Mechanisms of Molecular Carcinogenesis – Volume 2

Johannes Haybaeck
Editor

Mechanisms of Molecular Carcinogenesis – Volume 2

Springer

Editor
Johannes Haybaeck
Department of Pathology
Medical Faculty
Otto von Guericke University Magdeburg
Magdeburg
Germany

Institute of Pathology
Medical University of Graz
Graz
Austria

ISBN 978-3-319-85209-6 ISBN 978-3-319-53661-3 (eBook)
DOI 10.1007/978-3-319-53661-3

Softcover reprint of the hardcover 1st edition 2017

Printed on acid-free paper

This Springer imprint is published by Springer Nature
The registered company is Springer International Publishing AG
The registered company address is: Gewerbestrasse 11, 6330 Cham, Switzerland

Preface

This book entitled *Mechanisms of Molecular Carcinogenesis* addresses the latest developments in the assessment of molecular carcinogenesis. Mechanistic insights gained by various model systems in vitro and in vivo need to be validated in patients in order to find their way into clinics. Although the clinical relevance of model systems sometimes is not obvious, drug development is increasingly based on their mechanisms of action, and targeted drugs are a first step toward individualized medicine.

Patient-derived model systems that faithfully recapitulate human cancer are critical for the identification and validation of innovative drug targets and particular drugs and, thus, the basic understanding of cancer. Traditional approaches most often fail late in drug development (i.e., clinical phase II/III) due to substantial limitations of currently available preclinical models which inappropriately predict tumor plasticity and heterogeneity in the human patient. Scientists seek to overcome these limitations in cancer research by utilizing a panel of clinically well-characterized tumor tissues for the generation of different patient-derived 3D cell culture models (PD3D) containing either tumor cells alone or in combination with cancer-associated fibroblasts, as well as xenograft mouse models (PDX). The in-depth comparison of various models with regard to the stability of gene expression and their response toward chemotherapy poses a critical challenge in applied cancer research. Therefore, well-described mutations and translocations of particular tumor entities are nowadays characterized in the original patient tissue by next-generation sequencing, whereas their transcriptome is often analyzed using RNAseq.

Patient-derived tissues, as well as models, are often implemented into high-content-analysis and screening platforms for high-throughput drug discovery taking into account the influence of the tumor stroma on drug treatment efficacy. Model systems like transgenic or knockout animals, PD3D models, or cell lines are assumed to allow for compound profiling with high precision on both mRNA expression patterns and protein levels of novel targets, as well as the dissection and discovery of signaling pathways.

Novel tools allowing for the construction of computer-based models and simulation of biological processes are based on the emerging field of computational pathology, which is of high clinical relevance.

In-depth knowledge about the relevant molecular mechanisms of carcinogenesis is increasingly important for targeted molecular therapy in the framework of personalized medicine and patient care. Thus, the purpose of this book is to provide the reader with up-to-date insights into molecular and cellular mechanisms of cancer onset and progression, spread of cancer cells, and metastasis. It intends to fill the gap between basic cancer research and daily clinical practice where the prescription and advancement of routinely applied treatment strategies and targeted drugs can only be accomplished by individuals with a deeper understanding of the mode of action of the respective medications. Notably, the more advanced the tools for fighting cancer, the greater the need for a mechanistic understanding of medical approaches becomes. Therefore, this book deals with molecular diagnostics and their usability, as well as with targeted and genetic therapies. It draws a parallel to modern technology platforms and gives an overview of future developments.

This book aims at bridging the gap between basic and applied cancer research and the clinics, thereby trying to transfer knowledge from bench to bedside. A mechanistic understanding of carcinogenic events might be fundamental to the future of cancer research and treatment. Prognostic and predictive tumor biomarkers are extremely important and, thus, are highlighted in various chapters of this work.

Few medical areas have undergone such dramatic changes as did molecular pathology over the last few years. Thus, it is reasonable to have a look at this fascinating and very fast growing scientific field from different angles. Although excellent books on molecular technologies, diagnostic approaches, and therapeutic algorithms, are already available, a book addressing all these areas and simultaneously shedding light on the molecular mechanisms related to tumorigenesis is lacking.

This book not only provides a summary of basic knowledge but also, more importantly, gives an overview of the of recent advances in basic cancer research. We hope that it will serve as a comprehensive and concise source of knowledge, providing the reader with new developments and insights into carcinogenesis.

We hope that this book stimulates our readers, and that they will be fascinated by this exciting and scientifically, as well as clinically relevant, emerging topic.

Magdeburg, Germany Johannes Haybaeck
Graz, Austria

Contents

Molecular Classification of Breast Cancer 1

Iva Brčić and Marija Balić

Contents

Abstract

Breast cancer is the most common cancer in women, and it is one of the most intensively studied cancers. Out of all diagnosed breast cancer in women, only a small proportion develops in a familial setting, and for the large majority of women, the risk of developing breast cancer is less known. In this chapter, we describe in situ and invasive breast cancers in detail, referring to different immunohistochemical

I. Brčić (✉)
Institute of Pathology, Medical University of Graz, Auenbruggerplatz 25, 8036 Graz, Austria
e-mail: iva.brcic@medunigraz.at

M. Balić
Division of Oncology, Department of Internal Medicine, Medical University of Graz, Auenbruggerplatz 15, 8036 Graz, Austria
e-mail: marija.balic@medunigraz.at

J. Haybaeck (ed.), *Mechanisms of Molecular Carcinogenesis – Volume 2*,
DOI 10.1007/978-3-319-53661-3_1

and molecular events, as well as the molecular classification of this heterogeneous disease. We address most common and well-established biomarkers currently performed in breast cancer diagnostics. Finally, in the last part, we describe the implications of current knowledge on the further directions and address possible future perspectives in breast carcinogenesis.

1.1 Introduction

Breast cancer (BC) is the most common cancer in women, accounting for more than 1,500,000 cases worldwide annually. In 2012, the incidence rate was 1,676,000 [1]. BC constitutes a heterogeneous group of different tumors characterized by variable clinical and morphological features with distinct molecular alterations/aberrations. The traditional classification of BC is based upon the assessment of histological type and grade. Patient's age, tumor size, and lymph node status are also included in risk stratification and clinical treatment decisions. Immunohistochemical (IHC) analysis of receptor status on routinely processed tissue samples, namely, (1) expression of estrogen receptor α (ERα), (2) progesterone receptor (PR), and (3) overexpression/amplification of the human epidermal growth factor receptor 2 (HER2), also termed erythroblastosis oncogene B2 (ERBB2), is an imperative for complete pathological assessment, and their expression is the most important determinant of systemic treatment. The scoring of hormone receptors gives a better insight into potential response to endocrine treatment, and treatment with the humanized monoclonal antibody trastuzumab is largely dependent on the presence of HER2 either by immunohistochemistry or by in situ hybridization methods. Nonetheless, because of the genetic heterogeneity of breast cancers, histologically similar tumors may show different behaviors and responses to systemic therapies.

In the last two decades, different techniques have been used to collect molecular data on BC. By means of microarray-based gene expression profiling, the concept of heterogeneity among BCs has been established, demonstrating the need for distinct therapeutic approaches even further. More recently, next-generation sequencing (NGS) has brought new insights into BC classification. Prognostic and predictive subgroups have been suggested in order to enable individualized therapy [2].

Here, we review histological features of ductal carcinoma in situ (DCIS) and histological subtypes of invasive BCs and current state of their molecular classifications and discuss the biological processes that are likely to play a role in BC carcinogenesis. Special subtypes and familial BCs will also be discussed, as well as future perspectives.

1.2 Histopathological Classification of In Situ and Invasive Breast Cancer

The human breast is composed of a branching duct system of 15–20 lobes separated by fatty tissue. The terminal duct lobular units (TDLU) are the functional units of the breast and consist of the intralobular duct, ductules, and lobules that are lined by

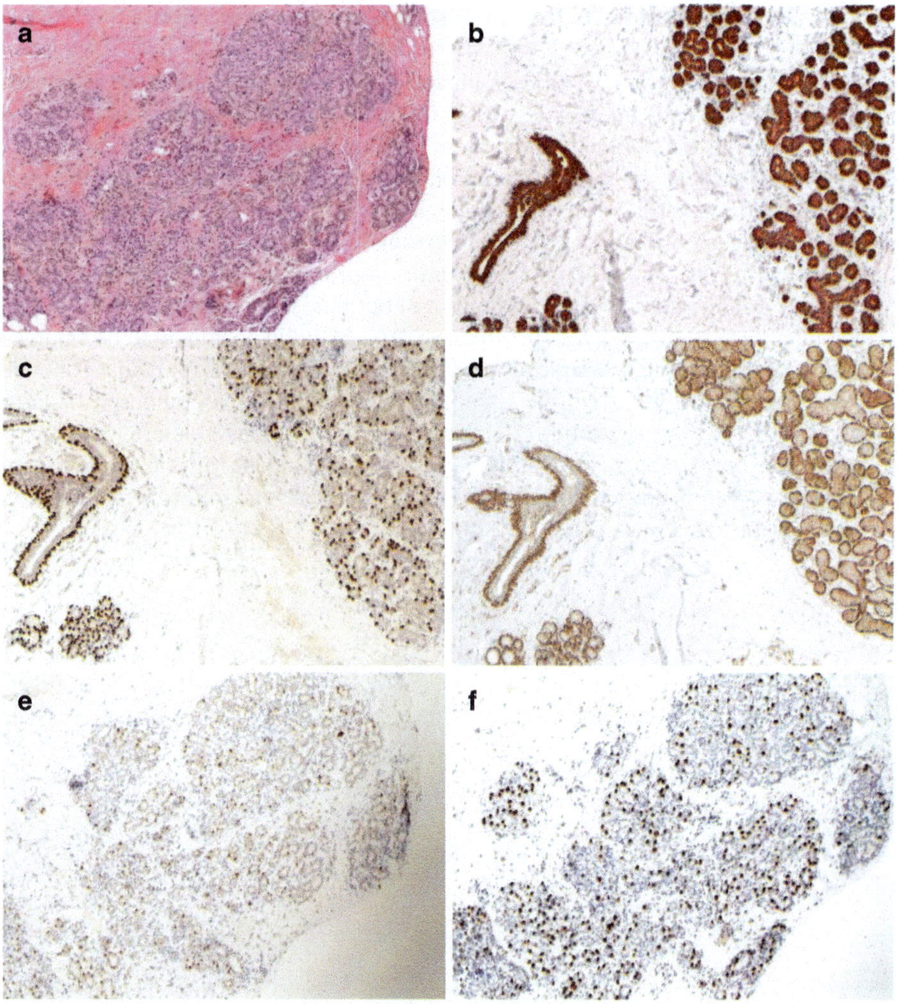

Fig. 1.1 Normal breast tissue. The terminal duct lobular unit (TDLU) is composed of the intralobular duct, ductules, and lobules lined by a single layer of luminal epithelium resting on a basal/myoepithelial layer (**a**). Luminal epithelium immunohistochemically highlighted by CK 8/18 (**b**). Basal layer immunohistochemically highlighted by p63 as nuclear staining (**c**) and SMA as cytoplasmic staining (**d**). Immunohistochemical expression of ER (**e**) and PR (**f**) seen as brown nuclear staining

a single layer of cuboidal epithelium (luminal epithelium) supported by underlying basal/myoepithelial cells. The IHC profile of these cells differs: luminal epithelial cells can be highlighted by low molecular weight keratins, such as CK8, 18, and 19, whereas basal/myoepithelial cells stain with antibodies against p63, smooth muscle actin (SMA), smooth muscle myosin, calponin, S100-protein, CD10, CK5/6, etc. (Fig. 1.1). Upon IHC, ERα, PR, and androgen receptors (AR) usually show a heterogeneous reaction in luminal cells, but are almost always negative in basal/myoepithelial cells [3].

1.2.1 Histological Features of Ductal Carcinoma In Situ

Ductal carcinoma in situ (DCIS) is characterized by proliferation of neoplastic luminal epithelial cells lined by a layer of myoepithelial cells and surrounded by an intact basement membrane, confined to the mammary ductal-lobular system. According to nuclear grade, differentiation, and presence of necrosis, they are classified into three grades: low, intermediate, and high grade [4]. Low-grade DCIS is composed of small, monomorphic cells showing micropapillary, cribriform, or solid growth pattern (Fig. 1.2). The nuclei are uniform in size and shape and have regular chromatin. Microcalcifications can be found, but mitosis and necrosis are uncommon. Intermediate-grade DCIS is characterized by mild to moderately enlarged cells; nuclei have variably coarse chromatin. Microcalcifications, punctuate or comedo necrosis, and mitosis can be found. High-grade DCIS is composed of large atypical cells growing in solid, cribriform, or micropapillary pattern. Nuclei are pleomorphic, with coarse chromatin and prominent nucleoli. Mitotic figures, comedo necrosis with necrotic debris in lumina of the ducts, and amorphous microcalcifications are frequently seen.

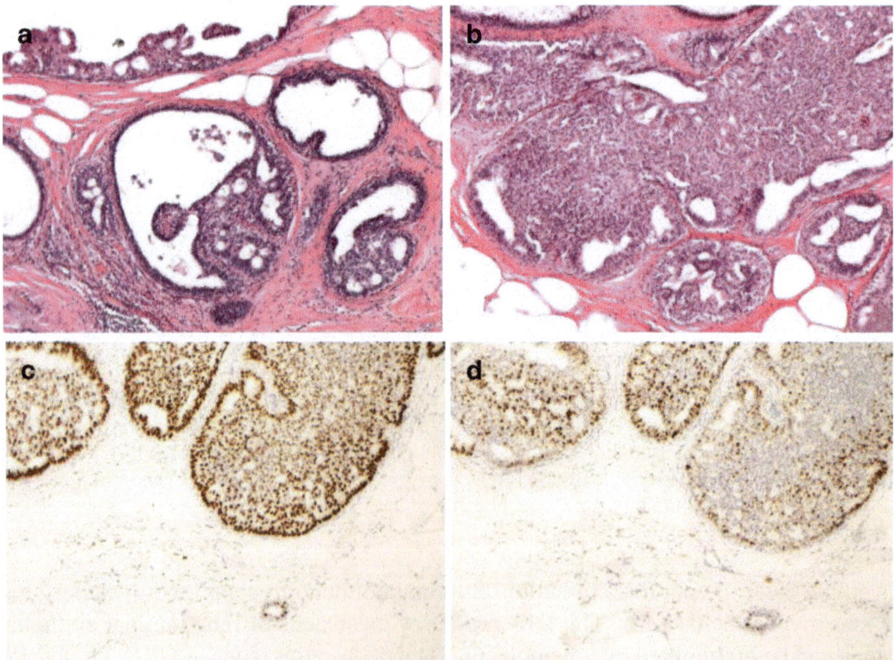

Fig. 1.2 Low-grade ductal carcinoma in situ. Cribriform (**a**) and solid pattern (**b**) of growth characterized by small, uniform cells. Immunohistochemical expression of ER (**c**) and PR (**d**) seen as brown nuclear staining

1.2.2 Histological Types and Grade of Invasive Breast Cancers

The World Health Organization (WHO) classification recognizes ten main histological types of invasive BC with their subtypes and various rare entities (Table 1.1). The most common type of BC is the so-called invasive carcinoma of no special type (IC-NST), previously called invasive ductal carcinoma not otherwise specified (IDC-NOS). It comprises a heterogeneous group of carcinomas that do not exhibit specific features of any other special type. The TDLU is regarded to be the site of

Table 1.1 WHO classification of invasive breast carcinomas, without papillary lesions and epithelial-myoepithelial tumors [8]

Invasive breast carcinoma (types)	Classification
Invasive carcinoma of no special type (NST)	8500/3
Pleomorphic carcinoma	8522/3
Carcinoma with osteoclast-like stromal giant cells	8035/3
Carcinoma with choriocarcinomatous features	
Carcinoma with melanotic features	
Invasive lobular carcinoma	8520/3
Classic lobular carcinoma	
Solid lobular carcinoma	
Alveolar lobular carcinoma	
Pleomorphic lobular carcinoma	
Tubulolobular carcinoma	
Mixed lobular carcinoma	
Tubular carcinoma	8211/3
Cribriform carcinoma	8201/3
Mucinous carcinoma	8480/3
Carcinoma with medullary features	
Medullary carcinoma	8510/3
Atypical medullary carcinoma	8513/3
Invasive carcinoma NST with medullary features	8500/3
Carcinoma with apocrine differentiation	
Carcinoma with signet-ring cell differentiation	
Invasive micropapillary carcinoma	8507/3
Metaplastic carcinoma of no special type	8575/3
Low-grade adenosquamous carcinoma	8570/3
Fibromatosis-like metaplastic carcinoma	8572/3
Squamous cell carcinoma	8070/3
Spindle cell carcinoma	8032/3
Metaplastic carcinoma with mesenchymal differentiation	
Chondroid differentiation	8571/3

(continued)

Table 1.1 (continued)

Invasive breast carcinoma (types)	Classification
Osseous differentiation	8571/3
Other types of mesenchymal differentiation	8575/3
Mixed metaplastic carcinoma	8575/3
Myoepithelial carcinoma	8982/3
Rare types	
Carcinoma with neuroendocrine features	8246/3
Neuroendocrine tumor, well differentiated	8041/3
Neuroendocrine carcinoma poorly differentiated (small cell carcinoma)	8574/3
Carcinoma with neuroendocrine differentiation	8502/3
Secretory carcinoma	8503/3
Invasive papillary carcinoma	8550/3
Acinic cell carcinoma	8430/3
Mucoepidermoid carcinoma	8525/3
Polymorphous carcinoma	8290/3
Oncocytic carcinoma	8314/3
Lipid-rich carcinoma	8315/3
Glycogen-rich clear cell carcinoma	8410/3
Sebaceous carcinoma	
Salivary gland/skin adnexal type tumors	8200/0
Cylindroma	8402/0
Clear cell hidradenoma	

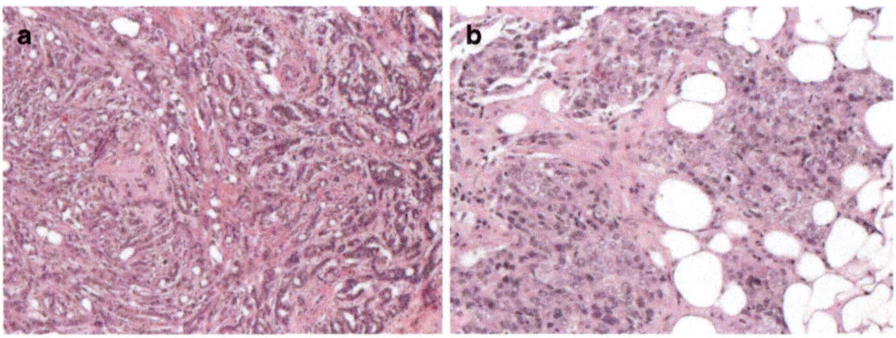

Fig. 1.3 Invasive breast carcinoma of no special type (IC-NST). Grade I tumor, characterized by tubule formations composed of small, monomorphic cells (**a**). Grade III tumor consists of sold sheets of large atypical cells with pleomorphic nuclei and prominent nucleoli (**b**)

origin of these tumors. Macroscopic appearance can vary: they can be irregular, stellate, or nodular, with poor-defined or pushing invasive margin. They are usually firm on palpation, and the cut surface is gray-white with yellow streaks. Morphological features can vary considerably: tumor cells can be arranged in tubules, clusters, cords, and trabeculae or have solid pattern of growth (Fig. 1.3) [5].

Tumor cells can have abundant or eosinophilic cytoplasm; nuclei may be uniform or pleomorphic with prominent nucleoli. Mitotic figures can be absent or numerous. Necrosis may be present or absent, as well as DCIS.

All invasive BCs are routinely classified according to histological grade [6]. Grade is assessed by the degree of differentiation (tubule and gland formation and nuclear pleomorphism) and mitotic count of a tumor (Fig. 1.3). Each parameter has a score of 1–3. Total scores of 3–5, 6–7, and 8–9 correspond to tumor grades I, II, and III (well to poorly differentiated), respectively. It is to be noted that multiple prognostic algorithms have incorporated histological grade for determination of therapy protocols. The most commonly used are the Nottingham Prognostic Index and Adjuvant! Online [7].

Assessment of ER, PR, HER2, and Ki67 by semiquantitative IHC analysis on routinely prepared, formalin-fixed, paraffin-embedded tissue sections is essential for the determination of appropriate therapy in routine practice. Two main ER variants have been described, ERα encoded by ESR1 gene and ERβ encoded by ESR2 gene. Upon immunohistochemistry, ERα and PR are expressed as nuclear staining of tumor cells. The proportion and intensity of positive tumor cells are measured and stated as a percentage of positive cells. ER- and PR-positive tumors are recommended to be the tumors that express at least 1% positive tumor nuclei [9] (Fig. 1.4). Different IHC semiquantitative scoring systems have been proposed and used in everyday work. The Remmele score [10] is assessed by the percentage of positive cells (0–4) and intensity of staining (0–3) with a range of 0–12: 0–1 being regarded as negative and 2–3, 4–8, and 9–12 as mild, moderate, and strongly positive, respectively (Table 1.2). The most established score is the Allred score [11], also based on the proportion of positive cells (0–5) and intensity of staining (0–3) (Table 1.2). The proportion and intensity are summed to produce total scores of 0–8 (0–2 is regarded as negative, while 3–8 as positive) with possibility to predict the respond to endocrine therapy [12].

HER2 status is determined by immunohistochemistry and/or in situ hybridization (ISH). According to guidelines from 2013 [13], HER2-positive tumors are those tumors that either express strong circumferential membrane staining of >10% of invasive tumor cells (IHC 3+ staining) or are amplified (HER2/CEP17 ratio by in situ hybridization (ISH) based on counting at least 20 cells of >2.2 or average HER2 gene copy number >6 signals/nucleus) (Fig. 1.5). Equivocal results are found in tumors that on IHC show either strong circumferential complete membrane staining of <10% of tumor cells or >10% of cells with weak to moderate circumferential incomplete membrane staining (IHC 2+) or have ISH ratio of 1.8–2.2 or average HER2 gene copy number 4–6 signals/nucleus. HER2-negative tumors show either no or little protein expression (IHC 1+ or 0) or have ISH ratio of <1.8 or average HER2 gene copy number of <4 signals/nucleus (Fig. 1.4e, f).

IHC detection of the Ki67 antigen is used to assess the growth fraction in BC by expressing the percentage of Ki67-positive nuclei. Several studies have shown that baseline Ki67 has prognostic value, and different cutoff points have been suggested to distinguish between tumors with low and high proliferative activity. However, different methods for counting Ki67 are used in daily practices with high intraobserver variability. It has been recommended that at least three high-power

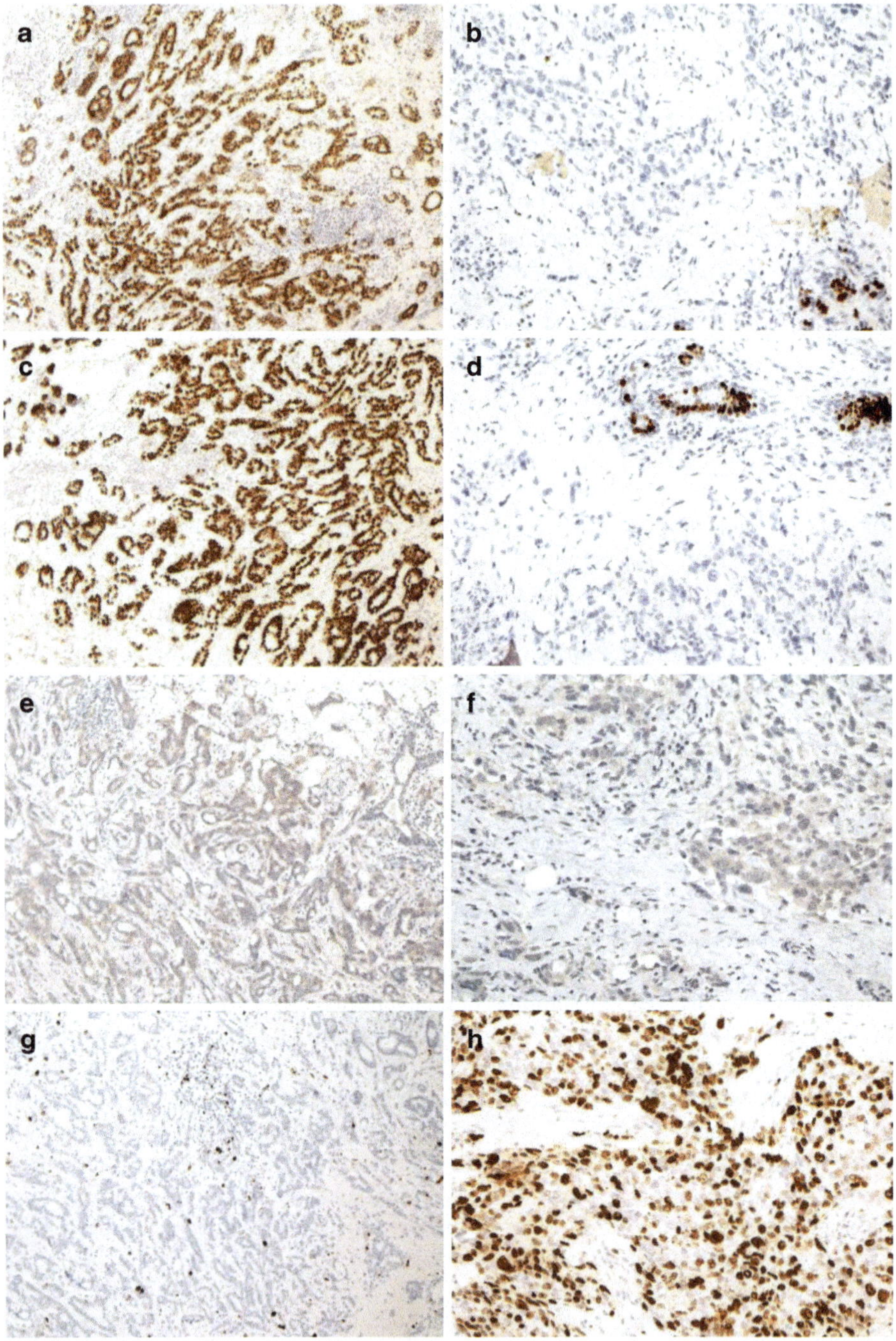

Fig. 1.4 Examples of ER (**a**, **b**), PR (**c**, **d**), HER2 (**e**, **f**), and Ki67 (**g**, **h**) staining in invasive breast carcinoma of no special type (IC-NST) in grade I tumor (**a**, **c**, **e**, **g**) and grade III tumor (**b**, **d**, **f**, **h**). We can appreciate brown nuclear staining in ER, PR, and Ki-67. HER2 is in both tumors negative

Table 1.2 Immunoreactive Allred and Remmele scores

Allred			Remmele		
% of positive cells	Intensity of staining	Score (effect on tamoxifen or letrozole therapy)	% of positive cells	Intensity of staining	Score
0 = 0%	0 = negative	0–1 = no effect	0 = 0%	0 = negative	0–1 = negative
1 = ≤ 1%	1 = weak/mild	2–3 = small (20%) chance of benefit	1 = < 10%	1 = weak/mild	2–3 = mildly positive
2 = 1–10%	2 = moderate	4–6 = moderate (50%) chance of benefit	2 = 10–50%	2 = moderate	4–8 = moderately positive
3 = 11–33%	3 = strong	7–8 = good (75%) chance of benefit	3 = 51–80%	3 = strong	9–12 = strongly positive
4 = 34–66%			4 = >80%		
5 = 67–100%					

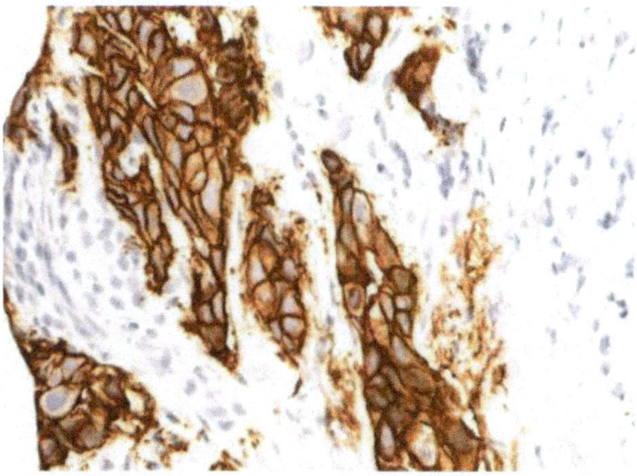

Fig. 1.5 HER2 immunohistochemical staining. Tumor cells express strong circumferential membrane staining of >10% of invasive tumor cells, regarded as IHC 3+ staining (HER2 positive)

(×40 objective) fields should be selected and the invasive edge of the tumor should be scored, counting of at least 500 malignant invasive cells [14]. In 2009, Cheang et al. [15] reported that the best cutoff value for Ki67 is 14%. More recently, Cserni et al. [16] showed that more realistic values for Ki67 ending with 0 or 5 should be introduced in routine practice. According to the newest guidelines, the cutoff is set at 20% (see more details under Sect. 1.3.1).

ARs have been known to be expressed in a proportion of BC, but due to recent advances in understanding their role, ARs regained a great amount of interest. The growing body of evidence indicates the importance of AR activity in the absence of estrogenic signaling. The AR is inhibitory to ERα, counteracting its oncogenic activity. It has been shown that BCs more often express AR than ERα and PR [17–19]. Therefore, AR has been considered a potential therapeutic target in ER-positive and ER-negative tumors that retain AR [20–23]. Furthermore, the potential of AR for therapy in triple-negative BC, which is currently without any targeted options, has been suggested, and the inclusion of AR expression in routine practice has been argued [24, 25].

1.3 Molecular Classification of Invasive Breast Carcinoma

Key genomic alterations in BC were identified more than two decades ago using loss of heterozygosity analysis (LOH) and comparative genomic hybridization (CGH). More recently, molecular characterization of BC has been done using "expression profiling" by measuring the entire transcriptome by microarray hybridization. This has revealed phenotypic subtypes of BC and diagnostic patterns of gene expression found useful in clinical research and increasingly in routine clinical practice [26–28].

1.3.1 Gene Expression Profiling in Breast Cancer

In 2000, Perou et al. [27] characterized variation in patterns of gene expression in 36 invasive ductal carcinomas, 2 invasive lobular carcinomas, 1 DCIS, 1 fibroadenoma, and 3 normal breast samples. They employed cDNA microarrays representing 8102 human genes and found the so-called intrinsic breast cancer subtypes arising from at least two distinct cell types (basal-like and luminal epithelial cells): luminal like, basal like, HER2 enriched, and normal like. This classification is based on the expression of three receptors: ERα, PR, and ERBB2 (HER2/NEU). The existence of described subtypes was confirmed by other authors [26, 28, 29]. Messenger RNA (mRNA) profiles of these subtypes showed that all of them are mainly driven by the expression of ER and ER-related genes, proliferation-related genes, and HER2 and genes mapping to the region of the HER2 amplicon on chromosome 17 [26–29].

Luminal tumors are ER positive and are subdivided into two categories. Based on the gene expression profiling, luminal A tumors cluster based on low expression of proliferation genes, whereas luminal B tumors cluster by higher number of genes associated with higher proliferation activity. By means of IHC, it has become increasingly accepted to define these two subtypes as suggested here: luminal A tumors are low grade, PR positive, and HER2 negative and show low proliferation activity (Ki-67 $< 20\%$). Luminal B tumors are higher grade, have higher proliferation index (Ki-67 $> 20\%$), and can be PR and HER2 either positive or negative [15]. Patients with luminal A tumors have better survival when compared to other described groups [12, 15, 28–30].

The other intrinsic subtypes, namely, basal like, HER2 positive, and normal like, are all ER negative. The basal-like group comprises multiple subtypes within the group and shows great diversity regarding specific morphological features, mutation profiles, metastatic behavior, response to chemotherapy, and clinical outcome [24, 30–35]. Upon histology, they are in most cases high grade and have pushing borders, central necrotic zones, conspicuous lymphocytic infiltrate, metaplastic areas, and medullary features [36].

The HER2-positive subtype is characterized by high expression of HER2 and a specific subset of related genes [27]. However, an incomplete overlap between the HER2 3+ IHC expression or HER2 gene amplification and the molecularly defined HER2-positive tumors has been described [37, 38]. HER2-amplified tumors that fall into the HER2-enriched subtype have worse prognosis when compared to the HER2-amplified tumors that fall into luminal B subtype [37].

The normal-like subtype gene expression pattern is defined by high expression of genes associated with adipose tissue and basal epithelial cells and low expression of genes characteristic of luminal epithelial cells [27]. Normal breast-like cancers consistently cluster with normal breast samples and fibroadenoma. It is suggested that this group should be regarded as normal tissue "contamination" rather than being a real intrinsic subtype [37].

In the last decade, different authors identified several more subtypes including claudin-low, molecular apocrine and interferon-related groups. The claudin-low tumors are in most of the cases triple negative; they are characterized by the low or absent expression of luminal markers, show enrichment for epithelial-to-mesenchymal transition markers and immune response genes, and have features suggestive of a "cancer stem cell-like" phenotype [39]. Tumors with medullary-like features and metaplastic carcinoma belong to this group.

Tumors with apocrine histological features that are ER negative or AR positive and may show amplification of HER2 are characteristic for the molecular apocrine subtype [22]. Using CGH, amplifications at 17q12 were found in 70% of these tumors [18]. Tumors in molecular apocrine group often recur (as well as basal-like tumors), but have a good response to neoadjuvant therapy [18].

The interferon-related group is characterized by the high expression of interferon-regulated genes, including STAT1, which is thought to be the transcription factor responsible for mediating interferon regulation of gene expression [26, 40]. It should be noted that the significance of these additional subtypes remains to be determined [41].

Meanwhile, there are several molecular platforms to perform molecular classification of BC, including Prediction Analysis of Microarray (PAM) 50 gene expression assay and MammaPrint/BluePrint and Endopredict [42–45]. It is clear that no substantial benefit can be expected just by better molecular classification of patients, and the goal of molecular tests is to help assessing patients' prognosis and predict therapeutic benefit. This approach is still controversial and not performed on routine basis, mostly due to different availabilities of the tests [46]. In the United States, the most prevalent test performed for the determination of patients who may benefit from chemotherapy is Oncotype DX, being evaluated in the adjuvant TAILORx trial. PAM-50 is a ceroid-based prediction method, designed as the 50-gene test developed to identify intrinsic breast cancer subtypes [37]. Risk of recurrence is derived from the expression profile based on 50 genes evaluated by PAM-50, with special weighting given to a set of proliferation-associated genes with a function of tumor size added. It can be assessed on formalin-fixed, paraffin-embedded tissues

[47]. The main restriction of all these tests is the lack of comparability of their results. However, with ongoing prospective clinical trials evaluating their value in stratification of patients for adjuvant treatment, their role may become clearer in the next few years.

1.3.2 Immunohistochemical Surrogates for Intrinsic Subtypes

In everyday practice, IHC surrogates for the previously described expression profiling-based intrinsic subtypes have been established. Panel of four markers (ER, PR, HER2, and Ki-67), used as IHC4 score, was found to be useful in clinics for the determination of the therapeutic strategies based upon biomarker expression patterns, as this panel has been validated against a gene expression profiling-defined intrinsic subtype classification [48, 49].

Nielsen et al. [47] showed that basal markers (CK5/6 and EGFR) improve identification of basal-like tumors. Recently, the Nottingham group has tried to identify key clinical phenotypes of breast cancer using a panel of ten protein biomarkers: ER, PR, CK5/6, CK7/8, EGFR, HER2, HER3, HER4, p53, and Mucin 1 [50]. These authors have divided BCs into novel subdivisions of luminal and basal tumors as follows: (1) luminal subtype into luminal A, luminal B, and luminal N (HER3 and HER4+); (2) basal subtype into basal-p53 altered (p53+) and basal-p53 normal (p53−); and (3) HER2 enriched into ER+ and ER−. Their study showed that both the luminal and basal BC phenotypes are heterogeneous and contain distinct subgroups.

1.3.3 Genome Profiling in Breast Cancer

The main goal of genome profiling of BC was to discover different alterations/mutations in the DNA of breast tumor cells [51]. In 2009, Stephens et al. published [52] complex landscapes of somatic rearrangement in human BC genomes. Since then, different technology platforms have been used to unravel a more detailed molecular characterization of BC.

The genomic alterations and the expression of specific genes vary across different molecular subtypes. Main alterations found in BC involve mutations in PIK3CA, PIK3R1, TP53, GATA3, PTEN, MAP3K1, MAP2K4, MLL3, AKT1, CDH1, and RB1 and amplifications of HER2, MYC, FGFR1/ZNF703, MDM2, and CCND1 [51, 53]. Luminal A group showed the largest number of mutated genes, including PIK3CA, which was found in 45% [51, 53]. Luminal B group showed mutations of TP53 and PIK3CA. Further mutations found in luminal ER-positive tumors include the following genes: GATA3, FOXA1, MAP3KI, and MAP2K4 as well as amplification of CCND1 and high expression of ESR1, XBP1, and MYB protein [26, 51]. The HER2-enriched subtype was found to have high expression of the 17q12-21 amplicon genes (HER2/ERBB2 and GRB7), FGFR4, TMEM45B, and GPR160 [26, 37]. Basal-like tumors in 80% display TP53 mutation. They also express

CK5/6, CK14, EGFR, c-KIT, FOXC1, caveolin 1&2, P-cadherin, and epidermal growth factor receptor (EGFR), usually have high Ki-67 index, and behave aggressively [26, 27, 54]. In addition, loss of RB1 and BRCA1 and amplification of MYC can be found [51].

1.3.4 Special Subtypes

The special subtypes comprise up to 25% of all BC, with lobular type being the most frequent. Data on expression profiling of these tumors are scarce.

Based on the ER status, we can divide these tumors into two groups: (1) ER positive (classic invasive lobular, tubular, micropapillary, mucinous, and neuroendocrine carcinomas) and (2) ER negative (apocrine, adenoid cystic, secretory, pleomorphic invasive lobular, metaplastic, and medullary carcinomas).

The most common special subtype of BC is invasive lobular carcinoma (ILC). They usually present as ill-defined mass, best detected by magnetic resonance imaging. Upon histology, classic ILC is characterized by diffuse growth of monomorphic cells, without cohesion, arranged in single-file pattern [55] (Fig. 1.6a). Several other variants have been described in this group showing different growth patterns (like pleomorphic lobular carcinoma). However, all these tumors are characterized by the loss of expression of transmembrane intercellular adhesion glycoprotein E-cadherin, encoded by CDH1, mapped on chromosome 16q22.1 (Fig. 1.6b). Different mechanisms of CDH1 gene inactivation have been reported, like CDH1 gene promotor methylation and mutations, LOH on 16q22, and deletion of 16q [56–59]. In addition, these tumors frequently harbor recurrent gains on 1q and are found to have region of amplification localized to chromosome 11, with FGF3 and CCND1 genes [60, 61], the latter one being a potential candidate for targeted therapy using EglN2 hydroxylase inhibitors [62].

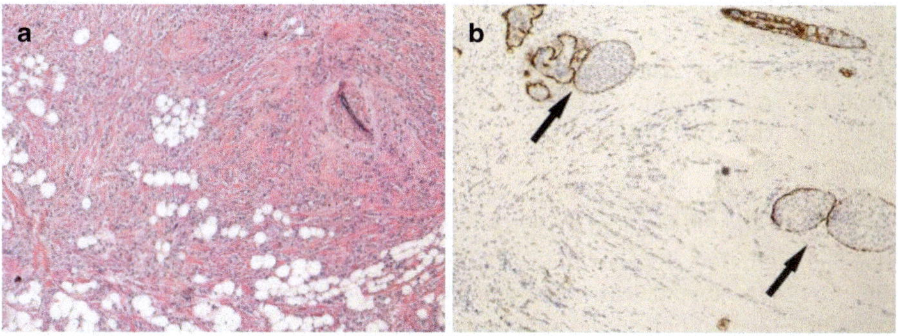

Fig. 1.6 Classic invasive lobular carcinoma with typical morphology: diffuse growth of monomorphic cells arranged in single-file pattern (**a**). Immunohistochemical staining with E-cadherin shows negative staining of tumor cells in invasive and in situ lobular component (*arrows*) of the tumor (**b**)

Microarray-based analysis of 11 special subtypes of BC showed that each of them, except for apocrine carcinomas, falls into one molecular subtype. Tubular, mucinous, and neuroendocrine carcinomas displayed a luminal phenotype and adenoid cystic, metaplastic, and medullary carcinomas a basal-like phenotype [34, 63]. Secretory carcinomas have indolent behavior, show triple-negative and basal-like phenotype, and are found to have the *t*(12;15) translocation that leads to the transformation of the ETV6-NTRK3 fusion gene [64]. Adenoid cystic carcinomas display basal-like phenotype and were found to consistently harbor the *t*(6;9) MYB/NFIB translocation [63]. PTEN and TOP2A genes, as well as DNA repair pathways, are downregulated in metaplastic carcinomas [65]. Furthermore, when compared to IC-NST cancers and basal-like subtype, metaplastic carcinomas were found to show higher expression of genes related to myoepithelial differentiation and epithelial-to-mesenchymal transition [65].

1.3.5 Familial Breast Cancer

The majority of the BCs are sporadic. However, about 5–10% of BC and ovarian cancers, usually found in younger women, occur in the familial setting. Most of them are associated with germline mutations in BRCA1 or BRCA2 genes [66, 67], tumor suppressor genes, normally expressed in the cells where they are involved in the repair of damaged DNA, and their mutations increase the risk for BC. BRCA1 is localized to chromosome 17q21 and BRCA2 to chromosome 13q12-q13. These mutations are present in 16–25% of high-risk familial BC [68]. For the remaining 75–84% of non-informative (BRCAX) patients, no responsible genes have yet been identified.

Tumors in patients with BRCA1 and BRCA2 mutations show different morphological features when compared to each other and when compared to sporadic BCs. Features associated with BRCA1 mutations are continuous pushing margins, lymphocytic infiltrate, trabecular growth pattern, necrosis, and a high mitotic count [69, 70]. BCs that harbor BRCA1 mutation are found to be more frequently aneuploid and have higher tumor cell proliferation rates compared with sporadic BCs [69, 71, 72]. According to the same authors, medullary or atypical medullary carcinomas were more common in BRCA1 group. Tumors from BRCA1 mutation carriers do not show ERBB2 amplification, are mostly basal-like subtype, and are therefore associated with no expression of ER and poor prognosis [29, 73].

Tumors from BRCA2 mutation carriers are predominantly of the luminal B subtype, have usually continuous pushing margins, and are more likely to be ER positive and high grade by showing less tubule formation when compared to sporadic BCs [69, 74]. Nonetheless, no differences between these two groups were found regarding pleomorphism and mitotic count. Age at diagnosis is found to be an important variable, and grade is identified as being more informative than ER status for BRCA2 mutation carrier prediction [75]. Specific histology type in this group of patients is still to be identified.

1.4 Molecular Classification of Ductal Carcinoma In Situ

DCIS is a heterogeneous group of diseases with distinct histological, immunohistochemical, and biological features. DCIS is described as the non-obligate precursor lesion of invasive ductal carcinoma [76]. This is supported by an observation that DCIS is found adjacent to invasive component in the vast majority of patients [77], usually presenting with concordant nuclear grade in both components.

During the last 20 years, linear multistep model of development from normal cells to invasive carcinoma was suggested. Several authors showed that the majority of transcriptomic changes occur during transition from normal luminal epithelial cells to DCIS [78, 79]. This process starts as flat epithelial atypia (FEA), progresses from atypical ductal hyperplasia (ADH) into DCIS, and evolves to invasive BC. Data suggest two distinct pathways: (1) from FEA, ADH, and low-grade DCIS to low-grade invasive carcinoma and (2) from high-grade DCIS to high-grade invasive carcinoma [80]. Molecular studies revealed that low-grade DCIS are ER positive, express BCL2, and frequently harbor chromosomal 16q and 17p loss [80–83]. In contrast, high-grade DCIS is characterized by TP53 mutations, high frequency of HER2 expression, chromosomal 8p and 13q loss, and 1q gain [82, 84–87].

Gene expression profiling showed differences in expression of a subset of genes between low- and high-grade DCIS [78, 88, 89]. Specifically, Ma et al. [78] revealed extensive similarities at the transcriptome level among the distinct stages of progression to BC. They suggested that gene expression alterations harboring potential for invasive growth are already present in preinvasive lesions. These authors showed that ADH, low-grade DCIS, and low-grade invasive ductal BC share similar gene expression signatures involving genes encountered in the ER phenotype. Gene expression profile consisting of genes associated with cell cycle processes and mitotic activity was found in high-grade DCIS and high-grade invasive BC. Furthermore, they also suggested that the ribonucleotide reductase M2 (RRM2) gene may play a role in both rapid cell proliferation and invasive growth behavior.

In 2008, Vincent-Salomon et al. [82] performed combined phenotypic and genomic analysis of 57 DCIS with gene expression profiling in 26 of the 57 cases. This group of authors showed that DCIS already displays molecular heterogeneity found in invasive BC and demonstrated that DCIS and invasive ductal BC share similar gene expression patterns. Further on, several other studies confirmed extensive genetic heterogeneity and presented evidence in support of clonal selection during the transition from DCIS to invasive BC [90–92]. They suggested that the progression process to invasive lesion may constitute an "evolutionary bottleneck."

1.4.1 Molecular Subtypes in DCIS

Molecular subtypes encountered in DCIS are similar to the major molecular subtypes present in invasive BC, but are found at different frequencies [34, 82, 93–95]. When DCIS is compared to invasive ductal BC, luminal A phenotype is significantly lower in frequency, whereas the luminal B and HER2-enriched phenotypes are significantly higher [96]. Further on, a correlation between DCIS grade and molecular subtypes is also observed: low-grade DCIS is ER positive and associated with luminal phenotype. In contrast, high-grade DCIS is ER negative, HER2 positive, and associated with basal-like phenotypes [96]. Overexpression of HER2 in high-grade DCIS suggests that HER2 plays an important role in tumor development [97].

In 2006, Hicks et al. identified three characteristic genomic subtypes (patterns) in diploid breast tumors [98]. The first pattern, called "simplex," was found in up to 60% of tumors investigated. This pattern showed broad segments of duplication and deletion, most frequently affecting entire chromosomes or chromosome arms, as seen in luminal A cancers. The other two patterns are complex. The second pattern, termed "complex-sawtooth," is characterized by multiple narrow segments of duplication and deletion, usually involving all chromosomes, and is associated with triple-negative, basal-like cancers. The third pattern is characterized by multiple closely spaced amplicons ("firestorms") confined to single chromosome arms. These complex high-level amplifications highly correlate with aggressive behavior and poor survival and are observed mostly in luminal B and HER2 subtypes.

For better understanding of DCIS, additional potential biomarkers have been investigated, including cell cycle regulation and apoptotic markers, proliferation markers, cell adhesion molecules, EGFR family receptors, angiogenesis-related proteins, extracellular markers proteins, and cyclooxygenase type 2 [99]. However, according to the National Comprehensive Cancer Network practice guidelines, ER is the only biomarker validated in the routine clinical practice [100]. The majority of patients with DCIS undergo surgical removal of lesions and/or microcalcifications found on mammography with radiation and/or prophylactic systemic therapy. After surgical resection, the likelihood of recurrence and/or progression to invasive cancer is low in DCIS lesions that are ER and/or PR positive, are HER2 negative, have normal expression of p53, and have low Ki-67 index. On the other hand, DCIS lesions with negative hormone status, HER2 amplification, TP53 mutations, and higher Ki-67 index show faster progression to invasive disease and have high recurrence potential [101].

1.5 Future Perspectives

Traditional histopathological features, ER, PR, HER2 status, and Ki-67, as well as TNM stage, are still used as basis for adjuvant therapy in patients with invasive BC. The role of AR in the BC pathogenesis remains to be elucidated. Better understanding of AR function in BC will hopefully lead to another potential therapeutic target. Furthermore, discovering clinically useful biomarkers to differentiate between low- and high-risk patients with DCIS of developing invasive BC remains an important goal. The employment of novel technologies, like depth massively parallel sequencing and single cell analysis, is a next step in discovering genetic alterations occurring in this transition process. Described molecular classification systems are currently of limited clinical value and are not embedded in routine practice. In the future, better understanding of molecular changes will provide the opportunity for more rational and personalized therapeutic options for patients with DCIS and/or invasive BC.

Conclusions

In situ and invasive BCs are a heterogeneous group of diseases that present with different morphologies and express different immunohistochemical and molecular markers. The progression from normal breast to invasive BC is a complex phenomenon. A better understanding of this molecular process will provide better prevention and enable guided treatment in individual patients. Elucidating the role of AR in the different subtypes of BCs will lead to improved treatment options.

References

1. Torre LA, Bray F, Siegel RL, Ferlay J, Lortet-Tieulent J, Jemal A. Global cancer statistics, 2012. CA Cancer J Clin. 2015;65:87–108. doi:10.3322/caac.21262.
2. Rakha EA, Ellis IO. Modern classification of breast cancer: should we stick with morphology or convert to molecular profile characteristics. Adv Anat Pathol. 2011;18:255–67. doi:10.1097/PAP.0b013e318220f5d1.
3. Allred C, Miller K, Viale G, Brogi E, Isola J. Molecular testing for estrogen receptor, progesterone receptor, and HER2. In: Lakhani SR, Ellis IO, Schnitt SJ, Tan PH, van de Vijver MJ, editors. World Health Organisation classification of tumours of the breast. Lyon: IARC; 2012. p. 22–3.
4. Schnitt SJ, Allred C, Britton P, Ellis IO, Lakhani SR, Morrow M, et al. Ductal carcinoma in situ. In: Lakhani SR, Ellis IO, Schnitt SJ, Tan PH, van de Vijver MJ, editors. World Health Organisation classification of tumours of the breast. Lyon: IARC; 2012. p. 90–4.
5. Ellis IO, Collins L, Ichihara S, MacGrogan G. Invasive carcinoma of no special type. In: Lakhani SR, Ellis IO, Schnitt SJ, Tan PH, van de Vijver MJ, editors. World Health Organisation classification of tumours of the breast. Lyon: IARC; 2012. p. 33–4.
6. Elston CW, Ellis IO. Pathological prognostic factors in breast cancer. I. The value of histological grade in breast cancer: experience from a large study with long-term follow-up. Histopathology. 1991;19:403–10.
7. Mook S, Schmidt MK, Rutgers EJ, van de Velde AO, Visser O, Rutgers SM, et al. Calibration and discriminatory accuracy of prognosis calculation for breast cancer with the online Adjuvant! Program: a hospital-based retrospective cohort study. Lancet Oncol. 2009;10:1070–6.
8. Lakhani SR, Ellis IO, Schnitt SJ, Tan PH, van de Vijver MJ. World Health Organisation classification of tumours of the breast. Lyon: IARC; 2012.
9. Hammond ME, Hayes DF, Dowsett M, Allred DC, Hagerty KL, Badve S, et al. American Society of Clinical Oncology/College Of American Pathologists guideline recommendations for immunohistochemical testing of estrogen and progesterone receptors in breast cancer. J Clin Oncol. 2010;28:2784–95. doi:10.1200/JCO.2009.25.6529.
10. Remmele W, Stegner HE. Recommendation for uniform definition of an immunoreactive score (IRS) for immunohistochemical estrogen receptor detection (ER-ICA) in breast cancer tissue. Pathologe. 1987;8:138–40.
11. Allred DC, Harvey JM, Berardo M, Clark GM. Prognostic and predictive factors in breast cancer by immunohistochemical analysis. Mod Pathol. 1998;11:155–68.
12. Brouckaert O, Laenen A, Vanderhaegen J, Wildiers H, Leunen K, Amant F, et al. Applying the 2011 St Gallen panel of prognostic markers on a large single hospital cohort of consecutively treated primary operable breast cancers. Ann Oncol. 2012;23:2578–84.

13. Wolff AC, Hammond ME, Hicks DG, Dowsett M, McShane LM, Allison KH, et al. Recommendations for human epidermal growth factor receptor 2 testing in breast cancer: American Society of Clinical Oncology/College of American Pathologists clinical practice guideline update. J Clin Oncol. 2013;31:3997–4013. doi:10.1200/JCO.2013.50.9984.
14. Dowsett M, Nielsen TO, A'Hern R, Bartlett J, Coombes RC, Cuzick J, et al. Assessment of Ki67 in breast cancer: recommendations from the International Ki67 in Breast Cancer working group. J Natl Cancer Inst. 2011;103:1656–64. doi:10.1093/jnci/djr393.
15. Cheang MC, Chia SK, Voduc D, Gao D, Leung S, Snider J, et al. Ki67 index, HER2 status, and prognosis of patients with luminal B breast cancer. J Natl Cancer Inst. 2009;101:736–50. doi:10.1093/jnci/djp082.
16. Cserni G, Vörös A, Liepniece-Karele I, Bianchi S, Vezzosi V, Grabau D, et al. Distribution pattern of the Ki67 labelling index in breast cancer and its implications for choosing cut-off values. Breast. 2014;23:259–63. doi:10.1016/j.breast.2014.02.003.
17. Collins LC, Cole KS, Marotti JD, Hu R, Schnitt SJ, Tamimi RM. Androgen receptor expression in breast cancer in relation to molecular phenotype: results from the Nurses' Health Study. Mod Pathol. 2011;24:924–31. doi:10.1038/modpathol.2011.54.
18. Guedj M, Marisa L, de Reynies A, Orsetti B, Schiappa R, Bibeau F, et al. A refined molecular taxonomy of breast cancer. Oncogene. 2012;31:1196–206. doi:10.1038/onc.2011.301.
19. Qi JP, Yang YL, Zhu H, Wang J, Jia Y, Liu N, et al. Expression of the androgen receptor and its correlation with molecular subtypes in 980 Chinese breast cancer patients. Breast Cancer (Auckl). 2012;6:1–8. doi:10.4137/BCBCR.S8323.
20. Chia K, O'Brien M, Brown M, Lim E. Targeting the androgen receptor in breast cancer. Curr Oncol Rep. 2015;17:4. doi:10.1007/s11912-014-0427-8.
21. Cochrane DR, Bernales S, Jacobsen BM, Cittelly DM, Howe EN, D'Amato NC, et al. Role of the androgen receptor in breast cancer and preclinical analysis of enzalutamide. Breast Cancer Res. 2014;16:R7. doi:10.1186/bcr3599.
22. Farmer P, Bonnefoi H, Becette V, Tubiana-Hulin M, Fumoleau P, Larsimont D, et al. Identification of molecular apocrine breast tumours by microarray analysis. Oncogene. 2005;24:4660–71. doi:10.1038/sj.onc.1208561.
23. McNamara KM, Sasano H. Androgen and breast cancer: an update. Curr Opin Endocrinol Diabetes Obes. 2016;23(3):249–56.
24. Fulford LG, Reis-Filho JS, Ryder K, Jones C, Gillett CE, Hanby A, Easton D, Lakhani SR. Basal-like grade III invasive ductal carcinoma of the breast: patterns of metastasis and long-term survival. Breast Cancer Res. 2007;9:R4. doi:10.1186/bcr1636.
25. Giannos A, Filipits M, Zagouri F, Brandstetter A, Tsigginou A, Sotiropoulou M, et al. Expression of ARs in triple negative breast cancer tumors: a potential prognostic factor? Onco Targets Ther. 2015;8:1843–7. doi:10.2147/OTT.S78254.
26. Hu Z, Fan C, Oh DS, Marron JS, He X, Qaqish BF, et al. The molecular portraits of breast tumors are conserved across microarray platforms. BMC Genomics. 2006;7:96. doi:10.1186/1471-2164-7-96.
27. Perou CM, Sørlie T, Eisen MB, van de Rijn M, Jeffrey SS, Rees CA, et al. Molecular portraits of human breast tumours. Nature. 2000;406:747–52.
28. Sorlie T, Perou CM, Tibshirani R, Aas T, Geisler S, Johnsen H, et al. Gene expression patterns of breast carcinomas distinguish tumor subclasses with clinical implications. Proc Natl Acad Sci U S A. 2001;98:10869–74. doi:10.1073/pnas.191367098.
29. Sorlie T, Tibshirani R, Parker J, Hastie T, Marron JS, Nobel A, et al. Repeated observation of breast tumor subtypes in independent gene expression data sets. Proc Natl Acad Sci U S A. 2003;100:8418–23. doi:10.1073/pnas.0932692100.
30. Voduc KD, Cheang MC, Tyldesley S, Gelmon K, Nielsen TO, Kennecke H. Breast cancer subtypes and the risk of local and regional relapse. J Clin Oncol. 2010;28:1684–91. doi:10.1200/JCO.2009.24.9284.
31. Banerjee S, Reis-Filho JS, Ashley S, Steele D, Ashworth A, Lakhani SR, et al. Basal-like breast carcinomas: clinical outcome and response to chemotherapy. J Clin Pathol. 2006;59:729–35. doi:10.1136/jcp.2005.033043.

32. Fulford LG, Easton DF, Reis-Filho JS, Sofronis A, Gillett CE, Lakhani SR, et al. Specific morphological features predictive for the basal phenotype in grade 3 invasive ductal carcinoma of breast. Histopathology. 2006;49:22–34.
33. Kennecke H, Yerushalmi R, Woods R, Cheang MC, Voduc D, Speers CH, et al. Metastatic behavior of breast cancer subtypes. J Clin Oncol. 2010;28:3271–7. doi:10.1200/JCO.2009.25.9820.
34. Livasy CA, Karaca G, Nanda R, Tretiakova MS, Olopade OI, Moore DT, et al. Phenotypic evaluation of the basal-like subtype of invasive breast carcinoma. Mod Pathol. 2006;19:264–71.
35. Shah SP, Roth A, Goya R, Oloumi A, Ha G, Zhao Y, et al. The clonal and mutational evolution spectrum of primary triple-negative breast cancers. Nature. 2012;486:395–9. doi:10.1038/nature10933.
36. Livasy CA, Perou CM, Karaca G, Cowan DW, Maia D, Jackson S, et al. Identification of a basal-like subtype of breast ductal carcinoma in situ. Hum Pathol. 2007;38:197–204.
37. Parker JS, Mullins M, Cheang MC, Leung S, Voduc D, Vickery T, et al. Supervised risk predictor of breast cancer based on intrinsic subtypes. J Clin Oncol. 2009;27:1160–7. doi:10.1200/jco.2008.18.1370.
38. Rouzier R, Perou CM, Symmans WF, Ibrahim N, Cristofanilli M, Anderson K, et al. Breast cancer molecular subtypes respond differently to preoperative chemotherapy. Clin Cancer Res. 2006;11:5678–85. doi:10.1158/1078- 0432.ccr-04-2421.
39. Prat A, Parker JS, Karginova O, Fan C, Livasy C, Herschkowitz JI, et al. Phenotypic and molecular characterization of the claudin-low intrinsic subtype of breast cancer. Breast Cancer Res. 2010;12:R68. doi:10.1186/bcr2635.
40. Bromberg JF, Horvath CM, Wen Z, Schreiber RD, Darnell Jr JE. Transcriptionally active Stat1 is required for the antiproliferative effects of both interferon alpha and interferon gamma. Proc Natl Acad Sci U S A. 1996;93:7673–8.
41. Weigelt B, Geyer FC, Reis-Filho JS. Histological types of breast cancer: how special are they? Mol Oncol. 2010;4:192–208. doi:10.1016/j.molonc.2010.04.004.
42. Bastien RR, Rodriguez-Lescure A, Ebbert MT, Prat A, Munárriz B, Rowe L, et al. PAM50 breast cancer subtyping by RT- qPCR and concordance with standard clinical molecular markers. BMC Med Genet. 2012;5:44. doi:10.1186/1755-8794-5-44.:44-5.
43. Bayraktar S, Royce M, Stork-Sloots L, de Snoo F, Glück S. Molecular subtyping predicts pathologic tumor response in early-stage breast cancer treated with neoadjuvant docetaxel plus capecitabine with or without trastuzumab chemotherapy. Med Oncol. 2014;31:163. doi:10.1007/s12032-014-0163-9.
44. Reis-Filho JS, Pusztai L. Gene expression profiling in breast cancer: classification, prognostication, and predilection. Lancet. 2011;378:1812–23. doi:10.1016/S0140-6736(11)61539-0.
45. van de Vijver MJ. Molecular tests as prognostic factors in breast cancer. Virchows Arch. 2014;464:283–91. doi:10.1007/s00428-014-1539-0.
46. Coates AS, Winer EP, Goldhirsch A, Gelber RD, Gnant M, Piccart-Gebhart M, et al. Tailoring therapies-improving the management of early breast cancer: St Gallen International Expert Consensus on the Primary Therapy of Early Breast Cancer 2015. Ann Oncol. 2015;26:1533–46. doi:10.1093/annonc/mdv221.
47. Nielsen TO, Parker JS, Leung S, Voduc D, Ebbert M, Vickery T, et al. A comparison of PAM50 intrinsic subtyping with immunohistochemistry and clinical prognostic factors in tamoxifen-treated estrogen receptor-positive breast cancer. Clin Cancer Res. 2010;16:5222–32. doi:10.1158/1078-0432.CCR-10-1282.
48. Cuzick J, Dowsett M, Pineda S, Wale C, Salter J, Quinn E, et al. Prognostic value of a combined estrogen receptor, progesterone receptor, Ki-67, and human epidermal growth factor receptor 2 immunohistochemical score and comparison with the Genomic Health recurrence score in early breast cancer. J Clin Oncol. 2011;29:4273–8. doi:10.1200/JCO.2010.31.2835.
49. Dowsett M, Sestak I, Lopez-Knowles E, Sidhu K, Dunbier AK, Cowens JW, et al. Comparison of PAM50 risk of recurrence score with oncotype DX and IHC4 for predicting risk of distant

recurrence after endocrine therapy. J Clin Oncol. 2013;31:2783–90. doi:10.1200/jco.2012.46.1558.
50. Green AR, Powe DG, Rakha EA, Soria D, Lemetre C, Nolan CC, et al. Identification of key clinical phenotypes of breast cancer using a reduced panel of protein biomarkers. Br J Cancer. 2013;109:1886–94. doi:10.1038/bjc.2013.528.
51. The Cancer Genome Atlas Network. Comprehensive molecular portraits of human breast tumours. Nature. 2012;490:61–70. doi:10.1038/nature11412.
52. Stephens PJ, McBride DJ, Lin ML, Varela I, Pleasance ED, Simpson JT, et al. Complex landscapes of somatic rearrangement in human breast cancer genomes. Nature. 2009;462:1005–10. doi:10.1038/nature08645.
53. Curtis C, Shah SP, Chin SF, Turashvili G, Rueda OM, Dunning MJ, et al. The genomic and transcriptomic architecture of 2,000 breast tumours reveals novel subgroups. Nature. 2012;486:346–52. doi:10.1038/nature10983.
54. Badve S, Dabbs DJ, Schnitt SJ, Baehner FL, Decker T, Eusebi V, et al. Basal-like and triple-negative breast cancers: a critical review with an emphasis on the implications for pathologists and oncologists. Mod Pathol. 2012;24:157–67. doi:10.1038/modpathol.2010.200.
55. Lakhani SR, Rakha E, Simpson PT. Invasive lobular carcinoma. In: Lakhani SR, Ellis IO, Schnitt SJ, Tan PH, van de Vijver MJ, editors. World Health Organisation classification of tumours of the breast. IARC: Lyon; 2012. p. 40–2.
56. Droufakou S, Deshmane V, Roylance R, Hanby A, Tomlinson I, Hart IR. Multiple ways of silencing E-cadherin gene expression in lobular carcinoma of the breast. Int J Cancer. 2001;92:404–8.
57. Rakha EA, Green AR, Powe DG, Roylance R, Ellis IO. Chromosome 16 tumor-suppressor genes in breast cancer. Genes Chromosomes Cancer. 2006;45:527–35.
58. Reis Filho JS, Simpson PT, Gale T, Lakhani SR. The molecular genetics of breast cancer: the contribution of comparative genomic hybridization. Pathol Res Pract. 2005;201:713–25.
59. Stange DE, Radlwimmer B, Schubert F, Traub F, Pich A, Toedt G, et al. High-resolution genomic profiling reveals association of chromosomal aberrations on 1q and 16p with histologic and genetic subgroups of invasive breast cancer. Clin Cancer Res. 2006;12:345–52.
60. Gruel N, Lucchesi C, Raynal V, Rodrigues MJ, Pierron G, Goudefroye R, et al. Lobular invasive carcinoma of the breast is a molecular entity distinct from luminal invasive ductal carcinoma. Eur J Cancer. 2010;46:2399–407. doi:10.1016/j.ejca. 2010.05.013.
61. Reis-Filho JS, Simpson PT, Turner NC, Lambros MB, Jones C, Mackay A, et al. FGFR1 emerges as a potential therapeutic target for lobular breast carcinomas. Clin Cancer Res. 2006;12:6652–62.
62. Zhang Q, Gu J, Li L, Liu J, Luo B, Cheung HW, et al. Control of cyclin D1 and breast tumorigenesis by the EglN2 prolyl hydroxylase. Cancer Cell. 2009;16:413–24. doi:10.1016/j.ccr.2009.09.029.
63. Weigelt B, Horlings HM, Kreike B, Hayes MM, Hauptmann M, Wessels LF, et al. Refinement of breast cancer classification by molecular characterization of histological special types. J Pathol. 2008;216:141–50. doi:10.1002/path.2407.
64. Tognon C, Knezevich SR, Huntsman D, Roskelley CD, Melnyk N, Mathers JA, et al. Expression of the ETV6-NTRK3 gene fusion as a primary event in human secretory breast carcinoma. Cancer Cell. 2002;2:367–76.
65. Weigelt B, Kreike B, Reis-Filho JS. Metaplastic breast carcinomas are basal-like breast cancers: a genomic profiling analysis. Breast Cancer Res Treat. 2009;117:273–80. doi:10.1007/s10549-008-0197-9.
66. Miki Y, Swensen J, Shattuck-Eidens D, Futreal PA, Harshman K, Tavtigian S, et al. A strong candidate for the breast and ovarian cancer susceptibility gene BRCA1. Science. 1994;266:66–71.
67. Wooster R, Bignell G, Lancaster J, Swift S, Seal S, Mangion J, et al. Identification of the breast cancer susceptibility gene BRCA2. Nature. 1995;378:789–92.
68. Stratton MR, Rahman N. The emerging landscape of breast cancer susceptibility. Nat Genet. 2008;40:17–22.

69. Lakhani SR, Jacquemier J, Sloane JP, Gusterson BA, Anderson TJ, van de Vijver MJ, et al. Multifactorial analysis of differences between sporadic breast cancers and cancers involving BRCA1 and BRCA2 mutations. J Natl Cancer Inst. 1998;90:1138–45.
70. Lakhani SR. The pathology of familial breast cancer: morphological aspects. Breast Cancer Res. 1999;1:31–5.
71. Marcus JN, Watson P, Page DL, Narod SA, Lenoir GM, Tonin P, et al. Hereditary breast cancer: pathobiology, prognosis, and BRCA1 and BRCA2 gene linkage. Cancer. 1996;77:697–709.
72. Pathology of familial breast cancer: differences between breast cancers in carriers of BRCA1 or BRCA2 mutations and sporadic cases. Breast Cancer Linkage Consortium. Lancet. 1997;349:1505–10.
73. van 't Veer LJ, Dai H, van de Vijver MJ, He YD, Hart AA, Mao M, et al. Gene expression profiling predicts clinical outcome of breast cancer. Nature. 2002;415:530–6.
74. Larsen MJ, Kruse TA, Tan Q, Laenkholm AV, Bak M, Lykkesfeldt AE, et al. Classifications within molecular subtypes enables identification of BRCA1/BRCA2 mutation carriers by RNA tumor profiling. PLoS One. 2013;8:e64268.
75. Spurdle AB, Couch FJ, Parsons MT, McGuffog L, Barrowdale D, Bolla MK, et al. Refined histopathological predictors of BRCA1 and BRCA2 mutation status: a large-scale analysis of breast cancer characteristics from the BCAC, CIMBA, and ENIGMA consortia. Breast Cancer Res. 2014;16:3419. doi:10.1186/s13058-014-0474-y.
76. Bombonati A, Sgroi DC. The molecular pathology of breast cancer progression. J Pathol. 2011;223:307–17. doi:10.1002/path.2808.
77. Evans AJ, Pinder SE, Snead DR, Wilson AR, Ellis IO, Elston CW. The detection of ductal carcinoma in situ at mammographic screening enables the diagnosis of small, grade 3 invasive tumours. Br J Cancer. 1997;75:542–4.
78. Ma XJ, Salunga R, Tuggle JT, Gaudet J, Enright E, McQuary P, et al. Gene expression profiles of human breast cancer progression. Proc Natl Acad Sci U S A. 2003;100:5974–9.
79. Porter D, Lahti-Domenici J, Keshaviah A, Bae YK, Argani P, Marks J, et al. Molecular markers in ductal carcinoma in situ of the breast. Mol Cancer Res. 2003;1:362–75.
80. Buerger H, Otterbach F, Simon R, Poremba C, Diallo R, Decker T, et al. Comparative genomic hybridization of ductal carcinoma in situ of the breast-evidence of multiple genetic pathways. J Pathol. 1999;187:396–402.
81. Quinn CM, Ostrowski JL, Harkins L, Rice AJ, Loney DP. Loss of bcl-2 expression in ductal carcinoma in situ of the breast relates to poor histological differentiation and to expression of p53 and c-erbb-2 proteins. Histopathology. 1998;33:531–366.
82. Vincent-Salomon A, Lucchesi C, Gruel N, Raynal V, Pierron G, Goudefroye R, et al. Integrated genomic and transcriptomic analysis of ductal carcinoma in situ of the breast. Clin Cancer Res. 2008;14:370–8. doi:10.1158/1078-0432.CCR-07-1465.
83. Zafrani B, Leroyer A, Fourquet A, Laurent M, Trophilme D, Validire P, et al. Mammographically-detected ductal in situ carcinoma of the breast analyzed with a new classification. A study of 127 cases: correlation with estrogen and progesterone receptors, p53 and c-erbb-2 proteins, and proliferative activity. Semin Diagn Pathol. 1994;11:208–14.
84. Buerger H, Mommers EC, Littmann R, Diallo R, Brinkschmidt C, Poremba C, et al. Correlation of morphologic and cytogenetic parameters of genetic instability with chromosomal alterations in in situ carcinomas of the breast. Am J Clin Pathol. 2000;114:854–9.
85. Hwang ES, DeVries S, Chew KL, Moore 2nd DH, Kerlikowske K, Thor A, et al. Patterns of chromosomal alterations in breast ductal carcinoma in situ. Clin Cancer Res. 2004;10:5160–7.
86. Simpson JF, Page DL. The p53 tumor suppressor gene in ductal carcinoma in situ of the breast. Am J Pathol. 2000;156:5–6.
87. van de Vijver MJ, Peterse JL, Mooi WJ, Wisman P, Lomans J, Dalesio O, et al. Neu-protein overexpression in breast cancer. Association with comedo-type ductal carcinoma in situ and limited prognostic value in stage II breast cancer. N Engl J Med. 1988;319:1239–45.
88. Adeyinka A, Emberley E, Niu Y, Snell L, Murphy LC, Sowter H, et al. Analysis of gene expression in ductal carcinoma in situ of the breast. Clin Cancer Res. 2002;8:3788–95.

89. Hannemann J, Velds A, Halfwerk JB, Kreike B, Peterse JL, van de Vijver MJ. Classification of ductal carcinoma in situ by gene expression profiling. Breast Cancer Res. 2006;8:R61.
90. Cowell CF, Weigelta B, Sakrb RA, Nga CKY, Hicksc J, Kingb TA, et al. Progression from ductal carcinoma in situ to invasive breast cancer: revisited. Mol Oncol. 2013;7:859–69. doi:10.1016/j.molonc.2013.07.005.
91. Hernandez L, Wilkerson PM, Lambros MB, Campion-Flora A, Rodrigues DN, Gauthier A, et al. Genomic and mutational profiling of ductal carcinomas in situ and matched adjacent invasive breast cancers reveals intra-tumour genetic heterogeneity and clonal selection. J Pathol. 2012;227:42–52.
92. Heselmeyer-Haddad K, Berroa Garcia LY, Bradley A, Ortiz-Melendez C, Lee WJ, Christensen R, et al. Single-cell genetic analysis of ductal carcinoma in situ and invasive breast cancer reveals enormous tumor heterogeneity yet conserved genomic imbalances and gain of MYC during progression. Am J Pathol. 2012;181:1807–22.
93. Bryan BB, Schnitt SJ, Collins LC. Ductal carcinoma in situ with basal-like phenotype: a possible precursor to invasive basal-like breast cancer. Mod Pathol. 2006;19:617–21.
94. Clark SE, Warwick J, Carpenter R, Bowen RL, Duffy SW, Jones JL. Molecular subtyping of DCIS: heterogeneity of breast cancer reflected in pre-invasive disease. Br J Cancer. 2011;104:120–7. doi:10.1038/sj.bjc.6606021.
95. Muggerud AA, Hallett M, Johnsen H, Kleivi K, Zhou W, Tahmasebpoor S, et al. Molecular diversity in ductal carcinoma in situ (DCIS) and early invasive breast cancer. Mol Oncol. 2010;4:357–68. doi:10.1016/j.molonc.2010.06.007.
96. Tamimi RM, Baer HJ, Marotti J, Galan M, Galaburda L, Fu Y, et al. Comparison of molecular phenotypes of ductal carcinoma in situ and invasive breast cancer. Breast Cancer Res. 2008;10:R67. doi:10.1186/bcr2128.
97. Roses RE, Paulson EC, Sharma A, Schueller JE, Nisenbaum H, Weinstein S, et al. HER-2/neu overexpression as a predictor for the transition from in situ to invasive breast cancer. Cancer Epidemiol Biomark Prev. 2009;18:1386–69. doi:10.1158/1055-9965.EPI-08-1101.
98. Hicks J, Krasnitz A, Lakshmi B, Navin NE, Riggs M, Leibu E, et al. Novel patterns of genome rearrangement and their association with survival in breast cancer. Genome Res. 2006;16:1465–79.
99. Lari SA, Kuerer HM. Biological markers in DCIS and risk of breast recurrence: a systematic review. J Cancer. 2011;2:232–61.
100. Gradishar WJ, Anderson BO, Balassanian R, Blair SL, Burstein HJ, Cyr A, et al. Invasive Breast Cancer Version 1.2016, NCCN clinical practice guidelines in oncology. J Natl Compr Cancer Netw. 2016;14:324–54.
101. Tsikitis VL, Chung MA. Biology of ductal carcinoma in situ classification based on biologic potential. Am J Clin Oncol. 2006;29:305–10.

Nonmelanoma Skin Cancer Actinic Keratosis and Squamous Cell Carcinoma

2

Birgit Aigner, Franz Legat, and Wolfgang Weger

Contents

B. Aigner (✉) • F. Legat • W. Weger
Medical University of Graz, University Clinic of Dermatology and Venerology, Auenbruggerplatz 8, 8036 Graz, Austria
e-mail: birgit.aigner@medunigraz.at; franz.legat@medunigraz.at; Wolfgang.weger@medunigraz.at

J. Haybaeck (ed.), *Mechanisms of Molecular Carcinogenesis – Volume 2*,
DOI 10.1007/978-3-319-53661-3_2

Abstract
Nonmelanoma skin cancer is the most common malignant tumor in Caucasians. Its pathogenesis is multifactorial. With a special focus on actinic keratosis and squamous cell carcinoma, we will give a short overview concerning basic knowledge and current histological classifications. Further, we discuss tumorigenesis and influencing factors, such as UV radiation, the inflammatory environment, hypoxia, presence of viruses, and the roles of the innate and adaptive immune system. Moreover, we give a short overview about targeted therapy of actinic keratosis and squamous cell carcinoma.

2.1 Introduction

Nonmelanoma skin cancer (NMSC) is the most common malignant tumor in Caucasians, comprising mainly basal cell carcinoma and cutaneous squamous cell carcinoma (SCC), among others [1–6]. Its etiology is multifactorial; environmental factors (e.g., cumulative lifetime sun exposure), genotype (recognized especially in genetic syndromes of impaired DNA repair, e.g., xeroderma pigmentosum syndrome) and phenotype (e.g., fair-skinned people), and viral infections, as well as suppression of the immune system, are of major impact [3, 5, 7–10].

The precursor lesion of SCC is the actinic keratosis (AK) [5, 11]. In Europe, a prevalence of 15% in men and 6% in women has been estimated. Over the age of 70 years, the prevalence increases to 34% in males and 18% in females [12]. Epidemiologic, clinical, histological, and genetic evidence exists for conversion of AKs to SCCs. Clinical studies suggest that between 0.025 and 16% of AKs progress to invasive SCCs, with extrapolation studies suggesting the risk of progression at approximately 10% [5, 13]. SCCs have an estimated lifetime risk of up to 11% [14]. Despite the fact that the overall 5-year cure rate is greater than 90% for SCC, this tumor can metastasize depending on its differentiation, location, and depth of invasion [10], and the high and still rising incidence substantially contributes to morbidity and mortality [7, 10, 15].

2.2 Histological Classifications

AKs are histologically classified based on the degree of epidermal keratinocytic atypia (keratinocytic intraepidermal neoplasia = KIN) (Table 2.1) [5]. In grade I (KIN I), atypical keratinocytes are present in the basal and suprabasal layer, whereas in grade II (KIN II) atypical keratinocytes cover the lower two-thirds of the epidermis. Grade III (KIN III) is defined by full epidermal atypia; this stage is equivalent to a SCC in situ [5, 16].

SCC is histologically characterized by epidermal proliferation of atypical keratinocytes invading the dermis, and areas of detachment may be present. Atypical keratinocytes have an eosinophilic cytoplasm and enlarged nuclei; mitotic figures are frequently observed. The WHO classification lists various subtypes based on the respective histomorphology (Table 2.2). Furthermore, SCCs are histologically graded based on the

Table 2.1 Clinical and histological grading of AKs

Grade	Clinical	Degree of epidermal involvement
Keratinocyte intraepidermal neoplasia type I (KIN I)	Subclinical lesions or a flat macule in sun-damaged skin	Basal layer crowding with hyperchromasia, subtle crowding, lack of hyperkeratosis or parakeratosis
Keratinocyte intraepidermal neoplasia type II (KIN II)	Pink to red papule or plaque with a rough surface	Atypical keratinocytes involving the lower two-thirds of the epidermis, alternating parakeratosis and orthokeratosis
Keratinocyte intraepidermal neoplasia type III (KIN III)	Red, scaly plaque	Atypical keratinocytes involving the entire epidermis, equivalent to squamous cell carcinoma in situ

Table 2.2 WHO histological classification of SCCs (see also [17])

Histological subtypes of SCCs	
Spindle cell SCC	Acantholytic SCC
Verrucous SCC	Keratoacanthoma
Lymphoepithelioma-like carcinoma	Desmoplastic SCC

Table 2.3 Histopathological grading of SCCs

UICC 1987		Broders	Percentage of undifferentiated tumor cells
Gx	Grade of differentiation cannot be determined		
G1	Well differentiated	Grad I	75% keratinocytes are well differentiated
G2	Moderately differentiated	Grad II	> 50% keratinocytes are well differentiated
G3	Poorly differentiated	Grad III	> 25% keratinocytes are well differentiated
G4	Not differentiated—spindle cell variety	Grad IV	< 25% keratinocytes are well differentiated

degree of nuclear atypia and dedifferentiation, both the UICC (Union International Contre Cancer) and the Broders classification are currently used (Table 2.3).

2.3 Influencing Factors for Carcinogenesis

2.3.1 UV Radiation

The spectrum of UV radiation can be subdivided into UVA (320–400 nm), UVB (290–320 nm), and UVC (200–290 nm) [18]. UV radiation itself acts as a potent and complete carcinogen via inducing genetic mutations in keratinocytes and promoting tumor cell growth [18]. Atypical keratinocytes are characterized by high rates of loss of heterozygosity, specifically on chromosome arms 17p, 17q, 9p, 9q, and 13q [5]. UV radiation furthermore is capable of altering epidermal growth factor receptors (EGFR), influencing the transcription factor NFkB (which will be discussed later in detail), as well as possibly activating the oncogene *H-ras* [5]. Together, the altered signal transduction results in inflammatory processes.

UVB radiation results in direct mutagenic effects on DNA due to transition of C → T and CC → TT [19]. Mutations in the tumor suppressor gene *p53*, which is located on chromosome 17p13.1, are frequent and most likely UVB induced [5]. Those mutations result in failure of apoptosis and therefore in genomic instability. UVB radiation further leads to expression of the so-called *Rhob* (Ras homolog gene family, member B) GTPase gene, which plays a major role in regulating apoptosis, therefore supporting initiation of SCC [20]. Moreover, the expression of cyclooxygenase 2, a parameter present in inflammatory setting, in the cutis is induced via UVB [21].

UVA radiation penetrates in deeper layers of the skin and leads to T → G mutations, and UVA light is moreover capable of producing reactive oxygen species (ROS) itself, resulting in oxidative damaged cells and abnormal proliferation. Oxidative stress leads to a production of platelet-activating factor (PAF), which per se might result in an increase of ROS.

2.3.2 Inflammation and Cancer

Several lines of evidence exist for a significant impact of chronic and persistent inflammatory processes contributing to various aspects of carcinogenesis [22–26]. Indeed, decisive inflammatory effects are known for tumor development, including initiation, promotion, and metastasis [25]. It is the expression of various immune mediators and modulators, as well as the abundance, localization, and activation state of different cell types in the tumor, which dictate whether tumor-promoting inflammation or antitumor immunity will follow [25].

Essential mediators of inflammation, cytokines and chemokines, are involved in acute phase responses and are major modulators to the extent of inflammation [27]. Several authors hypothesized that these soluble mediators play a certain role in initiation and progression, as well as in invasion and metastatic spread of cancers [28, 29]. Indeed, a distinct set of cytokines and chemokines are present in most, if not all, cancers [24]. Besides, TNF-α, IL-1 and IL-6, IFN-γ, and IL-10 have been repeatedly described to promote carcinogenesis of different cancers [23, 24, 30, 31].

2.4 Immune Cells in Tumor Genesis

2.4.1 Innate Immune Cells

Cells of the innate immune system are known to contribute to early neoplastic development [32]. Innate immune cells found in tumor microenvironment are macrophages, neutrophils, mast cells, antigen-presenting cells, and natural killer cells [25], all of which are able to produce or lead to the production of cytokines.

The majority of those cells express encoded pattern-recognition Toll-like receptors (TLRs) [32]. Activation of TLRs per se leads to several cascades, including production of interferon alpha (IFN-α), a Th1-mediated response and activation of the transcription factor NFkB [32], which is a key regulator, involved in many stages of inflammation and carcinogenesis [30]. In tumor cells and unaffected epithelial cells, too, NFkB can activate the expression of inflammatory cytokines and various angiogenic factors [30] and regulates the expression of genes that suppress tumor cell death [33]. In tumor cells, it might induce the expression of antiapoptotic cells or proteins, like BCL2 [30]. However, it has to be stated that NFkb has the potential to exhibit pro- and antitumor functions. It is for sure one important factor influencing the balance between pro- and anti-tumorigenic settings [30].

Innate immune cells, moreover, can directly contribute to cancer development by induction of DNA damage by free radicals [34].

The complement system is part of the innate immune system and keeps microbes and necrotic cells under control. Complement factor H (CFH)—one of the main inhibitors of the alternative pathway of the complement system—is clearly overexpressed in SCC and is capable of protecting tumor cells from complement attacks. Interestingly, some inflammatory molecules, such as IL-1β, TNF-α, and IFN-γ, seem to increase the expression of CFH in SCC cells. CFH was suggested to be a "biomarker" for progression of AK to SCC and a possible therapeutic target for the future.

Communication within the innate immune system and between the innate and adaptive immune system is based on various mediators. Produced by innate cells, IL-12, IL-15, IL-18, TNF-α, IFN-γ, and other cytokines promote differentiation and activation of immature antigen-presenting cells, B cells, as well as T-cells [32]. Th1-polarized cytokines, such as IL-1, TNF-α, and IFN-γ, are believed to have pro-inflammatory functions, whereas Th2-polarized cytokines, like IL-1, IL-10, and IL-13, are considered to have anti-inflammatory properties [30, 35].

Certain Th1-polarized cytokines have crucial roles in inflammation and, therefore, are being discussed in more detail here.

TNF-α is directly involved in inflammatory tissue alteration and stimulates other pro-inflammatory molecules responsible for the initiation of inflammatory processes [23, 31, 36, 37]. TNF-a modulates proliferation, activation, and differentiation of other cells [38], and its tumor-promoting effects have been demonstrated previously [33]. By stimulating the production of genotoxic molecules, this can lead to DNA mutations [33]. TNF-α itself is likely to be expressed by tumors and thus could operate as an autocrine cancer promoter [23]. Via the induction of genes encoding NFkB-dependent antiapoptotic molecules, TNF-α is capable of supporting tumor cell survival [33]. In addition, TNF-α further stimulates angiogenesis and metastasis via the suppression of many T-cell responses and macrophage effects [33]. Interestingly, another member of the TNF family, Fas (CD95), an initiator for apoptosis, seems to be almost direct proportionally diminished during the transformation from AK to SCC [39].

IL-6 is a pro-inflammatory molecule that plays an important role in dermal and epidermal cell proliferation and differentiation [38]. The correlation between IL-6 and both tumor promotion and metastasis has been discussed previously [40, 41]. IL-6 is one of the major growth-promoting inflammatory cytokines and is believed to be further responsible for antiapoptotic behavior [24, 33].

IL-17 has been demonstrated as being a key player in inflammatory responses, inducing the production of TNF-α, IL-6, IL-IL-1β itself, cooperating with IL-6 and TNF-α to enhance pro-inflammatory molecules and initiating the recruitment of immune cells to peripheral tissues [33].

IL-1, especially IL-1β, is thought to be an important factor promoting tumor invasiveness and metastasis [23].

IFN-γ is a pro-inflammatory cytokine that is produced by activated T-lymphocytes and NK-cells and influences a multitude of cellular processes [42]. IFN-γ modulates proliferation and differentiation of epidermal cells [42]. Intratumoral expression of IFN-γ has been associated with a more aggressive type of skin cancer, e.g., malignant melanoma [43].

IL-10 works as an immunosuppressive molecule and is able to inhibit NFkB and the production of several inflammatory cytokines [33]. IL-10 has been referred to as an important mediator in tumor formation [44, 45]. As for TNF-α, IL-10 may be expressed by tumor cells themselves [45, 46].

However, production of cytokines can have a pro- and anti-inflammatory or tumorigenic role. It depends not only on the type of cytokines that are produced but also on the surrounding stroma and microenvironment and on whether pro- or anti-inflammatory mechanisms are present. Those cytokines that are believed to present more pro-tumorigenic properties are IL-6, IL-17, and IL-23 [25].

Besides cytokines, an important link between innate and adaptive immune systems is Fcγ receptors [22]. Andreu et al. stated that B cells and humoral immunity support de novo carcinogenesis by activating FcRys [22]. FcRys are able to perform interactions between circulating (auto)antibodies and innate immune cells [22, 34].

2.4.2 Adaptive Immune Cells

B and T cells are the major players in the adaptive immune system. B cells are important components of the adaptive immunity, influencing premalignant progression of early squamous carcinogenesis in HPV 16 mice. B cells further lead to recruitment of leucocytes and are able to mediate chronic inflammation via altering cytokine and chemokine levels and may produce CXL1, IL-4, IL-6, and IL-10 [47]. It has been stated that spontaneous activation of B cells promotes de novo epithelial carcinogenesis by induction of chronic inflammation [47]. Moreover, via an antibody production, the complement cascade may be started, and due to the potent inflammatory factors C3a and C5a, recruitment and activation of leucocytes are induced [47]. B cells further are capable of inhibiting Th1-mediated anti-tumorigenic stimuli [22].

However, the adaptive immune system may modulate carcinogenesis by antitumor cytotoxic T cells and cytokine-mediated lysis of cancer cells [34]. Andreu et al. stated that chronically activated leucocytes recruited to premalignant tissues might functionally contribute to cancer development [22]. Depending on the antibody production, innate immune cells can be activated, triggering the inflammatory cascade (Figs. 2.1 and 2.2).

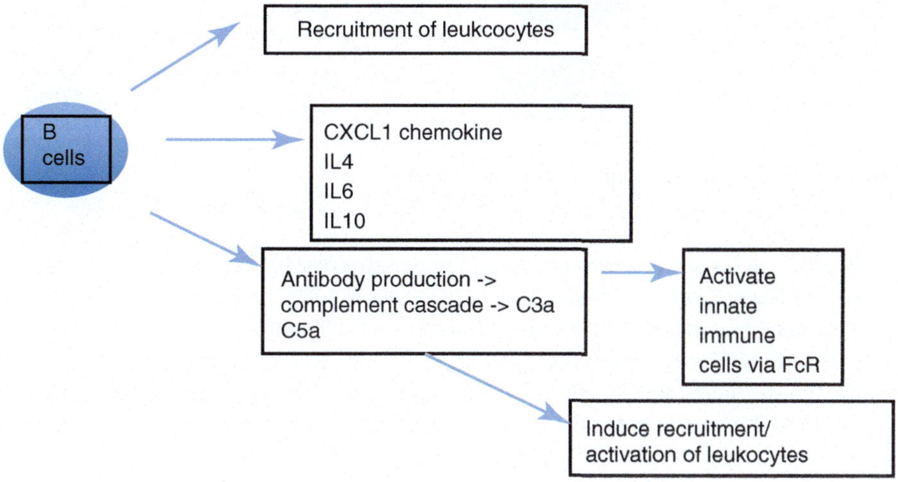

Fig. 2.1 B cells and selected roles

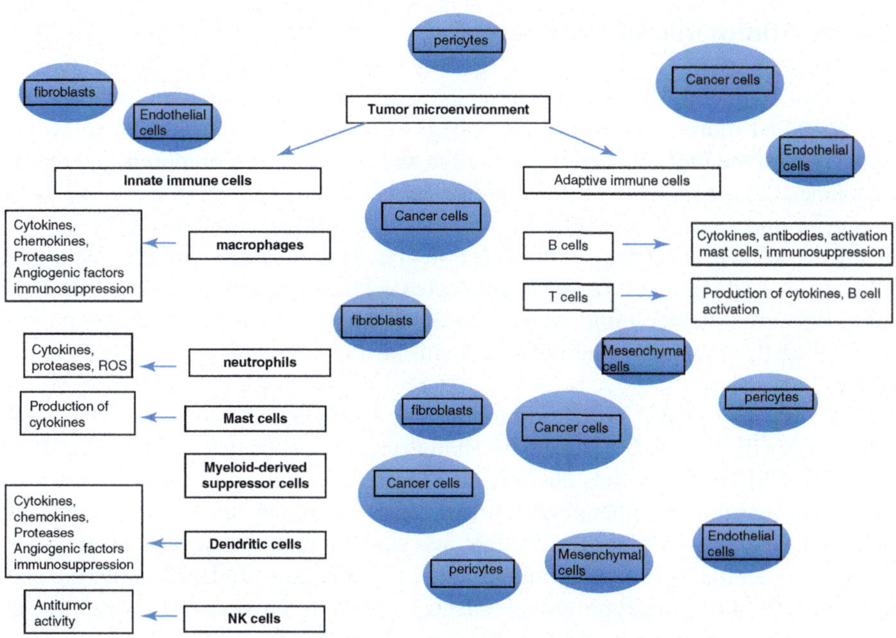

Fig. 2.2 Tumor microenvironment

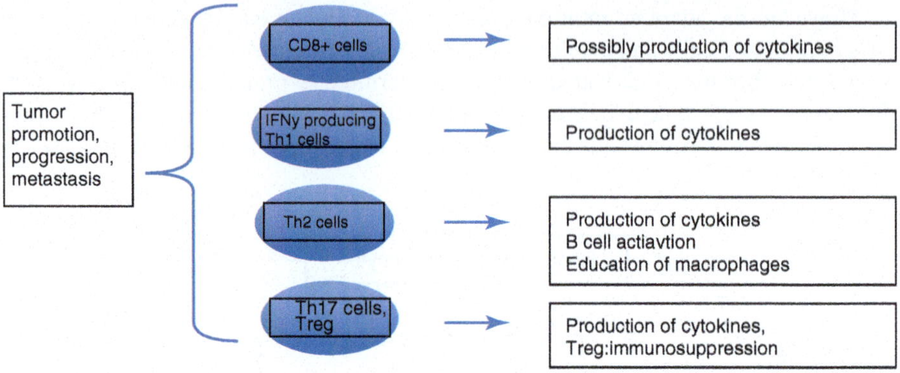

Fig. 2.3 T cells and roles with respect to tumorigenesis

T cells can have tumor-promoting and tumor-suppressing effects [25]. T cells consist of cytotoxic CD8+ cells and CD4+ cells, whereas CD4+ cells encompass T-helper, T-regulatory, and natural killer cells [25]. Interestingly, natural killer cells are the only ones that do not have a pro-tumorigenic role. T cells, which are predominantly found in the tumor microenvironment of solid tumors, are CD8+ and Th1, Th2, and Th17 cells [25]. CD8+ cells are capable of producing cytotoxic cytokines and may lead to direct lysis of cancer cells [25]. T-helper cells show pro-tumorigenic roles in producing cytokines, and Th2 cells lead to B-cell activation, too [25] (Fig. 2.3).

2.4.3 Tumor Microenvironment

The most frequently found immune cells within the tumor microenvironment are the so-called tumor-associated macrophages (TAMs). They are known to be key regulators of the link between inflammation and cancer and are important sources of cytokines [25, 48]. TAMs promote tumor growth, promote angiogenesis, are capable of remodeling tissue, and may suppress the adaptive immunity [25, 48, 49]. They are known to interact with cancer stem cells and cancer-initiating cells [48, 49], and a high level of TAMs is connected with poor prognosis [25, 49]. TAM products can influence tumorigenesis in many ways, as immunosuppressive cytokines, like IL-10 and pro-angiogenic growth factors, are produced by TAMs themselves [48].

The majority of the TAMs within the tumor microenvironment do have a M2 phenotype (IL-12 low, IL-23 low, IL-10 high) and are activated by IL-4 and IL-13 [25, 48]. CSF-1 and various chemokines, like CCl2, CCL18, and CCL17, can set macrophages in a M2 phenotype [49]. It has to be stated that M1 macrophages, which have an IL-12 high, IL-23 high, and IL-10 low phenotype, are believed to have the potential to exhibit antitumor activity and express Th1 cell-attracting chemokines [48, 49]. Cancer progression itself is known to be associated with a switch from the M1 phenotype to M2 [25, 49] (Figs. 2.4 and 2.5).

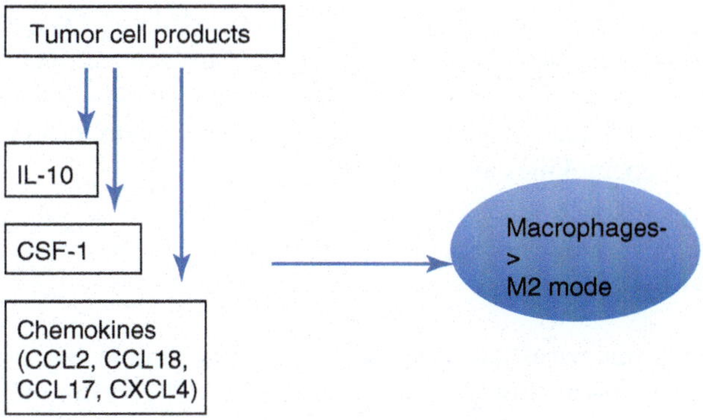

Fig. 2.4 Macrophages with a switch to M2 mode

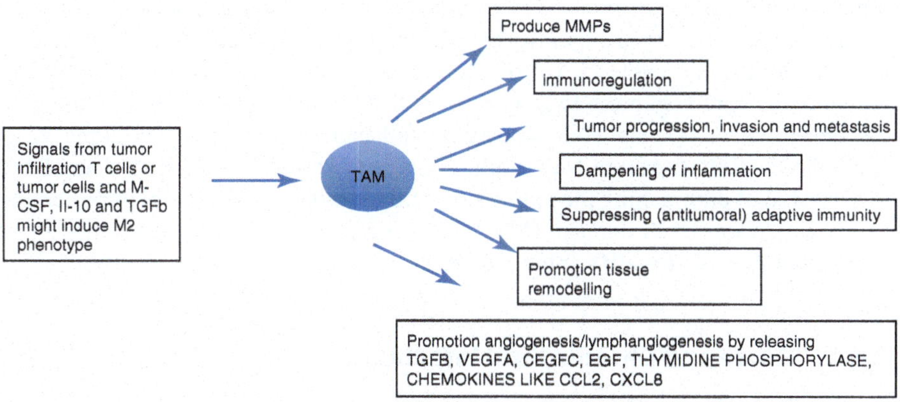

Fig. 2.5 TAMs and their effects

TAMs are further known to be capable of producing matrix metalloproteinases (MMPs) [48]. Two major MMPs, MMP-2 and MMP-9, functionally contribute to cancer development in different mechanisms, such as epithelial hyperproliferation, activation of angiogenesis, as well as tissue remodeling [34, 47]. Among others, complement factor H is known to be able to influence MMP-2 and MMP-3 [50]. A clear upregulation of MMP-1 and MMP-10 in SCC versus AK was further observed in a transcriptome profiling by Lambert et al. in 2014; both collagenases are believed to play certain parts in invasive behavior and therefore metastatic potential [7].

2.5 Hypoxia or Oxidative Stress and Inflammation

Hypoxia is known to significantly influence the innate and the adaptive immune system [51]. In tumors, concentrations of oxygen are frequently diminished, and high levels of hypoxia-inducible factors alpha and beta (HIFα, HIFβ) can be found

[51]. In hypoxia, inflammatory cytokines may be released by CD4+ cells. Moreover, Th1 polarization and T-cell effector activity are influenced. The two hypoxia-inducible factors dimerize and bind to a response element, which leads to transcription of NFkβ, and TLR may lead to angiogenesis, metabolism, proliferation, as well as production of chemokines and cytokines [51].

2.6 Viral Infection

It is well known that virus-associated chronic inflammation has pro-tumorigenic potential [25, 33]. Interestingly, more than 15% of all cancers may be linked to viruses [52]. Viral-associated skin cancers appear to depend on three major factors: a persisting virus, UV radiation, and the immune system [52]. Via oncoproteins that target cellular tumor suppressor genes, DNA oncoviruses can directly influence or induce malignant transformation and further control cell cycle, division, as well as escape mechanisms from apoptosis [52]. Moreover, viral-encoded oncoproteins may disrupt the cell cycle [52]. However, oncoviruses alone may not be sufficient for initiating carcinogenesis [52]. One virus that is known to be strongly linked to the development of cutaneous squamous cell cancer is human papilloma virus (HPV) [52]. With respect to NMSC, the so-called "E6-protein" of HPV 5 and HPV 8 is capable of diminishing the expression of IL-8, which leads to loss of protection to UV-dependent DNA damage [53]. Interestingly, patients with AK seemed to be more often positive for HPV infection [54].

2.6.1 Acute and Chronic Viral Infection

In general, during acute viral infection, naïve CD8+ cells are primed by antigens, expand, proliferate expansively, and differentiate into T effector cells, depending on the inflammatory setting [55, 56]. CD8+ T-cell response may lead to massive tissue damage [56]. Those CD8+ cells per se are able to control viral infection, as they recruit leucocytes, have cytotoxic activity, and produce cytokines themselves to eliminate infected cells [55]. Most of the effector cells will die, but up to 10% will become memory cells [57]. These memory T cells can produce cytokines, like TNF, IL-2, and IFN-γ. Due to an increase of the viral load, T cells become less (poly) functional [56]. However, some kind of self-renewal is possible and triggered by IL-7 and IL-15 [56]. In chronic viral infection, T cells may become exhausted and initially lose their ability of being cytotoxic, proliferation decreases, and they fail to differentiate into memory cells [56, 57]. The inhibitory receptor PD-1 seems to be able to modulate T-cell exhaustion [57]. Exhausted T cells overexpress several cell surface inhibitory receptors, like CTLA-4 and PD-1, and genes for chemotaxis, migration, and adhesion may be altered [57].

CD4+ cells are influenced by acute viral infection, too. Effector CD4+ cells are able to inhibit viral replication by producing antiviral cytokines and activate antigen-presenting cells and support CD8+ cell priming and promote the induction of

Table 2.4 Clinical grading for actinic keratosis (based on Werner et al. [61] S3 Guidelines for the Treatment of Actinic Keratosis)

Clinical grading of actinic keratosis		
G1	Mild	Slight palpability of AK
G2	Moderate	Easily palpable and visible AK
G3	Severe	Thick AK

Table 2.5 Subgroups concerning distribution and amount of AK as well as recommended therapeutic interventions based on S3 Guidelines [61] by Werner et al. and Dreno et al. [59])

Subgroup	Definition of subgroup	Recommended therapy
Single AK lesions	1–5 AK per affected body region or field	Cryotherapy, surgery, curettage, and electrocoagulation
Multiple AK	>6 AK in affected body region or field	Laser, (daylight) photodynamic therapy, topical treatments: 5-FU, imiquimod, COX inhibitors, ingenol mebutate
Field cancerization	>6 AK and chronic sun damage adjacent to various fields/body regions	(Daylight) Photodynamic therapy, 5-FUO, imiquimod, ingenol mebutate, chemical peeling

memory cells [55]. Via IL-2 CD4+ cells can greatly influence differentiation of CD8+ effector cells and differentiation of memory T cells [55]. IL-2 is further necessary for sufficient CD4+ memory differentiation [55]. Moreover, it has been stated that CD4+ cells are key components for long-term control of viruses [58]. Virally affected B cells are capable of producing antiviral neutralizing antibodies [56, 58].

2.6.2 Treatment: Actinic Keratosis

Concerning the treatment of actinic keratosis, several guidelines or consensus statements recommend to subdivide actinic keratosis according to their amount, grade, and distribution [59–61]. In addition to the histological grading (described above), a clinical grading exists. From a clinical aspect, actinic keratoses present themselves as flat to thick, mainly red, rough lesions showing a broad variety in thickness [62] (Tables 2.4 and 2.5).

2.7 Targeted Therapies in AK

2.7.1 5 Fluorouracil (5-FU)

5-FU is a pyrimidine analog, inhibiting DNA and RNA synthesis and therefore resulting in diminishing growth of atypical cells [59]. Topical 5-FU is used on a regular basis for treatment of thin to keratotic AKs on large areas once or twice a day for up to 4 weeks [59–61].

2.7.2 Imiquimod

Imiquimod is a Toll-like receptor 7 agonist and influences immune responses, above all stimulating apoptosis [59, 60]. It is used in patients suffering from non-hyperkeratotic multiple AKs. Two different treatment regimens, based on the concentration, may be followed: (1) imiquimod 5% cream which should be applied three times a week for 4 weeks (a second cycle may follow after a 4-week break) [59–61] and (2) imiquimod 3.75% cream that can be applied to the affected face or scalp once daily in the evening for 2 weeks, followed by a 2-week treatment break, and another treatment cycle of 2 weeks.

2.7.3 Cyclooxygenase (COX)-2 Inhibitors

COX 1 and COX 2 are enzymes that oxidize arachidonic acid to prostaglandin. COX 1 is the isoform expressed on a regular basis; COX 2, on the other hand, is synthesized in the presence of inflammation [63]. Its overexpression itself has already been linked to various tumors [21, 63]. An increase of COX 2 seems to be proportional to the development from AK to SCC [21, 63], and therefore its inhibition is used as a therapeutic target preventing AK transformation to SCC. For the treatment of AKs, a combination of topical diclofenac gel and hyaluronic acid is on the market [60, 61]. It is recommended for superficial and non-hyperkeratotic AKs and shall be administered twice a day for 8–12 weeks [59, 60].

2.7.4 Ingenol Mebutate

Ingenol mebutate gel is a newer substance derived from the so-called *Euphorbia peplus* plant [59]. It seems to lead not only to cytotoxicity but also to expression of inflammatory cytokines and may be used in non-hyperkeratotic areas of less than 25 cm^2; a lower dose is applied once daily for three consecutive days on the scalp and face, whereas on the trunk, a higher dose is to be administered for 2 days only [59, 61].

2.7.5 Photodynamic Therapy (PDT)

The photodynamic reaction depends on the simultaneous presence of a photosensitizer, visible light, and oxygen. In the treatment of actinic keratosis, a cream containing 5-aminolevulinic acid (5 ALA), or its methyl ester, is applied to the target lesions, eventually resulting in the accumulation of the photosensitizer protoporphyrin IX in tumor cells. After 3 h of incubation, the irradiation with visible light (blue or red light), in the presence of oxygen within the tumor, induces reactive oxygen species, resulting in cytotoxicity [60, 62]. PDT is ideal for patients with multiple actinic keratosis or field cancerization [61]; however, it can be painful during irradiation. Less painful is the so-called daylight PDT, whereby daylight is used instead of red light to induce the photodynamic reaction. It has been shown that treatment responses appear to be similar to conventional PDT for AK grade I and II.

2.7.6 Treatment: Cutaneous Squamous Cell Cancer

As actinic keratoses are precursor lesions of SCC, it is not surprising that the clinical appearance may be very similar. SCCs have a broad range of clinical appearance; usually they present either as hyperkeratotic plaque, crusty, or ulcerous lesion [17]. In any case where a clinical differentiation between hyperkeratotic AK and invasive SCC is not possible, a histological evaluation (see Tables 2.2 and 2.3) is recommended [17, 61]. Clinical staging should be based on the TNM staging, as in other invasive tumors. The latest TNM classification for cutaneous SCCs was published in 2010 by AJCC (Table 2.6) [64]. Breuninger et al. called for a more detailed histological description of SCCs: (1) histological grading, (2) statement concerning resection margins, (3) maximum longitudinal diameter, and (4) vertical tumor diameter to guarantee a guideline-specific treatment [17].

Clearly surgical excision is the first-line therapy; in those tumors that cannot be resected completely, radiotherapy should be performed [17]. If a vertical tumor thickness more than 2 mm is present, a lymph node ultrasound should be performed [17]. If metastases are present, a clinical staging with additional imaging tests and systemic treatment should be done [65]. To date, no clear recommendations can be

Table 2.6 AJCC TNM staging

AJCC TNM classification (based on Breuninger et al. [17] and Bonerandi et al. [62])	
T classification	
T1	Tumor <2 cm at largest horizontal width
T2	Tumor <2 cm at largest horizontal width
	+2–5 high-risk features
	Or tumor >2 cm at largest horizontal width
T3	Infiltration of facial and cranial bone
T4	Infiltration of skeletal bone or skull base
N classification	
N1	No regional lymph node metastases
N2a	Solitary, ipsilateral lymph node metastasis, maximum diameter >3 cm to max. 6 cm
N2b	Multiple, ipsilateral lymph node metastases, all with a maximum diameter 6 cm
N2c	Multiple, ipsilateral or contralateral lymph node metastases, all with a maximum diameter 6 cm
N3	Lymph node metastasis, diameter >6 cm
M classification	
M0	No distant metastases
M1	Distant metastases

made for systemic invasive metastasized SCC, and patients should be treated in centers under clinical studies once the patient is informed and has given consent [62, 65]. A polychemotherapy with cisplatin and 5 FUO is called the first line by Breuninger et al.; however, various chemotherapeutic agents (platins, 5 FUO, bleomycin, methotrexate, gemcitabine, etc.) have been used for the treatment of SCC, and no clear recommendation can be made from those reports [65].

2.8 Future (Targeted) Therapies or Treatment Options Used in Clinical Studies

2.8.1 Anti-epidermal Growth Factor Receptors (EGFR)

EGFR is a tyrosine kinase receptor, the activation of which may lead to keratinocyte proliferation, angiogenesis, and resistance to apoptosis and therefore leads to tumor progression [63, 66]. EGFR seems to be overexpressed in a multitude of SCCs and in all metastatic forms of it [66]. There exist various agents that are used for treatment. Cetuximab is a chimeric human-murine monoclonal antibody [62], and promising results for cutaneous squamous cell cancer have been published (mainly case reports) [65]. Both gefitinib and erlotinib inhibit EGFR and are currently under investigation [66]. Those inhibitors should be kept in mind as second-line treatment due to Stratigos et al. [65].

Insulin-like growth factor 1 receptor inhibition Picropodophyllin is a potent inhibitor of IGF-1R, and has been studied only in murine models [67], but showed promising results in combination with EGFR.

Conclusion

Actinic keratosis and cutaneous squamous cell cancer belong to the most common cancers all over the world, with a still rising incidence. Their clinical impact is mostly based on the amount of affected people, but a better understanding of the pathogenesis and insights into molecular processes are inevitable for future (pathogenetic and therefore therapeutic) concepts.

References

1. Diepgen TL, Mahler V. The epidemiology of skin cancer. Br J Dermatol. 2002;146(61):1–6.
2. Hoey SE, Devereux CE, Murray L, Catney D, Gavin A, Kumar S, Donnelly D, Dolan OM. Skin cancer trends in Northern Ireland and consequences for provision of dermatology services. Br J Dermatol. 2007;156:1301–7.
3. Madan V, Lear JT, Szeimies RM. Non melanoma skin cancer. Lancet. 2010;375:673–85.
4. Martinez JC, Otley CC. The management of melanoma and non-melanoma skin cancer: a review for the primary care physician. Mayo Clin Proc. 2001;76:1253–65.
5. Ratushny V, Gober MD, Hick R, Ridky TW, Sexkora JT. From keratinocyte to cancer: the pathogenesis and melding of cutaneous squamous cell cancer. J Clin Invest. 2012; 122(2):464–72.

6. Salasche SJ. Epidemiology of actinic keratosis and squamous cell carcinoma. J Am Acad Dermatol. 2000;42:4–7.
7. Lambert SR, Mladkova N, Gulati A, Hamoudi R, Purdie K, Cerio R, Leigh I, Proby C, Harwood CA. Key differences identified between actinic keratosis and cutaneous squamous cell carcinoma by transcriptome profiling. Br J Cancer. 2014;110:520–9.
8. Pfister DG, Harpern AC. Skin squamous cell cancer: the tie is right for greater involvement of the medical oncologist. J Clin Oncol. 2007;25(15):1953–4.
9. Rosso S, Zanetti R, Martinez C, Tormi MJ, Schraub S, Sancho-Garnier H, Franceschi S, Gafa L, Perea E, Navarro C, Laurent R, Schrameck C, Talamini R, Tumino R, Wechseler J. The multicenter south European study "Helios". II: Different sun exposure patterns in the aetiology of basal cell and squamous cell carcinomas of the skin. Br J Cancer. 1996;73(11):1447–54.
10. Yoong C, De'Ambrosis B. Cutaneous invasive squamous cell carcinoma: 10-year experience and recommendations for follow up. Australas J Dermatol. 2009;50:261–5.
11. Ackerman AB. Solar keratosis is squamous cell carcinoma. Arch Dermatol. 2003;139:1216–7.
12. Memon AA, Tomenson JA, Bothwell J, Friedmann PS. Prevalence of solar damage and actinic keratosis in a Merseyside population. Br J Dermatol. 2000;142:1154–9.
13. Glogau RG. The risk of progression to invasive disease. J Am Acad Dermatol. 2000;42:23–4.
14. Jensen V, Prasad AR, Smith A, Raju M, Wendel CS, Schmelz M, Leyva W, Warneke J, Krouse RS. Prognostic criteria for squamous cell cancer of the skin. J Surg Res. 2010;159:509–16.
15. Harwood CA, Leedham-Green M, Leigh IM, Proby CM. Low-dose retinoids in the prevention of cutaneous squamous cell carcinomas in organ transplant recipients: a 16-year retrospective study. Arch Dermatol. 2005;141:456–64.
16. Röwert-Huber J, Patel MJ, Forschner T, Ulrich C, Eberle J, Kerl H, Sterry W, Stockfleth E. Actinic keratosis is an early in situ squamous cell carcinoma: a proposal for reclassification. Br J Dermatol. 2007;156(Suppl 3):8–12.
17. Breuninger H, Eigentler T, Bootz F, Hauschild A, Kortmann RD, Wolff K, Stockfleth E, Szeimies RM, Rompel R, Garbe C, Grabbe S. Brief S2k guidelines- cutaneous squamous cell carcinoma. J Dtsch Dermatol Ges. 2013;11(3):37–45.
18. Berman B, Cockerell CJ. Pathobiology of actinic keratosis: ultraviolet-dependent keratinocyte proliferation. J Am Acad Dermatol. 2013;68(1):10–9.
19. Brash DE, Ziegler A, Jonason AS, Simon JA, Kunala S, Leffel DJ. Sunlight and sunburn in human skin cancer: p53, apoptosis, and tumor promotion. J Investig Dermatol Symp Proc. 1996;1:136–42.
20. Meyer N, Peyret-Lacombe A, Canguilhelm B, Medale-Giamarchi C, Mamouni K, Cristini A, Monferran S, Lamant L, Filleron T, Pradines A, Sordet O, Favre G. RhoB promotes cancer initiation by protecting keratinocytes from UVB- induced apoptosis but limits tumor aggressiveness. J Invest Dermatol. 2014;134:203–12.
21. Buckman SY, Gresham A, Hale P, Hruza G, Anast J, Masferrer J, Pentland AP. Cox2 expression s induced by UVB exposure in human skin: implications for the development of skin cancer. Carcinogenesis. 1998;19:723–9.
22. Andreu P, Johansson M, Affara NI, Pucci F, Tan T, Junankar S, Korets L, Lam J, Tawfik D, DeNardo DG, Naldini L, de Visser KE, De Palma M, Coussens LM. FcRγ activation regulates inflammation-associated squamous carcinogenesis. Cancer Cell. 2010;17:121–34.
23. Balkwill F, Charles KA, Mantovani A. Smoldering and polarized inflammation in the initiation and promotion of malignant disease. Cancer Cell. 2005;7:211–7.
24. Germano G, Allavena P, Mantovani A. Cytokines as a key component of cancer-related inflammation. Cytokine. 2008;43:374–9.
25. Grivennikov SI, Greten FR, Karin M. Immunity, inflammation and cancer. Cell. 2010;140:883–99.
26. Mantovani A, Pierotti MA. Cancer and inflammation: a complex relationship. Cancer Lett. 2008;267:180–1.
27. Gambichler T, Skrygan M, Hyun J, Bechara F, Tomi NS, Altmeyer P, Kreuter A. Cytokine mRNA expression in basal cell carcinoma. Arch Dermatol Res. 2006;298:139–41.

28. Szepietowski JC, Walker C, McKenna DB, Hunter JA, McKenzie RC. Leukemia inhibitory factor and interleukin-8 expression in non melanoma skin cancers. Clin Exp Dermatol. 2001;26:72–8.
29. Yen HAT, Chiang LC, Wen KH, Tsai CC, Yu CL, Yu HS. The expression of cytokines by an established basal cell carcinoma cell line (BCC-1/KMC) compared with cultured normal keratinocytes. Arch Dermatol Res. 1996;288:157–61.
30. Mantovani A, Allavena P, Sica A, Balkwill F. Cancer-related inflammation. Nature. 2008;454:436–44.
31. Schioppa T, Moore R, Thompson RG, Rosser EC, Kulbe H, Nedospasov S, Mauri C, Coussens LM, Balkwill FR. B regulatory cells and the tumor-promoting actions of TNF -α during squamous carcinogenesis. Proc Natl Acad Sci U S A. 2011;108(26):10662–1066.
32. De Visser KE, Coussens LM. The interplay between innate and adaptive immunity regulates cancer development. Cancer Immunol Immunother. 2005;54:1143–52.
33. Lin WW, Karin M. A cytokine-mediated link between innate immunity, inflammation and cancer. J Clin Invest. 2007;117(5):1175–83.
34. De Visser KE, Eichten A, Coussens LE. Paradoxical roles of the immune system during cancer development. Nature. 2006;6(1):24–37.
35. Coussens LM, Werb Z. Inflammation and cancer. Nature. 2002;420:860–7.
36. Bose F, Raeli L, Garutti C, Frigerio E, Cozzi A, Crimi M, Caprioli F, Scavelli R, Altomare G, Geginat J, Abrignani S, Reali E. Dual role of anti-TNF therapy: enhancement of TCR-mediated T cell activation in peripheral blood and inhibition of inflammation in target tissues. Clin Immunol. 2011;139:164–76.
37. Zaidi RM, Merlino G. The two faces of interferon-γ in cancer. Clin Cancer Res. 2011;17(19):6118–24.
38. Pietrzak AT, Zalewska A, Chodorowska G, Krasowska D, Michalak-Stoma A, Nockowski P, Osemlak P, Paskowski T, Rolinski JM. Cytokines and anticytokines in psoriasis. Clin Chim Acta. 2008;394:7–21.
39. Berhane T, Halliday GM, Cooke B, Barnetson RS. Inflammation is associated with progression of actinic keratosis to squamous cell carcinomas in humans. Br J Dermatol. 2002;146:810–5.
40. Hilbert DM, Kopf M, Mock BA, Köhler G, Rudikoff S. Interleukin 6 is essential for in vivo development of B lineage neoplasms. J Exp Med. 1995;182:243–8.
41. Smith CW, Chen Z, Dong G, Loukinova E, Pegram MY, Nicholas-Figueroa L, Van Waes C. The host environment promotes the development of primary and metastatic squamous cell carcinomas that constitutively express proinflammatory cytokines IL-1alpha, IL-6, GM-CSF, and KC. Clin Exp Metastasis. 1998;16(7):655–64.
42. Matsuura H, Sakaue M, Subbaramaiahi K, Kamitani H, Eling TE, Dannenbergi AJ, Tanabe T, Inoue H, Arata J, Jetten AM. Regulation of cyclooxygenase-2 by interferon g and transforming growth factor a in normal human epidermal keratinocytes and squamous carcinoma cells. J Biol Chem. 1999;274(41):29138–48.
43. Brocker EB, Zwaldo G, Holzmann B, Macher E, Sorg C. Inflammatory cell infiltrates in human melanoma at different stages of tumor progression. Int J Cancer. 1988;41(4):526–7.
44. Asadullah K, Döcke WD, Sabat R, Ebeling M, Volk HD, Sterry W. Interleukin-10 in der Dermatologie. Hautarzt. 1999;50:12–9.
45. Kim J, Modlin RL, Moy RL, Dubinett SM, McHugh T, Nickoloff BJ, Uyemura K. IL-10 production in cutaneous basal and squamous cell carcinomas. J Immunol. 1995;155:2240–7.
46. Huang SJ, Hijnen D, Murphy GF, Kupper TS, Calarese AW, Mollet IG, Schanbacher CF, Miller DM, Schmults D, Clark RA. Imiquimod enhances IFN-gamma production and effector function of T cells infiltrating human squamous cell carcinoma of the skin. J Invest Dermatol. 2009;126:2676–85.
47. De Visser KE, Korets LV, Coussens LM. De novo carcinogenesis promoted by chronic inflammation is B lymphocyte dependent. Cancer Cell. 2005;7:411–23.
48. Mantovani A, Biswas SK, GAldiero MR, Sica A, Locati M. Macrophage plasticity and polarization in tissue repair and remodelling. J Pathol. 2013;229:176–85.

49. Sica A, Mantovani A. Macrophage plasticity and polarization: in vivo veritas. J Clin Invest. 2012;122(3):787–95.
50. Riihila PM, Nissinen LM, Ala-aho R, Kallajoki M, Grenman R, Meri S, Peltonen S, Peltonen J, Kähäri VM. Complement factor H: a biomarker for progression of cutaneous squamous cell cancer. J Invest Dermatol. 2014;134(2):498–506.
51. Eltzschig HK, Carmeliet P. Hypoxia and inflammation. N Engl J Med. 2011;364:656–65.
52. Arron ST, Jennings L, Nindl I, Rosl F, Bavinck BB, Seckin D, Trakatelli M, Murphy GM. Viral oncogenesis and its role in nonmelanoma skin cancer. Br J Dermatol. 2011;164:1201–13.
53. Akgul B, Bostanci N, Westphal K, Nindl I, Navsaria H, Storey A, Pfister H. Human papillomavirus 5 and 8 E6 downregulate interleukin-8 secretion in primary human keratinocytes. J Gen Virol. 2010;91:888–92.
54. Madkan VK, Cook-Norris RH, Steadman MC, Arora A, Mendoza N, Tyring SK. The oncogenic potential of human papillomaviruses: a review on the role of host genetics and environmental cofactors. Br J Dermatol. 2007;57:228–41.
55. Kaech SM, Wherry EJ. Heterogeneity and cell-fate decisions in effector and memory CD8+ T cell differentiation during viral infection. Immunity. 2007;27:393–405.
56. Virgin HW, Wherry EJ, Ahmed R. Redefining chronic viral infection. Cell. 2009;138:30–50.
57. Wherry JE, Ha SJ, Kaech SM, Haining WN, Sarkar S, Kalia V, Subramaniam S, Blattman JN, Barber DL, Ahmed R. Molecular signature of CD8+ T cell exhaustion during chronic viral infection. Immunity. 2007;27:670–84.
58. Klenerman P, Hill A. T cells and viral persistence: lessons from diverse infections. Nat Immunol. 2005;6(9):873–9.
59. Dreno B, Amici JM, Basset-Seguin N, Cribier B, Claudel JP, Richard MA. Management of actinic keratosis: a practical report and treatment algorithm from AKTeam expert clinicians. J Eur Acad Dermatol Venereol. 2014;28:1141–9.
60. Stockfleth E, Terhorst D, Hausschild A, Zouboulis CC, SChlaeger M, Oster-Schmidt C, Diiepgen T, Gollnick H, Korting HC, Merk H, Sterry W, Kerl K, Berking C. S1 Letilinie zur Behandlung der aktinischen Keratosen. 2011. AWMF Link: http://www.awmf.org/uploads/tx_szleitlinien/013-041l_S1_Aktinische_Keratose_2012-verlaengert.pdf.
61. Werner RN, Stockfleth E, Connolly SM, Correia O, Erdmann R, Foley P, Gupta AK, Jacobs A, Kerl H, Lim HW, Martin G, Paquet M, Pariser DM, Rosumeck S, Röwert-Huber HJ, Sahota A, Sangueza OP, Shumack S, Sporbeck B, Swanson NA, Torezan L, Nast A. Evidence- and consensus-based (S3) Guidelines for the Treatment of Actinic Keratosis—International League of Dermatological Societies in cooperation with the European Dermatology Forum—Short version. J Eur Acad Dermatol Venereol. 2015;29(11):2069–79. doi:10.1111/jdv.13180.
62. Bonerandi JJ, Beauvillain CL, Chassagne JF, Chaussade V, Clavere P, Desouches C, Garnier F, Grossin M, Jourdain A, Lemonnier JY, Maillard H, Ortonna N, Rio E, Simon E, Sei JF, Grob JJ, Martin L. Guidelines for the diagnosis and treatment of cutaneous squamous cell carcinoma and precursor lesions. J Eur Acad Dermatol Venereol. 2011;25(5):1–51.
63. Spallone G, Botti E, Costanzo A. Targeted therapy in non melanoma skin cancers. Cancers. 2011;3:2255–73.
64. Edge SB, Byrd DR, Compton CC, et al. AJCC cancer staging manual. 7th ed. New York: Springer; 2010.
65. Stratigos A, Garbe C, Lebbe C, Malvehy J, del Marmol V, Pehamberger H, Peris K, Becker JC, Zalaudek I, Saiag P, Middleton MR, Bastholt L, Testori A, Grob JJ. Diagnosis and treatment of invasive squamous cell carcinoma of the skin: European consensus-based interdisciplinary guideline. Eur J Cancer. 2015;51(14):1989–2007.
66. Rudnick EW, Thareja S, Cherpelis B. Oral therapy for nonmelanoma skin cancer in patients with advanced disease and large tumor burden: a review of the literature with focus on a new generation of targeted therapies. Int J Dermatol. 2015;55(3):249–58. doi:10.1111/ijd.12961.
67. Galer CE, Corey CL, Wang Z, Younes MN, Gomez-Rivera F, Sa J, Ludwig DL, El-Naggar AK, Weber RS, Myers JN. Dual inhibition of epidermal growth factor receptor and insulin- like growth factor receptor 1 reduces angiogenesis and tumor growth in cutaneous squamous cell carcinoma. Head Neck. 2011;33:189–98.

Molecular Pathogenesis of Bone Tumours

3

Maria A. Smolle and Johannes Haybaeck

Contents

M.A. Smolle (✉)
Institute of Pathology, Medical University of Graz, Graz, Austria

Department of Orthopaedics and Trauma, Medical University of Graz, Graz, Austria
e-mail: maria.smolle@stud.medunigraz.at

J. Haybaeck, M.D., Ph.D.
Department of Pathology, Medical Faculty, Otto von Guericke University Magdeburg, 39120 Magdeburg, Germany

Institute of Pathology, Medical University of Graz, 8036 Graz, Austria
e-mail: johannes.haybaeck@med.ovgu.de

J. Haybaeck (ed.), *Mechanisms of Molecular Carcinogenesis – Volume 2*,
DOI 10.1007/978-3-319-53661-3_3

Abstract

Bone tumours constitute a rare tumour entity, thus comprising many subtypes, such as cartilage tumours, haematopoietic tumours, Ewing sarcomas, giant cell tumours, chordomas and tumours with undefined neoplastic features. Various genes and derived proteins are involved in the development of bone tumours and sarcomas. Some genes are unique for one neoplasm, whereas others are altered in different subtypes. Most bone sarcomas arise de novo. Dedifferentiation of benign bone neoplasms towards sarcomas has been observed, especially in hereditary tumour syndromes. These syndromes are characterised by mutations of specific genes that are involved in the development of bone neoplasms.

The following chapter will give an insight into the molecular pathogenesis of bone tumours by covering different genes, their regular function and their possible role during tumorigenesis. Different tumour types and hereditary tumour syndromes accompanied by increased likelihood of developing bone neoplasms will be discussed as well.

3.1 Introduction

Information about benign bone tumours is rather limited, as research has mainly concentrated on their malignant counterparts. Bone sarcomas are a rare tumour entity, as they account for only 0.2% of all malignancies. The most common representative is osteosarcoma, present in nearly 35% of cases, followed by chondrosarcoma (25%) and Ewing sarcoma (16%) [1]. A typical pattern of incidence is present in most entities of bone sarcomas: Around the second decade, the risk of developing osteosarcoma and Ewing sarcoma is the highest, followed by a second peak between 60 and 70 years of age. Chondrosarcomas, however, present with an increasing incidence up to the age of 75. For most bone sarcomas, a preferred region of appearance can be found: Whilst osteosarcomas mainly arise in the metaphysis, Ewing sarcomas are more common in the diaphysial area.

Although most bone sarcomas arise de novo, there is increasing evidence that some of them emerge from benign precursor lesions. This phenomenon can be observed in hereditary tumour syndromes, such as *Maffucci syndrome*, where patients present with multiple enchondromas and haemangiomas. These patients are at a high risk of developing haemangiosarcomas and chondrosarcomas [2].

The molecular pathogenesis of bone malignancies is very heterogeneous, as is the histological variety of bone tumour entities. A cascade of activating and inactivating mutations is necessary to promote development of bone tumours. Moreover, specific chromosomal translocations contribute to tumorigenesis of some bone malignancies. The *FISH* (fluorescence-in-situ-hybridisation) method is used to detect these translocations in order to differentiate between tumour entities. Microscopically, different bone and soft tissue tumours are quite difficult to differentiate. Therefore, immunohistochemistry and *FISH* constitute useful tools to define a tumour subtype.

3.2 Tumour Types

3.2.1 Cartilage Tumours

About 30% of all malignant bone neoplasms are of cartilage origin. The incidence of benign cartilage-forming bone tumours is not clear, as most of them remain undetected [3].

The most common benign cartilage-forming bone tumour is enchondroma, typically located in the metaphyseal area of long bones of the hand. Multifocal appearance of enchondromas is associated with *Ollier disease* and *Maffucci syndrome*. Especially in these syndromes, transformations from benign enchondromas to chondrosarcomas have been observed [3].

Osteochondromas constitute another benign cartilage-forming tumour entity, predominantly arising around the knee joint and pelvis. These tumours are composed of a bony “trunk” and a cartilage cap growing away from the joints [4]. Patients affected by the multiple hereditary osteochondromatosis disease present with multifocal appearance of osteochondromas [5].

Chondromyxoid fibroma and chondroblastoma are rather uncommon tumour entities. The former has to be distinguished from giant cell tumour of bone, and the latter is sometimes associated with aneurysmal bone cysts [6].

The second most common primary malignant bone tumour is chondrosarcoma, predominantly arising in the pelvis, femur and shoulder girdle [7]. Although most enchondromas can be found in the hand, chondrosarcomas in this area are uncommon. Whilst osteosarcomas have the highest incidence in children and young adults, chondrosarcomas predominantly arise in the sixth decade of life [3].

3.2.2 Osteogenic Tumours

Enostosis, osteoma, osteoblastoma and osteoid osteoma constitute the group of benign bone-forming tumours [8].

Osteomas arise from the periosteum and are localised on the bone cortex. They grow very slowly and can be found in the final stage of fibrous dysplasia, though the detailed aetiology is still unknown [8]. Most osteomas are detected in the third to fifth decades of life with a slight preference to the male gender.

The enostosis, also known as bone island, can be found in the spongiosa of the bone. Enostoses are often incidental findings; therefore, the incidence is not known [9].

Osteoid osteomas account for about 13% of all benign osteogenic tumours [10]. Patients typically report pain at night that can be relieved by salicylates. Osteoid osteomas are composed of a centrally located nidus that is surrounded by a sclerotic area [11]. The highest incidence can be found in young adults, with males affected three times as often as females [8].

Osteoblastomas were described as "giant osteoid osteomas" in the past, as they resemble osteoid osteomas despite a larger nidus (over 1.5 cm in size) [12]. They constitute a very rare entity of benign osteogenic tumours. The clinical presentation of osteoid osteomas is asymptomatic, though dull pain and pain-related scoliotic posture are often present [13].

Osteosarcomas are the most common primary malignant bone tumours. The highest incidence can be found in children and young adolescents, followed by another peak in the fifth decade [14]. The majority of osteosarcomas consist of osteoblastic cells, with varying chondroblastic and fibroblastic-fibrohistiocytic tissue [15]. A huge amount of hereditary syndromes is associated with increased incidence of developing an osteosarcoma, given the relatively low incidence of this sarcoma in the general population [16].

3.2.3 Ewing Sarcoma

Ewing sarcoma is the second most common bone malignancy in children and young adults after osteosarcoma. The origin of Ewing sarcoma is neurogenic tissue, which is why tumour cells exhibit morphological features as rosette formations typical for neural differentiation [17]. Eighty-five percent of Ewing sarcomas exhibit the *EWS-FLI1* fusion gene, leading to the production of oncogenic proteins [18]. *FLI1*, together with *ETV1*, *ETV4* and *FEV*, belongs to the ETS family of transcription factors [19]. The EWS amino-terminal domain has a ubiquitously activated promoter and an amino-terminal domain with transactivational potential [20].

3.2.4 Haematopoietic Tumours

The plasma cell myeloma, also known as plasmocytoma, is characterised by bone pain, osteolytic lesions, hypercalcemia, deposition of amyloid and a monoclonal gammopathy [21]. In 99% of patients, monoclonal proteins can be found in serum and urine of patients. In 50% of cases, IgG is present, followed by IgA in 20–25% and IgM, IgE or IgD in a few cases. Moreover, about 75% of patients have monoclonal light chains in their serum, known as Bence-Jones proteins [22].

The risk for progression to multiple myeloma is raised in solitary plasmocytoma. Therefore, surgical excision followed by chemotherapy is considered as a treatment option [23].

Though the aetiology is not known in detail, chronic infections, such as osteomyelitis and rheumatoid arthritis, are thought to trigger the development of plasmocytoma [22]. The incidental peak is around the sixth decade of life, with both sexes affected equally [24]. Whilst patients with low-grade extramedullary plasmocytoma develop multiple myeloma about 120 months after treatment, the time in high-grade tumours is significantly lower (26 months) [25].

Malignant lymphoma can affect the bone, especially at sites where persistent red marrow is present. Patients can be of any age group, though more cases have been reported in the later decades of life [26]. Interestingly, B symptoms are often absent in malignant lymphoma of bone, whereas pain is the predominant symptom. Neurological symptoms may occur due to involvement of the spine [27].

3.2.5 Giant Cell Tumours

Giant cell tumours of bone (GTCB) predominantly arise in the third and fourth decades of life. Typical localisations are the knee joint, axial skeleton and wrist. Patients usually present with painful swellings located in the epiphyses of long bones. However, joint involvement may be present, indicating a more aggressive type [28].

Histologically, GCTB are composed of bone erosions lined by mononuclear cells originating from macrophages, multiloculated giant cells and surrounding stroma. The latter is thought to be the malignant part of GCTB [29].

The receptor activator of nuclear factor kappa-B ligand (RANKL) plays a critical role in the development of GCTB, as it is overexpressed. This leads to the production of giant cells by fusion of monocytes. The giant cells work as osteoclasts and cause lysis of the bone [30]. The monoclonal RANKL inhibitor Denosumab is used in the treatment of GCTB where surgery cannot be performed [31].

3.2.6 Notochordal Tumours (Chordoma)

Chordomas constitute a rare tumour entity, accounting for only 1–4% of all primary bone malignancies [32]. This low-grade and slow-growing malignancy arises from embryonic remnants of the notochord and presents with an epithelial-mesenchymal presentation. Chordomas are most commonly found around the sacrum, followed by the skull base, cervical and thoracolumbar vertebrae [32]. Metastases arise in patients with advanced disease to the lungs, lymph nodes, skin and soft tissues, though surgical techniques have improved outcome [33].

3.2.7 Tumours of Undefined Neoplastic Nature

3.2.7.1 Aneurysmal Bone Cysts

Aneurysmal bone cysts (ABCs) are composed of intraosseous cavities without endothelial membranes that are filled with blood. ABCs predominantly occur around the knee joint, but also the pelvis, humerus and vertebral bodies can be affected [34]. Primary ABCs arise de novo, whereas secondary ABCs are associated with other benign or malignant bone neoplasms [35]. The ABCs are often painful, and radiologically, they present as expansile and lytic masses with defined margins not crossing the growth plates [36].

3.2.7.2 Simple Bone Cyst

Simple bone cysts, also known as juvenile bone cysts, are composed of a single serous or sero-sanguineous fluid-filled cavity. The preferred sites of occurrence are the same as for ABCs. As in ABCs, patients present with painful swellings, but the pain is often caused by pathological fracture [37].

3.2.7.3 Fibrous Dysplasia

Activating mutations in the *GNAS1* gene cause fibrous dysplasia. It can be present in a monostotic or polyostotic form [38]. Patients may present with pain, though pathological fractures are sometimes the cause for admission to hospital. The polyostotic form, together with café-au-lait spots and endocrine abnormalities, is called *McCune-Albright syndrome* [39].

3.2.7.4 Osteofibrous Dysplasia

Osteofibrous dysplasia (OFD) most commonly affects the middle third of the tibia and is often diagnosed in children [28]. OFD arises in the cortex of the affected bone and leads to a deformation of the affected limb. The prognosis is good, as the lesion stops growing around the age of 15 and heals [38].

3.2.7.5 Langerhans Cell Histiocytosis

The Langerhans cell histiocytosis, also named eosinophilic granuloma, is a very rare bone neoplasm, accounting for less than 1% of all bone tumours. It arises in every age group, predominantly affecting the calvarium, femur, mandible and pelvis [40]. Immunohistochemically, Langerhans cells express the S100 protein and membrane-based CD1a but lack CD45 [41]. Patients present with a swelling and pain around the lesion. The prognosis is good for the monostotic and polyostotic form [40].

3.2.7.6 Erdheim-Chester Disease

The Erdheim-Chester disease (ECD) is diagnosed between the fifth and seventh decades of life. Long bones of the limbs, but also flat bones, can be affected by the infiltration of lipid-loaded histiocytes causing osteosclerosis and fibrosis [42]. Extraskeletal involvement is present in 50% of cases, leading to renal, pulmonary, CNS and cardiovascular complications [43].

3.2.7.7 Chest Wall Hamartoma

Chest wall hamartomas present as masses of the thorax and are composed of mesenchymal tissue. The lesion is very rare, and most cases diagnosed affected toddlers. Predominantly, the lesion consists of cartilage tissue, mixed with elements of ABCs. The prognosis is excellent, as surgical removal of the lesion results in cure [44].

3.3 Molecular Pathogenesis

Gene	Pattern	Osteo-sarcoma	Chondro-sarcoma	Ewing's Sarcoma	Remarks
AEG-1	Overexpression	Yes			
ATRX	Underexpression	Yes			
CTGF	Overexpression	Yes	Yes		
ET-1	Overexpression	Yes			
EXT1	Underexpression		Yes		
GNAS	Overexpression	Yes			In parosteal osteosarcoma 55% exclusively!
IDH1	Underexpression		Yes		
MDM2	Overexpression	Yes			Not in periosteal osteosarcoma and high-grade osteo-sarcoma without low-grade osteosarcoma precursor
MEF2D	Overexpression	Yes			
MET	Overexpression	Yes		Yes	
NKD2	Underexpression	Yes			
p16/CDKN2A	Underexpression		Yes		
PRIM1	Overexpression	Yes			
RUNX2	Overexpression	Yes			
RUNX3	Overexpression			Yes	
SATB2	Overexpression	Yes			
TP53	Overexpression	Yes			Mutant type
WIF-1	Underexpression	Yes			

3.3.1 AEG-1

The astrocyte-elevated gene-1 (*AEG-1*) encodes for the AEG-1 protein, also known as MTDH (metadherin) [45]. AEG-1 is known to promote invasion of different tumour cells, such as Hela cells, glioma cells and non-small cell lung cancer [46].

In osteosarcoma, AEG-1 is overexpressed and correlates with the clinical parameters of the tumour [47]. Osteosarcoma cell invasion is promoted by AEG-1 through the JNK/c-Jun/MMP-2 pathway [48]. This is caused by the fact that matrix metalloproteinases (MMPs) are important for the destruction of the extracellular matrix in order to allow tumour growth, angiogenesis and invasion [49].

Moreover, AEG-1 regulates endothelin-1 (ET-1) expression at the transcriptional level [50]. Over the ETAR receptor, ET-1 then mediates the functions of AEG-1 on osteosarcoma cell invasion, as ET-1/ETAR signalling is located downstream of PI3K [51].

3.3.2 ATRX

The *ATRX* gene, located at Xq13.3, encodes for ATRX protein [52]. ATRX consists of two highly conserved regions, the N-terminal ADD domain and the C-terminal ATPase/helicase, resembling Swi2-/Snf2-like chromatin remodellers [53]. Therefore, the protein acts as a chromatin remodeller, but is also a repressor of the alternative lengthening of telomeres (ALT) pathway [54]. Inactivating mutations of *ATRX* are commonly found in paediatric osteosarcoma and positively correlate with ALT expression. Tumour cells with mutations of *ATRX* have considerably longer telomeres, indicating their immortalisation [55].

3.3.3 CDKN2A

The *CDKN2A* gene locus encodes for both the p16 and p14 proteins [56]. MDM2 acts as a negative regulator of p53 and is inhibited by p14 [57]. The p16 protein stops cyclin-dependent kinase-mediated phosphorylation of retinoblastoma (RB) and therefore cell-cycle progression [58]. A loss of *CDKN2A* therefore leads to unimpeded cellular growth, as both the p53 and RB inhibitors are absent [56].

The p16 mutation constitutes an important event in the pathogenesis of chondrosarcomas, as loss of p16/*CDKN2A* is present in approximately two thirds of patients with high-grade chondrosarcomas. In low-grade chondrosarcomas and enchondromas, however, no aberrant copy number of p16/*CDKN2A* is present [59].

3.3.4 CTGF

The *CTGF* gene encodes for the connective tissue growth factor (CTGF) protein, also known as CCN2 [60]. Belonging to the CCN family, CTGF—together with CCN1 and CCN3—is a matrix signalling modulator [61]. It is important for chondrogenesis, angiogenesis and wound healing, as it promotes cell proliferation, differentiation, migration and survival [62].

In osteosarcoma, overexpression of CTGF is associated with increased cell migration [63]. Moreover, CTGF promotes migration of chondrosarcoma cells by upregulating the MMP-13, FAK, MEK, NF-κB and ERK pathways [64].

3.3.5 ET-1

Endothelin-1 (ET-1), encoded by the *ET-1* gene, is expressed in various cells and promotes tumour cell survival and proliferation over the ET A receptor (ETAR) as an autocrine/paracrine growth factor. ET-1, as well as its receptors ETAR and ETBR, is also expressed in osteosarcoma tissue [65].

In osteosarcoma cells, the ET-1/ETAR signalling pathway induces cell survival and invasion against cisplatin-induced apoptosis, as the production of matrix metalloproteinases is enhanced, resulting in increased tumour cell invasion [66]. Moreover, ET-1 and the astrocyte-elevated gene-1 (AEG-1) interact with each other and promote tumour cell invasion over MMP-2 [47].

3.3.6 EXT1

The *EXT1* gene is located on chromosome 8 and encodes for the exostosin-1 (EXT) protein. It combines the functions of a glucuronic acid sugar-based transfer enzyme and an N-acetyl glucosamine glycosyltransferase [67]. The second function enhances the polymerisation of heparan sulphate (HS) chains, which is an important component of extracellular matrix glycoproteins [68]. HS controls cell adhesion, receptor ligand binding processes and signal transduction on the surface of cells [69].

EXT1 belongs to the EXT family, which comprises six proteins (exostosin (EXT) 1, 2 and 3; exostosin-like (EXTL) 1, 2 and 3) and plays a critical role in the pathogenesis of *hereditary multiple exostosis* (HME) and solitary osteochondromas [70]. Mutations of the tumour suppressor gene *EXT1* and *EXT2* cause impaired synthesis of HS in the growth plates, leading to an intracellular accumulation of HS core proteins. Therefore, inactivating mutations of *EXT1* and *EXT2* can be found in solitary, as well as in hereditary osteochondromas and in chondrosarcomas [71].

3.3.7 GNAS

The *GNAS* (guanine nucleotide-binding protein/alpha-subunit) gene encodes for the GNAS protein. Germline mutations of *GNAS*—predominantly at codon 201—cause the *McCune-Albright syndrome* and another variant, the *Mazabraud syndrome* [72]. These mutations lead to activation of the G protein-dependent adenylate cyclase, resulting in accumulated levels of intracellular cAMP [73].

One study discovered *GNAS* mutations in 55% of cases with parosteal osteosarcoma. This could be related to the fact that parosteal osteosarcoma histologically resembles malignant fibro-osseous tumours of the bone [72].

3.3.8 IDH1

The isocitrate dehydrogenase 1 (*IHD1*) gene is commonly mutated in acute myeloid leukaemia, gliomas and secondary glioblastomas [74, 75]. These mutations lead to specific amino acid changes near the active sites of the enzyme, resulting in a disability to convert isocitrate to α-ketoglutarate [76]. Moreover, the enzyme gains a new function leading to an accumulation of δ-2-hydroxyglutarate, and as an oncometabolite, this product enhances tumour-inducing processes [77].

IDH1 mutations are present in about 50% of cartilaginous tumours, including enchondromas, periosteal chondromas and chondrosarcomas. In other connective tissue neoplasms, however, no mutations of *IDH1* are present [78]. Interestingly, 90% of acral-based tumours present with an *IDH1* mutation, compared to 53% of long-bone tumours and only 35% of tumours in flat bones [78].

3.3.9 MDM2

The *mouse double minute 2 homolog (MDM2)* encodes for the protein MDM2. Amplification of *MDM2* is often accompanied by amplification of *cyclin-dependent kinase 4 (CDK4)* and *DNA primase 1 (PRIM1)*, leading to deregulation of the cell cycle [79]. MDM2 is a negative regulator of the tumour suppressor p53 and an inactivator of the tumour suppressor protein pRB [80, 81].

In low-grade osteosarcoma (including parosteal osteosarcoma, dedifferentiated low-grade osteosarcoma and low-grade central osteosarcoma), overexpression of MDM2 and CDK4 is present [82]. On the contrary, in periosteal osteosarcoma, MDM2 and CDK4 are underexpressed, implicating the particular role of these two proteins in the tumorigenesis of osteosarcomas [83]. Moreover, in high-grade osteosarcoma, overexpression co-expression of the two genes indicates low-grade osteosarcoma as the precursor lesion [84].

3.3.10 MEF2D

Four different genes named *MEF2A*, *MEF2B*, *MEF2C* and *MEF2D* encode for the myocyte enhancer factor-2 (MEF2) family proteins. These transcription factors are important for cellular division, differentiation and death by interacting with various co-factors of transcription [85]. The MEF2C protein is important for bone development by inducing expression of osterix and Runx2 in osteoblasts [86]. MEF2D promotes cellular proliferation by inducing the *c-jun* promoter, which is a positive regulator of cell-cycle progression [87].

In osteosarcoma, high expression levels of MEF2D are associated with poor prognosis. This is caused by enhancement of the G2/M transition in the cell cycle through suppression of CDKN1A and RPRM [88].

3.3.11 MET

The *MET* gene on chromosome 7 encodes for the receptor tyrosine kinase MET [89]. The protein consists of three subunits: the tyrosine kinase domain, a juxta-membrane domain and an extracellular sema domain (a region with homology to semaphorins). The only known MET ligand is the hepatocyte growth factor (HGF) [90]. Other functions of MET include fast tyrosine kinase phosphorylation, activation of the phosphatidylinositol 3 (PI3K)/Akt kinase pathway and the extracellular signal-regulated kinase (Erk) pathway [91].

High expression levels of MET can be found in Ewing sarcoma. The outcome for patients with Ewing sarcoma is poorer, when more membranous staining of the protein is present [91]. This could be related to the fact that only the MET receptors at the cell membrane are involved in signal transduction [91].

Also in osteosarcoma, MET is usually overexpressed [92]. The protein may transform osteo-progenitor cell populations lacking the ability of self-renewal into cell lines with the potential of self-renewal [93].

3.3.12 NKD2

The naked cuticle homolog (NKD) family includes the NKD1 and NKD2 protein. They are involved in regulation of the transforming growth factor (TGF)-α and Wnt pathways [94]. NKD2 usually represses tumour growth and metastatic spread in osteosarcoma cell lines. Therefore, downregulation of NKD2 is associated with decreased overall survival [B]. Although NKD2 is an important inhibitor of the Wnt pathway, inactivating mutations of alternative Wnt inhibitors (e.g. Wnt5a) seem to be necessary for enhanced activation of the Wnt pathway. In metastatic osteosarcoma, increased Wnt activity is present, though the detailed mechanisms by which Wnt enhances metastatic spread are still unknown [94].

3.3.13 PRIM1

The *DNA primase 1 (PRIM1)* gene is located on the short arm of chromosome 12 and encodes for PRIM1 [95]. It is necessary for initiation of DNA synthesis as it catalyses the elongation of primers. Therefore, together with CDK4 and MDM2, PRIM1 plays a role in the cell cycle by promoting transition from G1 into S phase [96]. Its mRNA expression is regulated during the cell cycle by transcriptional activation/repression and posttranscriptional degradation [97].

In about 40% of osteosarcoma cells, an overexpression of PRIM1 is present, indicating its potential role in tumorigenesis. However, it is unknown whether the co-amplification of *PRIM1*, *CDK4* and *MDM2* might have a synergistic effect on cell-cycle transition [96].

3.3.14 RUNX Family

The RUNX family consists of three proteins. They constitute the DNA-binding α-components of the core-binding factor (CBF) complex [98]. The CBF-β is necessary for a proper function of the RUNX proteins [99]. RUNX1 is important for haematopoiesis; RUNX2 is involved in osteogenesis, whereas neurogenesis and GI development are influenced by RUNX3 [100].

3.3.14.1 RUNX2

The *RUNX2* gene is located on chromosome 6 and encodes for RUNX2. This protein plays a key role in osteoblast differentiation by regulating gene expression during osteogenesis [101]. RUNX2 enhances gene transcription mainly in T-lymphocytes during development of the thymus. The protein is important for osteoblast differentiation during osteogenesis, as it interacts with various activating and repressing proteins [102]. Expression of *RUNX2* during bone formation is stimulated by different factors, such as ETS1, ELK1 and SP1 [103].

Overexpression of RUNX2 is present in osteosarcoma, and high levels of the protein are associated with poor response to neoadjuvant chemotherapy (<90% necrosis) [104].

3.3.14.2 RUNX3

The RUNX3 protein has oncogenic and tumour suppressing functions, depending on the type of malignancy. In colorectal, lung and breast cancer RUNX3 acts as a tumour suppressor [105, 106]. In basal cell carcinoma, head and neck squamous cell carcinoma and ovarian cancer, oncogenic functions have been described [107, 108]. RUNX3 and RUNX2 interact with the EWS-FLI fusion protein, which is characteristic of Ewing sarcoma. In this specific tumour type, RUNX3 acts as an oncogene, as suppression of RUNX3 leads to reduced cellular growth [109].

3.3.15 SATB2

The special AT-rich-binding protein 2 (SATB2) belongs to the SATB transcription factor family. It regulates chromatin organisation and remodelling through binding to AT-rich DNA sequences [110]. SATB2 induces osteoblast differentiation by repressing *Hoxa2* and interacting with ATF4 and RUNX2 [111].

In over 90% of osteosarcomas, high expression levels of SATB2 are present, whereas in other sarcoma types, only 1% of tumours are positive for this protein [112]. It is involved in the actin cytoskeleton dynamics by regulating specific genes and proteins, like EPLIN. As a result, SATB2 promotes cell adhesion and invasion in osteosarcoma tissue [112].

3.3.16 TP53

The *TP53* gene is located on chromosome 17 and encodes for a wild type and a mutant type of p53. Whilst the wild type acts as a tumour suppressor, the mutant type has oncogenic functions [113]. The latter promotes tumorigenesis by inducing cell proliferation and transformation and has anti-apoptotic effects [114]. The wild-type p53 induces expression of growth inhibition genes by interacting with the p53-binding site. The mutant type lacks these abilities [115].

In patients with osteosarcoma, high levels of mutant p53 positively correlate with a poor prognosis and negatively correlate with the apoptotic index (AI) [116].

3.3.17 WIF-1

The Wnt signalling pathway consists of a variety of Wnt ligands, frizzled receptors and lipoprotein receptor-related proteins (LPRPs), leading to the accumulation of Wnt effectors. High levels of beta-catenin in the nucleus induce expression of oncogenes important for the development of osteosarcoma, such as *c-Myc*, *Cyclin D1* and *c-Met* [117].

The Wnt inhibitory factor-1 (WIF-1) protein directly binds to ligands of the Wnt pathway. WIF-1 inhibits the Wnt signalling pathway; thus, it constitutes a tumour suppressor protein [118]. In osteosarcoma tissue, *WIF-1* is commonly downregulated, resulting in loss of inhibition of the Wnt signalling pathway [117]. On the contrary, experimental overexpression of WIF-1 significantly reduces the number of transcriptional targets of the Wnt pathway, such as metalloproteinases [117]. These proteins are important for tumorigenesis by inducing cellular migration and metastasis [119].

3.4 Syndromes

3.4.1 Li-Fraumeni Syndrome

The *Li-Fraumeni syndrome* (LFS) is the classic cancer predisposition syndrome, caused by a germline mutation in the *TP53* gene [120]. It is inherited in an autosomal-dominant way. Patients with LFS are at high risk of developing bone and soft tissue sarcomas, but also breast cancer, leukaemia, adrenocortical carcinoma and brain tumours [121].

About 95% of *TP53* mutations can be detected by sequencing the whole coding region from exon 2 to 11 [122]. Mutations lead to the production of a mutant type of p53 with oncogenic functions [113].

3.4.2 Mazabraud Syndrome

The combination of polyostotic fibrous dysplasia and multiple myxomas is called *Mazabraud syndrome* [123]. It is often associated with the *McCune-Albright syndrome*, where patients present with café-au-lait spots, endocrine abnormalities and fibrous dysplasia. Both conditions are caused by a mutation in the *GNAS1* gene [124].

The incidence of *Mazabraud syndrome* is twice as high in females than in males. The myxomas may be painful and limit movement, but they can be clinically inapparent as well [123].

3.4.3 McCune-Albright Syndrome

The *McCune-Albright syndrome* (MAS) is caused by an activating mutation in the *GNAS1* gene. *GNAS1* encodes for the G-protein alpha stimulator subunit and is located on chromosome 20 [124]. MAS patients have polyostotic fibrous dysplasia and present with extraskeletal manifestations like café-au-lait spots and hyperfunctioning endocrinopathies [39]. Inactivating mutations of *GNAS1* can be found in endocrine disorders with hormone resistance, whereas activating mutations are present in fibrous dysplasia and associated hyperactive endocrinopathies [125].

3.4.4 Ollier Disease (Enchondromatosis)

Ollier disease, also known as enchondromatosis, is characterised by asymmetrically distributed multiple enchondromas [126]. Patients present with painless masses predominantly located in the phalanges and metacarpals, whereas involvement of carpal bones is a rarity [127].

The pathophysiology of the syndrome is complicated and multifactorial. Mutations in the isocitrate dehydrogenases 1 and 2 (IDH1, IDH2) are associated with *Ollier disease* [128]. Parathyroid-related peptide type 1 receptor (PTHR1) mutations may be the cause for the development of enchondromas [129].

3.4.5 Paget's Disease of Bone

The characteristic of Paget's disease of bone (PDB) is disturbed bone turnover at one or more skeletal sites [130]. Three molecules, namely, receptor activator of nuclear factor kappa-B (RANK), RANK ligand (RANKL) and osteoprotegerin (OPG), regulate osteoclast differentiation. RANK is encoded by the *TNFRSF11A* gene, the RANKL by *TNSF11* and OPG by the *TNFRSF11B* gene [131]. Usually, the binding of RANKL to its receptor induces differentiation of osteoclasts, and bone resorption takes place. OPG acts as the counterpart by constituting a decoying receptor for RANKL [132].

In PDB, osteoclasts are more sensitive to RANKL, resulting in increased bone resorption. Increased bone formation follows the resorption, leading to abnormal new bone matrix with high risk of pathological fracture [133].

Activating mutations in the *TNFRSF11A* gene lead to early development of familial PDB and inactivating mutations on the gene of OPG cause juvenile PDB [132].

3.4.6 Retinoblastoma

The *RB1* gene encodes for the nuclear phosphoprotein (pRB), acting as a tumour suppressor. It plays a critical role in the regulation of the cell cycle between the G1 and S phase [134]. Germline mutations of *RB1* cause the *retinoblastoma* syndrome. Patients affected by this syndrome are at an extremely high risk of developing retinoblastoma, a malignant tumour of the eye [135].

Defective *RB1* leads to an increase in osteosarcoma incidence [136]. It has been reported that *retinoblastoma* patients are at higher risk of developing an osteosarcoma. This could be caused by the local radiotherapy administered due to a retinoblastoma [137]. Moreover, the risk of developing other malignancies, such as soft tissue sarcomas, brain tumours, cutaneous melanomas and breast or lung cancer, is elevated in these patients [138].

3.4.7 Rothmund-Thomson Syndrome

The *Rothmund-Thomson syndrome* is caused by a mutation in the *RECQL4* gene, located on chromosome 8 [139]. Patients affected by the syndrome present with poikiloderma, skeletal abnormalities, sparse facial hair, juvenile cataracts and a short stature [140]. The risk of developing cancer is raised, as for cutaneous neoplasms and osteosarcoma. Osteosarcomas occur earlier than in the general population, though they do not differ from sporadic cases of osteosarcoma [141].

Altogether, there are five RECQ helicases involved in the DNA replication, and mutations of the respective genes cause different syndromes. The *RECQL4* gene encodes for the RECQ protein-like 4 (RECQL4), an ATP-dependent DNA helicase domain similar to the RECQ helicase of *Escherichia coli* [142]. RECQL4 is overexpressed in the S phase of the cell cycle and after exposure to reactive oxygen species (ROS). This indicates that RECQL4 plays a role in the regulation of DNA expression as well as involvement in the reaction to oxidative stress [143, 144].

3.4.8 Werner Syndrome

The gene encoding the RECQ helicase 3, *RECQ3*, is mutated in *Werner syndrome*. Although the underlying genetic defect is the same in *Rothmund-Thomson syndrome* and *Werner syndrome*, the clinical presentation is rather different [140].

The *Werner syndrome* is an autosomal recessive disorder, leading to premature ageing of the affected patients. Clinical features are atrophic and tight skin, short stature, development of cataract and premature ageing [145].

Patients are at higher risk of developing various malignancies, such as soft tissue sarcomas, thyroid carcinomas, meningiomas, melanomas and osteosarcomas [146].

References

1. Siegel RL, Miller KD, Jemal A. Cancer statistics, 2015. CA Cancer J Clin. 2015;65:5–29.
2. Verdegaal SH, Bovee JV, Pansuriya TC, Grimer RJ, Ozger H, Jutte PC, San Julian M, Biau DJ, Van Der Geest IC, Leithner A, Streitburger A, Klenke FM, Gouin FG, Campanacci DA, Marec-Berard P, Hogendoorn PC, Brand R, Taminiau AH. Incidence, predictive factors, and prognosis of chondrosarcoma in patients with Ollier disease and Maffucci syndrome: an international multicenter study of 161 patients. Oncologist. 2011;16:1771–9.
3. Streitbuerger A, Hardes J, Gebert C, Ahrens H, Winkelmann W, Gosheger G. Cartilage tumours of the bone. Diagnosis and therapy. Orthopade. 2006;35:871–81. quiz 882
4. Glick R, Khaldi L, Ptaszynski K, Steiner GC. Dysplasia epiphysealis hemimelica (Trevor disease): a rare developmental disorder of bone mimicking osteochondroma of long bones. Hum Pathol. 2007;38:1265–72.
5. Schmale GA, Conrad 3rd EU, Raskind WH. The natural history of hereditary multiple exostoses. J Bone Joint Surg Am. 1994;76:986–92.
6. Bloem JL, Mulder JD. Chondroblastoma: a clinical and radiological study of 104 cases. Skelet Radiol. 1985;14:1–9.
7. Lee FY, Mankin HJ, Fondren G, Gebhardt MC, Springfield DS, Rosenberg AE, Jennings LC. Chondrosarcoma of bone: an assessment of outcome. J Bone Joint Surg Am. 1999;81:326–38.
8. Heuck A, Stabler A, Wortler K, Steinborn M. Benign bone-forming tumors. Radiologe. 2001;41:540–7.
9. Onitsuka H. Roentgenologic aspects of bone islands. Radiology. 1977;123:607–12.
10. Becce F, Jovanovic B, Guillou L, Theumann N. Painful fingertip swelling of the middle finger. Osteoid osteoma of the distal phalanx of the middle finger. Skelet Radiol. 2011;40(1479–80):1501–2.
11. Klein MH, Shankman S. Osteoid osteoma: radiologic and pathologic correlation. Skelet Radiol. 1992;21:23–31.
12. Dahlin DC, Johnson Jr EW. Giant osteoid osteoma. J Bone Joint Surg Am. 1954;36-A:559–72.
13. Kroon HM, Schurmans J. Osteoblastoma: clinical and radiologic findings in 98 new cases. Radiology. 1990;175:783–90.
14. Dorfman HD, Czerniak B. Bone cancers. Cancer. 1995;75:203–10.
15. Dahlin DC. Pathology of osteosarcoma. Clin Orthop Relat Res. 1975:23–32.
16. Kansara M, Thomas DM. Molecular pathogenesis of osteosarcoma. DNA Cell Biol. 2007;26:1–18.
17. Ozaki T. Diagnosis and treatment of Ewing sarcoma of the bone: a review article. J Orthop Sci. 2015;20:250–63.
18. Delattre O, Zucman J, Plougastel B, Desmaze C, Melot T, Peter M, Kovar H, Joubert I, de Jong P, Rouleau G, et al. Gene fusion with an ETS DNA-binding domain caused by chromosome translocation in human tumours. Nature. 1992;359:162–5.
19. Berg T, Kalsaas AH, Buechner J, Busund LT. Ewing sarcoma-peripheral neuroectodermal tumor of the kidney with a FUS-ERG fusion transcript. Cancer Genet Cytogenet. 2009;194:53–7.

20. Aman P, Panagopoulos I, Lassen C, Fioretos T, Mencinger M, Toresson H, Hoglund M, Forster A, Rabbitts TH, Ron D, Mandahl N, Mitelman F. Expression patterns of the human sarcoma-associated genes FUS and EWS and the genomic structure of FUS. Genomics. 1996;37:1–8.
21. Ellis PA, Colls BM. Solitary plasmacytoma of bone: clinical features, treatment and survival. Hematol Oncol. 1992;10:207–11.
22. Salmon SE, Seligmann M. B-cell neoplasia in man. Lancet. 1974;2:1230–3.
23. Soutar R, Lucraft H, Jackson G, Reece A, Bird J, Low E, Samson D, Working Group of the UK Myeloma Forum, British Committee for Standards in Haematology, British Society for Haematology. Guidelines on the diagnosis and management of solitary plasmacytoma of bone and solitary extramedullary plasmacytoma. Clin Oncol (R Coll Radiol). 2004;16:405–13.
24. Devesa SS, Silverman DT, Young Jr JL, Pollack ES, Brown CC, Horm JW, Percy CL, Myers MH, Mckay FW, Fraumeni Jr JF. Cancer incidence and mortality trends among whites in the United States, 1947–84. J Natl Cancer Inst. 1987;79:701–70.
25. Susnerwala SS, Shanks JH, Banerjee SS, Scarffe JH, Farrington WT, Slevin NJ. Extramedullary plasmacytoma of the head and neck region: clinicopathological correlation in 25 cases. Br J Cancer. 1997;75:921–7.
26. Heyning FH, Hogendoorn PC, Kramer MH, Hermans J, Kluin-Nelemans JC, Noordijk EM, Kluin PM. Primary non-Hodgkin's lymphoma of bone: a clinicopathological investigation of 60 cases. Leukemia. 1999;13:2094–8.
27. Moses AM, Spencer H. Hypercalcemia in Patients with Malignant Lymphoma. Ann Intern Med. 1963;59:531–6.
28. Campanacci M, Baldini N, Boriani S, Sudanese A. Giant-cell tumor of bone. J Bone Joint Surg Am. 1987;69:106–14.
29. Lacey DL, Boyle WJ, Simonet WS, Kostenuik PJ, Dougall WC, Sullivan JK, San Martin J, Dansey R. Bench to bedside: elucidation of the OPG-RANK-RANKL pathway and the development of denosumab. Nat Rev Drug Discov. 2012;11:401–19.
30. Cowan RW, Singh G. Giant cell tumor of bone: a basic science perspective. Bone. 2013;52:238–46.
31. Singh AS, Chawla NS, Chawla SP. Giant-cell tumor of bone: treatment options and role of denosumab. Biologics. 2015;9:69–74.
32. Mcmaster ML, Goldstein AM, Bromley CM, Ishibe N, Parry DM. Chordoma: incidence and survival patterns in the United States, 1973-1995. Cancer Causes Control. 2001;12:1–11.
33. Kaiser TE, Pritchard DJ, Unni KK. Clinicopathologic study of sacrococcygeal chordoma. Cancer. 1984;53:2574–8.
34. Rapp TB, Ward JP, Alaia MJ. Aneurysmal bone cyst. J Am Acad Orthop Surg. 2012;20:233–41.
35. Lopez-Barea F, Rodriguez-Peralto JL, Burgos-Lizaldez E, Alvarez-Linera J, Sanchez-Herrera S. Primary aneurysmal cyst of soft tissue. Report of a case with ultrastructural and MRI studies. Virchows Arch. 1996;428:125–9.
36. Kransdorf MJ, Sweet DE. Aneurysmal bone cyst: concept, controversy, clinical presentation, and imaging. AJR Am J Roentgenol. 1995;164:573–80.
37. Margau R, Babyn P, Cole W, Smith C, Lee F. MR imaging of simple bone cysts in children: not so simple. Pediatr Radiol. 2000;30:551–7.
38. Cohen Jr MM. Fibrous dysplasia is a neoplasm. Am J Med Genet. 2001;98:290–3.
39. Cremonini N, Graziano E, Chiarini V, Sforza A, Zampa GA. Atypical McCune-Albright syndrome associated with growth hormone-prolactin pituitary adenoma: natural history, long-term follow-up, and SMS 201-995--bromocriptine combined treatment results. J Clin Endocrinol Metab. 1992;75:1166–9.
40. Wester SM, Beabout JW, Unni KK, Dahlin DC. Langerhans' cell granulomatosis (histiocytosis X) of bone in adults. Am J Surg Pathol. 1982;6:413–26.

41. Nakajima T, Watanabe S, Sato Y, Shimosato Y, Motoi M, Lennert K. S-100 protein in Langerhans cells, interdigitating reticulum cells and histiocytosis X cells. Gan. 1982;73:429–32.
42. Caparros-Lefebvre D, Pruvo JP, Remy M, Wallaert B, Petit H. Neuroradiologic aspects of Chester-Erdheim disease. AJNR Am J Neuroradiol. 1995;16:735–40.
43. Veyssier-Belot C, Cacoub P, Caparros-Lefebvre D, Wechsler J, Brun B, Remy M, Wallaert B, Petit H, Grimaldi A, Wechsler B, Godeau P. Erdheim-Chester disease. Clinical and radiologic characteristics of 59 cases. Medicine (Baltimore). 1996;75:157–69.
44. Ayala AG, Ro JY, Bolio-Solis A, Hernandez-Batres F, Eftekhari F, Edeiken J. Mesenchymal hamartoma of the chest wall in infants and children: a clinicopathological study of five patients. Skelet Radiol. 1993;22:569–76.
45. Brown DM, Ruoslahti E. Metadherin, a cell surface protein in breast tumors that mediates lung metastasis. Cancer Cell. 2004;5:365–74.
46. Emdad L, Sarkar D, Su ZZ, Randolph A, Boukerche H, Valerie K, Fisher PB. Activation of the nuclear factor kappaB pathway by astrocyte elevated gene-1: implications for tumor progression and metastasis. Cancer Res. 2006;66:1509–16.
47. Wang F, Ke ZF, Sun SJ, Chen WF, Yang SC, Li SH, Mao XP, Wang LT. Oncogenic roles of astrocyte elevated gene-1 (AEG-1) in osteosarcoma progression and prognosis. Cancer Biol Ther. 2011;12:539–48.
48. Wang F, Ke ZF, Wang R, Wang YF, Huang LL, Wang LT. Astrocyte elevated gene-1 (AEG-1) promotes osteosarcoma cell invasion through the JNK/c-Jun/MMP-2 pathway. Biochem Biophys Res Commun. 2014;452:933–9.
49. Chambers AF, Matrisian LM. Changing views of the role of matrix metalloproteinases in metastasis. J Natl Cancer Inst. 1997;89:1260–70.
50. Ying Z, Li J, Li M. Astrocyte elevated gene 1: biological functions and molecular mechanism in cancer and beyond. Cell Biosci. 2011;1:36.
51. Liu B, Wu Y, Peng D. Astrocyte elevated gene-1 regulates osteosarcoma cell invasion and chemoresistance via endothelin-1/endothelin A receptor signaling. Oncol Lett. 2013;5:505–10.
52. Gibbons RJ, Picketts DJ, Villard L, Higgs DR. Mutations in a putative global transcriptional regulator cause X-linked mental retardation with alpha-thalassemia (ATR-X syndrome). Cell. 1995;80:837–45.
53. Conte D, Huh M, Goodall E, Delorme M, Parks RJ, Picketts DJ. Loss of Atrx sensitizes cells to DNA damaging agents through p53-mediated death pathways. PLoS One. 2012;7:e52167.
54. Bryan TM, Englezou A, Gupta J, Bacchetti S, Reddel RR. Telomere elongation in immortal human cells without detectable telomerase activity. EMBO J. 1995;14:4240–8.
55. Chen X, Bahrami A, Pappo A, Easton J, Dalton J, Hedlund E, Ellison D, Shurtleff S, Wu G, Wei L, Parker M, Rusch M, Nagahawatte P, Wu J, Mao S, Boggs K, Mulder H, Yergeau D, Lu C, Ding L, Edmonson M, Qu C, Wang J, Li Y, Navid F, Daw NC, Mardis ER, Wilson RK, Downing JR, Zhang J, Dyer MA, St. Jude Children's Research Hospital–Washington University Pediatric Cancer Genome Project. Recurrent somatic structural variations contribute to tumorigenesis in pediatric osteosarcoma. Cell Rep. 2014;7:104–12.
56. Quelle DE, Zindy F, Ashmun RA, Sherr CJ. Alternative reading frames of the INK4a tumor suppressor gene encode two unrelated proteins capable of inducing cell cycle arrest. Cell. 1995;83:993–1000.
57. Momand J, Zambetti GP, Olson DC, George D, Levine AJ. The mdm-2 oncogene product forms a complex with the p53 protein and inhibits p53-mediated transactivation. Cell. 1992;69:1237–45.
58. Serrano M, Hannon GJ, Beach D. A new regulatory motif in cell-cycle control causing specific inhibition of cyclin D/CDK4. Nature. 1993;366:704–7.
59. Amary MF, Ye H, Forbes G, Damato S, Maggiani F, Pollock R, Tirabosco R, Flanagan AM. Isocitrate dehydrogenase 1 mutations (IDH1) and p16/CDKN2A copy number change in conventional chondrosarcomas. Virchows Arch. 2015;466:217–22.

60. Dhar A, Ray A. The CCN family proteins in carcinogenesis. Exp Oncol. 2010;32:2–9.
61. Perbal B. CCN proteins: multifunctional signalling regulators. Lancet. 2004;363:62–4.
62. Luft FC. CCN2, the connective tissue growth factor. J Mol Med (Berl). 2008;86:1–3.
63. Tsai HC, Su HL, Huang CY, Fong YC, Hsu CJ, Tang CH. CTGF increases matrix metalloproteinases expression and subsequently promotes tumor metastasis in human osteosarcoma through down-regulating miR-519d. Oncotarget. 2014;5:3800–12.
64. Tan TW, Lai CH, Huang CY, Yang WH, Chen HT, Hsu HC, Fong YC, Tang CH. CTGF enhances migration and MMP-13 up-regulation via alphavbeta3 integrin, Fak, Erk, and NF-kappaB-dependent pathway in human chondrosarcoma cells. J Cell Biochem. 2009;107:345–56.
65. Nelson J, Bagnato A, Battistini B, Nisen P. The endothelin axis: emerging role in cancer. Nat Rev Cancer. 2003;3:110–6.
66. Felx M, Guyot MC, Isler M, Turcotte RE, Doyon J, Khatib AM, Leclerc S, Moreau A, Moldovan F. Endothelin-1 (ET-1) promotes MMP-2 and MMP-9 induction involving the transcription factor NF-kappaB in human osteosarcoma. Clin Sci (Lond). 2006;110:645–54.
67. Busse M, Feta A, Presto J, Wilen M, Gronning M, Kjellen L, Kusche-Gullberg M. Contribution of EXT1, EXT2, and EXTL3 to heparan sulfate chain elongation. J Biol Chem. 2007;282:32802–10.
68. Lind T, Tufaro F, Mccormick C, Lindahl U, Lidholt K. The putative tumor suppressors EXT1 and EXT2 are glycosyltransferases required for the biosynthesis of heparan sulfate. J Biol Chem. 1998;273:26265–8.
69. Busse-Wicher M, Wicher KB, Kusche-Gullberg M. The exostosin family: proteins with many functions. Matrix Biol. 2014;35:25–33.
70. Liu H, Wu S, Duan L, Zhu W, Zhang S, Hu X, Jia W, Yang G, Liu C, Li W, Yang L, Guo L, Lin Y, Wang Y, He M, Yang Z, He Y, Cai Z, Wang D. Identification of a novel EXT1 mutation in patients with hereditary multiple exostosis by exome sequencing. Oncol Rep. 2015;33:547–52.
71. Hameetman L, David G, Yavas A, White SJ, Taminiau AH, Cleton-Jansen AM, Hogendoorn PC, Bovee JV. Decreased EXT expression and intracellular accumulation of heparan sulphate proteoglycan in osteochondromas and peripheral chondrosarcomas. J Pathol. 2007;211:399–409.
72. Carter JM, Inwards CY, Jin L, Evers B, Wenger DE, Oliveira AM, Fritchie KJ. Activating GNAS mutations in parosteal osteosarcoma. Am J Surg Pathol. 2014;38:402–9.
73. Masters SB, Miller RT, Chi MH, Chang FH, Beiderman B, Lopez NG, Bourne HR. Mutations in the GTP-binding site of GS alpha alter stimulation of adenylyl cyclase. J Biol Chem. 1989;264:15467–74.
74. Abbas S, Lugthart S, Kavelaars FG, Schelen A, Koenders JE, Zeilemaker A, Van Putten WJ, Rijneveld AW, Lowenberg B, Valk PJ. Acquired mutations in the genes encoding IDH1 and IDH2 both are recurrent aberrations in acute myeloid leukemia: prevalence and prognostic value. Blood. 2010;116:2122–6.
75. Yan H, Parsons DW, Jin G, Mclendon R, Rasheed BA, Yuan W, Kos I, Batinic-Haberle I, Jones S, Riggins GJ, Friedman H, Friedman A, Reardon D, Herndon J, Kinzler KW, Velculescu VE, Vogelstein B, Bigner DD. IDH1 and IDH2 mutations in gliomas. N Engl J Med. 2009;360:765–73.
76. Zhao S, Lin Y, Xu W, Jiang W, Zha Z, Wang P, Yu W, Li Z, Gong L, Peng Y, Ding J, Lei Q, Guan KL, Xiong Y. Glioma-derived mutations in IDH1 dominantly inhibit IDH1 catalytic activity and induce HIF-1alpha. Science. 2009;324:261–5.
77. Dang L, White DW, Gross S, Bennett BD, Bittinger MA, Driggers EM, Fantin VR, Jang HG, Jin S, Keenan MC, Marks KM, Prins RM, Ward PS, Yen KE, Liau LM, Rabinowitz JD, Cantley LC, Thompson CB, Vander Heiden MG, Su SM. Cancer-associated IDH1 mutations produce 2-hydroxyglutarate. Nature. 2009;462:739–44.
78. Amary MF, Bacsi K, Maggiani F, Damato S, Halai D, Berisha F, Pollock R, O'donnell P, Grigoriadis A, Diss T, Eskandarpour M, Presneau N, Hogendoorn PC, Futreal A, Tirabosco

R, Flanagan AM. IDH1 and IDH2 mutations are frequent events in central chondrosarcoma and central and periosteal chondromas but not in other mesenchymal tumours. J Pathol. 2011;224:334–43.
79. Pedeutour F, Maire G, Sirvent N, Groupe francophone de cytogénétique oncologique. From cytogenetics to cytogenomics of adipose tissue tumors: 2. Malignant adipose tissue tumors. Bull Cancer. 2004;91:317–23.
80. Rayburn E, Zhang R, He J, Wang H. MDM2 and human malignancies: expression, clinical pathology, prognostic markers, and implications for chemotherapy. Curr Cancer Drug Targets. 2005;5:27–41.
81. Zheng L, Lee WH. The retinoblastoma gene: a prototypic and multifunctional tumor suppressor. Exp Cell Res. 2001;264:2–18.
82. Dujardin F, Binh MB, Bouvier C, Gomez-Brouchet A, Larousserie F, Muret A, Louis-Brennetot C, Aurias A, Coindre JM, Guillou L, Pedeutour F, Duval H, Collin C, de Pinieux G. MDM2 and CDK4 immunohistochemistry is a valuable tool in the differential diagnosis of low-grade osteosarcomas and other primary fibro-osseous lesions of the bone. Mod Pathol. 2011;24:624–37.
83. Righi A, Gambarotti M, Benini S, Gamberi G, Cocchi S, Picci P, Bertoni F. MDM2 and CDK4 expression in periosteal osteosarcoma. Hum Pathol. 2015;46:549–53.
84. Yoshida A, Ushiku T, Motoi T, Beppu Y, Fukayama M, Tsuda H, Shibata T. MDM2 and CDK4 immunohistochemical coexpression in high-grade osteosarcoma: correlation with a dedifferentiated subtype. Am J Surg Pathol. 2012;36:423–31.
85. Mckinsey TA, Zhang CL, Olson EN. MEF2: a calcium-dependent regulator of cell division, differentiation and death. Trends Biochem Sci. 2002;27:40–7.
86. Stephens AS, Stephens SR, Hobbs C, Hutmacher DW, Bacic-Welsh D, Woodruff MA, Morrison NA. Myocyte enhancer factor 2c, an osteoblast transcription factor identified by dimethyl sulfoxide (DMSO)-enhanced mineralization. J Biol Chem. 2011;286:30071–86.
87. Han TH, Prywes R. Regulatory role of MEF2D in serum induction of the c-jun promoter. Mol Cell Biol. 1995;15:2907–15.
88. Ma L, Liu J, Liu L, Duan G, Wang Q, Xu Y, Xia F, Shan J, Shen J, Yang Z, Bie P, Cui Y, Bian XW, Prieto J, Avila MA, Qian C. Overexpression of the transcription factor MEF2D in hepatocellular carcinoma sustains malignant character by suppressing G2-M transition genes. Cancer Res. 2014;74:1452–62.
89. Peruzzi B, Bottaro DP. Targeting the c-Met signaling pathway in cancer. Clin Cancer Res. 2006;12:3657–60.
90. Ma PC, Jagadeeswaran R, Jagadeesh S, Tretiakova MS, Nallasura V, Fox EA, Hansen M, Schaefer E, Naoki K, Lader A, Richards W, Sugarbaker D, Husain AN, Christensen JG, Salgia R. Functional expression and mutations of c-Met and its therapeutic inhibition with SU11274 and small interfering RNA in non-small cell lung cancer. Cancer Res. 2005;65:1479–88.
91. Fleuren ED, Roeffen MH, Leenders WP, Flucke UE, Vlenterie M, Schreuder HW, Boerman OC, VAN DER Graaf WT, Versleijen-Jonkers YM. Expression and clinical relevance of MET and ALK in Ewing sarcomas. Int J Cancer. 2013;133:427–36.
92. Ferracini R, Di Renzo MF, Scotlandi K, Baldini N, Olivero M, Lollini P, Cremona O, Campanacci M, Comoglio PM. The Met/HGF receptor is over-expressed in human osteosarcomas and is activated by either a paracrine or an autocrine circuit. Oncogene. 1995;10:739–49.
93. Dani N, Olivero M, Mareschi K, van Duist MM, Miretti S, Cuvertino S, Patane S, Calogero R, Ferracini R, Scotlandi K, Fagioli F, Di Renzo MF. The MET oncogene transforms human primary bone-derived cells into osteosarcomas by targeting committed osteo-progenitors. J Bone Miner Res. 2012;27:1322–34.
94. Yan D, Wallingford JB, Sun TQ, Nelson AM, Sakanaka C, Reinhard C, Harland RM, Fantl WJ, Williams LT. Cell autonomous regulation of multiple Dishevelled-dependent pathways by mammalian Nkd. Proc Natl Acad Sci U S A. 2001;98:3802–7.

95. Cloutier S, Hamel H, Champagne M, Yotov WV. Mapping of the human DNA primase 1 (PRIM1) to chromosome 12q13. Genomics. 1997;43:398–401.
96. Yotov WV, Hamel H, Rivard GE, Champagne MA, Russo PA, Leclerc JM, Bernstein ML, Levy E. Amplifications of DNA primase 1 (PRIM1) in human osteosarcoma. Genes Chromosomes Cancer. 1999;26:62–9.
97. Miyazawa H, Izumi M, Tada S, Takada R, Masutani M, Ui M, Hanaoka F. Molecular cloning of the cDNAs for the four subunits of mouse DNA polymerase alpha-primase complex and their gene expression during cell proliferation and the cell cycle. J Biol Chem. 1993;268:8111–22.
98. Levanon D, Negreanu V, Bernstein Y, Bar-Am I, Avivi L, Groner Y. AML1, AML2, and AML3, the human members of the runt domain gene-family: cDNA structure, expression, and chromosomal localization. Genomics. 1994;23:425–32.
99. Miller J, Horner A, Stacy T, Lowrey C, Lian JB, Stein G, Nuckolls GH, Speck NA. The core-binding factor beta subunit is required for bone formation and hematopoietic maturation. Nat Genet. 2002;32:645–9.
100. Kagoshima H, Shigesada K, Kohara Y. RUNX regulates stem cell proliferation and differentiation: insights from studies of *C. elegans*. J Cell Biochem. 2007;100:1119–30.
101. Komori T. Regulation of bone development and extracellular matrix protein genes by RUNX2. Cell Tissue Res. 2010;339:189–95.
102. Lian JB, Javed A, Zaidi SK, Lengner C, Montecino M, van Wijnen AJ, Stein JL, Stein GS. Regulatory controls for osteoblast growth and differentiation: role of Runx/Cbfa/AML factors. Crit Rev Eukaryot Gene Expr. 2004;14:1–41.
103. Zhang Y, Hassan MQ, Xie RL, Hawse JR, Spelsberg TC, Montecino M, Stein JL, Lian JB, van Wijnen AJ, Stein GS. Co-stimulation of the bone-related Runx2 P1 promoter in mesenchymal cells by SP1 and ETS transcription factors at polymorphic purine-rich DNA sequences (Y-repeats). J Biol Chem. 2009;284:3125–35.
104. Sadikovic B, Thorner P, Chilton-Macneill S, Martin JW, Cervigne NK, Squire J, Zielenska M. Expression analysis of genes associated with human osteosarcoma tumors shows correlation of RUNX2 overexpression with poor response to chemotherapy. BMC Cancer. 2010;10:202.
105. Chimge NO, Frenkel B. The RUNX family in breast cancer: relationships with estrogen signaling. Oncogene. 2013;32:2121–30.
106. Soong R, Shah N, Peh BK, Chong PY, Ng SS, Zeps N, Joseph D, Salto-Tellez M, Iacopetta B, Ito Y. The expression of RUNX3 in colorectal cancer is associated with disease stage and patient outcome. Br J Cancer. 2009;100:676–9.
107. Salto-Tellez M, Peh BK, Ito K, Tan SH, Chong PY, Han HC, Tada K, Ong WY, Soong R, Voon DC, Ito Y. RUNX3 protein is overexpressed in human basal cell carcinomas. Oncogene. 2006;25:7646–9.
108. Tsunematsu T, Kudo Y, Iizuka S, Ogawa I, Fujita T, Kurihara H, Abiko Y, Takata T. RUNX3 has an oncogenic role in head and neck cancer. PLoS One. 2009;4:e5892.
109. Bledsoe KL, Mcgee-Lawrence ME, Camilleri ET, Wang X, Riester SM, Van Wijnen AJ, Oliveira AM, Westendorf JJ. RUNX3 facilitates growth of Ewing sarcoma cells. J Cell Physiol. 2014;229:2049–56.
110. Dobreva G, Dambacher J, Grosschedl R. SUMO modification of a novel MAR-binding protein, SATB2, modulates immunoglobulin mu gene expression. Genes Dev. 2003;17:3048–61.
111. Dobreva G, Chahrour M, Dautzenberg M, Chirivella L, Kanzler B, Farinas I, Karsenty G, Grosschedl R. SATB2 is a multifunctional determinant of craniofacial patterning and osteoblast differentiation. Cell. 2006;125:971–86.
112. Seong BK, Lau J, Adderley T, Kee L, Chaukos D, Pienkowska M, Malkin D, Thorner P, Irwin MS. SATB2 enhances migration and invasion in osteosarcoma by regulating genes involved in cytoskeletal organization. Oncogene. 2015;34:3582–92.
113. Hermeking H. p53 enters the microRNA world. Cancer Cell. 2007;12:414–8.

114. Gudkov, A. V., Komarova, E. A.. Dangerous habits of a security guard: the two faces of p53 as a drug target. Hum Mol Genet. 2007;16 (Spec No 1):R67–72.
115. Pakos EE, Kyzas PA, Ioannidis JP. Prognostic significance of TP53 tumor suppressor gene expression and mutations in human osteosarcoma: a meta-analysis. Clin Cancer Res. 2004;10:6208–14.
116. Wu X, Cai ZD, Lou LM, Zhu YB. Expressions of p53, c-Myc, BCL-2 and apoptotic index in human osteosarcoma and their correlations with prognosis of patients. Cancer Epidemiol. 2012;36:212–6.
117. Rubin EM, Guo Y, Tu K, Xie J, Zi X, Hoang BH. Wnt inhibitory factor 1 decreases tumorigenesis and metastasis in osteosarcoma. Mol Cancer Ther. 2010;9:731–41.
118. Kawano Y, Kypta R. Secreted antagonists of the Wnt signalling pathway. J Cell Sci. 2003;116:2627–34.
119. Itoh Y. MT1-MMP: a key regulator of cell migration in tissue. IUBMB Life. 2006;58:589–96.
120. Malkin D, Li FP, Strong LC, Fraumeni Jr JF, Nelson CE, Kim DH, Kassel J, Gryka MA, Bischoff FZ, Tainsky MA, et al. Germ line p53 mutations in a familial syndrome of breast cancer, sarcomas, and other neoplasms. Science. 1990;250:1233–8.
121. Li FP, Fraumeni Jr JF, Mulvihill JJ, Blattner WA, Dreyfus MG, Tucker MA, Miller RW. A cancer family syndrome in twenty-four kindreds. Cancer Res. 1988;48:5358–62.
122. Yamada H, Shinmura K, Yamamura Y, Kurachi K, Nakamura T, Tsuneyoshi T, Yokota N, Maekawa M, Sugimura H. Identification and characterization of a novel germline p53 mutation in a patient with glioblastoma and colon cancer. Int J Cancer. 2009;125:973–6.
123. Zoccali C, Teori G, Prencipe U, Erba F. Mazabraud's syndrome: a new case and review of the literature. Int Orthop. 2009;33:605–10.
124. Weinstein LS, Shenker A, Gejman PV, Merino MJ, Friedman E, Spiegel AM. Activating mutations of the stimulatory G protein in the McCune-Albright syndrome. N Engl J Med. 1991;325:1688–95.
125. Spiegel AM, Weinstein LS, Shenker A. Abnormalities in G protein-coupled signal transduction pathways in human disease. J Clin Invest. 1993;92:1119–25.
126. Auyeung J, Mohanty K, Tayton K. Maffucci lymphangioma syndrome: an unusual variant of Ollier's disease, a case report and a review of the literature. J Pediatr Orthop B. 2003;12:147–50.
127. Kumar A, Jain VK, Bharadwaj M, Arya RK. Ollier Disease: Pathogenesis, Diagnosis, and Management. Orthopedics. 2015;38:e497–506.
128. Pansuriya TC, Van Eijk R, d'Adamo P, Van Ruler MA, Kuijjer ML, Oosting J, Cleton-Jansen AM, van Oosterwijk JG, Verbeke SL, Meijer D, van Wezel T, Nord KH, Sangiorgi L, Toker B, Liegl-Atzwanger B, San-Julian M, Sciot R, Limaye N, Kindblom LG, Daugaard S, Godfraind C, Boon LM, Vikkula M, Kurek KC, Szuhai K, French PJ, Bovee JV. Somatic mosaic IDH1 and IDH2 mutations are associated with enchondroma and spindle cell hemangioma in Ollier disease and Maffucci syndrome. Nat Genet. 2011;43:1256–61.
129. Couvineau A, Wouters V, Bertrand G, Rouyer C, Gerard B, Boon LM, Grandchamp B, Vikkula M, Silve C. PTHR1 mutations associated with Ollier disease result in receptor loss of function. Hum Mol Genet. 2008;17:2766–75.
130. Tan A, Ralston SH. Paget's disease of bone. QJM. 2014;107:865–9.
131. Ralston SH, Langston AL, Reid IR. Pathogenesis and management of Paget's disease of bone. Lancet. 2008;372:155–63.
132. Ralston SH, Layfield R. Pathogenesis of Paget disease of bone. Calcif Tissue Int. 2012;91:97–113.
133. Naot D, Bava U, Matthews B, Callon KE, Gamble GD, Black M, Song S, Pitto RP, Cundy T, Cornish J, Reid IR. Differential gene expression in cultured osteoblasts and bone marrow stromal cells from patients with Paget's disease of bone. J Bone Miner Res. 2007;22:298–309.

134. Zhang J, Schweers B, Dyer MA. The first knockout mouse model of retinoblastoma. Cell Cycle. 2004;3:952–9.
135. Abramson DH, Schefler AC. Update on retinoblastoma. Retina. 2004;24:828–48.
136. Gonzalez-Vasconcellos I, Domke T, Kuosaite V, Esposito I, Sanli-Bonazzi B, Nathrath M, Atkinson MJ, Rosemann M. Differential effects of genes of the Rb1 signalling pathway on osteosarcoma incidence and latency in alpha-particle irradiated mice. Radiat Environ Biophys. 2011;50:135–41.
137. Wong FL, Boice Jr JD, Abramson DH, Tarone RE, Kleinerman RA, Stovall M, Goldman MB, Seddon JM, Tarbell N, Fraumeni Jr JF, Li FP. Cancer incidence after retinoblastoma. Radiation dose and sarcoma risk. JAMA. 1997;278:1262–7.
138. Schulz CJ, Riddle MP, Valdimirsdottir HB, Abramson DH, Sklar CA. Impact on survivors of retinoblastoma when informed of study results on risk of second cancers. Med Pediatr Oncol. 2003;41:36–43.
139. Kitao S, Shimamoto A, Goto M, Miller RW, Smithson WA, Lindor NM, Furuichi Y. Mutations in RECQL4 cause a subset of cases of Rothmund-Thomson syndrome. Nat Genet. 1999;22:82–4.
140. Larizza L, Roversi G, Volpi L. Rothmund-Thomson syndrome. Orphanet J Rare Dis. 2010;5:2.
141. Wang LL, Levy ML, Lewis RA, Chintagumpala MM, Lev D, Rogers M, Plon SE. Clinical manifestations in a cohort of 41 Rothmund-Thomson syndrome patients. Am J Med Genet. 2001;102:11–7.
142. Sharma S, Doherty KM, Brosh Jr RM. Mechanisms of RecQ helicases in pathways of DNA metabolism and maintenance of genomic stability. Biochem J. 2006;398:319–37.
143. Sengupta S, Shimamoto A, Koshiji M, Pedeux R, Rusin M, Spillare EA, Shen JC, Huang LE, Lindor NM, Furuichi Y, Harris CC. Tumor suppressor p53 represses transcription of RECQ4 helicase. Oncogene. 2005;24:1738–48.
144. Woo LL, Futami K, Shimamoto A, Furuichi Y, Frank KM. The Rothmund-Thomson gene product RECQL4 localizes to the nucleolus in response to oxidative stress. Exp Cell Res. 2006;312:3443–57.
145. Huang S, Lee L, Hanson NB, Lenaerts C, Hoehn H, Poot M, Rubin CD, Chen DF, Yang CC, Juch H, Dorn T, Spiegel R, Oral EA, Abid M, Battisti C, Lucci-Cordisco E, Neri G, Steed EH, Kidd A, Isley W, Showalter D, Vittone JL, Konstantinow A, Ring J, Meyer P, Wenger SL, von Herbay A, Wollina U, Schuelke M, Huizenga CR, Leistritz DF, Martin GM, Mian IS, Oshima J. The spectrum of WRN mutations in Werner syndrome patients. Hum Mutat. 2006;27:558–67.
146. Goto M, Miller RW, Ishikawa Y, Sugano H. Excess of rare cancers in Werner syndrome (adult progeria). Cancer Epidemiol Biomark Prev. 1996;5:239–46.

Molecular and Cellular Mechanisms of Carcinogenesis in the Large Bowel

4

Iva Brčić, Catarina Callé, and Cord Langner

Contents

Abstract

Colorectal cancer is one of the most intensively studied cancers with well-documented precursor lesions. The acquisition of genomic instability plays a central role in its development. In the majority of cases, tumor growth results from different combinations of sporadic genetic events and epigenetic alterations, resulting in increased cell proliferation and decreased cell death. Three main pathways have been identified: chromosomal instability (CIN)

I. Brčić • C. Langner (✉)
Institute of Pathology, Medical University of Graz, Auenbruggerplatz 25, 8036 Graz, Austria
e-mail: iva.brcic@medunigraz.at; cord.langner@medunigraz.at

C. Callé
Department of Pathology, Portuguese Oncology Institute Francisco Gentil, EPE, Lisbon, Portugal

J. Haybaeck (ed.), *Mechanisms of Molecular Carcinogenesis – Volume 2*,
DOI 10.1007/978-3-319-53661-3_4

pathway, microsatellite instability (MSI) pathway, and CpG island hypermethylation phenotype (CIMP) pathway. Within these pathways, somatic *BRAF* and/or *KRAS* mutations have been identified as major players. Up to 5% of colorectal cancers develop in the setting of inherited syndromes, such as Lynch syndrome, familial adenomatous polyposis, *MUTYH*-associated polyposis, and certain hamartomatous polyposis conditions, including Peutz-Jeghers syndrome and juvenile polyposis syndrome. In this chapter, we describe the above-mentioned pathways and syndromes in detail, referring to different molecular events and different precursor lesions. In the last part, we address possible future perspectives in colorectal carcinogenesis.

4.1 Introduction

Colorectal cancer (CRC) ranks the third most frequent cancer in men (after lung and prostate cancer) and second in women (after breast cancer), representing approximately 9.7% of all new cancer cases diagnosed worldwide [1, 2]. In 2012, an estimated 746.000 men and 614.000 women were diagnosed with CRC, and 694.0000 died of the disease [1, 2]. In the last decade (2001–2010), the global incidence rate decreased by approximately 3% per year [3].

On the molecular level, CRC is one of the most intensively studied cancers. The existence of well-documented precursor lesions indicates multistep cancer development. In fact, this type of cancer represents a very heterogeneous disease regarding the clinical presentation, likelihood of cure, pattern of extension, and response to treatment [4]. The acquisition of genomic instability plays a central role in its development. In the majority of cases, CRC results from different combinations of sporadic genetic events and epigenetic alterations, resulting in increased cell proliferation and decreased cell death [5]. Kindred and twin studies, also studies based on family clusters, estimated that approximately 30% of all CRC cases are inherited [6, 7].

In the last decade, a growing body of scientific evidence demonstrated that different CRC subtypes can be separated based upon combinations of different genetic markers. Three major signaling pathways have been recognized, all characterized by specific precursor lesions, mechanisms of carcinogenesis, and natural history: the chromosomal instability (CIN) pathway, the microsatellite instability (MSI) pathway, and the CpG island hypermethylation phenotype (CIMP) pathway [5, 8] (Fig. 4.1). Within these pathways, somatic *BRAF* and/or *KRAS* mutations have been identified as major players [5]. Up to 5% of CRCs develop in the setting of inherited syndromes like Lynch syndrome (LS), familial adenomatous polyposis (FAP), *MUTYH*-associated polyposis (MAP), and certain hamartomatous polyposis conditions [9].

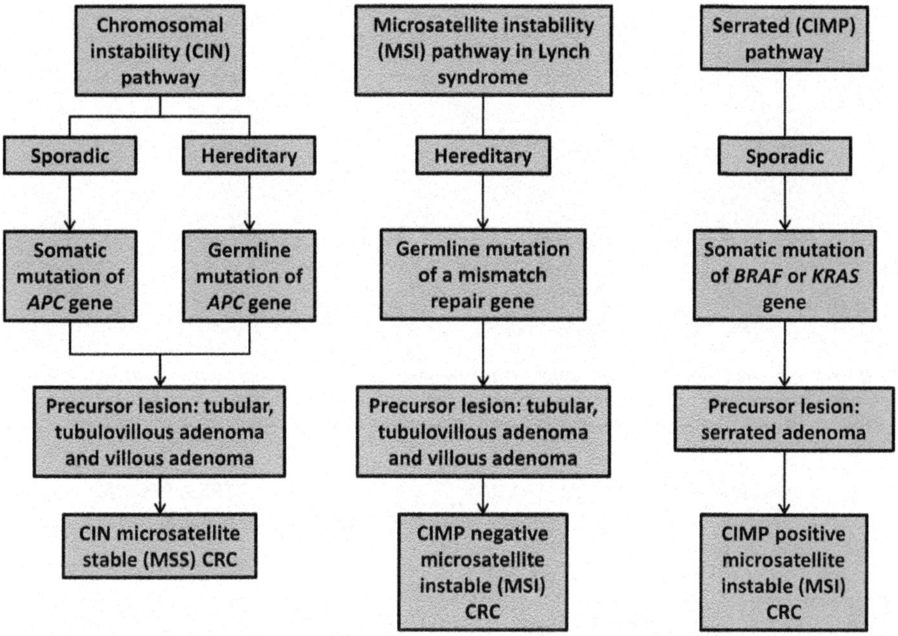

Fig. 4.1 Three major carcinogenic pathways have been identified in colorectal cancer (*CRC*): chromosomal instability (*CIN*), microsatellite instability (*MSI*), and CpG island methylator phenotype (*CIMP*)

4.2 Molecular Classification of Colorectal Cancer

In this chapter, we will describe the three major pathways responsible for CRC: chromosomal instability (CIN), microsatellite instability (MSI), and CpG island hypermethylation phenotype (CIMP). We will also refer to the MAP and to hamartomatous polyposis syndrome, such as Peutz-Jeghers syndrome (PJS) and juvenile polyposis syndrome (JPS), and will finally address possible future perspectives.

4.2.1 The Chromosomal Instability (CIN) Pathway

The CIN pathway, also called the "traditional pathway," is characterized by imbalance in chromosomal number (aneuploidy), subsequent loss of heterozygosity (LOH) of genes, and subchromosomal genetic amplifications [10]. The time of tumor development via this pathway is approximately 10–15 years. The initial lesion in this pathway is the dysplastic aberrant crypt focus (ACF) [11]. It is a microscopic mucosal lesion that develops into conventional adenomas, i.e., tubular

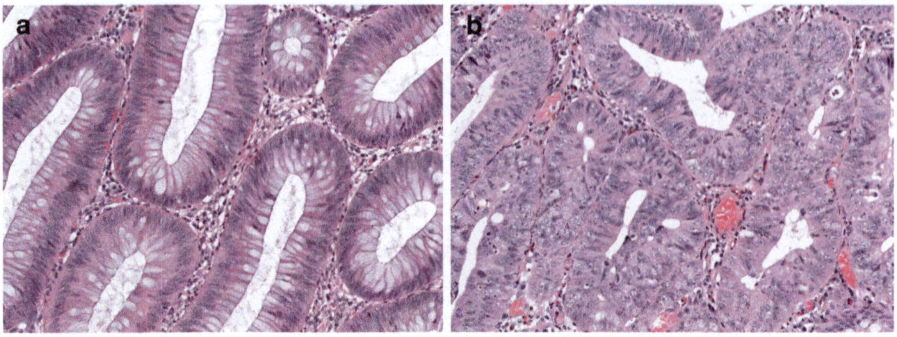

Fig. 4.2 Tubular colorectal adenoma with low-grade dysplasia, characterized by well-formed glands and pseudostratified, polarized, hyperchromatic nuclei (**a**). High-grade dysplasia characterized by increased architectural complexity and more severe atypia with loss of nuclear polarity (**b**)

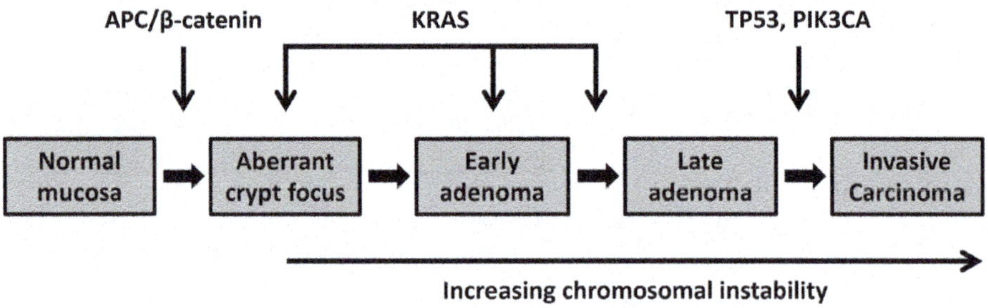

Fig. 4.3 Multistep genetic model of colorectal carcinogenesis (adenoma-carcinoma sequence): chromosomal instability is observed in benign adenomas and increases in conjunction with tumor progression (from [12], S. Karger AG, with permission)

adenomas, tubulovillous, and villous adenomas (Fig. 4.2), which are the macroscopically discernable precursor lesions of sporadic CRCs arising via this pathway [12]. It is of note that traditional adenomas are also considered to be the precursor lesions in the hereditary cancers, namely, LS and FAP [12, 13].

Already in 1990, Fearon and Vogelstein [14] proposed a multistep model of sequential genetic alterations, responsible for adenoma and ultimately carcinoma formation within the colorectum (Fig. 4.3). Mutation in the adenomatous polyposis coli (*APC*) tumor-suppressor gene located on chromosome 5q21 has been identified as the first step of this model [15]. Both copies of the *APC* gene must be functionally inactivated for adenomas to develop. Specifically, *APC* mutation interferes with phosphorylation of β-catenin, a component of the *Wnt* signaling pathway that regulates apoptosis, growth, and differentiation. Consequently, β-catenin is not ubiquitinated and destroyed by the proteasome. It accumulates in the cytosol and is translocated to the nucleus, where it interacts with T-cell factor (TCF)/lymphoid enhancer factor (LEF), converting these effectors into transcriptional activators [16]. Activation of the *Wnt* pathway is present in up to 80% of adenomas [11].

The second molecular event is an activating mutation of Kirsten-rat sarcoma 2 viral oncogene (*KRAS*), which is, however, not unique for this pathway. This

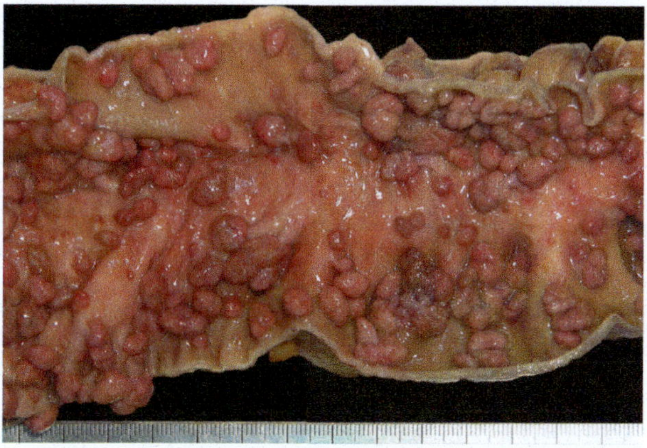

Fig. 4.4 Gross presentation of familial adenomatous polyposis (FAP): the colectomy specimen shows numerous adenomatous polyps

mutation is found in approximately 45% of CRCs and constitutively activates the MAPK signaling pathway [17]. The genomic change in adenoma-carcinoma sequence also includes LOH of chromosome 18q, which is present in up to 60% of tumors [18]. Many important tumor-suppressor genes are located at 18q21.1—*DCC*, *SMAD2*, and *SMAD4*, the latter being involved in TGF-β signaling, responsible for regulation of growth and apoptosis. Mutational inactivation of the tumor-suppressor *TP53* (17p13) occurs as a late event (at the transition from high-grade adenoma to invasive cancer) in up to 80% of CRC [17]. Mutational activation of phosphatidylinositol-4, 5-bisphosphate 3-kinase catalytic subunit alpha (PIK3CA) occurs also in the late phase, but in a small number of cases [10].

Recognition of the central role of *APC* mutations in tumorigenesis has improved our understanding of FAP, the second most common inherited CRC syndrome. APC-associated polyposis conditions also include attenuated FAP, Gardner syndrome (FAP with epidermoid cysts, osteomas, dental anomalies, and/or desmoid tumors), and Turcot syndrome (colonic polyps with central nervous system tumors) [9, 19]. FAP is characterized by the development of hundreds to thousands of conventional adenomas beginning in early adolescence (Fig. 4.4). The disease inevitably leads to CRC, thereby prompting prophylactic colectomy. This syndrome accounts for only <1% of all CRCs. The neoplastic polyps are distributed among the colorectum and can also be observed in the stomach and small bowel, in particular the duodenum. Attenuated FAP is a less severe form, characterized by <100 colonic adenomatous polyps with tendency for proximal location. In individuals with attenuated FAP, adenoma and CRC development is delayed by 15 years when compared to classic FAP [20].

4.2.2 Microsatellite Instability (MSI) Pathway and Lynch Syndrome (LS)

Errors that occur during DNA replication are corrected by the mismatch repair (MMR) system, which includes the following proteins: MLH1, PMS2, MSH2, and MSH6. This system is necessary for maintaining genomic stability. During mismatch repair, the MMR proteins form heterodimers, that is, MLH1 builds a complex with

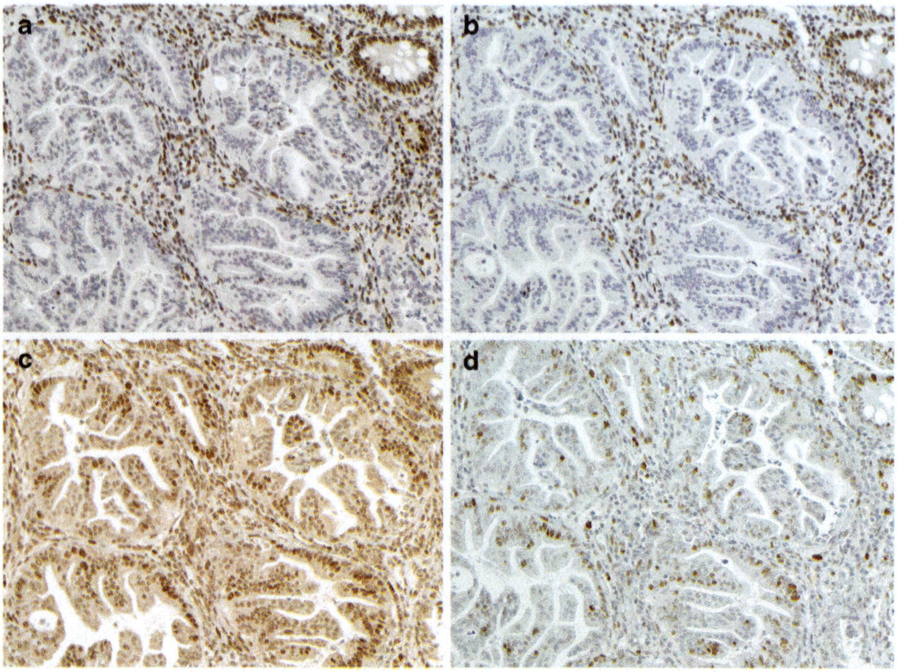

Fig. 4.5 Mismatch repair (MMR) protein expression in a cancer with high-level microsatellite instability (MSI-H): MLH1 (**a**) and PMS2 (**b**) staining is lost in the tumor cell nuclei, while the expression of MSH2 (**c**) and MSH6 (**d**) is retained. Nonneoplastic stromal and inflammatory cells serve as internal positive control (serial sections)

PMS2, and MSH2 builds another with MSH6 [21, 22]. It is well known that the MLH1 and MSH2 proteins are the dominant components of their heterodimers. In consequence, mutations in *MLH1* or *MSH2* gene lead to proteolytic degradation of the corresponding dimer and subsequent loss of both, the main and the auxiliary partner proteins (Fig. 4.5) [23]. If a mutation occurs in one of the auxiliary genes, i.e., *PMS2* or *MSH6*, this results in a loss of the respective PMS2 or MSH6 protein, but does not cause secondary loss of the dominant protein, that is, MLH1 or MSH2 [3].

When the MMR system does not function properly, the cells accumulate genetic errors. These may happen also in so-called microsatellites, that is, repetitive segments of DNA (two to five nucleotides in length), which are found scattered throughout the genome in the noncoding regions between genes or within genes [24]. MSI is defined as a change of any length of these repeating units, due to deletion or insertion [25].

For MSI testing, different panels of microsatellite markers have been used. The first consensus of the National Cancer Institute (NCI) recommended the use of a panel of five markers for MSI testing [26]. This included two mononucleotide repeats (BAT-25 and BAT-26) and three dinucleotide repeats (D5S346, D2S123, and D17S250) [27]. Other panels are solely based upon mononucleotide repeat markers, which can be amplified and analyzed in a single assay [28, 29]. Tumors with instability in two or more of the five markers qualify for high-level MSI (MSI-H; Fig. 4.6), whereas those with instability at one repeat qualify for low-level MSI

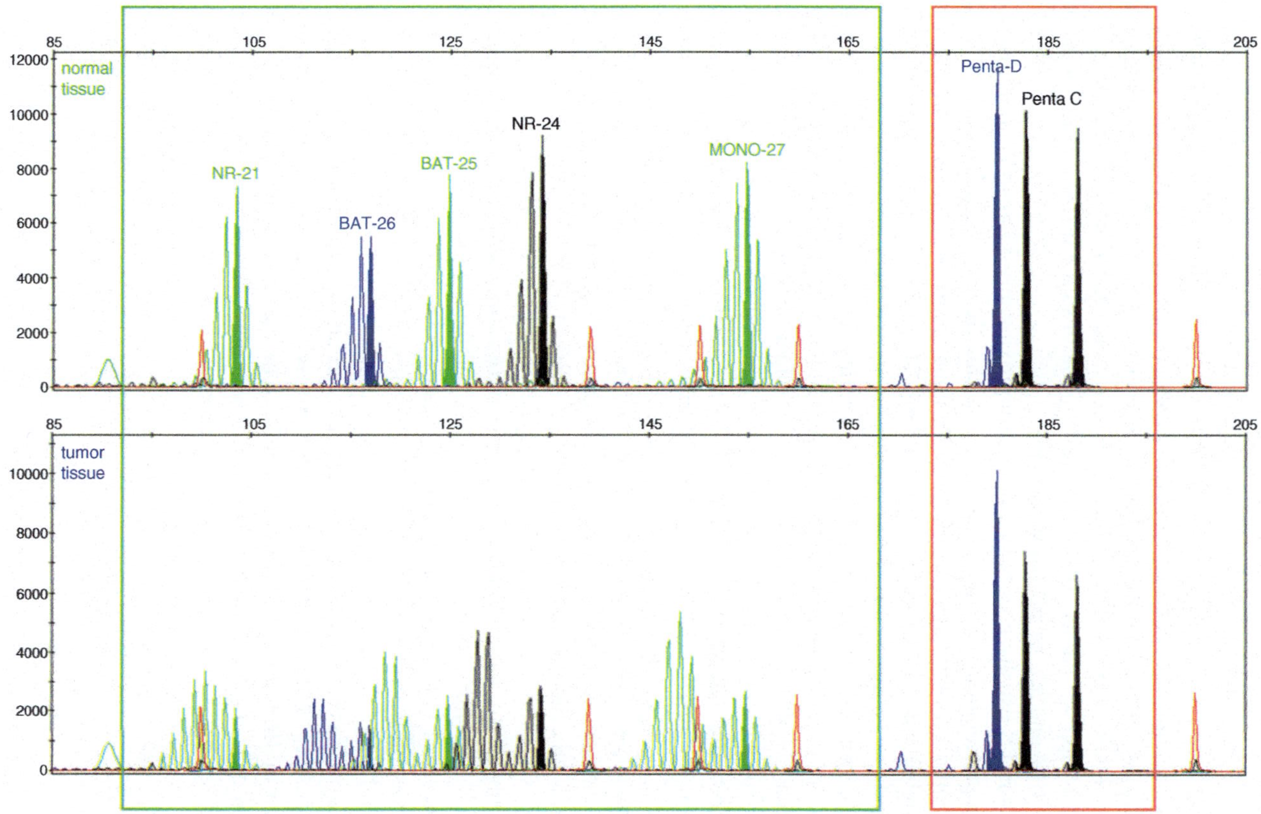

Fig. 4.6 Example of cancer with high-level microsatellite instability (MSI-H): The MSI profile is assessed by a panel of five monomorphic mononucleotide repeats. Instability for all markers is observed, as shown by additional alleles (allelic shifts). Two polymorphic pentanucleotide repeats (Penta C and Penta D) are included for sample identification

(MSI-L). When all markers are stable, the lesion is called microsatellite stable (MSS) [4, 28].

Approximately 15% of CRCs are genetically instable due to MSI [30]. The majority of these tumors (80%) are sporadic and arise due to hypermethylation of the *MLH1* gene promoter [31]. Other 20% of tumors are inherited, that is, caused by germ line mutation in one of the MMR genes and associated with LS [32]. This syndrome follows an autosomal dominant trait of inheritance and accounts for 2–4% of all CRCs [33, 34]. Specifically, mutations in *MLH1* and *MSH2* account for most cases (approximately 40% each), while mutations in *MSH6* and *PMS2* account for only 10% and 5%, respectively [33, 35].

CRCs in LS usually occur at early age (approximately 45 years) and are right-sided (approximately 70% proximal to the splenic flexure) [33]. In addition, they may be multifocal with syn- and/or metachronous tumor development, and there is also a higher risk for extracolonic tumors [36]. These mainly include endometrial, ovary, and gastric tumors [9].

The lifetime risk of cancer in LS is depending on sex and the mutated MMR gene [37–44]. In patients with *MLH1* or *MSH2* mutation, the risk of CRC has been calculated 27–74% for males and 22–53% for females, respectively, with mean age at diagnosis varying from 27 to 46 years (69 years for sporadic cancers). The risk of endometrium cancer is 14–54% [45]. When *MSH6* is mutated, the CRC risk appears to be lower (18%), while the endometrium cancer risk is not changed. Smaller studies reported a lower *PMS2* mutation penetrance for CRC and endometrium cancer as compared with *MLH1* and *MSH2* mutation carriers and similar or even lower risks as compared with *MSH6* mutation carriers [46]. A large European cohort recently reported a cumulative risk of CRC of 19% for males and 11% for females, while the risk of endometrium cancer was 12%. In this cohort, the mean age at diagnosis for both CRC and endometrium cancer was higher as compared with *MLH1* or *MSH2* mutation carriers. When compared with *MSH6*, the mean age at diagnosis of CRC was lower, and the mean age at diagnosis of endometrium cancer was similar [46].

Several tools are available to assist the clinical diagnosis of LS, including analyses of family history, tumor testing, mutation prediction models, and genetic testing. The Amsterdam criteria were created first in 1990 and then reestablished in 1999 as Amsterdam criteria II defining clinical criteria needed for the diagnosis of HNPCC [45, 47–51]. These criteria include individual patient and family history of colonic and extracolonic tumors. They are listed in Table 4.1.

The revised Bethesda guidelines are a third set of clinicopathologic criteria developed to identify individuals that should be investigated for LS by evaluation of MSI and/or immunohistochemical (IHC) testing (Table 4.2) [52].

Table 4.1 Amsterdam criteria I and II for the diagnosis of Lynch syndrome [45, 47–51]

Amsterdam criteria I
1. Three or more relatives with histologically verified CRC, one of which is a first-degree relative of the other two
2. Two or more generations should be affected
3. One or more patients with CRC should be diagnosed before the age of 50 years
4. Familial adenomatous polyposis (FAP) should be excluded
Amsterdam criteria II
1. Three or more relatives with histologically verified Lynch syndrome-associated cancer (CRC, cancer of the endometrium, small bowel, ureter, or renal pelvis), one of which is a first-degree relative of the other two
2. Two or more generations should be affected
3. One or more cancer patients should be diagnosed before the age of 50 years
4. Familial adenomatous polyposis (FAP) should be excluded

Table 4.2 The revised Bethesda Guidelines [45, 49–52]. Colorectal cancers (CRCs) should be tested for microsatellite instability (MSI) in the following settings:

1. CRC diagnosed in a patient who is less than 50 years of age
2. Presence of synchronous or metachronous CRC or other Lynch syndrome-associated tumor[a], regardless of age
3. CRC with MSI-H histology diagnosed in a patient who is less than 60 years of age
4. Patient with CRC and CRC or Lynch syndrome-associated tumor[a] diagnosed in at least one first-degree relative less than 50 years of age
5. Patient with CRC and CRC or Lynch syndrome-associated tumor[a] diagnosed in two first-degree or second-degree relatives, regardless of age

[a]Lynch syndrome-associated tumors: cancers of the colorectum, endometrium, stomach, ovary, pancreas, biliary tract, small bowel, ureter, and renal pelvis; brain tumors (usually glioblastoma as seen in Turcot syndrome); sebaceous gland adenomas; and keratoacanthomas (in Muir-Torre syndrome)

Adenomas and CRCs in LS arise earlier and at more proximal location when compared to sporadic neoplasm. The rate of adenoma development is similar to the rate of adenoma development in the sporadic setting, but progression to cancer occurs at increased rate. This is in contrast to FAP, which has an increased rate of adenoma formation, while progression to cancer is believed to occur at a similar rate to that of sporadic adenomas. In LS, the germ line inactivation of one of the mismatch repair genes, coupled with somatic inactivation of the remaining allele, increases the mutation rate and, subsequently, the rate of progression from adenoma to cancer (Fig. 4.7) [12, 53].

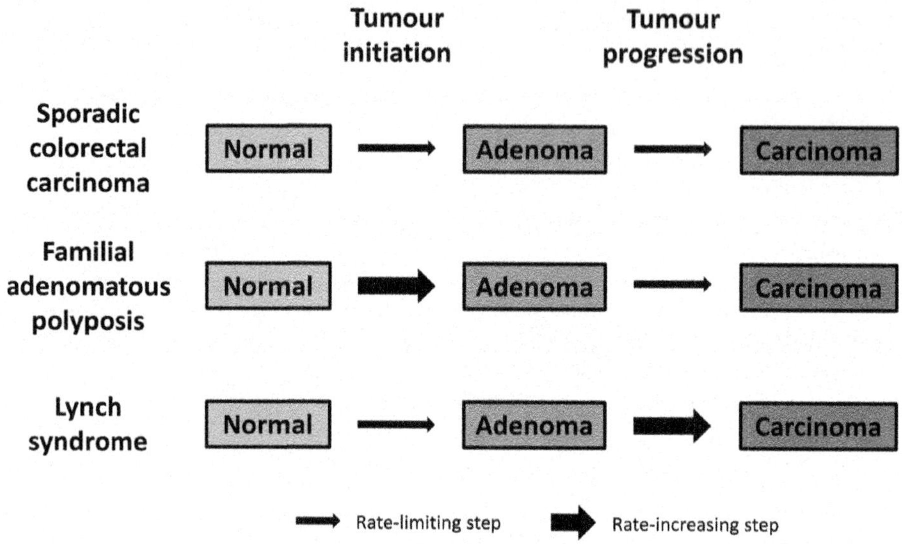

Fig. 4.7 Relative effects of germ line mutations on the rate of tumor initiation and progression: In sporadic tumors, adenoma formation and cancer development are rate-limiting steps. In familial adenomatous polyposis, adenoma formation occurs at an increased rate, while adenomas progress to cancer at a rate similar to the sporadic setting. The mutation rate within adenomatous polyps and, subsequently, the rate of progression from adenoma to cancer are increased in Lynch syndrome (from [12], S. Karger AG, with permission)

4.2.3 CpG Island Methylator Phenotype (CIMP) and Serrated Pathway

CpG dinucleotides (cytosine nucleotide followed by a guanine nucleotide) are uncommon in the human genome. However, in the promoter region of about half of all genes, clusters of these nucleotides, called CpG islands, are found [54]. Aberrant (hyper)methylation of CpG-rich promoters leads to epigenetic silencing of tumor-suppressor genes and ultimately cancer. The methylation status of the tumor can be assessed according to the degree of methylation as CIMP high, CIMP low, or CIMP negative [55]. However, molecular analysis of CIMP and classification of methylation level appear to be poorly standardized. Hence, up to date, no precise definition of CIMP and no consensus recommendation are available [3].

Sporadic MSI-H CRCs occur in patients without germ line mutation in a MMR gene. These tumors occur preferably in the right colon. They are diagnosed more commonly in women, often at advanced age [56, 57]. These cancers develop from serrated precursor lesions [31] through the CIMP or "serrated pathway" (Fig. 4.8), characterized by *BRAF* mutation (characteristically V600E) and hypermethylation in CpG-rich gene promoters, which leads to silencing of distinct tumor-suppressor genes, including the MMR gene *MLH1*, as well as *p16*, *MGMT*, and *IGFB7* [58–62].

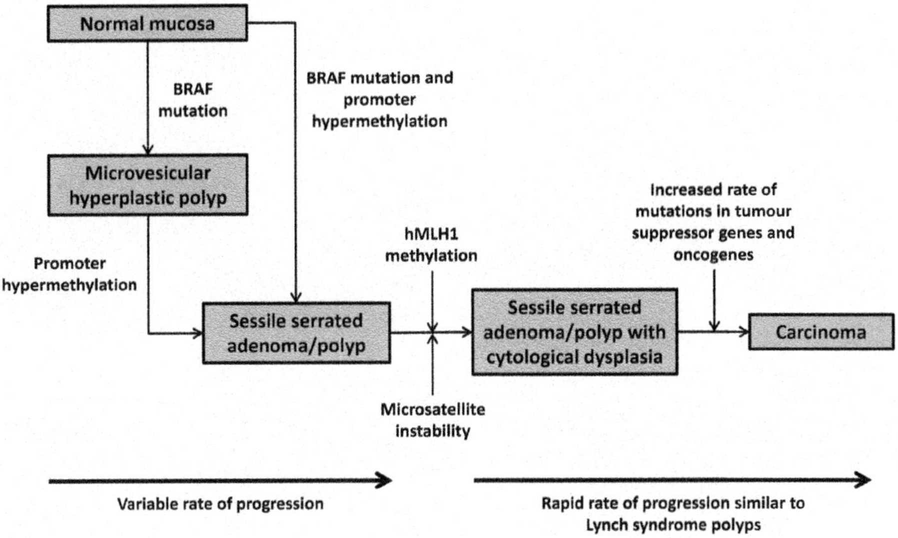

Fig. 4.8 Colorectal carcinogenesis according to the "serrated (CIMP) pathway": sporadic colorectal adenocarcinomas with high-level microsatellite instability (MSI-H) develop from serrated precursor lesions due to promoter hypermethylation of the *MLH1* gene (from [12], S. Karger AG, with permission)

Sessile serrated adenomas/polyps (SSA/P) are considered to be the main precursor lesions of the serrated pathway. They account for approximately 5–25% of all serrated lesions occurring in the colorectum [13, 63, 64]. They may arise from large microvesicular hyperplastic polyps or develop de novo from normal colonic mucosa. Uncomplicated SSA/Ps do not show dysplasia. Dysplasia may, however, occur during neoplastic progression (Fig. 4.9). There appears to be a histological continuum from non-dysplastic ACF to microvesicular hyperplastic polyps to SSA/P to SSA/P with cytological dysplasia and ultimately to invasive ("serrated") adenocarcinoma [12].

Serrated lesions can also occur in familial setting. Serrated polyposis syndrome is a rare condition characterized by multiple and/or large serrated polyps of the colon. Guarinos et al. [65] identified *BRAF* mutations in 63% and *KRAS* mutations in 10% of lesions occurring in this syndrome; 43% of the lesions were CIMP high. A single per patient analysis showed that all patients had *BRAF* or *KRAS* mutation in more than 25% of the polyps, and 84.8% of patients had a mutation in *BRAF* or *KRAS* in more than 50% of their polyps [65]. Germ line loss-of-function mutations in oncogene-induced senescence pathways may play an additional role in the disease [66].

Traditional serrated adenomas (TSAs) are much less common than the other serrated lesions, accounting for approximately 1% of colorectal polyps (Fig. 4.10). The majority of lesions are detected in the distal colon [12]. TSAs may originate from preexisting non-dysplastic serrated polyps, including hyperplastic polyps and SSA/Ps.

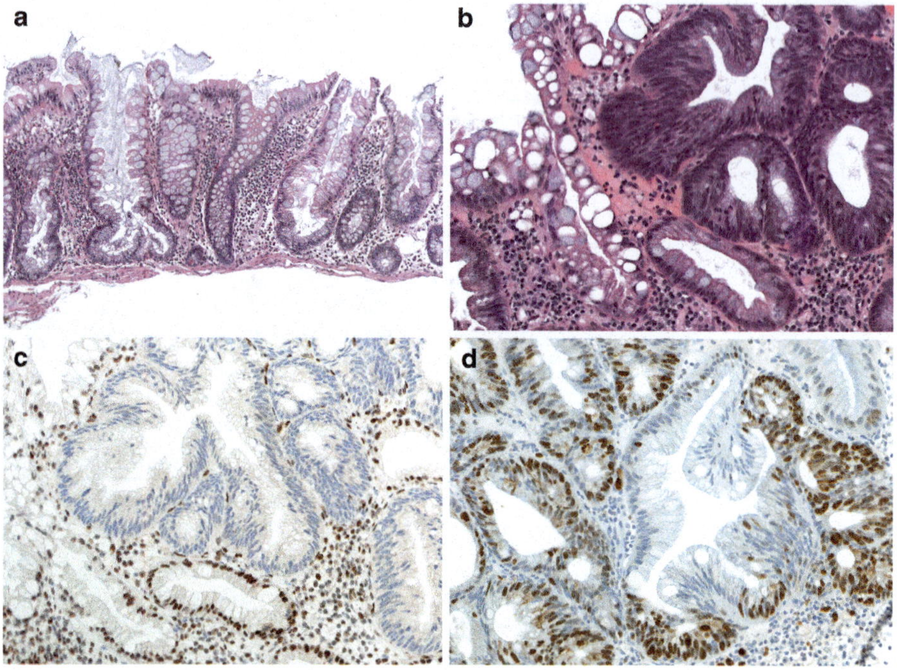

Fig. 4.9 Sessile serrated adenoma/polyp (SSA/P) with increased serration of non-dysplastic crypts, with T-shaped ("anchor") crypts and mature goblet cells at the crypt bases (**a**). Cytological dysplasia is not present in uncomplicated SSA/P, but develops with progression toward carcinoma (**b**), often in conjunction with promoter hypermethylation of the *MLH1* gene, as illustrated by loss of nuclear MLH1 expression (**c**). The proliferation rate (MIB-1) is markedly increased in the dysplastic glands (**d**)

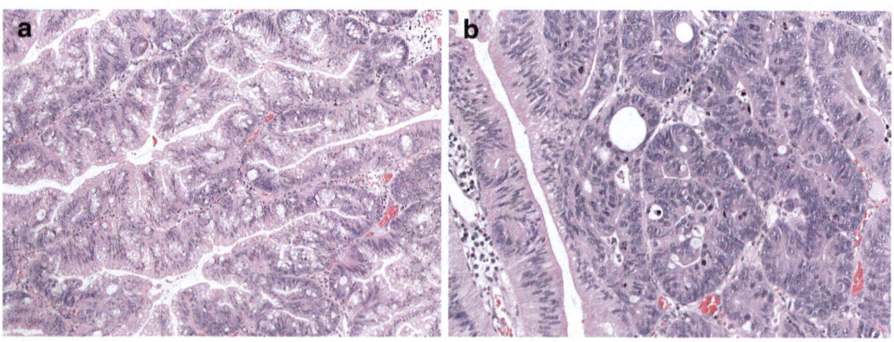

Fig. 4.10 Traditional serrated adenoma (TSA) with slit-like serration, cytoplasmic eosinophilia, and proliferative "ectopic crypts" (**a**). In high-grade dysplasia, marked architectural complexity and nuclear atypia with increased nuclear/cytoplasmic ratio are observed (**b**)

On the molecular level, these lesions are characterized by *BRAF* mutations, giving rise to *BRAF*-mutated MSS CRCs [67, 68]. TSAs may alternatively develop de novo. These lesions mainly show mutations in the *KRAS* gene. Malignant progression occurs via *TP53* mutation and *Wnt* pathway activation regardless of mutation status [67–69].

4.2.4 MUTYH-Associated Polyposis (MAP)

MAP is a hereditary condition caused by biallelic germ line mutations in *MUTYH* gene and has an autosomal recessive pattern of inheritance [70]. It is characterized by the development of multiple neoplastic polyps in the colorectum and increased risk of CRC [9]. The colonic phenotype of MAP mimics FAP—however, in addition to multiple adenomatous polyps, hyperplastic polyps and SSA/Ps can also be found [71].

The *MUTYH* gene product is involved in the base-excision repair pathway and protects against oxidative DNA damage. Individuals with >10 colorectal adenomas who do not have mutation in *APC* should undergo genetic testing for MAP [9].

4.2.5 Hamartomatous Polyposis Conditions

Hamartomatous polyposis conditions include PJS, JPS, hereditary mixed polyposis syndrome, Cowden syndrome, Bannayan-Riley-Ruvalcaba syndrome, and Cronkhite-Canada syndrome [72]. This group of disorders is characterized by the development of multiple benign-appearing polyps in the gastrointestinal tract. Affected individuals bear an increased risk of cancer, not only in the gastrointestinal tract but also in other organs [9]. Carcinogenesis, that is, progression of the hamartomatous polyps to cancer or cancer development de novo, is still largely unclear, suggesting different pathways from adenomatous polyposis. In this chapter, we will concentrate on the two most common conditions, that is, PJS and JPS. They can both be sporadic or familial, in the hereditary setting having an autosomal dominant pattern of inheritance.

In PJS the key clinical features are hyperpigmentation (melanosis) of the lips, mouth, and oral mucosa and polyposis of the small intestine. Affected individuals harbor a mutation in the *STK11* gene [73]. Lifetime cancer risk is as high as 81–93%, with 50% risk for breast cancer and 39% risk for colon cancer [74].

JPS is caused by germ line mutations in either *MADH4* (*SMAD4, DPC4)* or *BMPR1A*, which can be found in 18.2% or 20.8% of affected individuals, respectively [75]. The condition is characterized by multiple hamartomatous polyps, most commonly arising in the colon but rarely also in the stomach, duodenum, and small bowel. For both sporadic and familial JPS, mean age of CRC diagnosis is 37 years [76]. Lifetime cancer risk has been estimated 38% for colonic and 21% for upper GI cancers, including the stomach, pancreas, and small bowel [77].

4.3 Future Perspectives

Recent data indicate an even greater complexity of cancer development in the colorectum. Thus, germ line exonuclease domain mutations (EDMs) of *POLE* and *POLD1* have been shown to confer high risk of multiple colorectal adenomas and carcinoma, a condition named polymerase proofreading-associated polyposis (PPAP). Somatic *POLE* EDMs have also been found in sporadic CRCs and endometrial cancers. It is believed that both the germ line and the somatic mutations cause impair polymerase proofreading resulting in "ultramutated," yet microsatellite stable (MSS), tumors [78].

In addition, Guinney et al. [79] reported four "consensus molecular subtypes" (CMS) of colorectal cancer: CMS1 (MSI immune, 14%), hypermutated, microsatellite unstable, and strong immune activation; CMS2 (canonical, 37%), epithelial, marked WNT, and MYC signaling activation; CMS3 (metabolic, 13%), epithelial and evident metabolic dysregulation; and CMS4 (mesenchymal, 23%), prominent transforming growth factor-β activation, stromal invasion, and angiogenesis. It is of note that 13% of samples tested showed mixed features, which could be explained by intratumoral heterogeneity or by a transition phenotype. The significance of this "consensus" publication is, however, currently unclear.

Conclusion

Different molecular and cellular mechanisms of carcinogenesis have been identified in the large bowel. These mainly include CIN, MSI, and CIMP pathways. Familial cancers may arise within FAP, LS, and MAP syndromes. Hamartomatous polyposis syndromes likewise harbor increased cancer risk. Four consensus molecular subtypes (CMS1-CMS4) have been described recently, awaiting validation by other groups.

References

1. Cancer fact sheet n°297. WHO Media Center. 2015. http://www.who.int/mediacentre/factsheets/fs297/en/. Accessed 2 Dec 2015.
2. GLOBOCAN. Estimated cancer incidence, mortality and prevalence in 2012 (2012) International Agency for Research on Cancer–WHO. 2012. http://globocan.iarc.fr/Pages/fact_sheets_cancer.aspx. Accessed 2 Dec 2015.
3. Setaffy L, Langner C. Microsatellite instability in colorectal cancer: clinicopathological significance. Pol J Pathol. 2015;66:203–21.
4. Bosman FT, Yua P. Molecular pathology of colorectal cancer. Pol J Pathol. 2014;65:257–66.
5. Samadder NJ, Vierkant RA, Tillmans LS, et al. Associations between colorectal cancer molecular markers and pathways with clinicopathologic features in older women. Gastroenterology. 2013;145:348–56. doi:10.1053/j.gastro.2013.05.001.
6. Lichtenstein P, Holm NV, Verkasalo PK, et al. Environmental and heritable factors in the causation of cancer—analyses of co-horts of twins from Sweden, Denmark, and Finland. N Engl J Med. 2000;343:78–85.
7. Grady WM. Genetic testing for high-risk colon cancer patients. Gastroenterology. 2003;124:1574–94.

8. Leggett B, Whitehall V. Role of the serrated pathway in colorectal cancer pathogenesis. Gastroenterology. 2010;138:2088–100. doi:10.1053/j.gastro.2009.12.066.
9. Jasperson KW, Tuohy TM, Neklason DW. Hereditary and familial colon cancer. Gastroenterology. 2010;138:2044–58. doi:10.1053/j.gastro.2010.01.054.
10. Pino MS, Chung DC. The chromosomal instability pathway in colon cancer. Gastroenterology. 2009;138:2059–72. doi:10.1053/j.gastro.2009.12.065.
11. Takayama T, Ohi M, Hayashi T, et al. Analysis of K-ras, APC, and beta-catenin in aberrant crypt foci in sporadic adenoma, cancer, and familial adenomatous polyposis. Gastroenterology. 2001;121:599–611.
12. Langner C. Serrated and non-serrated precursor lesions of colorectal cancer. Dig Dis. 2015;33:28–37. doi:10.1159/000366032.
13. Bettington M, Walker N, Clouston A, et al. The serrated pathway to colorectal carcinoma: current concepts and challenges. Histopathology. 2013;62:367–86. doi:10.1111/his.12055.
14. Fearon ER, Vogelstein B. A genetic model for colorectal tumorigenesis. Cell. 1990;61:759–67.
15. Bodmer WF, Bailey CJ, Bodmer J, et al. Localization of the gene for familial adenomatous polyposis on chromosome 5. Nature. 1987;328:614–6.
16. White BD, Chien AJ, Dawson DW. Dysregulation of Wnt/β-catenin signaling in gastrointestinal cancers. Gastroenterology. 2012;142:219–32. doi:10.1053/j.gastro.2011.12.001.
17. Leslie A, Carey FA, Pratt NR, et al. The colorectal adenoma-carcinoma sequence. Br J Surg. 2002;89:845–60.
18. Vogelstein B, Fearon ER, Hamilton SR, et al. Genetic alterations during colorectal-tumor development. N Engl J Med. 1988;319:525–32.
19. Claes K, Dahan K, Tejpar S, et al. The genetics of familial adenomatous polyposis (FAP) and MutYH-associated polyposis (MAP). Acta Gastroenterol Belg. 2011;74:421–6.
20. Burt RW, Leppert MF, Slattery ML, et al. Genetic testing and phenotype in a large kindred with attenuated familial adenomatous polyposis. Gastroenterology. 2004;127:444–51.
21. Acharya S, Wilson T, Gradia S, et al. hMSH2 forms specific mispair-binding complexes with hMSH3 and hMSH6. Proc Natl Acad Sci U S A. 1996;93:13629–34.
22. Kadyrov FA, Dzantiev L, Constantin N, et al. Endonucleolytic function of MutLalpha in human mismatch repair. Cell. 2006;126:297–308.
23. Young J, Simms LA, Biden KG, et al. Features of colorectal cancers with high-level microsatellite instability occurring in familial and sporadic settings: parallel pathways of tumorigenesis. Am J Pathol. 2001;159:2107–16.
24. Sideris M, Papagrigoriadis S. Molecular biomarkers and classification models in the evaluation of the prognosis of colorectal cancer. Anticancer Res. 2014;34:2061–8.
25. Boland CR, Koi M, Chang DK, et al. The biochemical basis of microsatellite instability and abnormal immunohistochemistry and clinical behavior in Lynch syndrome: from bench to bedside. Familial Cancer. 2008;7:41–52.
26. Boland CR, Thibodeau SN, Hamilton SR, et al. A National Cancer Institute Workshop on Microsatellite Instability for cancer detection and familial predisposition: development of international criteria for the determination of microsatellite instability in colorectal cancer. Cancer Res. 1998;58:5248–57.
27. Mead LJ, Jenkins MA, Young J, et al. Microsatellite instability markers for identifying early-onset colorectal cancers caused by germ-line mutations in DNA mismatch repair genes. Clin Cancer Res. 2007;13:2865–9.
28. Suraweera N, Duval A, Reperant M, et al. Evaluation of tumor microsatellite instability using five quasimonomorphic mononucleotide repeats and pentaplex PCR. Gastroenterology. 2002;123:1804–11.
29. Patil DT, Bronner MP, Portier BP, et al. A five-marker panel in a multiplex PCR accurately detects microsatellite instability-high colorectal tumors without control DNA. Diagn Mol Pathol. 2012;21:127–33. doi:10.1097/PDM.0b013e3182461cc3.
30. Jass JR. Classification of colorectal cancer based on correlation of clinical, morphological and molecular features. Histopathology. 2007;50:113–30.

31. Boland CR, Goel A. Microsatellite instability in colorectal cancer. Gastroenterology. 2010;138:2073–87. doi:10.1053/j.gastro.2009.12.064.
32. Lynch HT, Smyrk T, Lynch JF. Molecular genetics and clinical-pathology features of hereditary nonpolyposis colorectal carcinoma (Lynch syndrome): historical journey from pedigree anecdote to molecular genetic confirmation. Oncology. 1998;55:103–8.
33. Lynch HT, de la Chapelle A. Hereditary colorectal cancer. N Engl J Med. 2003;348:919–32.
34. Hampel H, Frankel WL, Martin E, et al. Feasibility of screening for Lynch syndrome among patients with colorectal cancer. J Clin Oncol. 2008;26:5783–8. doi:10.1200/JCO.2008.17.5950.
35. Cicek MS, Lindor NM, Gallinger S, et al. Quality assessment and correlation of microsatellite instability and immunohistochemical markers among population- and clinic-based colorectal tumors results from the Colon Cancer Family Registry. J Mol Diagn. 2011;13:271–81. doi:10.1016/j.jmoldx.2010.12.004.
36. Bellizzi AM, Frankel WL. Colorectal cancer due to deficiency in DNA mismatch repair function: a review. Adv Anat Pathol. 2009;16:405–17. doi:10.1097/PAP.0b013e3181bb6bdc.
37. Dunlop MG, Farrington SM, Carothers AD, et al. Cancer risk associated with germline DNA mismatch repair gene mutations. Hum Mol Genet. 1997;6:105–10.
38. Quehenberger F, Vasen HF, van Houwelingen HC. Risk of colorectal and endometrial cancer for carriers of mutations of the hMLH1 and hMSH2 gene: correction for ascertainment. J Med Genet. 2005;42:491–6.
39. Jenkins MA, Baglietto L, Dowty JG, et al. Cancer risks for mismatch repair gene mutation carriers: a population-based early onset case-family study. Clin Gastroenterol Hepatol. 2006;4:489–98.
40. Alarcon F, Lasset C, Carayol J, et al. Estimating cancer risk in HNPCC by the GRL method. Eur J Hum Genet. 2007;15:831–6.
41. Senter L, Clendenning M, Sotamaa K, et al. The clinical phenotype of Lynch syndrome due to germline PMS2 mutations. Gastroenterology. 2008;135:419–28. doi:10.1053/j.gastro.2008.04.026.
42. Choi YH, Cotterchio M, McKeown-Eyssen G, et al. Penetrance of colorectal cancer among MLH1/MSH2 carriers participating in the colorectal cancer familial registry in Ontario. Hered Cancer Clin Pract. 2009;7:14. doi:10.1186/1897-4287-7-14.
43. Baglietto L, Lindor NM, Dowty JG, et al. Risks of Lynch syndrome cancers for MSH6 mutation carriers. J Natl Cancer Inst. 2010;102:193–201. doi:10.1093/jnci/djp473.
44. Bonadona V, Bonaïti B, Olschwang S, et al. Cancer risks associated with germline mutations in MLH1, MSH2, and MSH6 genes in Lynch syndrome. JAMA. 2011;305:2304–10. doi:10.1001/jama.2011.743.
45. Giardiello FM, Allen JI, Axilbund JE, et al. Guidelines on genetic evaluation and management of Lynch syndrome: a consensus statement by the US Multi-Society Task Force on colorectal cancer. Gastroenterology. 2014;147:502–26. doi:10.1053/j.gastro.2014.04.001.
46. ten Broeke SW, Brohet RM, Tops CM, et al. Lynch syndrome caused by germline PMS2 mutations: delineating the cancer risk. J Clin Oncol. 2015;33:319–25. doi:10.1200/JCO.2014.57.8088.
47. Vasen HFA, Mecklin J-P, Meera Khan P, et al. The International Collaborative Group on hereditary non-polyposis colorectal cancer (ICG-HNPCC). Dis Colon Rectum. 1991;34:424–5.
48. Vasen HF, Watson P, Mecklin JP, et al. New clinical criteria for hereditary nonpolyposis colorectal cancer (HNPCC, Lynch syndrome) proposed by the International Collaborative group on HNPCC. Gastroenterology. 1999;116:1453–6.
49. Giardiello FM, Allen JI, Axilbund JE, et al. Guidelines on genetic evaluation and management of Lynch syndrome: a consensus statement by the U.S. Multi-Society Task Force on Colorectal Cancer. Gastrointest Endosc. 2014;80:197–220. doi:10.1016/j.gie.2014.06.006.
50. Giardiello FM, Allen JI, Axilbund JE, et al. Guidelines on genetic evaluation and management of Lynch syndrome: a consensus statement by the US Multi-Society Task Force on Colorectal Cancer. Dis Colon Rectum. 2014;57:1025–48.
51. Giardiello FM, Allen JI, Axilbund JE, et al. Guidelines on genetic evaluation and management of Lynch syndrome: a consensus statement by the US Multi-society Task Force on colorectal cancer. Am J Gastroenterol. 2014;109:1159–79. doi:10.1038/ajg.2014.186.

52. Umar A, Boland CR, Terdiman JP, et al. Revised Bethesda Guidelines for hereditary nonpolyposis colorectal cancer (Lynch syndrome) and microsatellite instability. J Natl Cancer Inst. 2004;96:261–8.
53. Fearon ER. Molecular genetics of colorectal cancer. Annu Rev Pathol. 2011;6:479–507. doi:10.1146/annurev-pathol-011110-130235.
54. Bird AP. CpG-rich islands and the function of DNA methylation. Nature (London). 1986;321:209–13.
55. Issa JP. CpG island methylator phenotype in cancer. Nat Rev Cancer. 2004;4:988–93.
56. Lee S, Cho NY, Choi M, et al. Clinicopathological features of CpG island methylator phenotype-positive colorectal cancer and its adverse prognosis in relation to KRAS/BRAF mutation. Pathol Int. 2008;58:104–13. doi:10.1111/j.1440-1827.2007.02197.x.
57. Yamamoto E, Suzuki H, Yamano HO, et al. Molecular dissection of premalignant colorectal lesions reveals early onset of the CpG island methylator phenotype. Am J Pathol. 2012;181:1847–61. doi:10.1016/j.ajpath.2012.08.007.
58. Chirieac LR, Shen L, Catalano PJ, et al. Phenotype of microsatellite-stable colorectal carcinomas with CpG island methylation. Am J Surg Pathol. 2005;29:429–36.
59. Juo YY, Johnston FM, Zhang DY, et al. Prognostic value of CpG island methylator phenotype among colorectal cancer patients: a systematic review and meta-analysis. Ann Oncol. 2014;25:2314–27. doi:10.1093/annonc/mdu149.
60. Kriegl L, Neumann J, Vieth M. Up and downregulation of p16(Ink4a) expression in BRAF-mutated polyps/adenomas indicates a senescence barrier in the serrated route to colon cancer. Mod Pathol. 2011;24:1015–22. doi:10.1038/modpathol.2011.43.
61. Wajapeyee N, Serra RW, Zhu X, et al. Oncogenic BRAF induces senescence and apoptosis through pathways mediated by the secreted protein IGFBP7. Cell. 2008;132:363–74. doi:10.1016/j.cell.2007.12.032.
62. Guarinos C, Sánchez-Fortún C, Rodríguez-Soler M, et al. Serrated polyposis syndrome: molecular, pathological and clinical aspects. World J Gastroenterol. 2012;18:2452–61. doi:10.3748/wjg.v18.i20.2452.
63. Rex DK, Ahnen DJ, Baron JA, et al. Serrated lesions of the colorectum: review and recommendations from an expert panel. Am J Gastroenterol. 2012;107:1315–29. doi:10.1038/ajg.2012.161.
64. Fernando WC, Miranda MS, Worthley DL, et al. The CIMP phenotype in BRAF mutant serrated polyps from a prospective colonoscopy patient cohort. Gastroenterol Res Pract. 2014;2014:374926. doi:10.1155/2014/374926.
65. Guarinos C, Sánchez-Fortún C, Rodríguez-Soler M, et al. Clinical subtypes and molecular characteristics of serrated polyposis syndrome. Clin Gastroenterol Hepatol. 2013;11:705–11. doi:10.1016/j.cgh.2012.12.045.
66. Gala MK, Mizukami Y, Le LP, et al. Germline mutations in oncogene-induced senescence pathways are associated with multiple sessile serrated adenomas. Gastroenterology. 2014;146: 520–19.
67. Bettington ML, Chetty R. Traditional serrated adenoma: an update. Hum Pathol. 2015;46:933–8. doi:10.1016/j.humpath.2015.04.002.
68. Bettington ML, Walker NI, Rosty C, et al. A clinicopathological and molecular analysis of 200 traditional serrated adenomas. Mod Pathol. 2015;28:414–27. doi:10.1038/modpathol.2014.122.
69. Bettington M, Walker N, Rosty C, et al. Clinicopathological and molecular features of sessile serrated adenomas with dysplasia or carcinoma. Gut. 2017;66(1):97–106. doi:10.1136/gutjnl-2015-310456.
70. Al-Tassan N, Chmiel NH, Maynard J, et al. Inherited variants of MYH associated with somatic G:C → T:A mutations in colorectal tumors. Nat Genet. 2002;30:227–32.
71. Boparai KS, Dekker E, Van Eeden S, et al. Hyperplastic polyps and sessile serrated adenomas as a phenotypic expression of MYH-associated polyposis. Gastroenterology. 2008;135:2014–8. doi:10.1053/j.gastro.2008.09.020.
72. Calva D, Howe JR. Hamartomatous polyposis syndromes. Surg Clin North Am. 2008;88:779–817. doi:10.1016/j.suc.2008.05.002.

73. Jenne DE, Reimann H, Nezu J-I, et al. Peutz-Jeghers syndrome is caused by mutations in a novel serine threonine kinase. Nat Genet. 1998;18:38.
74. McGarrity TJ, Amos C. Peutz-Jeghers syndrome: clinicopathology and molecular alterations. Cell Mol Life Sci. 2006;63:2135–44.
75. Howe JR, Sayed MG, Ahmed AF, et al. The prevalence of MADH4 and BMPR1A mutations in juvenile polyposis and absence of BMPR2, BMPR1B, and ACVR1 mutations. J Med Genet. 2004;41:484.
76. Giardiello FM, Hamilton SR, Kern SE, et al. Colorectal neoplasia in juvenile polyposis or juvenile polyps. Arch Dis Child. 1991;66:971.
77. Howe JR, Mitros FA, Summers RW. The risk of gastrointestinal carcinoma in familial juvenile polyposis. Ann Surg Oncol. 1998;5:751.
78. Briggs S, Tomlinson I. Germline and somatic polymerase ε and δ mutations define a new class of hypermutated colorectal and endometrial cancers. J Pathol. 2013;230:148–53. doi:10.1002/path.4185.
79. Guinney J, Dienstmann R, Wang X, et al. The consensus molecular subtypes of colorectal cancer. Nat Med. 2015;21:1350–6. doi:10.1038/nm.3967.

Pancreatic Carcinogenesis

5

V. Taucher and Johannes Haybaeck

Contents

V. Taucher (✉)
Institute of Pathology, Medical University of Graz, Graz, Austria
e-mail: valentin.taucher@stud.medunigraz.at

J. Haybaeck, M.D., Ph.D.
Department of Pathology, Medical Faculty, Otto von Guericke University Magdeburg, 39120 Magdeburg, Germany

Institute of Pathology, Medical University of Graz, 8036 Graz, Austria
e-mail: johannes.haybaeck@med.ovgu.de

J. Haybaeck (ed.), *Mechanisms of Molecular Carcinogenesis – Volume 2*,
DOI 10.1007/978-3-319-53661-3_5

Abstract

Pancreatic cancer (PC) is a lethal disease with a devastating prognosis. This is due to the late occurrence of clinical symptoms in advanced cases, where operative cure is seldom possible. PC is divided into subtypes, which include pancreatic ductal adenocarcinoma (PDAC), colloid carcinoma (CC), pancreatoblastoma, cystadenocarcinoma, and histologically-mixed types. PDAC is the most common and aggressive subtype, comprising 95% of PC cases. It evolves by a series of molecular aberrations that lead to malignant transformation. The precursor lesions pancreatic intraepithelial neoplasia (PanIN), intraductal papillary mucinous neoplasm (IPMN), and mucinous cystic neoplasm (MCN) can evolve into PDAC. In most cases, however, PanINs are the precursors of PDAC. IPMN progresses to either PDAC or CC, depending on the histologic subtype of the underlying IMPN. The microscopically visible PanINs are well characterized molecularly and genetically, which has helped in determining the sequence of changes that occur during pancreatic carcinogenesis. IPMNs and MCNs are macroscopic lesions that are seen with standard imaging techniques and often present with a cystic morphology and symptoms attributable to ductal obstruction. Activating *KRAS* mutations are the most common mutation in PanINs and IPMNs. Studies have shown that constitutively active KRAS leads to changes that involve not only increased growth of tumor cells but also gives precursor lesions a survival advantage by promoting tumor desmoplasia, anabolic metabolism, and immunologic tumor tolerance. Furthermore, the occurrence of *TP53* mutations in PanINs with mutant *KRAS* has been shown to be essential for the transformation to PDAC. Multiple pathways are altered in pancreatic carcinogenesis. These include MAPK/ERK, PI3K/AKT, Wnt/β-catenin, TGF-β, hedgehog, chromatin regulation, autophagy, and G1/S-phase transition. Moreover, noncoding RNA has been shown to be involved in almost all steps of pancreatic carcinogenesis.

5.1 Pancreatic Carcinogenesis

Pancreatic cancer (PC) is a lethal disease with a dismal prognosis. Annually, PC is the cause of more than 40,000 deaths in the European Union (EU) and is the fourth most common cancer death in the United States and the EU [1, 2]. The late occurrence of symptoms often leads to the correct diagnosis in late clinical stages, when curative therapy is impossible [3]. The most common type of PC is pancreatic ductal adenocarcinoma (PDAC), accounting for over 95% of cases [3]. The second most common type of PC is acinar-cell carcinoma (ACC), which accounts for 2–3% of cases in adults, but is more common in the pediatric population [4]. Other types include colloid carcinoma (CC), pancreatoblastoma, cystadenocarcinoma, and overlapping phenotypes, such as mixed acinar-ductal adenocarcinoma and mixed acinar-neuroendocrine-ductal carcinoma [5]. Although a minor increase of overall survival of PDAC patients has been achieved in recent years, inoperable PDAC still conveys an expected survival of less than 1 year after diagnosis [6].

5.1.1 Pancreatic Ductal Adenocarcinoma (PDAC)

PDAC accounts for the vast majority of PC cases and is also the most aggressive subtype [3]. Although it represents the most intensively investigated subtype of PC, the survival of PDAC patients has not increased substantially in recent years [3]. Macroscopically, PDAC presents as a firm fibrotic tumor, with a white-yellow appearance and loss of normal pancreatic architecture [7]. Microscopically, infiltrative PDAC is characterized by glandular components with intraluminal necrosis in the pancreatic parenchyma without abiding to the normal glandular architecture [7]. Furthermore, invasion into extrapancreatic structures, such as perineural or vascular invasion, is often found and is diagnostic of PDAC [8, 9]. It has been recognized for some time that PDAC develops out of precursor lesions, a process that will be discussed thoroughly in this chapter.

5.1.2 Risk Factors for PDAC

Although risk factors for PDAC, such as chronic pancreatitis (CP) and hepatitis B virus (HBV) infection, have been acknowledged, most cases develop in persons without obvious risk factors [10–12]. Research has shown that the risk factors most strongly associated with PDAC are CP, diabetes, smoking, helicobacter pylori infection, and longterm alcohol consumption over 30 gramms per day [13, 14]. Studies have shown that these risk factors also apply to early- and very-early-onset PDAC, i.e., before the age of 60 or 45 years, respectively [15, 16]. Allergic diseases, such as asthma and allergic rhinitis, decrease the risk of PDAC significantly, highlighting the role of the immune response in pancreatic carcinogenesis [14]. Genetic predisposition to PDAC is caused by germline mutations of *BRCA1*, *BRCA2*, *CDKN2A*, *ATM*, *PALB*, and mismatch-repair genes, leading to familial pancreatic cancer syndromes in some cases if transmission to the next generation occurs. Moreover, a number of single-nucleotide polymorphisms (SNPs) are associated with increased risk for PDAC, including *ABO*, *PDX1*, *NR5A2*, and *KLF5* [17]. In patients with PDAC, *BRCA2* germline mutations are found in approximately 10% [18], highlighting the importance of familial pancreatic cancer to the overall incidence of PDAC cases.

5.1.3 Acinar-Cell Carcinoma (ACC)

Compared to PDAC, this type of cancer is characterized by different mutations and alterations. Interestingly, over 30% of ACCs do not have any somatic mutations, but are solely characterized by chromosomal aberrations. This is demonstrated by a high value of fractional allelic losses (FAL) of 0.27 (range, 0–0.89), indicating that 27% of information carrying chromosomal arms are affected by loss of heterozygosity [19]. Rigaud et al. have demonstrated that 12 chromosomes show loss of heterozygosity in over 50% of cases, with chromosomes 4q, 1p, and 17p in over 70% of cases. Amplification of GATA5 and the mTOR-associated LKB1 has also been reported [20]. ACCs commonly harbor changes in β-catenin-related signaling, such as adenomatous polyposis coli (APC), and express pancreatic exocrine enzymes, such as trypsinogen [3, 4]. Alterations in the APC/β-catenin pathway commonly include APC deletions and promoter hypermethylation and result in underexpression in 58% of cases [21]. Recent mouse-model-based data have shown that the mTOR pathway may be crucial for ACC development [22]. *TP53* mutations or deletions have been found in roughly 50% of cases and are found especially in metastases and are therefore thought to come along with malignant change [23].

5.1.4 Colloid Carcinoma

Colloid Carcinoma (CC), also referred to as mucinous non-cystic carcinoma, is a variant of pancreatic cancer that is characterized by substantial mucin production within large parenchymal pools. Floating tumor material is found within these pools, which represent a barrier for metastasis, explaining the relatively benign nature of the disease in comparison with PDAC [24]. Moreover, the tumor cells may have a diverging appearance [25]. CCs are largely located in the head of the pancreas, but may also affect all anatomic regions of the pancreas in a multinodular fashion [25]. Whereas the average 5-year survival of PDAC patients does not exceed 20% [26], the average 5-year survival of CC patients is >50% [25]. CCs are thought to arise from intraductal papillary mucinous neoplasms (IPMNs), although mainly from intestinal-type IPMNs and possibly from mucinous cystic neoplasms (MCNs) [25, 27]; however, the data is still inconclusive in this matter.

5.1.5 Precursors of PDAC

PDACs are thought to develop by a series of genetic and epigenetic alterations from nonneoplastic pancreatic epithelium (NNE), to precursor lesions, to invasive cancer [28]. These lesions include pancreatic intraepithelial neoplasia (PanIN), IPMN, and MCN [29–31]. PanIN has been shown to be the most important precursor lesion in the development of most PDACs [32]. PanINs are categorized into three grades, according to their histologic degree of cellular atypia into PanIN-1A, PanIN-1B, PanIN-2, and PanIN-3 lesions [33]. PanIN-1 lesions are found frequently in normal human pancreata, whereas higher-grade PanINs are rarely found in the absence of PDACs and IPMN-associated carcinomas [28, 34]. Moreover, case reports have shown that PanIN-3 lesions may advance to PDACs [28, 35, 36].

5.1.5.1 Pancreatic Intraepithelial Neoplasia (PanIN)

PanINs are the most important and best investigated PDAC precursor lesions that are categorized into three histologic grades, according to their degree of cellular atypia [33]. PanINs are flat microscopic intraepithelial lesions that develop out of acinar cells in a process termed acinar-to-ductal metaplasia (ADM) Conversely, ductal epithelium is rarely the origin of PanINs [37, 38]. PanINs are characterized by varying degrees of intracellular mucin, pseudo-papillary or papillary structures and nuclear atypia, according to the histologic grade of PanIN. In the ADM model, chronic inflammatory

Table 5.1 Histopathologic grades of PanIN

PanIN-1A	Flat lesion with tall columnar mucinous epithelium with normally shaped nuclei
PanIN-1B	Papillary lesion, otherwise same morphology as PanIN-1A
PanIN-2	Papillary lesion showing nuclear pseudostratification and crowding, focal loss of polarity, and sometimes cellular abnormalities such as hyperchromatic nuclei or enlargement can be present
PanIN-3	Pronouncedly papillary lesion with diffuse loss of polarity and cribriform structures. Enlarged nuclei with increased nuclear to cytoplasm ratio and polymorphisms similar to a well-differentiated adenocarcinoma

stresses, such as smoking, HBV infection, and recurring flares of CP, lead to metaplastic acinar cells by the acquisition of activating *KRAS* mutations, which then transform into PanINs by a series of genetic and epigenetic alterations [37, 39]. However, PanINs often harbor different variants of *KRAS* mutations, indicating low clonality of these precursors, while established PDACs normally harbor the similar mutations [40]. Although most people are thought to harbor low-grade PanINs in a considerable frequency, high-grade PanIN-3 lesions are found mainly in the surrounding area of a PDAC [34, 41]. With respect to IPMN, PanIN progresses exclusively to PDAC [33, 42]. Furthermore, recent studies have shown that PanIN-associated PDACs are more aggressive than IPMN-associated PDACs [43] (Tables 5.1).

5.1.5.2 Intraductal Papillary Mucinous Neoplasm (IPMN)

IPMNs are rare macroscopic tumors that can progress to tubular adenocarcinoma (i.e., PDAC) and colloid type in 50% of cases, respectively [33, 42]. With respect to PanINs, IPMNs are symptomatic in up to 45% of cases because of the substantial tumor size, leading to compression of local anatomical structures, such as the common bile duct or pancreatic duct [44]. Common symptoms therefore include jaundice, abdominal pain, weight loss, steatorrhea, new-onset diabetes, and acute pancreatitis [4, 44–46]. IPMN is graded into three grades according to the degree of intraepithelial atypia: IPMN with low-grade dysplasia, IPMN with intermediate-grade dysplasia, and IMPN with high-grade dysplasia [33]. Microscopically, IPMNs are characterized by vast papillary formations and tall epithelial cells with abundant apical mucin [33]. In addition, IPMNs arise in different histologic phenotypes, which comprise gastric, intestinal, oncocytic, and pancreatobiliary phenotypes [33]. Interestingly, tumors in the main pancreatic duct commonly harbor infiltrative compounds, i.e., PC at higher frequency than branch-duct IPMNs [47]. The most commonly mutated genes in IPMN, in descending order, are *KRAS*, *GNAS*, *TP53*, *RNF43*, *SMAD4*, *SMARCA4*, and *ARID1A* [42]. However, the frequencies of different gene mutations and alterations of protein expression diverge significantly between the different histologic subtypes [42]. Some authors have listed *RNF43* mutations as being the third most frequently mutated gene in IPMNs [48]. Evidence shows that there are significant statistical correlations between the prevalence of IPMNs and extrapancreatic diseases, especially autoimmune diseases, such as inflammatory bowel disease, systemic lupus erythematosus, and rheumatoid arthritis [49]. Moreover, IPMNs may be associated with an increased incidence for extrapancreatic carcinomas of the colon, prostate, stomach, bile duct, and breast [50–53].

Table 5.2 Histopathologic grades of IPMN

IPMN with low-grade dysplasia	Tall columnar epithelium with abundant apical mucin, usually gastric type. More commonly found in branch ducts
IPMN with intermediate-grade dysplasia	Complex architecture only focal loss of polarity. Nuclear enlargement and hyperchromasia. Commonly found in main ducts
IPMN with high-grade dysplasia	Cribriform/pseudopapillary lesion with severe loss of polarity. Budding of cells into the lumen. Severely dysplastic cells with nuclear polymorphisms and intestinal-type epithelium. Commonly found in main ducts

However, a recent systematic review has questioned the conclusiveness of the respective data [54] (Table 5.2).

5.1.5.3 Mucinous Cystic Neoplasm (MCN)

MCNs are even rarer than IPMNs and are hitherto poorly investigated. They represent macroscopic lesions that are characterized by intensive mucin production, a highly mucinous columnar epithelium, and an ovarian-type stroma underlying the dysplastic epithelium and may progress to mucinous cystadenocarcinoma [33, 55]. MCNs are more common in woman than in men and are associated with cystadenocarcinoma in about 20% of cases [55].

5.1.5.4 Molecular Alterations in PDAC

PDACs harbor hundreds of mutations in genes involved in diverse pathways and cellular functions; however, some alterations have proven essential, while others represent passenger mutations that are not considered crucial for pancreatic carcinogenesis [56–59]. A good example is mutations in *TTN* encoding titin, which are present in up to 20% of cases, although some cytoskeletal proteins, such as paladin and α-catenin, which interact with titin, have been shown to have influence on PDAC cells [60]. The most commonly mutated genes in PDAC are *KRAS*, *CDKN2A*, *TP53*, *SMAD4*, *MLL3*, *ARID1A*, and *SF3B1* [58]. In ACC, over 30% of cases do not harbor any specific mutation. Instead, this type of cancer is characterized by severe chromosomal changes [19]. Other common aberrations are *SMAD4* (26%), *TP53* (13%), and *JAK1* (17%) [19].

5.1.5.5 Common Mutations in PDACs, PanINs, MCNs, and IPMNs

Many mutations occur during pancreatic carcinogenesis to PDAC. Whole-exome sequencing-studies have found diverging frequencies of several mutations, which may be due to insufficient case numbers in some studies [40, 58, 61, 62]. However, the frequencies of some major mutations are largely consistent among different studies.

Activating mutations (e.g., G12D or G12V) in *KRAS* on chromosome 12 are found in most cases of PanINs, with frequencies up to 90% in PanIN-3 lesions. These mutations lead to increased MAPK/ERK and PI3K/AKT signaling and are thought to be initiating events in the progression from NNE to PanIN [40, 63]. Epigenetic mechanisms, such as the downregulation of KRAS-suppressing microRNAs, have also been acknowledged [64]. For example, Lohr et al. have reported frequencies of KRAS mutations of 36%, 44%, and 87% of PanIN-1a, PanIN-1b, and PanIN-2 and PanIN-3 lesions, respectively, which represents a significant increase in progression toward higher-grade lesions ($p < 0.001$) [63]. Although *KRAS* mutations belong to the most frequent mutations in IPMN, they are significantly less common than in PanINs [63, 65]. However, data suggests a gradual increase of *KRAS* mutation in the progression of IMPN to PDAC. In IPMN with high-grade dysplasia and IPMN with associated carcinoma, *KRAS* mutations are found in 40–57% and 50–57%, respectively [66, 67]. These data suggest that *KRAS* mutations are associated with a more malignant phenotype. Moreover, *KRAS* mutations are also more common in gastric- and pancreatobiliary-type IPMNs than other types [42]. Other authors have found little to no increase of *KRAS* mutations in the progression of IPMN, leading to the assumption that *KRAS* mutations are not related to progression but are early changes associated with the formation of the lesion [42, 68]. Hence, there is still disagreement in the scientific community about the significance and frequency of *KRAS* mutations in IMPN. *KRAS* mutations are also found in about one third of MCNs [69, 70]. *KRAS* mutations are thought to be a prerequisite for the progression of PanINs to PDACs, which is eventually caused by aberrations in *TP53*, *SMAD4*, and *CDKN2A* [71]. Collins et al. have demonstrated in a murine model that oncogenic KRAS is necessary for the formation and maintenance of PanINs [72, 73]. KRAS signaling leads to increased expression of cyclooxygenase 2 (COX2) by increased nuclear factor kappa B (NF-κB) signaling, which furthermore enhances KRAS activity [74]. *KRAS* mutations lead to a chronic state of inflammation by upregulation of inflammation-associated chemokines and reduced T-cell immunity against neoplastic cells [75, 76].

p53 overexpression is found in up to 57% of PanIN-3 lesions and 85% of PDACs, which correlates well with *TP53* mutation. Overexpression is caused by accumulation of the dysfunctional protein in the nucleus, leading to insufficient recognition by the ubiquitin-protein ligase Mdm2 and JNK [32, 77]. Deletion of *TP53* alleles, on the other hand, would not lead to accumulation, although in most cases of PDAC mutation is present [57, 78]. *TP53* mutations are believed to be an important factor in the development of a malignant phenotype and represent a variety of SNPs in different genomic regions and loss of heterozygosity [32, 40]. This is due to the tumor-suppressive function of p53, which allows the acquisition of additional genetic aberrations and leads to increased proliferation and reduced apoptosis. It may also be due to a gain-of-function character of *TP53* mutations, although this is still debated [79]. However, early studies have shown that mice with *TP53* mutations suffered more frequently from distant metastases than did mice with *TP53*-allele deletion, indicating a gain-of-function quality of *TP53* mutation [80, 81]. Studies have shown that the aggressive nature of p53-mutant tumors in mice is due to increased PDGF receptor-β signaling and that PDGF receptor-β expression is correlated to poor disease-free survival in PDAC patients postoperatively [82]. p53 dysfunction is also a cause for impaired function of p21, which leads to increasing G1/S-phase progression [32]. Germline *TP53* polymorphisms have been shown to correlate with increased risk for PDAC in men, but not in women [83]. Studies on murine models have shown that activating *KRAS* mutations and *TP53* mutations lead to the development of PDAC in 75% of cases when occurring in rapid sequence [84]. This highlights the importance of these two mutations in the development of PDAC. p53 overexpression is found in 17–38% of IPMNs with high-grade dysplasia [85–89] and in about 10% of IPMNs in total [65]. *TP53* mutation is predominantly found in IPMN with high-grade dysplasia and IPMN carcinoma, indicating that *TP53* mutations are associated with the development of a more malignant phenotype in IPMNs [42]. IPMN-associated PDACs and colloid-type carcinomas show p53 overexpression and *TP53* mutation in lower frequencies than "common," i.e., PanIN-associated PDACs [25, 42]. MCNs commonly do not exhibit p53 overexpression before invasive carcinoma has developed, indicating that *TP53* mutation is associated with malignant behavior in these lesions [90].

Downregulation of p16 is frequently observed in PanINs, with frequencies of 29%, 55%, and 71% in PanIN-1, PanIN-2, and PanIN-3, respectively [91]. PDACs do not express p16 on the protein level in almost 80–100% of cases [88, 91]. Therefore, downregulation of p16 is thought to be one of the intermediate events in pancreatic carcinogenesis after *KRAS* mutation and before or contemporaneously with *TP53* mutations. Downregulation of p16 is caused by point mutations, homozygous deletions, or epigenetic silencing in the CDKN2A promoter region [57, 92], leading to accelerated cell growth because of lack of p16-mediated repression of the G1/S-phase transition in the cell cycle [93]. Increasing loss of p16 expression is seen in the progression of IPMN to carcinoma. IPMN with low-grade dysplasia and IPMN with high-grade dysplasia display loss of p16 expression in 20% and 80–100% of cases, respectively [88, 94, 95]. In MCNs, loss of p16 expression has also been reported; however, only small study samples have been investigated [96].

SMAD4 mutations are among the late events in pancreatic carcinogenesis via the development of PanIN, and downregulation of SMAD4 is seen in >90% of PDAC cases [88]. The reported frequencies of *SMAD4* mutations have varied substantially among different studies. Deletion and point mutation cause loss of SMAD4 expression [57, 58, 61]. *SMAD4* inactivating mutations by point mutation or homozygous deletion lead to the switch from growth-inhibitory signaling of the TGF-β pathway to growth-promoting signaling in affected cells. Furthermore, in this context, TGF-β signaling also leads to promotion of invasive behavior and epithelial mesenchymal transition (EMT) [97, 98]. In PanINs, studies have consistently shown normal expression in PanIN-1 and PanIN-2 lesions and reduced or lost expression in 30–55% of PanIN-3 cases [88, 99]. This is in accordance with experimental data that show that *SMAD4* mutations are sufficient to initiate PDAC formation in the context of oncogenic KRAS [100]. IPMN with low- and intermediate-grade dysplasia consistently shows normal SMAD4 expression, while IPMN with high-grade dysplasia and IPMN carcinoma shows loss of SMAD4 expression in 7–75% of cases [94, 101]. A problem with reported expression frequencies is the use of different positivity criteria in the analysis of SMAD4 positivity by different study authors, a general problem in immunohistochemistry-based studies. However, it may indicate lower frequencies of *SMAD4* mutation or deletion in IPMN-associated PDACs. In MCNs, loss of SMAD4 is found frequently in cases with invasive carcinoma, but not in noninvasive lesions, pointing toward a causative role of SMAD4 mutations in the malignant transformation of MCNs [102, 103].

GNAS mutations are common in intestinal-type IPMN and colloid-type carcinoma compared to pancreatobiliary-type IPMN and PanIN-associated PDAC, reflecting a discrete pathway in pancreatic carcinogenesis [42, 61, 62, 69, 70]. In total, they are among the most common mutations in IPMNs, together with *KRAS* mutations [65]. The *GNAS* gene encodes the α-subunit of the stimulatory G-protein (Gαs), which is responsible for activating the enzyme protein kinase A (PKA) upon receptor activation and downstream signaling molecules, such as cAMP response element-binding protein 1 (CREB1). Thus, GNAS mutation leads to constitutive activation of G-protein-coupled receptor (GPCR) signaling [104]. Studies have shown that GNAS mutations concur with increased expression of MUC2 and MUC5AC, without increasing tumor cell proliferation [105]. In PanINs and PDAC, *GNAS* mutations are found only sporadically [70, 106]. In contrast, *GNAS* mutations are found in frequencies of 41–66% of IPMNs, without significant differences between grades [62, 69, 70, 107]. However, in cases with wild-type *KRAS*, *BRAF*, and *PI3KCA*, mutations of *GNAS* are much more common [61]. Some authors have demonstrated decreasing frequencies of *GNAS* mutations in the progression of IPMN, indicating that *GNAS* mutations may be important in the early stages of disease, but malignant change is less likely in these tumors [70].

Myeloid/lymphoid or mixed-lineage leukemia protein 3 (MLL3) is an enzyme encoded by the *MLL3* gene, which is member of a histone-methylation complex. In PDAC, it is inactivated by homozygous deletion or point mutations in roughly 10% of cases [57]. Mutations or deletions in *MLL* and *MLL2*, which are all involved in chromatin regulation, are found with lower frequency. Interestingly, mutations in the *MLL* genes do not overlap, indicating that alteration in one of these interacting molecules is sufficient for disruption of normal chromatin regulation [108].

Brahma-related gene 1 (Brg1), encoded by the *SMARCA4* gene, is a catalytic subunit of the SWI/SNF-chromatin remodeling complexes. It is underexpressed in 28% of IPMNs with low-grade dysplasia, in 52% with intermediate-grade dysplasia, and in 76% with high-grade dysplasia [109–111]. Mutations of *SMARCA4* are also found in 1–8% of PDAC cases [58], but with frequencies of up to 12–15% in IPMN-associated carcinomas [42]. Mouse model data have shown that loss of Brg1 leads to formation of IPMNs that progress to carcinoma in the presence of activating *KRAS* mutation. Brg1 inhibits dedifferentiation and malignant transformation [110, 111]. However, IPMN-associated PDAC exhibits higher levels of Brg1 expression, and in vitro data have shown that Brg1 leads to increased malignant behavior of tumor cells when overexpressed in PDAC. This shows a dual role of Brg1 in IPMN and IPMN-associated PDAC [42].

RNF43 is the third most frequently mutated gene in IPMNs and is also frequently mutated in MCNs [48], although some authors have listed *TP53* mutations as the third most common mutations in IPMN [42]. *RNF43* encodes a ubiquitin ligase that resides mostly in the endoplasmatic reticulum and the nuclear membrane and promotes tumor cell proliferation [112]. Springer et al. found *RNF43* mutations in 38% of IMPNs [48]. Similar results were obtained by Wu et al. [113]. Amato et al.

showed that *RNF43* mutations are more common in intestinal-type IPMNs, compared to pancreatobiliary or gastric-type IPMNs, suggesting that these mutations represent a discrete pathway in pancreatic carcinogenesis [65]. However, only *STK11* (serine/threonine kinase 11) mutations are infrequently found in PDAC but are more common in IPMNs [114, 115]. Germline *STK11* mutations lead to the Peutz-Jeghers syndrome, which is characterized by the predisposition to development of gastrointestinal polyps, pigmented mucocutaneous macules, and IPMNs [115]. Loss of *STK11* expression is more common in higher-grade IPMNs and in lesions with deletion of both alleles. *STK11* mutation is found in 27–32% of IPMNs, but only in 7% of PDAC cases [114].

MUC16 has been shown to be mutated in PanINs, prior to the progression to PDAC. In PDAC, MUC16 upregulation is found frequently and is associated with increased mTOR signaling. However, mutations of *MUC16* were not found frequently in next-generation sequencing studies. Jones et al. found *MUC16* mutations in only 5–8% of PDAC cases, overall ranking it the seventh most common mutation [57].

5.1.5.6 Chromosomal Aberrations

Apart from typical mutations, PDAC is characterized by genomic instability and chromosomal alterations, although less extensively than ACC. These alterations include structural aberrations, such as nonreciprocal translocations, deletions, and amplifications [19, 116, 117]. Mouse model data have shown that mutations in *KRAS* and *TP53* increase the likelihood of chromosome fusion in murine PDACs, without significantly reduced telomere length [118]. Common chromosomal alterations in ACC are found on chromosomes 11p, 15, 17p, 18q and 22 [19]. Commonly, gains of chromosome material are more frequent than deletions [116]. Telomere shortening by telomere breakage is one of the earliest changes in pancreatic carcinogenesis and is found in PanIN-1A lesions and even nonneoplastic epithelium in surrounding area of PanINs [119]. This may lead to chromosomal fusion, DNA fragmentation, or unstable ring chromosomes, promoting the acquisition of further chromosomal aberrations. Furthermore, chromosomal aberrations go along with telomerase expression [120], indicating that chromosomal instability may contribute to immortalization of tumor cells. Telomere shortening has also been shown to affect virtually all IPMN cases [121]. *EGFR*, *HER-2*, and topoisomerase II (*TOP2A*) amplification or, in some cases, deletion have been reported in PDAC [122–124]. IPMN shows different chromosomal aberrations with respect to PDAC. These include alterations of chromosomes 5, 6, 11, and 18 [125–127].

5.1.5.7 Signaling Pathways in PDAC

Many signaling pathways are altered in pancreatic carcinogenesis to PDAC. Jones et al. have determined that 14 signaling pathways are altered in PDAC, which are KRAS signaling, TGF-β signaling, apoptosis, JNK signaling, integrin signaling, Wnt/Notch signaling, hedgehog signaling, G1/S-phase transition, DNA damage control, small GTPase signaling, invasion and hemophilic cell adhesion, chromatin remodeling, SWI-SNF, and ROBO/SLIT [57, 128].

MAPK/ERK and PI3K/AKT signaling is one of the cornerstones of pancreatic carcinogenesis [63]. This is due to early occurrence of activating *KRAS* mutations and downregulation of several miRNAs that inhibit KRAS expression [63, 129, 130]. Moreover, oncogenic KRAS has been shown to alter activation of many intracellular pathways, such as PI3K/AKT, RalGDS/p38 MAPK, Rac and Rho, NF-κB, chromatin modification, and Rassf1 signaling [131, 132]. MAPK/ERK signaling is characterized by a cascade of sequential phosphorylation in which a RAS protein (e.g., KRAS or HRAS) cooperates with phosphorylate and activates MEK, which phosphorylates MAPK/ERK. MAPK activates transcription factors, such as c-myc and CREB, leading to increased transcription of growth-promoting factors and uncontrolled tumor proliferation in the case of PDAC [133]. In vitro studies with PanIN-derived cell lines have shown that the MAPK/ERK pathway is highly active at all stages, indicated by increased expression of p-MAPK, EGFR, and HER-2 [134]. Mouse model data have shown that activation of the MAPK/ERK pathway is caused by increased activation of HER-2 and not by EGFR. This is also suggested by the frequent finding of HER-2 overexpression compared to EGFR overexpression [118]. In addition, MUC4, which is overexpressed in >90% of PDACs and high-grade PanINs, leads to increased MAPK/ERK signaling by agonistic function on the HER-2 receptor [135]. Activating *KRAS* mutations also leads to increased PI3K/AKT signaling, which has been shown to considerably promote pancreatic carcinogenesis [136–138]. In this signaling pathway, receptor tyrosine kinase activation leads to phosphorylation and association of the enzyme subunits p85 and p110 to form the fully active PI3K molecule. PI3K phosphorylates PIP2 to PIP3 and promotes AKT phosphorylation by PDK1 kinase and mTOR. In the nucleus, AKT exerts its function by phosphorylating various transcription factors, which leads to a switch of gene expression towards expression of anti-apoptotic and growth- and invasion-promoting molecules [139]. Constitutively active PI3K leads to the formation of a large quantity of preneoplastic and neoplastic lesions in murine pancreata, from ADM and PanIN lesions to PDACs [136]. Eser et al. demonstrated in a murine model that overactive PI3K signaling, together with deactivation of cellular senescence inducing p16 expression, leads to the progression to PDAC [136].

Wnt/β-catenin signaling is not among the most strongly dysregulated pathways in PDAC, but contributes to tumor growth substantially [140, 141]. Murine models have shown that constitutive activation of the Wnt/β-catenin pathway inhibits the development of murine PanIN (mPanINs) lesions [142]. Conversely, most PDAC cases harbor mutations and altered promoter methylation in genes encoding members of the Wnt/β-catenin pathway [57, 143]. In PanINs, β-catenin alteration, including nuclear β-catenin expression and loss of membranous β-catenin, is a late change [141, 144], although other authors have demonstrated aberration of Wnt/β-catenin signaling in early-grade PanINs [134]. As seen in PDAC, these changes may be largely attributable to promoter hypermethylation of members of the Wnt/β-catenin pathway, such as the genes encoding Wnt ligands *WNT5A* and *WNT7A* or the cytoplasmic molecules *WNT9A* and *APC2*, *SOX1*, *SOX7*, *ROBO1*, *ROBO2*, *SLIT2*, *RNF43*, and *WIF-1* [143].

TGF-β signaling is altered in all cases of PDAC and has been shown to be crucial for pancreatic carcinogenesis [57]. TGF-β signaling is a highly important signaling pathway that is involved in numerous physiologic processes in both embryogenesis and adult cells. Receptor activation leads to phosphorylation of SMAD proteins that regulate downstream gene expression. [145]. In PDACs, members of the TGF-β pathway are mutated in 100% of cases, especially *SMAD3*, *SMAD4*, *TGFBR1*, and *TGFBR2* [57, 61]. In addition, the newly discovered cancer-associated gene *ACVR1B*, which is also part of the TGF-β pathway, is mutated in some PDACs [61]. The common mediator-SMAD SMAD4, which interacts with receptor-regulated SMADs, transduces growth-inhibitory stimuli in non-neoplastic epithelial cells. Point mutation and homozygous deletion abrogate proper SMAD4 signaling in PDACs [57], leading to increased G1/S-phase transition and loss of pro-apoptotic signaling [100, 146, 147]. Studies have shown that apoptosis is commonly inhibited in PDAC cells. Inactivation of at least one of the following genes that encode the apoptosis-associated molecules *CASP10*, *VCP*, *CAD*, and *HIP1* [57] has been reported *in PDAC*. Expression of anti-apoptotic molecules of the inhibitor of apoptosis family (IAP) is found in PanINs, such as cellular inhibitor of apoptosis protein 1 (cIAP1) and (cIAP2), which act by inhibition of caspases 3, 7, and 9 but also by inhibition of TRAIL receptors CD95 and collaboration with anti-apoptotic transcription factor NF-κB [148, 149]. PDAC cells are also insensitive to TRAIL and CD95 (TRAIL receptor)-mediated apoptosis, caused by increased activation of protein kinase C (PKC) and NF-κB, leading to inhibition of the mitochondrial apoptosis pathway by upregulation of BCL-XL [150, 151]. The anti-apoptotic molecule BCL-2 is regulated by intact p53 signaling. Its underexpression correlates with PDAC grade and poor survival [152]. However, in PDACs, *TP53* mutation commonly leads to p53 overexpression, which may be the cause of this paradoxical finding, as mutant p53 has been considered as having oncogenic function and not merely being dysfunctional [81]. NF-κB, STAT-3, and MAPK/ERK signaling also upregulate anti-apoptotic molecules BCL-2, BCL-XL, and MCL-1 in the majority of PDACs [153–155]. PI3K/AKT signaling is hyperactive in virtually all PDAC cases and also inhibits apoptosis by upregulation of anti-apoptotic molecules (e.g., BCL-XL) and downregulation of pro-apoptotic molecules (e.g., BAD) [156–158].

G1/S-phase transition is commonly dysregulated in PDAC cells. This is caused by various upstream signaling pathways that lead to increased activation of cyclin-dependent kinases (CDKs) [77, 91, 106, 159–161]. Cyclin D1 is overexpressed with increasing frequency in the progression of PanINs and in most PDAC cases. This is caused by gene amplification and overactive signaling pathways, such as KRAS [32, 162, 163]. Cyclin D1 is encoded by the proto-oncogene *CCND1* that regulates the G1/S-phase transition by interacting with CDK4 and CDK6, leading to phosphorylation and inactivation of retinoblastoma protein (Rb) [163, 164]. *CCND1* and *CDK4* amplifications have been found frequently in PDACs [61]. *CDKN2A*, which encodes for p16, counteracts cyclin D1 function by blocking its association with CDKs and phosphorylation of Rb [93]. Many PanINs and most PDAC exhibit p16 underexpression caused by homozygous deletion, point mutation, or promoter hypermethylation [57, 159, 165, 166]. PDACs harbor mutations in G1/S-phase transition-associated genes *FBXW7*, *CHD1*, and *APC2* [57]. The tumor suppressor FBXW7 is a recognition protein that aggregates to target molecules to propagate degradation by the Skp1-Cul1, F-box (SCF) ubiquitin-ligase complex. These molecules include growth-promoting molecules c-MYC, cyclin E, c-JUN, and Notch [167, 168]. In PDAC, FBXW7 has been shown to be present mostly in its inactive form, which is caused by phosphorylation at Thr205 by ERK-1, leading to ubiquitination and degradation in the proteasome [167]. The Wnt signaling member APC is not thoroughly investigated in PDAC. However, studies have shown that functioning APC may be needed in pancreatic carcinogenesis to PDAC [169]. This may be reflected by the low prevalence of APC mutations in PDACs [170].

DNA damage control is crucial for the integrity and function of the genome and inhibits tumorigenesis in normal cells [171]. In PDAC, many genes that are implicated in DNA damage control are mutated, including *ERCC4*, *RANBP2*, *EP300*, and *TP53* [57]. *TP53* mutations, as discussed above, are important factors that contribute to the malignant transformation of PanIN to PDAC and are present in most PDACs [32, 84]. P53 is activated upon cellular stresses, such as DNA damage or oxidative stress, leading to apoptosis or growth arrest and DNA repair [172]. Moreover, downstream molecules, such as p21 and associated miRNAs, most prominently miR-34, are altered in pancreatic carcinogenesis [32, 173–175]. *CDKN2A* mutation also leads to dysfunction of the alternate transcript p14ARF, which cooperates with p53 by inhibiting p53 degradation by Mdm2 [176]. Downregulation of p14ARF is a common finding in PDACs that correlates with the presence of distant metastases [177]. The histone acetyltransferase EP300 regulates gene transcription by chromatin remodeling and interacts with p53 to mediate growth arrest upon DNA damage by acetylation of the p53 C-terminal domain [178]. Downregulation of EP300 is found especially in highly metastatic tumors and is caused by increased expression of EP300-targeting miRNAs, such as miR-194, miR-200b, miR-200c, and miR-429 [179]. In PanINs, mutations of DNA damage control-associated genes *TOP2A*, *CHEK2*, *POLH*, and *APLF* have found to be mutated as well [40]. A recent study investigating PanIN genetics found that *ataxia telangiectasia mutated (ATM)* is mutated in PanINs, especially in cases with wild-type *TP53* [40], indicating that the disruption of DNA damage control is necessary for pancreatic carcinogenesis.

Chromatin remodelling is an important epigenetic mechanism that regulates gene expression and is substantially altered in PDAC [128, 180]. Transcriptional regulation is achieved by dynamic manipulation of chromatin structure, which is carried out by chromatin-modifying enzymes. Histone acetylation at lysine residues, which is an important part of chromatin regulation, is mediated by enzymes named histone acetyltransferases (HATs), while histone deacetylation is carried out by histone deacetylases (HDACs). Histone methylation works in concert with histone acetylation to regulate chromatin state [181]. Another important factor that affects transcriptional activity is gene promoter methylation by DNA methyltransferases, which cooperate with histone-modifying enzymes to regulate gene expression [181, 182]. Several genes encoding chromatin-remodeling enzymes are mutated in PDAC, such as *EPC1*, *ARID1A*, and the histone-methylating enzyme

MLL3 [58]. The Switch/sucrose nonfermentable (SWI/SNF) complex is a large tumor-suppressive multiunit complex containing more than ten molecules that regulates transcription of growth and differentiation-related genes by chromatin remodeling [183]. *ARID1A*, encoding one of the subunits of the SWI/SNF complex, is mutated in 4–8% of PDAC cases [58]. Another subunit, *SMARCA4*, is also mutated in a small proportion of cases [58]. High-mobility group A protein 2 (HMGA2) is a nuclear chromatin modifying factor that is overexpressed in PDAC tissue, which correlates negatively with the expression of the cell-adhesion molecule E-cadherin and positively with expression of EMT-associated molecule vimentin. HMGA2 has also been shown to be an effector molecule of the KRAS pathway [132].

Hedgehog signaling comprises a family of molecules that make up the three hedgehog signaling pathways: sonic hedgehog (SHH), Indian hedgehog (IHH), and desert hedgehog (DHH), which are named after their respective activating ligands and are utterly important in vertebrate organogenesis [184]. The stimulation of hedgehog ligands is transduced by a pair of co-activating receptors located on the cellular membrane. These include Patched (Ptc), which inhibits the activation of Smoothened (Smo) in the absence of ligand stimulation. Upon stimulation, Ptc is inactivated, leading to the translocation of Gli proteins into the nucleus to propagate the expression of target genes [185]. Hedgehog signaling has been shown to be an important factor in the development of PDAC-associated fibrosis by stimulation of pancreatic stellate cells (PSCs), whereas it affects the PDAC cell compartment only to a small extent [186–188]. This finding has also been observed in PanINs, which stimulate stromal cells by the secretion of SHH as well [189]. In vitro and in vivo knockout of SHH leads only to minimal fibrotic tissue in comparison to cells with wild-type SHH, which show a massive desmoplastic reaction. This is explained by reduced proliferation of stromal cells and little to no production of type I collagen or fibronectin. Mechanistically, Bailey et al. found that knockout of SHH leads to the absence of paracrine stimulation of stromal cells by tumor cells with SHH. Additionally, the authors found that their results are attributable to impaired transformation and proliferation of PSCs [186]. Conversely, IHH does not promote the proliferation or transformation of PSCs, but instead leads to increasing motility of PSCs in Matrigel™ assays. This is explained by IHH-dependent inhibition of tissue inhibitor of metalloproteinases 2 (TIMP-2) and consecutive production of matrix metalloproteinase 14 (MMP14 or MT1-MMP) that cleaves extracellular ECM components, such as collagen, fibronectin, laminin, and proteoglycans [190].

Notch signaling has also been shown to play a role in pancreatic development as well as in early pancreatic carcinogenesis [191]. Notch signaling controls cell differentiation through lateral inhibition in embryogenesis. Thus, Notch signaling determines the differentiation of progenitor cells into endocrine or exocrine cells [192]. The Notch proteins are transmembrane receptors that share structural similarities with EGFR and include Notch-1 to Notch-4 [193]. Upon receptor activation, proteolytic release of the Notch intracellular domain (Notch ICD) and consecutive nuclear translocation leads to the activation of target genes [193]. Bailey et al. have reported frequent amplification of Notch signaling members, such as Notch-1 and E3 ubiquitin-protein ligase MIB1 (MIB1) [128]. Furthermore, multiple members of the Notch pathway, such as Notch-1, Delta-1, and presenilin-1, are overexpressed in PDACs [194, 195]. Experiments have shown that Notch-2 is especially needed for the development of PanINs, causing increased expression of c-myc and epithelial-to-mesenchymal transition (EMT) by upregulation of TGF-β. Conversely, knockout of Notch-1 was shown to have less influence on pancreatic carcinogenesis [191]. Miyamoto et al. have also shown that TGF-α-induced ADM is caused by upregulation of Notch signaling members, such as Hes-1 [196]. Therefore, the importance of Notch signaling seems to be highest in early stages of pancreatic carcinogenesis, although members of the Notch pathway are also expressed in PDACs [196].

5.1.5.8 Tumor Metabolism and Autophagy

Knowledge of the metabolism of malignant tumors has been shown to be utterly important for the understanding of tumors, which was accounted for in the new version of the "hallmarks of cancer" by D. Hanahan and R. Weinberg [197]. Normal cells produce energy by oxidative phosphorylation of adenosine monophosphate (AMP) or adenosine diphosphate (ADP) to adenosine triphosphate (ATP). Fast-growing tumors are characterized by increased glycolysis and uptake and production of nucleotides, lipids, and amino acids, which are required for cell-cycle progression [198]. Moreover, the synthesis of cell-membrane-bound lipids requires disproportionally high amounts of glucose to provide the necessary carbon atoms [198]. In PDAC, the excess glucose uptake is mediated by oncogenic KRAS and its stimulation of MAPK/ERK and mTOR signaling [199]. Oncogenic KRAS also promotes proliferation by increased production of DNA ribose backbone molecules in the pentose phosphate pathway. Energy production, however, is largely dependent on glutamine as fuel of the citric acid cycle [199]. As oncogenic *KRAS* mutations belong to the first mutations found in pancreatic carcinogenesis [63], it is clear that the metabolic derangement of PDAC is evident before malignant transformation.

Macroautophagy is a cellular process that leads to the disassembly and recycling of damaged intracellular structures, such as misfolded proteins or malfunctioning organelles, which are altered differentially in pancreatic carcinogenesis [200]. Macroautophagy enables the efficient production of energy, amino acids, nucleotides, and lipids in nutrient-deprived locations, such as the desmoplastic environment of PDACs [201]. In PanIN-3 lesions, macroautophagy has been observed in up to 80% of cases [200]. In PDACs, macroautophagy increases tumor proliferation and viability, while its inhibition leads to increased resistance to ionizing radiation and chemotherapeutics gemcitabine and 5-flourouracil [202, 203]. In addition, expression of macroautophagy-associated molecules correlates with stem-cell-like characteristics, poor overall survival, and decreased progression-free survival in PDAC patients [204]. Mouse model data shows that the progression of PanINs is dependent on functional macroautophagy, while loss of macroautophagy capability leads to cellular senescence and only sporadic progression to high-grade PanINs [205]. Conversely, loss of macroautophagy capability increases tumor growth in established PDACs in the context of *TP53* mutation, by stimulating glucose uptake, glycolysis, and activation of the citric acid and pentose phosphate pathway [205]. Furthermore, loss of macroautophagy may promote the acquisition of additional chromosomal and genetic aberrations [206].

Macropinocytosis describes a process in which extracellular fluids and molecules are imported into the cytosol in the form of vesicles, i.e., macropinosomes [207]. Macropinocytosis was shown to be induced in the presence of oncogenic KRAS in PanINs and PDACs, which leads to the increasing degradation of albumin and its degradation to amino acids [207].

Furthermore, amino acids have been shown to increase PDAC growth by direct activation of the mTOR pathway [208]. Moreover, increased serum levels of amino acids are found early even before the first diagnosis, which is caused by catabolic effects attributable to oncogenic KRAS [209]. In addition, overexpression of the c-myc-regulated L-type amino-acid transporter 1 (LAT1) is a typical finding in many tumors, especially PDAC [210, 211]. LAT1 is a transporter for large neutral amino acids, such as leucine, valine, phenylalanine, and histidine, which are overexpressed in over 50% of cases, indicating that the increased uptake of amino acids is dependent on expression of the respective transporters [212]. LAT1 is also overexpressed in PDAC precursor IPMN [213].

5.1.5.9 Tumor Macro- and Microenvironment

The pancreatic tumor environment in PDAC patients is characterized by progressive fibrosis and desmoplastic transformation, which makes up to 80% of the macroscopic tumor volume, and has increasingly been shown to be crucial for the viability, proliferation, and invasive capability of PDAC cells [214–216]. In PanINs, desmoplasia is not found in low-grade lesions, whereas it is found increasingly in PanIN-3 lesions [217]. Desmoplastic transformation is fundamentally dependent on lost growth-suppressive TGF-β signaling, whereas knockout of the TGF-β receptor (TGFBR) leads to suppression of tumor-associated fibrosis by proliferation of pancreatic cancer-associated fibroblasts (CAFs) and increased killing of PDAC cells by CD8+ T cells [218]. Hedgehog signaling has also been shown to be a central factor in cancer-associated pancreatic fibrosis by the stimulation of PSCs [186–190]. The inhibition of CAFs is an important factor in PDAC development and maintenance [219]. Pancreatic CAFs have been shown to have different cellular origins, predominantly PSCs, but may also originate from bone marrow or resident fibroblasts [220]. The role of PSCs has emerged as important part of PDAC biology in recent years [214, 220–222]. PSCs are cells similar to hepatic stellate cells that express mesenchymal and myogenic markers, such as vimentin and desmin. When activated, PSCs transform into myofibroblast-like cells that produce extracellular matrix (ECM) components, such as collagen, laminin, and fibronectin [216]. PSCs induce oncogenic behavior in PDAC cells and promote the expression of stem-cell-associated markers ABCG2, Nestin, and LIN28 [223]. PSCs also secrete growth-promoting factors, such as platelet-derived growth factor (PDGF), cyclooxygenase 2 (COX-2), trefoil factor family (TFF) peptides, fibroblast growth factor 2 (FGF2), and periostin [224–228]. PSCs are thought to suppress the development and progression of pancreatic neoplasms by the engulfment and degradation of damaged pancreatic cells [216]. In established PDACs, however, stellate cells substantially support the desmoplastic change and promote PDAC metastasis and peripheral nerve infiltration [188, 229]. The interaction of PSCs with PDAC cells leads to dose-dependent proliferation of PDAC cells in vitro and may promote PDAC cell invasion by the expression of matrix metalloproteinase 3 (MMP-3) and MMP-14 [190, 215, 216]. For example, studies have shown, as explained above, that PDAC cells recruit stellate cells by secreting SHH in a paracrine manner to increase tumor desmoplasia [186]. This has also been observed in murine PanINs, indicating that the interaction with the pancreatic stroma is important even in early pancreatic carcinogenesis [189]. Decreased occurrence of apoptosis is also found in PSC-rich tumors [224]. Studies have also shown that a subset of PSCs expressing surface marker CD10 are of particular importance and are increased in PDAC environment [215]. Interestingly, PSCs do not only facilitate metastasis of PDAC cells, but have been shown to metastasize along with PDAC cells into distant organs in a PDGF dependent manner [230].

The expression of vascular endothelial growth factor (VEGF) in PDAC cells has long since been recognized as marker of poor prognosis and increased frequency of distant metastasis [231]. Moreover, the level of VEGF expression correlates with microvessel density in PDACs, which has been linked to impaired survival following curative resection [232]. The hypoxic environment of PDACs leads to the expression of angiogenic growth factors and promotes the migratory and invasive behavior of PSCs and PDAC cells [233, 234]. Ide et al. have shown that hypoxia leads to expression of the transcription factors c-Met in PDAC cells and angiogenic hepatocyte growth factor (HGF) in stromal cells. The expression of HGF was shown to correlate significantly with c-Met expression, implying a stroma-induced activation of HGF/c-Met signaling [234]. Moreover, VEGF, MMP-2, MMP-7, and MMP-14 were also significantly overexpressed [234]. Studies have shown that PSCs promote the formation of in vitro tubes, which is an in vitro marker for tumor angiogenesis while this process is inhibited by the addition anti-VEGF antibodies [230]. In addition, increased angiogenesis was found in murine xenograft models under treatment with hypoxic culture mediums, which was determined to be due to increased VEGF secretion [233]. Conversely, other authors have shown that, when compared to other solid tumors, PDACs are characterized by high interstitial fluid pressure due to elevated hyaluronic acid production, which causes collapse of blood vessels and impairs the diffusion of chemotherapeutic drugs [235]. However, the fibrotic tumor environment has shown to restrict tumor progression in animal models as well [236–238], highlighting the complexity of the tumor stroma relationship in PDAC.

5.1.5.10 Immunomodulation by PDACs

Immunomodulation is found early in pancreatic carcinogenesis and leads to increased resistance of the tumor to the host's immune system. The three phases of the cancer immunomodulation: elimination, equilibrium, and escape are an important part of tumor evolution [239]. Moreover, the immune system switches to a tumor-promoting state in many tumors, especially in PDAC [240]. T-cell targeting of tumor cells is inhibited by different factors, including marked parenchymal and tumor invasion by immunosuppressive macrophages and leukocytes [241–244]. This accumulation of immunosuppressive CD11b+ and Gr-1+ myeloid cells and CD4+Foxp3+CD25+ regulatory T cells is already seen in PanINs in genetically engineered mouse models and inhibits the targeting of tumor cells by CD8+ T cells [76, 244]. This process has been shown to be dependent on tumor-derived granulocyte-macrophage colony-stimulating factor (GM-CSF) [245], and the secretion of other factors. The extremely desmoplastic tumor environment itself also contributes to the immune escape of PDAC cells [218]. Intratumoral release of cytokines, such as tumor necrosis factor-α (TNFα), interferon-γ (IFN-γ), and interleukin-6 (IL-6), is low in PDACs compared to other tumor entities [246]. Moreover, production of antibodies is also significantly lower than observed in other tumors [247]. Pylayeva-Gupta et al. have shown that activating *KRAS* mutations leads to upregulation of GM-CSF production as a result of increased activation of the MAPK and PI3K pathways [76]. In another study, Pylayeva-Gupta et al. showed that the attraction of IL-35-producing B-cells by CAFs, mediated by secretion of the chemoattractant CXCL13, leads to increased growth of murine PanINs. Moreover, overexpression of CXCL13 was noted in human PanINs as well [248]. Gunderson et al. showed that B cells lead to increased differentiation of T cells toward the TH2 phenotype, thereby reducing the production of CD8+ T cells that attack tumor cells. This is caused by action of the B-cell Bruton tyrosine kinase (BTK), which may be partially abrogated with the use of BTK inhibitors [249].

5.1.5.11 Noncoding RNA

Noncoding RNA (ncRNA) is a relatively young research field, but studies have demonstrated that the behavior and sometimes evolution of malignant tumors are substantially influenced by ncRNA. ncRNA comprises noncoding RNA transcripts of highly variable size, which is the cornerstone for the current classification system of ncRNA. Of these, micro RNA (miRNA) and long non-coding RNA (lncRNA) had the highest impact in scientific research in recent years. miRNAs regulate gene expression by targeting of certain "seed regions" on mRNA in the cytoplasm, leading to inhibition of translation. lncRNAs exert their function by altering gene expression in the nucleus in different ways. In the world of ncRNA, the importance of *KRAS* is indisputable as well. Recent research has shown that at least 500 miRNAs are abnormally regulated in PDAC [250]. Moreover, at least 107 miRNAs are abnormally regulated in PanINs, indicating that aberration of miRNAs is important in the earliest stages of pancreatic carcinogenesis [251]. Many miRNAs are induced by KRAS signaling, such as miR-155, miR-21, and miR-221 [252–254]. These miRNAs suppress the expression of various tumor-suppressive molecules, such as Foxo3a, TP53INP1, PTEN, p27, and p57 [253, 255–257]. Other miRNAs with tumor-suppressive function, such as miR-217, miR-143, miR-206, miR-145, and Let-7, are suppressed by active KRAS signaling [129, 130, 258, 259]. miRNAs also interfere in other pathways, such as TGF-β, STAT, PI3K/AKT, chromatin regulation, and DNA damage control [129, 260–262]. Knowledge of the functions and the significance of lncRNAs is still evolving. However, some lncRNAs have shown to influence PDAC cell growth substantially. At least 319 lncRNAs have shown to be dysregulated in PDACs, but only a handful have been investigated in detail [263]. In accordance with the highlighted significance of the MAPK/ERK pathway in PDAC, many lncRNAs have been shown to influence this important pathway as well [264]. The lncRNA HOTAIR has been shown to increase oncogenic behavior of PDAC cells by increasing gene expression of target genomic regions by chromatin regulation [265, 266]. The lncRNA H19 has been shown to increase malignant behavior of PDAC cells by disrupting the tumor-suppressive function of the miRNA let-7, thus increasing the expression of HMGA2 [261]. The only thoroughly investigated tumor-suppressive lncRNA is ENST00000480739, which inhibits HIF-1α expression [267].

5.2 Integrated Progression Model of PDAC

5.2.1 Pancreatic Intraepithelial Neoplasia

The first change that initiates PanIN-associated tumorigenesis of PDAC is ADM, caused by various exogenic and endogenic stresses, such as smoking and CP flares [37, 39]. These stresses lead to ADM through the repression of the GATA6 promoter, which is essential for proper differentiation of pancreatic cells [268]. ADM is accompanied by expression of ductal-specific markers CK19 and Sox9 and loss of p21 expression [269, 270]. However, the exact mechanisms are still unclear, although research has shown that Notch signaling is essential for this process [191, 196]. ADM may then progress to form PanINs in the context of oncogenic *KRAS* mutation [39, 88]. Murine models have shown that CP may lead to ADM and consecutive PanIN and PDAC development [271]. In PanINs, the first changes found are *KRAS* mutations, causing increased MAPK/ERK, STAT, and AKT/PI3K signaling. *KRAS* mutations have been shown to be crucial for PanIN formation and maintenance in experimental models [72, 73, 134]. MUC4 leads to stabilization of the HER-2 receptor and further increases KRAS signaling. HER-2 amplification and overexpression furthermore enhance KRAS signaling [124, 135]. *KRAS* mutation leads to upregulation of oncogenic proteins c-MYC, NFATc1, EGFR, and SPINK1 and downregulation of the senescence inducer p21 [32, 272–274]. Moreover, oncogenic KRAS promotes the expression of oncogenic microRNAs, such as miR-155, miR-21, and miR-31, which are all upregulated in PanINs and often target multiple tumor-suppressor proteins on mRNA level, such as the tumor suppressors PTEN, Foxo3a, and Bcl-2 [251, 252, 275]. Oncogenic KRAS also represses the expression of tumor-suppressive miRNAs [129, 261, 276, 277]. Hedgehog signaling induces the desmoplastic transformation of the neighboring stroma, and the proliferation of PSCs in PanINs, which becomes more pronounced in progression to PanIN-3 and PDAC [189]. These alterations significantly contribute to malignant behavior of tumor cells. Telomere shortening is found early in pancreatic carcinogenesis, which may promote the acquisition of chromosomal alterations, such as deletions or fusion of chromosomes [119]. Moreover, it may lead to selection pressure toward cells that express telomerase, thus promoting immortalization of tumor cells. Early PanIN lesions are characterized by increased proliferation, but no anchorage-independent growth, as are PDAC cells [134]. The expression of COX-2 as a mediator of local inflammation is already found in PanIN-1 and PanIN-2 lesions and functions by increasing KRAS activity further [278, 279]. Late changes that initiate invasive behavior and further increases of proliferation and cell division are *SMAD4* and *TP53* mutations [134]. Mutations in *SMAD4* lead to reprogramming of TGF-β signaling from growth repression to growth promotion [97]. *SMAD4* mutations also lead to increased Wnt/β-catenin signaling, which promotes EMT and proliferation [280] and is typical in high-grade PanINs and PDAC, which is indicated by reduced membranous β-catenin and E-cadherin staining [144]. *SMAD4* and *TP53* mutations probably occur together in a short time frame. Together with oncogenic KRAS signaling, *TP53* and *SMAD4* mutations have individually been shown to accelerate pancreatic carcinogenesis to PDAC [84, 100], indicating the importance of these two alterations as late steps in pancreatic carcinogenesis. PanINs show increased levels of macroautophagy, which increases PanIN and PDAC-cell viability in states of oxygen and nutrient deprivation [200, 202, 204]. Whereas it increases survival of PanINs and early PDAC, loss of macroautophagy eventually increases cellular anaerobic glucose-dependent metabolism and propels tumor proliferation [205]. The immune response increases PanIN proliferation by secretion of proinflammatory cytokines and chemokines, such as interleukin-35 (IL-35) [248].

5.2.2 Mucinous Cystic Neoplasm

Because of the relative rarity of MCNs and associated cancer, it is still poorly investigated. Common changes include EGFR overexpression, which is present in most MCN carcinomas (61.2%), but not in non-invasive MCNs [281]. *KRAS* mutations have been found in different frequencies in the progression of MCN lesions [282, 283]. Jimenez et al. found a gradual increase of *KRAS* mutations, with virtually all malignant cases being affected by *KRAS* mutation, indicating that tumorigenicity in MCNs is promoted by active KRAS signaling. *RNF43* mutations are found frequently in MCNs, similar to IPMNs [48]. Late changes associated with malignant behavior are *TP53* and *SMAD4* mutations, similarly to conventional PDAC [103, 282, 284]. Moreover, increased expression of the Notch signaling pathway-associated molecules Jagged1 and Hes1 is seen in high-grade MCNs [285]. Interestingly, genetically engineered mouse models have shown that activating *KRAS* mutations and heterozygous loss of *SMAD4* lead to the formation of different pancreatic lesions, including PanIN-3 lesions. However, many mice develop lesions that resemble human MCNs, which progress to PDAC in many cases [284]. These carcinomas are characterized by spontaneous mutations in *TP53* and *CDKN2A*, showing that tumorigenesis from established MCN lesions to PDAC may involve similar mechanisms and cellular pathways as in PanIN-associated tumorigenesis [284].

5.2.3 Intraductal Papillary Mucinous Neoplasm

In IPMNs, the first changes observed in intestinal-type IPMNs are *GNAS* mutations, resulting in constitutive activation of Gαs, while *KRAS* mutations are an early finding in pancreatobiliary-type IPMNs [70]. *KRAS* mutations are found early, but are not significantly increased in the progression [68]. This shows that increased MAKP/ERK signaling is typical for pancreatobiliary-type IPMNs, which more commonly progress to PDAC, while intestinal-type IPMNs and colloid-carcinoma are characterized by increased GPCR signaling [105]. Germline mutations of *STK11* convey an increased risk of IPMN development. Moreover, they are found in IPMNs in one third of patients that do not harbor *SKT11* mutations [114]. Chromatin remodeling pathway alterations are frequent in IPMN lesions. *SMARCA4* mutations and downregulation of the gene product Brg1 are associated with formation of IPMNs and cooperate with oncogenic KRAS to promote formation of carcinoma [109–111]. Telomere shortening with decreasing telomere length in progression to carcinoma is common, indicating a role in malignant transformation [121]. In contrast to PanINs, expression of S100P is a constant finding in IPMNs with low-grade dysplasia, implying a fundamental role of calcium-signaling in the formation of IPMN lesions [286, 287]. Reduced p16 staining is found frequently in IPMNs with low-grade dysplasia, similar to PanINs [85]. *RNF43*, which encodes a ubiquitin ligase that promotes tumor cell proliferation, is mutated in up to 38% of IPMNs [48, 112]. Telomere shortening is found with increasing frequency and severity in IPMNs [121], indicating that chromosomal damage is already present in early stages. This goes along with increased telomerase expression [121], indicating that immortalization of tumor cells occurs at early stages of the disease. IPMN with intermediate-grade dysplasia is characterized by increasing expression of c-myc and cyclin D1, indicating that increased cell-cycle progression is found at this grade [85, 88, 94, 95, 288]. Moreover, fist numerical chromosomal aberrations are commonly found in IPMN with intermediate-grade dysplasia [127]. Late changes that promote progression to carcinoma include *TP53* and *SMAD4* mutations, although there are differences with respect to the histologic type of IPMN [85, 101].

References

1. Malvezzi M, Bertuccio P, Levi F, La Vecchia C, Negri E. European cancer mortality predictions for the year 2014. Ann Oncol. 2014;25(8):1650–6.
2. Siegel R, Naishadham D, Jemal A. Cancer statistics, 2012. CA Cancer J Clin. 2012;62(1):10–29.
3. Ryan DP, Hong TS, Bardeesy N. Pancreatic adenocarcinoma. N Engl J Med. 2014;371(11):1039–49.
4. Klimstra DS. Nonductal neoplasms of the pancreas. Mod Pathol. 2007;20(Suppl 1):S94–112.
5. La Rosa S, Sessa F, Capella C. Acinar cell carcinoma of the pancreas: overview of clinicopathologic features and insights into the molecular pathology. Front Med (Lausanne). 2015;2:41.

6. Chiorean EG, Coveler AL. Pancreatic cancer: optimizing treatment options, new, and emerging targeted therapies. Drug Des Devel Ther. 2015;9:3529–45.
7. Hruban R, Klimstra D, Pitman M. Tumors of the pancreas. Atlas of tumor pathology. 6th Series ed. Washington, DC: American Registry of Pathology; 2007.
8. Hirai I, Kimura W, Ozawa K, Kudo S, Suto K, Kuzu H, et al. Perineural invasion in pancreatic cancer. Pancreas. 2002;24(1):15–25.
9. Kloppel G, Lingenthal G, von Bulow M, Kern HF. Histological and fine structural features of pancreatic ductal adenocarcinomas in relation to growth and prognosis: studies in xenografted tumours and clinico-histopathological correlation in a series of 75 cases. Histopathology. 1985;9(8):841–56.
10. Raimondi S, Maisonneuve P, Lohr JM, Lowenfels AB. Early onset pancreatic cancer: evidence of a major role for smoking and genetic factors. Cancer Epidemiol Biomark Prev. 2007;16(9):1894–7.
11. Ben Q, Li Z, Liu C, Cai Q, Yuan Y, Wang K, et al. Hepatitis B virus status and risk of pancreatic ductal adenocarcinoma: a case-control study from China. Pancreas. 2012;41(3):435–40.
12. Shi C, Klein AP, Goggins M, Maitra A, Canto M, Ali S, et al. Increased prevalence of precursor lesions in familial pancreatic cancer patients. Clin Cancer Res. 2009;15(24):7737–43.
13. Rosato V, Polesel J, Bosetti C, Serraino D, Negri E, La Vecchia C. Population attributable risk for pancreatic cancer in Northern Italy. Pancreas. 2015;44(2):216–20.
14. Maisonneuve P, Lowenfels AB. Risk factors for pancreatic cancer: a summary review of meta-analytical studies. Int J Epidemiol. 2015;44(1):186–98.
15. McWilliams RR, Maisonneuve P, Bamlet WR, Petersen GM, Li D, Risch HA, et al. Risk factors for early-onset and very-early-onset pancreatic adenocarcinoma: a pancreatic cancer case-control consortium (PanC4) analysis. Pancreas. 2016;45(2):311–6.
16. Piciucchi M, Capurso G, Valente R, Larghi A, Archibugi L, Signoretti M, et al. Early onset pancreatic cancer: risk factors, presentation and outcome. Pancreatology. 2015;15(2):151–5.
17. Childs EJ, Mocci E, Campa D, Bracci PM, Gallinger S, Goggins M, et al. Common variation at 2p13.3, 3q29, 7p13 and 17q25.1 associated with susceptibility to pancreatic cancer. Nat Genet. 2015;47(8):911–6.
18. Goggins M, Hruban RH, Kern SE. BRCA2 is inactivated late in the development of pancreatic intraepithelial neoplasia: evidence and implications. Am J Pathol. 2000;156(5):1767–71.
19. Jiao Y, Yonescu R, Offerhaus GJ, Klimstra DS, Maitra A, Eshleman JR, et al. Whole-exome sequencing of pancreatic neoplasms with acinar differentiation. J Pathol. 2014;232(4):428–35.
20. Rigaud G, Moore PS, Zamboni G, Orlandini S, Taruscio D, Paradisi S, et al. Allelotype of pancreatic acinar cell carcinoma. Int J Cancer. 2000;88(5):772–7.
21. Furlan D, Sahnane N, Bernasconi B, Frattini M, Tibiletti MG, Molinari F, et al. APC alterations are frequently involved in the pathogenesis of acinar cell carcinoma of the pancreas, mainly through gene loss and promoter hypermethylation. Virchows Arch. 2014;464(5):553–64.
22. Ding L, Han L, Li Y, Zhao J, He P, Zhang W. Neurogenin 3-directed cre deletion of Tsc1 gene causes pancreatic acinar carcinoma. Neoplasia. 2014;16(11):909–17.
23. La Rosa S, Bernasconi B, Frattini M, Tibiletti MG, Molinari F, Furlan D, et al. TP53 alterations in pancreatic acinar cell carcinoma: new insights into the molecular pathology of this rare cancer. Virchows Arch. 2016;468(3):289–96.
24. Baker ML, Seeley ES, Pai R, Suriawinata AA, Mino-Kenudson M, Zamboni G, et al. Invasive mucinous cystic neoplasms of the pancreas. Exp Mol Pathol. 2012;93(3):345–9.
25. Adsay NV, Pierson C, Sarkar F, Abrams J, Weaver D, Conlon KC, et al. Colloid (mucinous noncystic) carcinoma of the pancreas. Am J Surg Pathol. 2001;25(1):26–42.
26. Konstantinidis IT, Warshaw AL, Allen JN, Blaszkowsky LS, Castillo CF, Deshpande V, et al. Pancreatic ductal adenocarcinoma: is there a survival difference for R1 resections versus locally advanced unresectable tumors? What is a "true" R0 resection? Ann Surg. 2013;257(4):731–6.

27. Tada M, Kawabe T, Arizumi M, Togawa O, Matsubara S, Yamamoto N, et al. Pancreatic cancer in patients with pancreatic cystic lesions: a prospective study in 197 patients. Clin Gastroenterol Hepatol. 2006;4(10):1265–70.
28. Brat DJ, Lillemoe KD, Yeo CJ, Warfield PB, Hruban RH. Progression of pancreatic intraductal neoplasias to infiltrating adenocarcinoma of the pancreas. Am J Surg Pathol. 1998;22(2):163–9.
29. Kim KM, Lee JK, Shin JU, Lee KH, Lee KT, Sung JY, et al. Clinicopathologic features of intraductal papillary neoplasm of the bile duct according to histologic subtype. Am J Gastroenterol. 2012;107(1):118–25.
30. Nakanuma Y, Sasaki M, Ishikawa A, Tsui W, Chen TC, Huang SF. Biliary papillary neoplasm of the liver. Histol Histopathol. 2002;17(3):851–61.
31. Feldmann G, Beaty R, Hruban RH, Maitra A. Molecular genetics of pancreatic intraepithelial neoplasia. J Hepato-Biliary-Pancreat Surg. 2007;14(3):224–32.
32. Biankin AV, Kench JG, Morey AL, Lee CS, Biankin SA, Head DR, et al. Overexpression of p21(WAF1/CIP1) is an early event in the development of pancreatic intraepithelial neoplasia. Cancer Res. 2001;61(24):8830–7.
33. Bosman FT, Carneiro F, Hruban RH, Theise ND. WHO classification of tumours of the digestive system. 4th ed. Geneva: World Health Organization; 2010.
34. Andea A, Sarkar F, Adsay VN. Clinicopathological correlates of pancreatic intraepithelial neoplasia: a comparative analysis of 82 cases with and 152 cases without pancreatic ductal adenocarcinoma. Mod Pathol. 2003;16(10):996–1006.
35. Brockie E, Anand A, Albores-Saavedra J. Progression of atypical ductal hyperplasia/carcinoma in situ of the pancreas to invasive adenocarcinoma. Ann Diagn Pathol. 1998;2(5):286–92.
36. Takaori K, Kobashi Y, Matsusue S, Matsui K, Yamamoto T. Clinicopathological features of pancreatic intraepithelial neoplasias and their relationship to intraductal papillary-mucinous tumors. J Hepato-Biliary-Pancreat Surg. 2003;10(2):125–36.
37. Rooman I, Real FX. Pancreatic ductal adenocarcinoma and acinar cells: a matter of differentiation and development? Gut. 2012;61(3):449–58.
38. Ray KC, Bell KM, Yan J, Gu G, Chung CH, Washington MK, et al. Epithelial tissues have varying degrees of susceptibility to Kras(G12D)-initiated tumorigenesis in a mouse model. PLoS One. 2011;6(2):e16786.
39. Shi G, DiRenzo D, Qu C, Barney D, Miley D, Konieczny SF. Maintenance of acinar cell organization is critical to preventing Kras-induced acinar-ductal metaplasia. Oncogene. 2013;32(15):1950–8.
40. Murphy SJ, Hart SN, Lima JF, Kipp BR, Klebig M, Winters JL, et al. Genetic alterations associated with progression from pancreatic intraepithelial neoplasia to invasive pancreatic tumor. Gastroenterology. 2013;145(5):1098–1109.e1.
41. Cubilla AL, Fitzgerald PJ. Morphological lesions associated with human primary invasive nonendocrine pancreas cancer. Cancer Res. 1976;36(7 PT 2):2690–8.
42. Tan MC, Basturk O, Brannon AR, Bhanot U, Scott SN, Bouvier N, et al. GNAS and KRAS mutations define separate progression pathways in intraductal papillary mucinous neoplasm-associated carcinoma. J Am Coll Surg. 2015;220(5):845–54.
43. Flattet Y, Yamaguchi T, Andrejevic-Blant S, Halkic N. Pancreatic adenocarcinoma: the impact of preneoplastic lesion pattern on survival. Biosci Trends. 2016;9(6):402–6.
44. Minagawa N, Sato N, Mori Y, Tamura T, Higure A, Yamaguchi K. A comparison between intraductal papillary neoplasms of the biliary tract (BT-IPMNs) and intraductal papillary mucinous neoplasms of the pancreas (P-IPMNs) reveals distinct clinical manifestations and outcomes. Eur J Surg Oncol. 2013;39(6):554–8.
45. Grutzmann R, Post S, Saeger HD, Niedergethmann M. Intraductal papillary mucinous neoplasia (IPMN) of the pancreas: its diagnosis, treatment, and prognosis. Dtsch Arztebl Int. 2011;108(46):788–94.
46. Salvia R, Crippa S, Falconi M, Bassi C, Guarise A, Scarpa A, et al. Branch-duct intraductal papillary mucinous neoplasms of the pancreas: to operate or not to operate? Gut. 2007;56(8):1086–90.

47. Tanaka M, Chari S, Adsay V, Fernandez-del Castillo C, Falconi M, Shimizu M, et al. International consensus guidelines for management of intraductal papillary mucinous neoplasms and mucinous cystic neoplasms of the pancreas. Pancreatology. 2006;6(1–2):17–32.
48. Springer S, Wang Y, Dal Molin M, Masica DL, Jiao Y, Kinde I, et al. A combination of molecular markers and clinical features improve the classification of pancreatic cysts. Gastroenterology. 2015;149(6):1501–10.
49. Roch AM, Rosati CM, Cioffi JL, Ceppa EP, DeWitt JM, Al-Haddad MA, et al. Intraductal papillary mucinous neoplasm of the pancreas, one manifestation of a more systemic disease? Am J Surg. 2015;211(3):512–8.
50. Eguchi H, Ishikawa O, Ohigashi H, Tomimaru Y, Sasaki Y, Yamada T, et al. Patients with pancreatic intraductal papillary mucinous neoplasms are at high risk of colorectal cancer development. Surgery. 2006;139(6):749–54.
51. Riall TS, Stager VM, Nealon WH, Townsend CM,Jr, Kuo YF, Goodwin JS, et al. Incidence of additional primary cancers in patients with invasive intraductal papillary mucinous neoplasms and sporadic pancreatic adenocarcinomas. J Am Coll Surg. 2007;204(5):803–13; discussion 813–4.
52. Sugiyama M, Atomi Y. Extrapancreatic neoplasms occur with unusual frequency in patients with intraductal papillary mucinous tumors of the pancreas. Am J Gastroenterol. 1999;94(2):470–3.
53. Lubezky N, Ben-Haim M, Lahat G, Marmor S, Solar I, Brazowski E, et al. Intraductal papillary mucinous neoplasm of the pancreas: associated cancers, family history, genetic predisposition? Surgery. 2012;151(1):70–5.
54. Pugliese L, Keskin M, Maisonneuve P, D'Haese JG, Marchegiani G, Wenzel P, et al. Increased incidence of extrapancreatic neoplasms in patients with IPMN: fact or fiction? A critical systematic review. Pancreatology. 2015;15(3):209–16.
55. Wilentz RE, Albores-Saavedra J, Hruban RH. Mucinous cystic neoplasms of the pancreas. Semin Diagn Pathol. 2000;17(1):31–42.
56. Lawrence MS, Stojanov P, Polak P, Kryukov GV, Cibulskis K, Sivachenko A, et al. Mutational heterogeneity in cancer and the search for new cancer-associated genes. Nature. 2013;499(7457):214–8.
57. Jones S, Zhang X, Parsons DW, Lin JC, Leary RJ, Angenendt P, et al. Core signaling pathways in human pancreatic cancers revealed by global genomic analyses. Science. 2008;321(5897):1801–6.
58. Biankin AV, Waddell N, Kassahn KS, Gingras MC, Muthuswamy LB, Johns AL, et al. Pancreatic cancer genomes reveal aberrations in axon guidance pathway genes. Nature. 2012;491(7424):399–405.
59. Mann KM, Ward JM, Yew CC, Kovochich A, Dawson DW, Black MA, et al. Sleeping beauty mutagenesis reveals cooperating mutations and pathways in pancreatic adenocarcinoma. Proc Natl Acad Sci U S A. 2012;109(16):5934–41.
60. Beck MR, Otey CA, Campbell SL. Structural characterization of the interactions between palladin and alpha-actinin. J Mol Biol. 2011;413(3):712–25.
61. Witkiewicz AK, McMillan EA, Balaji U, Baek G, Lin WC, Mansour J, et al. Whole-exome sequencing of pancreatic cancer defines genetic diversity and therapeutic targets. Nat Commun. 2015;6:6744.
62. Furukawa T, Kuboki Y, Tanji E, Yoshida S, Hatori T, Yamamoto M, et al. Whole-exome sequencing uncovers frequent GNAS mutations in intraductal papillary mucinous neoplasms of the pancreas. Sci Rep. 2011;1:161.
63. Lohr M, Kloppel G, Maisonneuve P, Lowenfels AB, Luttges J. Frequency of K-ras mutations in pancreatic intraductal neoplasias associated with pancreatic ductal adenocarcinoma and chronic pancreatitis: a meta-analysis. Neoplasia. 2005;7(1):17–23.
64. Yu S, Lu Z, Liu C, Meng Y, Ma Y, Zhao W, et al. miRNA-96 suppresses KRAS and functions as a tumor suppressor gene in pancreatic cancer. Cancer Res. 2010;70(14): 6015–25.

65. Amato E, Molin MD, Mafficini A, Yu J, Malleo G, Rusev B, et al. Targeted next-generation sequencing of cancer genes dissects the molecular profiles of intraductal papillary neoplasms of the pancreas. J Pathol. 2014;233(3):217–27.
66. Z'graggen K, Rivera JA, Compton CC, Pins M, Werner J, Fernandez-del Castillo C, et al. Prevalence of activating K-ras mutations in the evolutionary stages of neoplasia in intraductal papillary mucinous tumors of the pancreas. Ann Surg. 1997;226(4):491–8; discussion 498–500.
67. Schonleben F, Qiu W, Remotti HE, Hohenberger W, Su GH. PIK3CA, KRAS, and BRAF mutations in intraductal papillary mucinous neoplasm/carcinoma (IPMN/C) of the pancreas. Langenbeck's Arch Surg. 2008;393(3):289–96.
68. Kitago M, Ueda M, Aiura K, Suzuki K, Hoshimoto S, Takahashi S, et al. Comparison of K-ras point mutation distributions in intraductal papillary-mucinous tumors and ductal adenocarcinoma of the pancreas. Int J Cancer. 2004;110(2):177–82.
69. Wu J, Matthaei H, Maitra A, Dal Molin M, Wood LD, Eshleman JR, et al. Recurrent GNAS mutations define an unexpected pathway for pancreatic cyst development. Sci Transl Med. 2011;3(92):92ra66.
70. Hosoda W, Sasaki E, Murakami Y, Yamao K, Shimizu Y, Yatabe Y. GNAS mutation is a frequent event in pancreatic intraductal papillary mucinous neoplasms and associated adenocarcinomas. Virchows Arch. 2015;466(6):665–74.
71. Serrano M, Lin AW, McCurrach ME, Beach D, Lowe SW. Oncogenic ras provokes premature cell senescence associated with accumulation of p53 and p16INK4a. Cell. 1997;88(5):593–602.
72. Collins MA, Bednar F, Zhang Y, Brisset JC, Galban S, Galban CJ, et al. Oncogenic Kras is required for both the initiation and maintenance of pancreatic cancer in mice. J Clin Invest. 2012;122(2):639–53.
73. Collins MA, Yan W, Sebolt-Leopold JS, Pasca di Magliano M. MAPK signaling is required for dedifferentiation of acinar cells and development of pancreatic intraepithelial neoplasia in mice. Gastroenterology. 2014;146(3):822–834.e7.
74. Daniluk J, Liu Y, Deng D, Chu J, Huang H, Gaiser S, et al. An NF-kappaB pathway-mediated positive feedback loop amplifies Ras activity to pathological levels in mice. J Clin Invest. 2012;122(4):1519–28.
75. Ancrile BB, O'Hayer KM, Counter CM. Oncogenic ras-induced expression of cytokines: a new target of anti-cancer therapeutics. Mol Interv. 2008;8(1):22–7.
76. Pylayeva-Gupta Y, Lee KE, Hajdu CH, Miller G, Bar-Sagi D. Oncogenic Kras-induced GM-CSF production promotes the development of pancreatic neoplasia. Cancer Cell. 2012;21(6):836–47.
77. Maitra A, Adsay NV, Argani P, Iacobuzio-Donahue C, De Marzo A, Cameron JL, et al. Multicomponent analysis of the pancreatic adenocarcinoma progression model using a pancreatic intraepithelial neoplasia tissue microarray. Mod Pathol. 2003;16(9):902–12.
78. Scarpa A, Capelli P, Mukai K, Zamboni G, Oda T, Iacono C, et al. Pancreatic adenocarcinomas frequently show p53 gene mutations. Am J Pathol. 1993;142(5):1534–43.
79. Oren M, Rotter V. Mutant p53 gain-of-function in cancer. Cold Spring Harb Perspect Biol. 2010;2(2):a001107.
80. Lang GA, Iwakuma T, Suh YA, Liu G, Rao VA, Parant JM, et al. Gain of function of a p53 hot spot mutation in a mouse model of Li-Fraumeni syndrome. Cell. 2004;119(6):861–72.
81. Morton JP, Timpson P, Karim SA, Ridgway RA, Athineos D, Doyle B, et al. Mutant p53 drives metastasis and overcomes growth arrest/senescence in pancreatic cancer. Proc Natl Acad Sci U S A. 2010;107(1):246–51.
82. Weissmueller S, Manchado E, Saborowski M, Morris 4th JP, Wagenblast E, Davis CA, et al. Mutant p53 drives pancreatic cancer metastasis through cell-autonomous PDGF receptor beta signaling. Cell. 2014;157(2):382–94.
83. Sonoyama T, Sakai A, Mita Y, Yasuda Y, Kawamoto H, Yagi T, et al. TP53 codon 72 polymorphism is associated with pancreatic cancer risk in males, smokers and drinkers. Mol Med Rep. 2011;4(3):489–95.

84. Bailey JM, Hendley AM, Lafaro KJ, Pruski MA, Jones NC, Alsina J, et al. p53 mutations cooperate with oncogenic Kras to promote adenocarcinoma from pancreatic ductal cells. Oncogene. 2016;35(32):4282–8.
85. Sasaki S, Yamamoto H, Kaneto H, Ozeki I, Adachi Y, Takagi H, et al. Differential roles of alterations of p53, p16, and SMAD4 expression in the progression of intraductal papillary-mucinous tumors of the pancreas. Oncol Rep. 2003;10(1):21–5.
86. Miyasaka Y, Nagai E, Yamaguchi H, Fujii K, Inoue T, Ohuchida K, et al. The role of the DNA damage checkpoint pathway in intraductal papillary mucinous neoplasms of the pancreas. Clin Cancer Res. 2007;13(15 Pt 1):4371–7.
87. Norfadzilah MY, Pailoor J, Retneswari M, Chinna K, Noor LM. P53 expression in invasive pancreatic adenocarcinoma and precursor lesions. Malays J Pathol. 2011;33(2):89–94.
88. Furukawa T, Fujisaki R, Yoshida Y, Kanai N, Sunamura M, Abe T, et al. Distinct progression pathways involving the dysfunction of DUSP6/MKP-3 in pancreatic intraepithelial neoplasia and intraductal papillary-mucinous neoplasms of the pancreas. Mod Pathol. 2005;18(8):1034–42.
89. Nishikawa N, Kimura Y, Okita K, Zembutsu H, Furuhata T, Katsuramaki T, et al. Intraductal papillary mucinous neoplasms of the pancreas: an analysis of protein expression and clinical features. J Hepato-Biliary-Pancreat Surg. 2006;13(4):327–35.
90. Zamboni G, Scarpa A, Bogina G, Iacono C, Bassi C, Talamini G, et al. Mucinous cystic tumors of the pancreas: clinicopathological features, prognosis, and relationship to other mucinous cystic tumors. Am J Surg Pathol. 1999;23(4):410–22.
91. Wilentz RE, Geradts J, Maynard R, Offerhaus GJ, Kang M, Goggins M, et al. Inactivation of the p16 (INK4A) tumor-suppressor gene in pancreatic duct lesions: loss of intranuclear expression. Cancer Res. 1998;58(20):4740–4.
92. Liggett Jr WH, Sidransky D. Role of the p16 tumor suppressor gene in cancer. J Clin Oncol. 1998;16(3):1197–206.
93. Rayess H, Wang MB, Srivatsan ES. Cellular senescence and tumor suppressor gene p16. Int J Cancer. 2012;130(8):1715–25.
94. Biankin AV, Biankin SA, Kench JG, Morey AL, Lee CS, Head DR, et al. Aberrant p16(INK4A) and DPC4/Smad4 expression in intraductal papillary mucinous tumours of the pancreas is associated with invasive ductal adenocarcinoma. Gut. 2002;50(6):861–8.
95. Miyasaka Y, Nagai E, Ohuchida K, Fujita H, Nakata K, Hayashi A, et al. Senescence in intraductal papillary mucinous neoplasm of the pancreas. Hum Pathol. 2011;42(12):2010–7.
96. Sorio C, Capelli P, Lissandrini D, Moore PS, Balzarini P, Falconi M, et al. Mucinous cystic carcinoma of the pancreas: a unique cell line and xenograft model of a preinvasive lesion. Virchows Arch. 2005;446(3):239–45.
97. Hahn SA, Hoque AT, Moskaluk CA, da Costa LT, Schutte M, Rozenblum E, et al. Homozygous deletion map at 18q21.1 in pancreatic cancer. Cancer Res. 1996;56(3):490–4.
98. Yang J, Weinberg RA. Epithelial-mesenchymal transition: at the crossroads of development and tumor metastasis. Dev Cell. 2008;14(6):818–29.
99. Wilentz RE, Su GH, Dai JL, Sparks AB, Argani P, Sohn TA, et al. Immunohistochemical labeling for dpc4 mirrors genetic status in pancreatic adenocarcinomas: a new marker of DPC4 inactivation. Am J Pathol. 2000;156(1):37–43.
100. Kojima K, Vickers SM, Adsay NV, Jhala NC, Kim HG, Schoeb TR, et al. Inactivation of Smad4 accelerates Kras(G12D)-mediated pancreatic neoplasia. Cancer Res. 2007;67(17):8121–30.
101. Iacobuzio-Donahue CA, Klimstra DS, Adsay NV, Wilentz RE, Argani P, Sohn TA, et al. Dpc-4 protein is expressed in virtually all human intraductal papillary mucinous neoplasms of the pancreas: comparison with conventional ductal adenocarcinomas. Am J Pathol. 2000;157(3):755–61.
102. Gerdes B, Wild A, Wittenberg J, Barth P, Ramaswamy A, Kersting M, et al. Tumor-suppressing pathways in cystic pancreatic tumors. Pancreas. 2003;26(1):42–8.
103. Iacobuzio-Donahue CA, Wilentz RE, Argani P, Yeo CJ, Cameron JL, Kern SE, et al. Dpc4 protein in mucinous cystic neoplasms of the pancreas: frequent loss of expression in invasive carcinomas suggests a role in genetic progression. Am J Surg Pathol. 2000;24(11):1544–8.

104. Lania AG, Mantovani G, Spada A. Mechanisms of disease: mutations of G proteins and G-protein-coupled receptors in endocrine diseases. Nat Clin Pract Endocrinol Metab. 2006;2(12):681–93.
105. Komatsu H, Tanji E, Sakata N, Aoki T, Motoi F, Naitoh T, et al. A GNAS mutation found in pancreatic intraductal papillary mucinous neoplasms induces drastic alterations of gene expression profiles with upregulation of mucin genes. PLoS One. 2014;9(2):e87875.
106. Kanda M, Matthaei H, Wu J, Hong SM, Yu J, Borges M, et al. Presence of somatic mutations in most early-stage pancreatic intraepithelial neoplasia. Gastroenterology. 2012;142(4):730–733.e9.
107. Kanda M, Knight S, Topazian M, Syngal S, Farrell J, Lee J, et al. Mutant GNAS detected in duodenal collections of secretin-stimulated pancreatic juice indicates the presence or emergence of pancreatic cysts. Gut. 2013;62(7):1024–33.
108. Sausen M, Phallen J, Adleff V, Jones S, Leary RJ, Barrett MT, et al. Clinical implications of genomic alterations in the tumour and circulation of pancreatic cancer patients. Nat Commun. 2015;6:7686.
109. Dal Molin M, Hong SM, Hebbar S, Sharma R, Scrimieri F, de Wilde RF, et al. Loss of expression of the SWI/SNF chromatin remodeling subunit BRG1/SMARCA4 is frequently observed in intraductal papillary mucinous neoplasms of the pancreas. Hum Pathol. 2012;43(4):585–91.
110. Roy N, Malik S, Villanueva KE, Urano A, Lu X, Von Figura G, et al. Brg1 promotes both tumor-suppressive and oncogenic activities at distinct stages of pancreatic cancer formation. Genes Dev. 2015;29(6):658–71.
111. von Figura G, Fukuda A, Roy N, Liku ME, Morris Iv JP, Kim GE, et al. The chromatin regulator Brg1 suppresses formation of intraductal papillary mucinous neoplasm and pancreatic ductal adenocarcinoma. Nat Cell Biol. 2014;16(3):255–67.
112. Sugiura T, Yamaguchi A, Miyamoto K. A cancer-associated RING finger protein, RNF43, is a ubiquitin ligase that interacts with a nuclear protein, HAP95. Exp Cell Res. 2008;314(7):1519–28.
113. Wu J, Jiao Y, Dal Molin M, Maitra A, de Wilde RF, Wood LD, et al. Whole-exome sequencing of neoplastic cysts of the pancreas reveals recurrent mutations in components of ubiquitin-dependent pathways. Proc Natl Acad Sci U S A. 2011;108(52):21188–93.
114. Sahin F, Maitra A, Argani P, Sato N, Maehara N, Montgomery E, et al. Loss of Stk11/Lkb1 expression in pancreatic and biliary neoplasms. Mod Pathol. 2003;16(7):686–91.
115. Sato N, Rosty C, Jansen M, Fukushima N, Ueki T, Yeo CJ, et al. STK11/LKB1 Peutz-Jeghers gene inactivation in intraductal papillary-mucinous neoplasms of the pancreas. Am J Pathol. 2001;159(6):2017–22.
116. Aguirre AJ, Brennan C, Bailey G, Sinha R, Feng B, Leo C, et al. High-resolution characterization of the pancreatic adenocarcinoma genome. Proc Natl Acad Sci U S A. 2004;101(24):9067–72.
117. Hahn SA, Seymour AB, Hoque AT, Schutte M, da Costa LT, Redston MS, et al. Allelotype of pancreatic adenocarcinoma using xenograft enrichment. Cancer Res. 1995;55(20):4670–5.
118. Hingorani SR, Wang L, Multani AS, Combs C, Deramaudt TB, Hruban RH, et al. Trp53R172H and KrasG12D cooperate to promote chromosomal instability and widely metastatic pancreatic ductal adenocarcinoma in mice. Cancer Cell. 2005;7(5):469–83.
119. van Heek NT, Meeker AK, Kern SE, Yeo CJ, Lillemoe KD, Cameron JL, et al. Telomere shortening is nearly universal in pancreatic intraepithelial neoplasia. Am J Pathol. 2002;161(5):1541–7.
120. Gisselsson D, Jonson T, Petersen A, Strombeck B, Dal Cin P, Hoglund M, et al. Telomere dysfunction triggers extensive DNA fragmentation and evolution of complex chromosome abnormalities in human malignant tumors. Proc Natl Acad Sci U S A. 2001;98(22):12683–8.
121. Hashimoto Y, Murakami Y, Uemura K, Hayashidani Y, Sudo T, Ohge H, et al. Telomere shortening and telomerase expression during multistage carcinogenesis of intraductal papillary mucinous neoplasms of the pancreas. J Gastrointest Surg. 2008;12(1):17–28; discussion 28–9.

122. Liang ZY, Wang WZ, Gao J, Wu SF, Zeng X, Liu TH. Topoisomerase IIalpha and HER2/neu gene alterations and their correlation in pancreatic ductal adenocarcinomas. Zhonghua Bing Li Xue Za Zhi. 2007;36(2):102–6.
123. Tsiambas E, Karameris A, Lazaris AC, Talieri M, Triantafillidis JK, Cheracakis P, et al. EGFR alterations in pancreatic ductal adenocarcinoma: a chromogenic in situ hybridization analysis based on tissue microarrays. Hepato-Gastroenterology. 2006;53(69):452–7.
124. Dancer J, Takei H, Ro JY, Lowery-Nordberg M. Coexpression of EGFR and HER-2 in pancreatic ductal adenocarcinoma: a comparative study using immunohistochemistry correlated with gene amplification by fluorescencent in situ hybridization. Oncol Rep. 2007;18(1):151–5.
125. Abe T, Fukushima N, Brune K, Boehm C, Sato N, Matsubayashi H, et al. Genome-wide allelotypes of familial pancreatic adenocarcinomas and familial and sporadic intraductal papillary mucinous neoplasms. Clin Cancer Res. 2007;13(20):6019–25.
126. Fritz S, Fernandez-del Castillo C, Mino-Kenudson M, Crippa S, Deshpande V, Lauwers GY, et al. Global genomic analysis of intraductal papillary mucinous neoplasms of the pancreas reveals significant molecular differences compared to ductal adenocarcinoma. Ann Surg. 2009;249(3):440–7.
127. Soldini D, Gugger M, Burckhardt E, Kappeler A, Laissue JA, Mazzucchelli L. Progressive genomic alterations in intraductal papillary mucinous tumours of the pancreas and morphologically similar lesions of the pancreatic ducts. J Pathol. 2003;199(4):453–61.
128. Bailey P, Chang DK, Nones K, Johns AL, Patch AM, Gingras MC, et al. Genomic analyses identify molecular subtypes of pancreatic cancer. Nature. 2016;531(7592):47–52.
129. Zhao WG, Yu SN, Lu ZH, Ma YH, Gu YM, Chen J. The miR-217 microRNA functions as a potential tumor suppressor in pancreatic ductal adenocarcinoma by targeting KRAS. Carcinogenesis. 2010;31(10):1726–33.
130. Keklikoglou I, Hosaka K, Bender C, Bott A, Koerner C, Mitra D, et al. MicroRNA-206 functions as a pleiotropic modulator of cell proliferation, invasion and lymphangiogenesis in pancreatic adenocarcinoma by targeting ANXA2 and KRAS genes. Oncogene. 2015;34(37):4867–78.
131. Pylayeva-Gupta Y, Grabocka E, Bar-Sagi D. RAS oncogenes: weaving a tumorigenic web. Nat Rev Cancer. 2011;11(11):761–74.
132. Watanabe S, Ueda Y, Akaboshi S, Hino Y, Sekita Y, Nakao M. HMGA2 maintains oncogenic RAS-induced epithelial-mesenchymal transition in human pancreatic cancer cells. Am J Pathol. 2009;174(3):854–68.
133. Schubbert S, Shannon K, Bollag G. Hyperactive Ras in developmental disorders and cancer. Nat Rev Cancer. 2007;7(4):295–308.
134. Shen R, Wang Q, Cheng S, Liu T, Jiang H, Zhu J, et al. The biological features of PanIN initiated from oncogenic Kras mutation in genetically engineered mouse models. Cancer Lett. 2013;339(1):135–43.
135. Singh AP, Moniaux N, Chauhan SC, Meza JL, Batra SK. Inhibition of MUC4 expression suppresses pancreatic tumor cell growth and metastasis. Cancer Res. 2004;64(2):622–30.
136. Eser S, Reiff N, Messer M, Seidler B, Gottschalk K, Dobler M, et al. Selective requirement of PI3K/PDK1 signaling for Kras oncogene-driven pancreatic cell plasticity and cancer. Cancer Cell. 2013;23(3):406–20.
137. Jimeno A, Tan AC, Coffa J, Rajeshkumar NV, Kulesza P, Rubio-Viqueira B, et al. Coordinated epidermal growth factor receptor pathway gene overexpression predicts epidermal growth factor receptor inhibitor sensitivity in pancreatic cancer. Cancer Res. 2008;68(8):2841–9.
138. Kennedy AL, Morton JP, Manoharan I, Nelson DM, Jamieson NB, Pawlikowski JS, et al. Activation of the PIK3CA/AKT pathway suppresses senescence induced by an activated RAS oncogene to promote tumorigenesis. Mol Cell. 2011;42(1):36–49.
139. Nicholson KM, Anderson NG. The protein kinase B/Akt signalling pathway in human malignancy. Cell Signal. 2002;14(5):381–95.

140. Lowy AM, Fenoglio-Preiser C, Kim OJ, Kordich J, Gomez A, Knight J, et al. Dysregulation of beta-catenin expression correlates with tumor differentiation in pancreatic duct adenocarcinoma. Ann Surg Oncol. 2003;10(3):284–90.
141. Pasca di Magliano M, Biankin AV, Heiser PW, Cano DA, Gutierrez PJ, Deramaudt T, et al. Common activation of canonical Wnt signaling in pancreatic adenocarcinoma. PLoS One. 2007;2(11):e1155.
142. Morris 4th JP, Cano DA, Sekine S, Wang SC, Hebrok M. Beta-catenin blocks Kras-dependent reprogramming of acini into pancreatic cancer precursor lesions in mice. J Clin Invest. 2010;120(2):508–20.
143. Vincent A, Omura N, Hong SM, Jaffe A, Eshleman J, Goggins M. Genome-wide analysis of promoter methylation associated with gene expression profile in pancreatic adenocarcinoma. Clin Cancer Res. 2011;17(13):4341–54.
144. Al-Aynati MM, Radulovich N, Riddell RH, Tsao MS. Epithelial-cadherin and beta-catenin expression changes in pancreatic intraepithelial neoplasia. Clin Cancer Res. 2004;10(4):1235–40.
145. Massague J. TGFbeta in cancer. Cell. 2008;134(2):215–30.
146. Jimeno A, Hidalgo M. Molecular biomarkers: their increasing role in the diagnosis, characterization, and therapy guidance in pancreatic cancer. Mol Cancer Ther. 2006;5(4):787–96.
147. Koorstra JB, Hustinx SR, Offerhaus GJ, Maitra A. Pancreatic carcinogenesis. Pancreatology. 2008;8(2):110–25.
148. Esposito I, Kleeff J, Abiatari I, Shi X, Giese N, Bergmann F, et al. Overexpression of cellular inhibitor of apoptosis protein 2 is an early event in the progression of pancreatic cancer. J Clin Pathol. 2007;60(8):885–95.
149. Bhanot U, Heydrich R, Moller P, Hasel C. Survivin expression in pancreatic intraepithelial neoplasia (PanIN): steady increase along the developmental stages of pancreatic ductal adenocarcinoma. Am J Surg Pathol. 2006;30(6):754–9.
150. Trauzold A, Wermann H, Arlt A, Schutze S, Schafer H, Oestern S, et al. CD95 and TRAIL receptor-mediated activation of protein kinase C and NF-kappaB contributes to apoptosis resistance in ductal pancreatic adenocarcinoma cells. Oncogene. 2001;20(31):4258–69.
151. Hinz S, Trauzold A, Boenicke L, Sandberg C, Beckmann S, Bayer E, et al. Bcl-XL protects pancreatic adenocarcinoma cells against CD95- and TRAIL-receptor-mediated apoptosis. Oncogene. 2000;19(48):5477–86.
152. Hu YX, Watanabe H, Ohtsubo K, Yamaguchi Y, Ha A, Motoo Y, et al. Bcl-2 expression related to altered p53 protein and its impact on the progression of human pancreatic carcinoma. Br J Cancer. 1999;80(7):1075–9.
153. Boucher MJ, Morisset J, Vachon PH, Reed JC, Laine J, Rivard N. MEK/ERK signaling pathway regulates the expression of Bcl-2, Bcl-X(L), and Mcl-1 and promotes survival of human pancreatic cancer cells. J Cell Biochem. 2000;79(3):355–69.
154. Greten FR, Weber CK, Greten TF, Schneider G, Wagner M, Adler G, et al. Stat3 and NF-kappaB activation prevents apoptosis in pancreatic carcinogenesis. Gastroenterology. 2002;123(6):2052–63.
155. Weichert W, Boehm M, Gekeler V, Bahra M, Langrehr J, Neuhaus P, et al. High expression of RelA/p65 is associated with activation of nuclear factor-kappaB-dependent signaling in pancreatic cancer and marks a patient population with poor prognosis. Br J Cancer. 2007;97(4):523–30.
156. Bondar VM, Sweeney-Gotsch B, Andreeff M, Mills GB, McConkey DJ. Inhibition of the phosphatidylinositol 3′-kinase-AKT pathway induces apoptosis in pancreatic carcinoma cells in vitro and in vivo. Mol Cancer Ther. 2002;1(12):989–97.
157. Datta SR, Dudek H, Tao X, Masters S, Fu H, Gotoh Y, et al. Akt phosphorylation of BAD couples survival signals to the cell-intrinsic death machinery. Cell. 1997;91(2):231–41.
158. Jones RG, Parsons M, Bonnard M, Chan VS, Yeh WC, Woodgett JR, et al. Protein kinase B regulates T lymphocyte survival, nuclear factor kappaB activation, and Bcl-X(L) levels in vivo. J Exp Med. 2000;191(10):1721–34.

159. Fukushima N, Sato N, Ueki T, Rosty C, Walter KM, Wilentz RE, et al. Aberrant methylation of preproenkephalin and p16 genes in pancreatic intraepithelial neoplasia and pancreatic ductal adenocarcinoma. Am J Pathol. 2002;160(5):1573–81.
160. Heilmann AM, Perera RM, Ecker V, Nicolay BN, Bardeesy N, Benes CH, et al. CDK4/6 and IGF1 receptor inhibitors synergize to suppress the growth of p16INK4A-deficient pancreatic cancers. Cancer Res. 2014;74(14):3947–58.
161. Liu F, Korc M. Cdk4/6 inhibition induces epithelial-mesenchymal transition and enhances invasiveness in pancreatic cancer cells. Mol Cancer Ther. 2012;11(10):2138–48.
162. Biankin AV, Kench JG, Biankin SA, Lee CS, Morey AL, Dijkman FP, et al. Pancreatic intraepithelial neoplasia in association with intraductal papillary mucinous neoplasms of the pancreas: implications for disease progression and recurrence. Am J Surg Pathol. 2004;28(9):1184–92.
163. Gansauge S, Gansauge F, Ramadani M, Stobbe H, Rau B, Harada N, et al. Overexpression of cyclin D1 in human pancreatic carcinoma is associated with poor prognosis. Cancer Res. 1997;57(9):1634–7.
164. Alao JP. The regulation of cyclin D1 degradation: roles in cancer development and the potential for therapeutic invention. Mol Cancer. 2007;6:24.
165. Hustinx SR, Leoni LM, Yeo CJ, Brown PN, Goggins M, Kern SE, et al. Concordant loss of MTAP and p16/CDKN2A expression in pancreatic intraepithelial neoplasia: evidence of homozygous deletion in a noninvasive precursor lesion. Mod Pathol. 2005;18(7):959–63.
166. Huang L, Goodrow TL, Zhang SY, Klein-Szanto AJ, Chang H, Ruggeri BA. Deletion and mutation analyses of the P16/MTS-1 tumor suppressor gene in human ductal pancreatic cancer reveals a higher frequency of abnormalities in tumor-derived cell lines than in primary ductal adenocarcinomas. Cancer Res. 1996;56(5):1137–41.
167. Ji S, Qin Y, Shi S, Liu X, Hu H, Zhou H, et al. ERK kinase phosphorylates and destabilizes the tumor suppressor FBW7 in pancreatic cancer. Cell Res. 2015;25(5):561–73.
168. Welcker M, Clurman BE. FBW7 ubiquitin ligase: a tumour suppressor at the crossroads of cell division, growth and differentiation. Nat Rev Cancer. 2008;8(2):83–93.
169. Strom A, Bonal C, Ashery-Padan R, Hashimoto N, Campos ML, Trumpp A, et al. Unique mechanisms of growth regulation and tumor suppression upon Apc inactivation in the pancreas. Development. 2007;134(15):2719–25.
170. Liberman E, Kraus S, Sagiv E, Dulkart O, Kazanov D, Arber N. The APC E1317Q and I1307K polymorphisms in non-colorectal cancers. Biomed Pharmacother. 2007;61(9):566–9.
171. Boschman CR, Stryker S, Reddy JK, Rao MS. Expression of p53 protein in precursor lesions and adenocarcinoma of human pancreas. Am J Pathol. 1994;145(6):1291–5.
172. Levine AJ. P53, the cellular gatekeeper for growth and division. Cell. 1997;88(3):323–31.
173. He L, He X, Lim LP, de Stanchina E, Xuan Z, Liang Y, et al. A microRNA component of the p53 tumour suppressor network. Nature. 2007;447(7148):1130–4.
174. Ji Q, Hao X, Zhang M, Tang W, Yang M, Li L, et al. MicroRNA miR-34 inhibits human pancreatic cancer tumor-initiating cells. PLoS One. 2009;4(8):e6816.
175. Liu C, Cheng H, Shi S, Cui X, Yang J, Chen L, et al. MicroRNA-34b inhibits pancreatic cancer metastasis through repressing Smad3. Curr Mol Med. 2013;13(4):467–78.
176. Buschmann T, Minamoto T, Wagle N, Fuchs SY, Adler V, Mai M, et al. Analysis of JNK, Mdm2 and p14(ARF) contribution to the regulation of mutant p53 stability. J Mol Biol. 2000;295(4):1009–21.
177. Yu GZ, Zhu MH, Ni CR, Li FM, Zheng JM, Gong ZJ. Expression of proteins in p53 (p14ARF-mdm2-p53-p21WAF/CIP1) pathway and their significance in exocrine pancreatic carcinoma. Zhonghua Bing Li Xue Za Zhi. 2004;33(2):130–4.
178. Yuan ZM, Huang Y, Ishiko T, Nakada S, Utsugisawa T, Shioya H, et al. Role for p300 in stabilization of p53 in the response to DNA damage. J Biol Chem. 1999;274(4):1883–6.
179. Mees ST, Mardin WA, Wendel C, Baeumer N, Willscher E, Senninger N, et al. EP300—a miRNA-regulated metastasis suppressor gene in ductal adenocarcinomas of the pancreas. Int J Cancer. 2010;126(1):114–24.

180. Singh SK, Ellenrieder V. Senescence in pancreatic carcinogenesis: from signalling to chromatin remodelling and epigenetics. Gut. 2013;62(9):1364–72.
181. Fahrner JA, Eguchi S, Herman JG, Baylin SB. Dependence of histone modifications and gene expression on DNA hypermethylation in cancer. Cancer Res. 2002;62(24):7213–8.
182. Dawson MA, Kouzarides T. Cancer epigenetics: from mechanism to therapy. Cell. 2012;150(1):12–27.
183. Narlikar GJ, Fan HY, Kingston RE. Cooperation between complexes that regulate chromatin structure and transcription. Cell. 2002;108(4):475–87.
184. Bai Y, Bai Y, Dong J, Li Q, Jin Y, Chen B, et al. Hedgehog Signaling in Pancreatic Fibrosis and Cancer. Medicine (Baltimore). 2016;95(10):e2996.
185. Watkins DN, Peacock CD. Hedgehog signalling in foregut malignancy. Biochem Pharmacol. 2004;68(6):1055–60.
186. Bailey JM, Swanson BJ, Hamada T, Eggers JP, Singh PK, Caffery T, et al. Sonic hedgehog promotes desmoplasia in pancreatic cancer. Clin Cancer Res. 2008;14(19):5995–6004.
187. Hwang RF, Moore TT, Hattersley MM, Scarpitti M, Yang B, Devereaux E, et al. Inhibition of the hedgehog pathway targets the tumor-associated stroma in pancreatic cancer. Mol Cancer Res. 2012;10(9):1147–57.
188. Li X, Wang Z, Ma Q, Xu Q, Liu H, Duan W, et al. Sonic hedgehog paracrine signaling activates stromal cells to promote perineural invasion in pancreatic cancer. Clin Cancer Res. 2014;20(16):4326–38.
189. Tian H, Callahan CA, DuPree KJ, Darbonne WC, Ahn CP, Scales SJ, et al. Hedgehog signaling is restricted to the stromal compartment during pancreatic carcinogenesis. Proc Natl Acad Sci U S A. 2009;106(11):4254–9.
190. Shinozaki S, Ohnishi H, Hama K, Kita H, Yamamoto H, Osawa H, et al. Indian hedgehog promotes the migration of rat activated pancreatic stellate cells by increasing membrane type-1 matrix metalloproteinase on the plasma membrane. J Cell Physiol. 2008;216(1):38–46.
191. Mazur PK, Einwachter H, Lee M, Sipos B, Nakhai H, Rad R, et al. Notch2 is required for progression of pancreatic intraepithelial neoplasia and development of pancreatic ductal adenocarcinoma. Proc Natl Acad Sci U S A. 2010;107(30):13438–43.
192. Apelqvist A, Li H, Sommer L, Beatus P, Anderson DJ, Honjo T, et al. Notch signalling controls pancreatic cell differentiation. Nature. 1999;400(6747):877–81.
193. Andersson ER, Sandberg R, Lendahl U. Notch signaling: simplicity in design, versatility in function. Development. 2011;138(17):3593–612.
194. Ruiz-Hidalgo MJ, Garces C, Laborda J. Notch-1 expression levels in 3 T3-L1 cells influence ras signaling and transformation by oncogenic ras. Int J Oncol. 1999;14(4):777–83.
195. Weijzen S, Rizzo P, Braid M, Vaishnav R, Jonkheer SM, Zlobin A, et al. Activation of Notch-1 signaling maintains the neoplastic phenotype in human Ras-transformed cells. Nat Med. 2002;8(9):979–86.
196. Miyamoto Y, Maitra A, Ghosh B, Zechner U, Argani P, Iacobuzio-Donahue CA, et al. Notch mediates TGF alpha-induced changes in epithelial differentiation during pancreatic tumorigenesis. Cancer Cell. 2003;3(6):565–76.
197. Hanahan D, Weinberg RA. Hallmarks of cancer: the next generation. Cell. 2011;144(5):646–74.
198. Vander Heiden MG, Cantley LC, Thompson CB. Understanding the Warburg effect: the metabolic requirements of cell proliferation. Science. 2009;324(5930):1029–33.
199. Ying H, Kimmelman AC, Lyssiotis CA, Hua S, Chu GC, Fletcher-Sananikone E, et al. Oncogenic Kras maintains pancreatic tumors through regulation of anabolic glucose metabolism. Cell. 2012;149(3):656–70.
200. Yang S, Wang X, Contino G, Liesa M, Sahin E, Ying H, et al. Pancreatic cancers require autophagy for tumor growth. Genes Dev. 2011;25(7):717–29.
201. Rabinowitz JD, White E. Autophagy and metabolism. Science. 2010;330(6009):1344–8.
202. Hashimoto D, Blauer M, Hirota M, Ikonen NH, Sand J, Laukkarinen J. Autophagy is needed for the growth of pancreatic adenocarcinoma and has a cytoprotective effect against anticancer drugs. Eur J Cancer. 2014;50(7):1382–90.

203. Mukubou H, Tsujimura T, Sasaki R, Ku Y. The role of autophagy in the treatment of pancreatic cancer with gemcitabine and ionizing radiation. Int J Oncol. 2010;37(4):821–8.
204. Yang MC, Wang HC, Hou YC, Tung HL, Chiu TJ, Shan YS. Blockade of autophagy reduces pancreatic cancer stem cell activity and potentiates the tumoricidal effect of gemcitabine. Mol Cancer. 2015;14(1):179. doi:10.1186/s12943-015-0449-3.
205. Rosenfeldt MT, O'Prey J, Morton JP, Nixon C, MacKay G, Mrowinska A, et al. P53 Status Determines the Role of Autophagy in Pancreatic Tumour Development. Nature. 2013;504(7479):296–300.
206. Mathew R, Kongara S, Beaudoin B, Karp CM, Bray K, Degenhardt K, et al. Autophagy suppresses tumor progression by limiting chromosomal instability. Genes Dev. 2007;21(11):1367–81.
207. Commisso C, Davidson SM, Soydaner-Azeloglu RG, Parker SJ, Kamphorst JJ, Hackett S, et al. Macropinocytosis of protein is an amino acid supply route in Ras-transformed cells. Nature. 2013;497(7451):633–7.
208. Liu KA, Lashinger LM, Rasmussen AJ, Hursting SD. Leucine supplementation differentially enhances pancreatic cancer growth in lean and overweight mice. Cancer Metab. 2014;2(1):6. doi:10.1186/2049-3002-2-6.
209. Mayers JR, Wu C, Clish CB, Kraft P, Torrence ME, Fiske BP, et al. Elevation of circulating branched-chain amino acids is an early event in human pancreatic adenocarcinoma development. Nat Med. 2014;20(10):1193–8.
210. Wang Q, Holst J. L-type amino acid transport and cancer: targeting the mTORC1 pathway to inhibit neoplasia. Am J Cancer Res. 2015;5(4):1281–94.
211. Hayashi K, Jutabha P, Endou H, Anzai N. c-Myc is crucial for the expression of LAT1 in MIA Paca-2 human pancreatic cancer cells. Oncol Rep. 2012;28(3):862–6.
212. Kaira K, Sunose Y, Arakawa K, Ogawa T, Sunaga N, Shimizu K, et al. Prognostic significance of L-type amino-acid transporter 1 expression in surgically resected pancreatic cancer. Br J Cancer. 2012;107(4):632–8.
213. Yanagisawa N, Ichinoe M, Mikami T, Nakada N, Hana K, Koizumi W, et al. High expression of L-type amino acid transporter 1 (LAT1) predicts poor prognosis in pancreatic ductal adenocarcinomas. J Clin Pathol. 2012;65(11):1019–23.
214. Kadaba R, Birke H, Wang J, Hooper S, Andl CD, Di Maggio F, et al. Imbalance of desmoplastic stromal cell numbers drives aggressive cancer processes. J Pathol. 2013;230(1):107–17.
215. Ikenaga N, Ohuchida K, Mizumoto K, Cui L, Kayashima T, Morimatsu K, et al. CD10+ pancreatic stellate cells enhance the progression of pancreatic cancer. Gastroenterology 2010;139(3):1041–51, 51.e1–8.
216. Apte MV, Pirola RC, Wilson JS. Pancreatic stellate cells: a starring role in normal and diseased pancreas. Front Physiol. 2012;3:344.
217. Korc M. Pancreatic cancer-associated stroma production. Am J Surg. 2007;194(4 Suppl):S84–6.
218. Principe DR, DeCant B, Mascarinas E, Wayne E, Diaz A, Akagi N, et al. TGFbeta signaling in the pancreatic tumor microenvironment promotes fibrosis and immune evasion to facilitate tumorigenesis. Cancer Res. 2016;76(9):2525–39.
219. Olumi AF, Grossfeld GD, Hayward SW, Carroll PR, Tlsty TD, Cunha GR. Carcinoma-associated fibroblasts direct tumor progression of initiated human prostatic epithelium. Cancer Res. 1999;59(19):5002–11.
220. Nielsen MF, Mortensen MB, Detlefsen S. Key players in pancreatic cancer-stroma interaction: Cancer-associated fibroblasts, endothelial and inflammatory cells. World J Gastroenterol. 2016;22(9):2678–700.
221. Hwang RF, Moore T, Arumugam T, Ramachandran V, Amos KD, Rivera A, et al. Cancer-associated stromal fibroblasts promote pancreatic tumor progression. Cancer Res. 2008;68(3):918–26.
222. Xu Z, Pothula SP, Wilson JS, Apte MV. Pancreatic cancer and its stroma: a conspiracy theory. World J Gastroenterol. 2014;20(32):11216–29.

223. Hamada S, Masamune A, Takikawa T, Suzuki N, Kikuta K, Hirota M, et al. Pancreatic stellate cells enhance stem cell-like phenotypes in pancreatic cancer cells. Biochem Biophys Res Commun. 2012;421(2):349–54.
224. Vonlaufen A, Joshi S, Qu C, Phillips PA, Xu Z, Parker NR, et al. Pancreatic stellate cells: partners in crime with pancreatic cancer cells. Cancer Res. 2008;68(7):2085–93.
225. Guppy NJ, El-Bahrawy ME, Kocher HM, Fritsch K, Qureshi YA, Poulsom R, et al. Trefoil factor family peptides in normal and diseased human pancreas. Pancreas. 2012;41(6):888–96.
226. Yoshida S, Ujiki M, Ding XZ, Pelham C, Talamonti MS, Bell Jr RH, et al. Pancreatic stellate cells (PSCs) express cyclooxygenase-2 (COX-2) and pancreatic cancer stimulates COX-2 in PSCs. Mol Cancer. 2005;4:27.
227. Liu Y, Du L. Role of pancreatic stellate cells and periostin in pancreatic cancer progression. Tumour Biol. 2015;36(5):3171–7.
228. Bachem MG, Schunemann M, Ramadani M, Siech M, Beger H, Buck A, et al. Pancreatic carcinoma cells induce fibrosis by stimulating proliferation and matrix synthesis of stellate cells. Gastroenterology. 2005;128(4):907–21.
229. Apte MV, Park S, Phillips PA, Santucci N, Goldstein D, Kumar RK, et al. Desmoplastic reaction in pancreatic cancer: role of pancreatic stellate cells. Pancreas. 2004;29(3):179–87.
230. Xu Z, Vonlaufen A, Phillips PA, Fiala-Beer E, Zhang X, Yang L, et al. Role of pancreatic stellate cells in pancreatic cancer metastasis. Am J Pathol. 2010;177(5):2585–96.
231. Seo Y, Baba H, Fukuda T, Takashima M, Sugimachi K. High expression of vascular endothelial growth factor is associated with liver metastasis and a poor prognosis for patients with ductal pancreatic adenocarcinoma. Cancer. 2000;88(10):2239–45.
232. Niedergethmann M, Hildenbrand R, Wostbrock B, Hartel M, Sturm JW, Richter A, et al. High expression of vascular endothelial growth factor predicts early recurrence and poor prognosis after curative resection for ductal adenocarcinoma of the pancreas. Pancreas. 2002;25(2):122–9.
233. Masamune A, Kikuta K, Watanabe T, Satoh K, Hirota M, Shimosegawa T. Hypoxia stimulates pancreatic stellate cells to induce fibrosis and angiogenesis in pancreatic cancer. Am J Physiol Gastrointest Liver Physiol. 2008;295(4):G709–17.
234. Ide T, Kitajima Y, Miyoshi A, Ohtsuka T, Mitsuno M, Ohtaka K, et al. Tumor-stromal cell interaction under hypoxia increases the invasiveness of pancreatic cancer cells through the hepatocyte growth factor/c-Met pathway. Int J Cancer. 2006;119(12):2750–9.
235. Provenzano PP, Cuevas C, Chang AE, Goel VK, Von Hoff DD, Hingorani SR. Enzymatic targeting of the stroma ablates physical barriers to treatment of pancreatic ductal adenocarcinoma. Cancer Cell. 2012;21(3):418–29.
236. Ijichi H, Chytil A, Gorska AE, Aakre ME, Bierie B, Tada M, et al. Inhibiting Cxcr2 disrupts tumor-stromal interactions and improves survival in a mouse model of pancreatic ductal adenocarcinoma. J Clin Invest. 2011;121(10):4106–17.
237. Ozdemir BC, Pentcheva-Hoang T, Carstens JL, Zheng X, Wu CC, Simpson TR, et al. Depletion of carcinoma-associated fibroblasts and fibrosis induces immunosuppression and accelerates pancreas cancer with reduced survival. Cancer Cell. 2014;25(6): 719–34.
238. Rhim AD, Oberstein PE, Thomas DH, Mirek ET, Palermo CF, Sastra SA, et al. Stromal elements act to restrain, rather than support, pancreatic ductal adenocarcinoma. Cancer Cell. 2014;25(6):735–47.
239. Dunn GP, Old LJ, Schreiber RD. The three Es of cancer immunoediting. Annu Rev Immunol. 2004;22:329–60.
240. Vonderheide RH, Bayne LJ. Inflammatory networks and immune surveillance of pancreatic carcinoma. Curr Opin Immunol. 2013;25(2):200–5.
241. De Monte L, Reni M, Tassi E, Clavenna D, Papa I, Recalde H, et al. Intratumor T helper type 2 cell infiltrate correlates with cancer-associated fibroblast thymic stromal lymphopoietin production and reduced survival in pancreatic cancer. J Exp Med. 2011;208(3):469–78.

242. Ostrand-Rosenberg S, Sinha P. Myeloid-derived suppressor cells: linking inflammation and cancer. J Immunol. 2009;182(8):4499–506.
243. Tassi E, Gavazzi F, Albarello L, Senyukov V, Longhi R, Dellabona P, et al. Carcinoembryonic antigen-specific but not antiviral CD4+ T cell immunity is impaired in pancreatic carcinoma patients. J Immunol. 2008;181(9):6595–603.
244. Clark CE, Beatty GL, Vonderheide RH. Immunosurveillance of pancreatic adenocarcinoma: insights from genetically engineered mouse models of cancer. Cancer Lett. 2009;279(1):1–7.
245. Bayne LJ, Beatty GL, Jhala N, Clark CE, Rhim AD, Stanger BZ, et al. Tumor-derived granulocyte-macrophage colony-stimulating factor regulates myeloid inflammation and T cell immunity in pancreatic cancer. Cancer Cell. 2012;21(6):822–35.
246. Garbe AI, Vermeer B, Gamrekelashvili J, von Wasielewski R, Greten FR, Westendorf AM, et al. Genetically induced pancreatic adenocarcinoma is highly immunogenic and causes spontaneous tumor-specific immune responses. Cancer Res. 2006;66(1):508–16.
247. Gnjatic S, Ritter E, Buchler MW, Giese NA, Brors B, Frei C, et al. Seromic profiling of ovarian and pancreatic cancer. Proc Natl Acad Sci U S A. 2010;107(11):5088–93.
248. Pylayeva-Gupta Y, Das S, Handler JS, Hajdu CH, Coffre M, Koralov S, et al. IL-35 producing B cells promote the development of pancreatic neoplasia. Cancer Discov. 2016;6(3):247–55.
249. Gunderson AJ, Kaneda MM, Tsujikawa T, Nguyen AV, Affara NI, Ruffell B, et al. Bruton tyrosine kinase dependent immune cell cross-talk drives pancreas cancer. Cancer Discov. 2016;6(3):270–85.
250. Ye S, Yang L, Zhao X, Song W, Wang W, Zheng S. Bioinformatics method to predict two regulation mechanism: TF-miRNA-mRNA and lncRNA-miRNA-mRNA in pancreatic cancer. Cell Biochem Biophys. 2014;70(3):1849–58.
251. Yu J, Li A, Hong SM, Hruban RH, Goggins M. MicroRNA alterations of pancreatic intraepithelial neoplasias. Clin Cancer Res. 2012;18(4):981–92.
252. du Rieu MC, Torrisani J, Selves J, Al Saati T, Souque A, Dufresne M, et al. MicroRNA-21 is induced early in pancreatic ductal adenocarcinoma precursor lesions. Clin Chem. 2010;56(4):603–12.
253. Basu A, Alder H, Khiyami A, Leahy P, Croce CM, Haldar S. MicroRNA-375 and MicroRNA-221: potential noncoding RNAs associated with antiproliferative activity of benzyl isothiocyanate in pancreatic cancer. Genes Cancer. 2011;2(2):108–19.
254. Wang P, Zhu CF, Ma MZ, Chen G, Song M, Zeng ZL, et al. Micro-RNA-155 is induced by K-Ras oncogenic signal and promotes ROS stress in pancreatic cancer. Oncotarget. 2015;6(25):21148–58.
255. Xu Q, Li P, Chen X, Zong L, Jiang Z, Nan L, et al. miR-221/222 induces pancreatic cancer progression through the regulation of matrix metalloproteinases. Oncotarget. 2015;6(16):14153–64.
256. Sarkar S, Dubaybo H, Ali S, Goncalves P, Kollepara SL, Sethi S, et al. Down-regulation of miR-221 inhibits proliferation of pancreatic cancer cells through up-regulation of PTEN, p27(kip1), p57(kip2), and PUMA. Am J Cancer Res. 2013;3(5):465–77.
257. Park JK, Lee EJ, Esau C, Schmittgen TD. Antisense inhibition of microRNA-21 or −221 arrests cell cycle, induces apoptosis, and sensitizes the effects of gemcitabine in pancreatic adenocarcinoma. Pancreas. 2009;38(7):e190–9.
258. Yan X, Chen X, Liang H, Deng T, Chen W, Zhang S, et al. miR-143 and miR-145 synergistically regulate ERBB3 to suppress cell proliferation and invasion in breast cancer. Mol Cancer. 2014;13:220. doi:10.1186/1476-4598-13-220.
259. Torrisani J, Bournet B, du Rieu MC, Bouisson M, Souque A, Escourrou J, et al. let-7 MicroRNA transfer in pancreatic cancer-derived cells inhibits in vitro cell proliferation but fails to alter tumor progression. Hum Gene Ther. 2009;20(8):831–44.
260. Hao J, Zhang S, Zhou Y, Hu X, Shao C. MicroRNA 483-3p suppresses the expression of DPC4/Smad4 in pancreatic cancer. FEBS Lett. 2011;585(1):207–13.
261. Ma C, Nong K, Zhu H, Wang W, Huang X, Yuan Z, et al. H19 promotes pancreatic cancer metastasis by derepressing let-7's suppression on its target HMGA2-mediated EMT. Tumour Biol. 2014;35(9):9163–9.

262. Gironella M, Seux M, Xie MJ, Cano C, Tomasini R, Gommeaux J, et al. Tumor protein 53-induced nuclear protein 1 expression is repressed by miR-155, and its restoration inhibits pancreatic tumor development. Proc Natl Acad Sci U S A. 2007;104(41):16170–5.
263. Wang Q, Jiang H, Ping C, Shen R, Liu T, Li J, et al. Exploring the Wnt pathway-associated LncRNAs and genes involved in pancreatic carcinogenesis driven by Tp53 mutation. Pharm Res. 2015;32(3):793–805.
264. Tahira AC, Kubrusly MS, Faria MF, Dazzani B, Fonseca RS, Maracaja-Coutinho V, et al. Long noncoding intronic RNAs are differentially expressed in primary and metastatic pancreatic cancer. Mol Cancer. 2011;10:141. doi:10.1186/1476-4598-10-141.
265. Kim K, Jutooru I, Chadalapaka G, Johnson G, Frank J, Burghardt R, et al. HOTAIR is a negative prognostic factor and exhibits pro-oncogenic activity in pancreatic cancer. Oncogene. 2013;32(13):1616–25.
266. Gupta RA, Shah N, Wang KC, Kim J, Horlings HM, Wong DJ, et al. Long non-coding RNA HOTAIR reprograms chromatin state to promote cancer metastasis. Nature. 2010;464(7291):1071–6.
267. Sun YW, Chen YF, Li J, Huo YM, Liu DJ, Hua R, et al. A novel long non-coding RNA ENST00000480739 suppresses tumour cell invasion by regulating OS-9 and HIF-1alpha in pancreatic ductal adenocarcinoma. Br J Cancer. 2014;111(11):2131–41.
268. Hermann PC, Sancho P, Canamero M, Martinelli P, Madriles F, Michl P, et al. Nicotine promotes initiation and progression of KRAS-induced pancreatic cancer via Gata6-dependent dedifferentiation of acinar cells in mice. Gastroenterology. 2014;147(5):1119–33.e4.
269. Kopp JL, von Figura G, Mayes E, Liu FF, Dubois CL, Morris 4th JP, et al. Identification of Sox9-dependent acinar-to-ductal reprogramming as the principal mechanism for initiation of pancreatic ductal adenocarcinoma. Cancer Cell. 2012;22(6):737–50.
270. Grabliauskaite K, Hehl AB, Seleznik GM, Saponara E, Schlesinger K, Zuellig RA, et al. P21 limits senescence and acinar-to-ductal metaplasia formation during pancreatitis. J Pathol. 2014;235(3):502–14.
271. Guerra C, Schuhmacher AJ, Canamero M, Grippo PJ, Verdaguer L, Perez-Gallego L, et al. Chronic pancreatitis is essential for induction of pancreatic ductal adenocarcinoma by K-Ras oncogenes in adult mice. Cancer Cell. 2007;11(3):291–302.
272. Chen NM, Singh G, Koenig A, Liou GY, Storz P, Zhang JS, et al. NFATc1 links EGFR signaling to induction of Sox9 transcription and acinar-ductal transdifferentiation in the pancreas. Gastroenterology. 2015;148(5):1024–1034.e9.
273. Schleger C, Verbeke C, Hildenbrand R, Zentgraf H, Bleyl U. c-MYC activation in primary and metastatic ductal adenocarcinoma of the pancreas: incidence, mechanisms, and clinical significance. Mod Pathol. 2002;15(4):462–9.
274. Ozaki N, Ohmuraya M, Hirota M, Ida S, Wang J, Takamori H, et al. Serine protease inhibitor Kazal type 1 promotes proliferation of pancreatic cancer cells through the epidermal growth factor receptor. Mol Cancer Res. 2009;7(9):1572–81.
275. Kent OA, Mendell JT, Rottapel R. Transcriptional regulation of miR-31 by oncogenic KRAS mediates metastatic phenotypes by repressing RASA1. Mol Cancer Res. 2016;14(3):267–77.
276. Khan S, Ebeling MC, Zaman MS, Sikander M, Yallapu MM, Chauhan N, et al. MicroRNA-145 targets MUC13 and suppresses growth and invasion of pancreatic cancer. Oncotarget. 2014;5(17):7599–609.
277. Kent OA, Chivukula RR, Mullendore M, Wentzel EA, Feldmann G, Lee KH, et al. Repression of the miR-143/145 cluster by oncogenic Ras initiates a tumor-promoting feed-forward pathway. Genes Dev. 2010;24(24):2754–9.
278. Maitra A, Ashfaq R, Gunn CR, Rahman A, Yeo CJ, Sohn TA, et al. Cyclooxygenase 2 expression in pancreatic adenocarcinoma and pancreatic intraepithelial neoplasia: an immunohistochemical analysis with automated cellular imaging. Am J Clin Pathol. 2002;118(2):194–201.
279. Philip B, Roland CL, Daniluk J, Liu Y, Chatterjee D, Gomez SB, et al. A high-fat diet activates oncogenic Kras and COX2 to induce development of pancreatic ductal adenocarcinoma in mice. Gastroenterology. 2013;145(6):1449–58.

280. White BD, Chien AJ, Dawson DW. Dysregulation of Wnt/beta-catenin signaling in gastrointestinal cancers. Gastroenterology. 2012;142(2):219–32.
281. Kirby RE, Lewandrowski KB, Southern JF, Compton CC, Warshaw AL. Relation of epidermal growth factor receptor and estrogen receptor-independent pS2 protein to the malignant transformation of mucinous cystic neoplasms of the pancreas. Arch Surg. 1995;130(1):69–72.
282. Jimenez RE, Warshaw AL, Z'graggen K, Hartwig W, Taylor DZ, Compton CC, et al. Sequential accumulation of K-ras mutations and p53 overexpression in the progression of pancreatic mucinous cystic neoplasms to malignancy. Ann Surg. 1999;230(4):501–9; discussion 509–11.
283. Kuboki Y, Shiratori K, Hatori T, Fujita I, Kimijima A, Yamamoto M, et al. Association of epidermal growth factor receptor and mitogen-activated protein kinase with cystic neoplasms of the pancreas. Mod Pathol. 2010;23(8):1127–35.
284. Izeradjene K, Combs C, Best M, Gopinathan A, Wagner A, Grady WM, et al. Kras(G12D) and Smad4/Dpc4 haploinsufficiency cooperate to induce mucinous cystic neoplasms and invasive adenocarcinoma of the pancreas. Cancer Cell. 2007;11(3):229–43.
285. Fukushima N, Sato N, Prasad N, Leach SD, Hruban RH, Goggins M. Characterization of gene expression in mucinous cystic neoplasms of the pancreas using oligonucleotide microarrays. Oncogene. 2004;23(56):9042–51.
286. Nakata K, Nagai E, Ohuchida K, Hayashi A, Miyasaka Y, Aishima S, et al. S100P is a novel marker to identify intraductal papillary mucinous neoplasms. Hum Pathol. 2010;41(6):824–31.
287. Liu H, Shi J, Wang HL, Zhang J, Brown RE, Wilkerson M, et al. Expression of von Hippel-Lindau gene product (pVHL) and S100P in cystic neoplasms of the pancreas—with an implication for their roles in tumorigenesis. Ann Clin Lab Sci. 2012;42(2):109–17.
288. Chetty R, Serra S, Salahshor S, Alsaad K, Shih W, Blaszyk H, et al. Expression of Wnt-signaling pathway proteins in intraductal papillary mucinous neoplasms of the pancreas: a tissue microarray analysis. Hum Pathol. 2006;37(2):212–7.

Hepatocellular Carcinogenesis

6

Nicole Golob-Schwarzl, Sonja Kessler, and Johannes Haybaeck

Contents

N. Golob-Schwarzl
Institute of Pathology, Medical University of Graz, Graz, Austria
e-mail: Nicole.GolobSchwarzl@cbmed.at

S. Kessler (✉)
Department of Pharmacy, Pharmaceutical Biology, Saarland University, Saarbruecken, Germany
e-mail: s.kessler@mx.uni-saarlagd.de

J. Haybaeck, M.D., Ph.D.
Department of Pathology, Medical Faculty, Otto von Guericke University Magdeburg, 39120 Magdeburg, Germany

Institute of Pathology, Medical University of Graz, 8036 Graz, Austria
e-mail: johannes.haybaeck@med.ovgu.de

J. Haybaeck (ed.), *Mechanisms of Molecular Carcinogenesis – Volume 2*,
DOI 10.1007/978-3-319-53661-3_6

6.1 Human HCC Research

Liver cancer is the second leading cause of cancer mortality worldwide, with approximately 600,000 cancer related deaths [1, 2]. The most common type of liver cancer is hepatocellular carcinoma (HCC). In the United States, its incidence has increased over the last few years and is expected to double in the next 10–20 years [3, 4]. Despite progressive achievements in management and diagnosis, the mean survival of patients with HCC is less than 8 months, and thus, HCC is still one of the most fatal cancers [2, 3]. The relevant treatment strategies for HCC are surgical resection, liver transplantation, and local ablation. Recurrence occurs in up to 70% of patients within 5 years after resection [2, 5, 6].

Two pronounced types of recurrence are known in HCC. The early recurrence develops from primary cancer cells spreading to the surrounding liver and is usually observed within the first 2 years after surgery. In comparison, late recurrence, which is typically monitored more than 2 years after surgery, appears to be a result of chronic liver damage, known as the field effect, creating de novo tumors irrespective of resected primary tumors [5]. The two types of recurrence follow different clinical courses and presumably appear in distinct biological contexts [7]. To improve disease treatment, it is therefore important to recapitulate and understand the biological characteristics of each type of recurrence and to generate distinct molecular prognostication systems that are capable of determining patients at high risk of developing either type. Treatment of recurrences is of utmost importance. Our knowledge of the genetic alterations of these two types, in particular late recurrence, is incomplete [2].

Inflammation and chronic injury are known to benefit from tumor development. HCC is one of the best examples, and more than 90% of HCCs arise from hepatic injury and inflammation [8, 9]. Chronically unresolved inflammation is often associated with continuing hepatic injury and simultaneous regeneration, which makes the liver susceptible to developing HCC. This process resembles a continuing wound-healing response, regardless of the differences among various etiological factors, such as alcohol, viruses, and fatty liver [10, 11]. However, the coexistence of cirrhosis and inflammation complicates early diagnosis of HCC. Therefore, biomarkers differentiating HCC from inflammation and cirrhosis are needed in order to improve prognosis of the respective patients. Besides, biomarkers may influence the design of novel chemopreventive strategies during HCC surveillance of patients with liver cirrhosis.

A perfect biomarker for HCC might be one that allows clinicians to reliably and robustly diagnose asymptomatic patients and can be largely used in a screening procedure. Usually, a biomarker useful for clinical use has to reach a level of sensitivity and specificity of ≥90% and is cost-effective and invasive to enable broad use. Therefore, the most worthwhile biomarker should be tumor specific and simply detectable in body fluids, such as plasma, serum, and bile [12].

6.2 Biomarkers in HCC

6.2.1 Alpha-Fetoprotein and Alpha-Fetprotein-L3

Alpha-fetoprotein (AFP) is considered the most valuable biomarker for HCC evaluation, since it was already found in serum of HCC patients in 1964 [12, 13]. AFP is used as a reference biomarker to screen and relief the diagnosis of HCC. There are three forms of AFP (AFP-L1, AFP-L2, and AFP-L3). AFP-L3 appears to differentiate HCC from chronic liver diseases and may be an indicator of HCC with total serum AFP levels ≥200 ng/mL [14]. For early HCC detection, AFP-L3 has been suggested as biomarker as it has higher specificity than AFP [12]. The use of AFP levels to diagnose HCC is based on a threshold value of 200 ng/mL, a specificity of 0.960, a sensitivity of 0.310 and an area under the curve (AUC) of 0.835 [15, 16].

6.2.2 Des-*y*-Carboxyprothrombin

Des-*y*-carboxyprothrombin (DCP) is a deviant prothrombin molecule that is increased in HCC, a defect in posttranslational carboxylation that leads to the production of DCP. In this context, DCP loses its normal prothrombin function, but can take on a major role in promoting malignant proliferation in HCC. The level of serum DCP in patients with benign and malignant liver diseases differs significantly from normal, and its diagnostic sensitivity is probably higher compared to AFP [17, 18].

At a 125 mAU/mL threshold, DCP has high sensitivity (89%), specificity (95%), and an AUC of 0.797 in the prediction of HCC. Altogether, for an early stage of HCC, DCP can be regarded as an excellent biomarker [14, 17].

6.2.3 Glypican-3

Glypican-3 (GPC3) is a cell surface-linked heparin sulfate proteoglycan and is involved in cell proliferation, survival, and tumor suppression, but is not expressed in healthy and nonmalignant hepatocytes. As GPC3 is detected in HCC cells but not in benign liver tissues, it has the capability of a biomarker for the diagnosis of early stage HCC [19, 20]. GPC3 is not limited in its capability as serum biomarker. GPC3 is also tested as an immune specific target. Corresponding immune responses and antitumor efficacy with high tolerance were shown in a phase I clinical trial of a GPC3 peptide vaccine for patients with advanced HCC [21].

6.2.4 Osteopontin

Osteopontin (OPN) is an integrin-binding glycophosphoprotein which is involved in many cellular functions, for example, in the regulation of survival, invasion, migration, and metastasis of tumor cells [22]. OPN is expressed in bile duct epithelium, stellate cells, and Kupffer cells, but not in hepatocytes [23]. Compared with cirrhosis, chronic hepatitis B, chronic hepatitis C, or healthy controls, OPN plasma levels were significantly increased in HCC patients [24].

6.2.5 Golgi Protein 73

Golgi protein 73 (GP73) is a Golgi-specific membrane protein and can be detected in serum of patients with liver disease, notably HCC [25]. Studies showed significantly higher serum GP73 levels in patients with HCC compared to healthy adults and hepatitis B virus carriers without hepatic diseases [26]. GP73 was considered as a potential biomarker also for early HCC diagnosis. Serum GP73 levels demonstrated improved sensitivity relative to AFP in detecting early stage HCC [27].

6.2.6 Squamous Cell Carcinoma Antigen

Squamous cell carcinoma antigen (SCCA) is highly expressed in epithelial tumors and has a role in protecting tumor cells from apoptosis. SCCA levels are increased in HCC patients with normal, physiological-AFP levels. This characteristic is helpful in the early detection and follow-up diagnoses for patients treated for HCC [14, 28]. At a threshold of 0.368 ng/mL, SCCA has a sensitivity of 84.2%, a specificity of 48.9%, and an AUC of 0.705 [29].

A potential alternative biomarker is the IgM immune complex. SCCA has been observed to bind to IgM (SCCA-IgM IC). Its expression was increased in the early phase of hepatocarcinogenesis. SCCA-IgM has a higher diagnostic performance than the free biomarker and was therefore undetectable in the serum of a healthy adult. SCCA-IgM IC was increased in patients with cirrhosis progressing toward HCC development, and the sensitivity was higher than AFP. Therefore, SCCA-IgM IC may be a useful serum marker for early HCC detection in some cases [30].

6.2.7 Annexin A2

Annexin A2 is upregulated in many tumor types and has multiple roles in tumorigenic processes, proliferation, apoptosis, cell migration, invasion and adhesion processes, which are essential for cancer metastasis [31, 32]. The serum concentration of annexin A2 in HCC was found to be frequently raised compared to healthy controls and individuals with benign liver disease or other malignant tumors [33]. In combination with AFP, it might be an important independent and discriminative serological biomarker for evaluating early stage HCC in patients with normal serum AFP.

6.2.8 Midkine

Midkine (MDK) is a heparin-binding growth factor, firstly identified as retinoic acid responsive gene. It plays a role in cell growth, migration, survival, angiogenesis, and carcinogenesis [34]. MDK levels were found to be higher in cases of HCC compared to cirrhosis or healthy controls [35]. Therefore, MDK can be used in HCC patients to diagnose tumor recurrence [36].

6.2.9 AXL

It is a receptor tyrosine kinase that is involved in the proliferation, chemoresistance and survival of many malignancies. An increased AXL expression has been identified as a poor prognostic factor for recurrence-free survival, as well as overall survival in colon and pancreatic cancer [37]. The diagnostic value in early stage diagnosis of HCC was analyzed in a multicenter study [38]. The sensitivity of AXL was found to be much higher than that of AFP in early stage HCC. AXL and AFP together have been reported to reach an extraordinarily high AUC in detecting early stage HCC, with a sensitivity of 80.8% and specificity of 92.3% [39].

6.2.10 Thioredoxins

Thioredoxins (TRXs) are involved in several biological processes, such as apoptosis and proliferation, regulation of protein states, and protection against oxidative stress [39]. The TRX expression is increased in many neoplasms and shows correlation with prognosis, specifically in colorectal and lung carcinoma [40] and is currently under investigation for detection of early stage HCC.

6.2.11 Nucleic Acids

The technology of microarray is a powerful tool to test nucleic acids for the identification of various clinically relevant molecular biomarkers and brings a new dimension to the diagnosis. Three genes were associated with HCC development, chemokine (C-X-C motif) receptor 2 (CXCR2), C-C chemokine receptor type 2 (CCR2), and E1A-binding protein P400 (EP400). Combined measurements of these three gene markers increased the accuracy in the detection of early stage HCC [12].

6.2.12 MicroRNA

MicroRNAs (miRNAs) are endogenous, small, noncoding RNAs that bind to complementary sequences in 3′-untranslated regions of target mRNAs to induce their degradation. miRNAs have been found to regulate diverse processes in worms, flies, and mammals, including humans [41]. Around 500 miRNA genes have been identified and found to be important components of complex functional pathways controlling important cellular processes, such as proliferation, differentiation, and apoptosis. In human cancer development, miRNAs have been determined to function both as oncogenes and as tumor suppressor genes. Each single type of miRNA is stable and can downregulate hundreds of genes at a time. Such diversity in functional roles enables miRNAs to be used as diagnostic tool for early cancer detection, risk and prognosis assessment, and as new therapeutic targets [42].

miRNAs connected with HCC development have been investigated as biomarkers to diagnose the disease. Several studies have indicated miR 200a and miR 200b, two members of the miR 200 family, as deregulated during the development of both HCC and liver fibrosis [43]. The increased levels of serum miR-21 have been used to characterize cases of HCC from chronic hepatitis and healthy controls. Also serum miR-15b and miR-130b are additional potential miRNA markers that are significantly upregulated in HCC. A panel of seven miRNAs (miR-122, miR-192, miR-21, miR-223, miR-26a, miR-27a, and miR-801) has been shown to have high diagnostic potential in the early diagnosis of HBV-related HCC [44].

In addition to their expression profiles, miRNAs are particularly attractive as potential biomarkers. One point is that miRNAs are highly stable and readily detected in serum and plasma in HCC patients. The second point is that miRNAs arise in the urine, which represents a noninvasive and easily obtainable resource for biomarkers. Five deregulated miRNAs (miR-625, miR-532, miR-618, miR-516-5P, and miR-650) have already been detected in urine and have been used for screening of high-risk patients for early detection of HCC [45]. Further investigations are necessary to establish specific circulating miRNA(s) as reliable biomarkers for detecting HCC at an early stage.

In conclusion, advances in technologies hold great promises for the identification of novel early diagnostic HCC biomarkers. Circulating miRNAs are particularly intriguing as a whole new class of biomarkers and may outperform traditional serum protein markers. Finally, novel biomarkers may provide important clues as to our understanding of hepatic oncogenesis and ultimately lead to establishing better treatment strategies. Simultaneous advancement in many medical disciplines will hopefully change the poor prognosis of HCC patients.

6.3 Animal Models in HCC Research

Due to the increasing incidence of HCC and its high mortality, there is a special need for successful animal models reflecting the human pathogenesis of HCC. Animal models provide the opportunity to mimic the multistep process of hepatocarcinogenesis. Researchers can study tumor-host interactions, perform drug screenings, and implement various therapeutic experiments. Thus, animal models are important for both basic as well as translational studies in HCC research.

However, the different risk factors mentioned above and the heterogeneity of HCC itself make it impossible to generate an ideal model for all purposes. Each animal model recaps only some aspects of hepatocarcinogenesis. Therefore, investigators are forced to select the specific model fitting best to their scientific hypothesis.

In general, there are four different groups of HCC models: xenograft models, models using carcinogens to induce HCC, genetically engineered models, and models of spontaneous tumor formation. Details of specific animal models are given in Tables 6.1, 6.2 and 6.3. There are also excellent reviews available focusing on either xenograft and genetically engineered models [46, 47] models to study tumor-host interactions [48], or HCC metastasis and treatment models [49], respectively.

Table 6.1 Subcutaneous and orthotopic xenograft/allograft models for HCC

Mouse strain	Cell lines	Total volume	Injection	Cell number	References
Subcutaneous xenografts					
Female SCID mice	PLC/PRF/5	200 μL	s.c.	5×10^6	[54]
Balb/c nude mice	SK-Hep	100 μL	s.c.	5×10^6	[55]
Balb/c nude mice	HepG2	100 μL	s.c.	1×10^6	[56]
Female BALB/c nude mice	BEL7404		s.c.	2×10^6	[57]
Balb/c nude mice	QGY-7703		s.c.	1×10^6	[58]
Male BALB/C nude mice	HCCLM6	100 μL	s.c.	3×10^6	[59]
Male NCr athymic mice	Huh7	100 μL	s.c.	1×10^6	[60]
Male athymic nude mice	PLC/PRF/5	100 μL	s.c.	1×10^6	[52]
Male BALB/c nude mice	Huh7	100 μL	s.c.	5×10^6	[61]
Male BALB/c nude mice	Hep3B	100 μL	s.c.	5×10^6	[61]
Female BALB/c nude mice	SMMC-7721		s.c.	1×10^7	[62]
Orthotopic xenografts					
Athymic nude mice	PLC/PRF/5	20 μL	Intrahepatic	1×10^6	[52]
Male BALB/c nude mice	Huh7		Intrahepatic	2×10^6	[61]

(continued)

Table 6.1 (continued)

Mouse strain	Cell lines	Total volume	Injection	Cell number	References
Male BALB/c nude mice	Huh7	50 μL	Intrahepatic	1×10^6	[63]
Male BALB/c nude mice	Hep3B	30–50 μL	Intrahepatic	2×10^6	[64]
Male nude mice	HepG2	25 μL	Intrahepatic	5×10^5	[65]
Male nude mice	SMMC7721	200 μL	Intrasplenic	5×10^6	[53]
Male BALB/c nude mice	QGY-7703	25 μL	Intrahepatic	2×10^6	[66]
Male nude mice	MHCC97L	30 μL	Intrahepatic	2×10^6	[67]
C57BL/6J	Hepa1.6	100 μL	Portal vein	2.5×10^6	[68]
AFP/βGal	HepaβGal	100 μL	Portal vein	2×10^6	[69]
C57BL/6	Hepatocytes from SV40 T-Ag tg MTD2 mice		Splenic vein	5×10^5	[70]
BALB/c	H22	30–50 μL	Intrahepatic	1×10^6	[71]
C3H/He	Hepa129	50 μL	Intrahepatic	1.25×10^5	[72]

Table 6.2 Chemically induced HCC models

Diet/carcinogene	Time to tumor formation (week)	References
Diethylnitrosamine (DEN) ± phenobarbital	20–40	[83–87]
Choline-deficient ethionine-supplemented diet (CDE)	30–35	[75, 88, 89]
Ciprofibrate	60	[84, 85, 90]
Thioacetamide (TAA)	43	[81, 91]
Carbon tetrachloride (CCl_4)	15–20	[79, 87, 92]
Aflatoxin B	52	[93]

6.3.1 Xenograft Models

There are two types of xenograft models for HCC. Subcutaneous xenografts are performed by implanting HCC cells or tissue fragments subcutaneously, usually into the flanks of immunodeficient mice. As a fast model, subcutaneous xenografts are widely used to test new drugs or genes affecting tumor growth. Details of cell lines used can be found in Table 6.1.

Table 6.3 Genetically engineered HCC models

Gene	Details of modification	Promoter	Incidence	Latency	References
Viral models					
Hepatitis B virus large envelope protein	BgIII-A fragment of HBV encoding large envelope protein	Albumin	72% males		[101, 102]
Hepatitis B virus X protein	EcoRI–BgIII fragment of HBV X	HBV X	84%	13–24 months	[103–105]
Hepatitis C virus	HCV core-E1–E2 transgenic and HCV core transgenic	Albumin/HBV X	+DEN: 100%	32 weeks	[96]
Hepatitis C virus	HCV core and HCV E1–E2	HBV X	Core transgenics: 32%	16–23 months	[96, 106]
Hepatitis C virus	HCV core-E1–E2 transgenic and the entire HCV transgenic	Albumin		13 months	[97]
Non-viral models					
Acox1–/–	Fatty acyl-CoA oxidase null (AOX–/–)		100%	15 months	[107]
Alpha-1-antitrypsin (AAT)	AAT Z genomic clones		82%	16–18 months	[108]
APC knockout liver-specific	ApcΔex14 knockout (–/–)		67%	9 months	[109]
β-Catenin exon 3 knockout and activated H-ras (H-rasG12V)	Catnblox(ex3) knockout and H-ras (Tglox(pA)H-ras*) double-transgenic	Human CMV	100%	6 months	[110]
c-myc	c-myc overexpression	Albumin/α1 antitrypsin	54%	15 months	[85, 111–115]
c-myc and E2F-1	Mouse c-myc and human E2F-1 overexpression	Albumin	100%	6–8 months	[85, 111, 116, 117]
c-myc and TGFα	c-myc overexpression; TGFα overexpression	Albumin/ metallothionein 1	18%	4 months	[85, 118]
E2F-1	E2F-1 overexpression	Albumin	33%	12 months	[85, 111, 119]

(continued)

Table 6.3 (continued)

Gene	Details of modification	Promoter	Incidence	Latency	References
EGF transgenic	Alb-DS4 encoding autocrine growth factor IgEGF crossed with AAT-myc mice			4.4–7.1 months	[120]
ELF+/− knockout	ELF+/− knockout mice		40%	15 months	[121, 122]
HGF transgenic	Mouse HGF expression	Metallothionein/ albumin		17 months	[123, 124]
HGF + c-myc	Mouse c-myc and human HGF	Albumin	c-myc: 60% HCC	16 months	[125]
HGF + TGFα	Mouse TGFα overexpression and human HGF	Metallothionein/ albumin	HGF/TGFα: 33%; TGFα:60%		[126]
H-ras	Mutant c-H-ras overexpression	Albumin	Hepatomegaly		[114]
IGF2 transgenic	IGF2 overexpression	Urinary protein	<10%	18–24 months	[127]
IGF2 knockout and TGFα transgenic	TGFα overexpression and IGF2 heterozygous knockout mice	Metallothionein	TGFα/IGF2wt/−: 100%	18 months	[128]
Lymphotoxin β receptor deficient	Lymphotoxin β receptor knockout		18%	12 months	[129]
Mdr-2	Mdr-2 gene knockout			6–12 months	[130, 131]
Met transgenic	Tetracycline-inducible expressing human Met	LAP	60%	12 months	[132, 133]
Met and β-catenin	Human MET, constitutively-activated mutated form of β-catenin (ΔN90-CTNNB1), and dominant-negative TCF-1 (DNHNF1)		74%	1 month	[132]
c-Met conditional knockout	c-Met conditional liver-specific knockout Met (c-metfl/f)	Albumin	100%	6 months	[134]
p53 germline knockout	p53 germline knockout mice expressing viral receptor TVA	Albumin	42%	4 months	[135, 136]
p53 and INK4a/ARF conditional mutant mice	p53 conditional mutant and INK4a/ARF conditional mutant	Albumin	>90%		[137]

p53 conditional expression	Oncogenic ras (Hras V12) and a tet-responsive P53 miRNA design short hairpin RNA				[138]
PDGF-C	Human PDGF-C expression	Albumin	80%	12 months	[139]
PTEN−/−	PTENlox/lox	Albumin	66% HCC	78 weeks	[140, 141]
SV40 T-antigen conditional and inducible expression	SV40 T-antigen expression	Albumin/urinary protein/ metallothionein 1	100%	3–7 months	[113, 114, 142–146]
mTERT−/− and p53+/− or WT	Germline mTERT and p53 knockout		+CCl4: 100%	50 weeks	[147]
TGFα	TGFα overexpression	Metallothionein 1	50–100% HCC	10–15 months	[98, 148, 149]
TGFβ	Porcine TGFβ overexpression	Albumin	+TAA: 100%		[150–152]

In orthotopic xenograft models, tumor cells are intrahepatically injected or implanted (Table 6.1). In contrast to subcutaneous xenografts, orthotopic xenograft tumor models allow to study the effects of tumor microenvironment on tumor development and growth. By using imaging methods, such as magnetic resonance imaging (MRI) [50], positron emission tomography (PET) [51], and bioluminescence or fluorescence imaging (BLI/FLI) [52, 53], primary tumors and metastasis can be monitored.

6.3.2 Chemically Induced HCC Models

Most of the HCC models using carcinogens (Table 6.2) act via a genotoxic mechanism [73]. Diethylnitrosamine (DEN) is the most commonly used carcinogen for modeling hepatocarcinogenesis. It was first used in the 1960s to induce HCC in rodents [74]. DEN is an alkylating agent, which is activated by enzymes of the cytochrome P (CYP) 450 family and leads to mutagenic DNA adducts. If injected into animals not older than 2 weeks, DEN acts as a complete carcinogen since hepatocytes are still proliferating in young mice [75]. Treatment of older mice needs the additional application of a promoting agent like phenobarbital to induce hepatocarcinogenesis [76]. For the DEN model, differences in the susceptibility caused by sex and/or strain differences have been reported [77, 78]. Compared to human hepatocarcinogenesis, DEN-induced HCC development follows a slow multistep sequential process. Histopathological changes from basophilic foci to hyperplastic nodules and adenomas finally leading to HCC occur [78].

The most commonly used compound for fibrotic liver disease is carbon tetrachloride (CCl_4), which can also be combined with DEN to induce cirrhosis-associated HCC. Similar to DEN, CCl_4 is metabolized by CYP450, leading to liver damage by reactive oxygen species and peroxidative degeneration of membrane phospholipids [79]. Other chemicals like thioacetamide or diets, such as the choline-deficient ethionine-supplemented diet (CDE), damage hepatocytes by inducing fibrotic changes and oxidative stress [80–82].

6.3.3 Genetically Engineered HCC Models

The development of transgenic mice and gene targeting approaches opened new possibilities for cancer research. Specific molecular events which happen during tumor development and progression can be studied in detail. Genetic models are important tools to understand interactions among genes and other factors to study HCC susceptibility, development, disease progression, and potential therapies.

Transgenic mice are established to overexpress oncogenes or mutated tumor suppressor genes using an ectopic promoter and enhancer elements [94, 95]. Nowadays, conditional gene targeting models are available enabling temporal control of gene expression, either as transgenic or knockout models. In addition, tissue-specific promoters, such as those for albumin [96, 97], metallothionein [98], transthyretin [99], and liver-activating protein [100], control expression restricted to the liver.

6.3.4 HCC Models of Spontaneous Tumor Formation

As is known for other tumor types, different genetic backgrounds carry a different susceptibility to spontaneous HCC development. For example, C3H and CBA mice are commonly recognized to be more prone, while LP, 129sv, DBA2, BALB/c, and C57BL6 are more resistant to spontaneous HCC development. The fatty liver Shionogi (FLS) mouse is a relatively new inbred strain, which develops steatosis, hepatic inflammation, and HCC mostly in males with an incidence of 40% within 16 months of age [153].

In general, spontaneous HCC models are not widely used due to their unpredictable and low tumor incidence.

Conclusion

Animal models are widely used in HCC research. Still, there is a need for additional models reflecting human hepatocarcinogenesis. Depending on the specific issue, the best fitting model has to be chosen. However, if used appropriately, the existing models display important predictive tools for HCC research.

References

1. International Agency for Research on Cancer. GLOBOCAN 2012: estimated cancer incidence, mortality and prevalence worldwide in 2012. Lyon: Ierational Agency for Research on Cancer; 2013.
2. Kim JH, Sohn BH, Lee HS, Kim SB, Yoo JE, et al. Genomic predictors for recurrence patterns of hepatocellular carcinoma: model derivation and validation. PLoS Med. 2014;11:e1001770.
3. El Serga HB. Hepatocellular carcinoma: recent trends in the United States. Gastroenterology. 2004;127:S27–34.
4. Davila JA, Morgan RO, Shaib Y, McGlynn KA, El Serage HB. Hepatitis C infection and the increasing incidence of hepatocellular carcinoma: a population-based study. Gastroenterology. 2004;127:1372–80.
5. Imamura H, Matsuyama Y, Tanaka E, Ohkuba T, Hasegawa K, et al. Risk factors contributing to early and late phase intrahepatic recurrence of hepatocellular carcinoma after hepatectomy. J Hepatol. 2003;38:200–7.
6. Llovet JM, Fuster J, Bruix J. Intention-to-treat analysis of surgical treatment for early hepatocellular carcinoma: resection versus transplantation. Hepatology. 1999;30:1434–40.
7. Poon RT, Fan ST, Ng IO, Lo CM, Lui CL, et al. Different risk factors and prognosis for early and late intrahepatic recurrence after resection of hepatocellular carcinoma. Cancer. 2000;89:500–7.
8. Mantovani A, Allavena P, Sica A, Balkwill F. Cancer-related inflammation. Nature. 2008;454:436–44.
9. Bosch FX, Ribes J, Cleries R, Diaz M. Epidemiology of hepatocellular carcinoma. Clin Liver Dis. 2005;9:191–211.

10. Sun B, Karin M. Inflammation and liver tumorigenesis. Front Med. 2013;7:242–54.
11. Dvorak HF. Tumors: wounds that do not heal. Similarities between tumor stroma generation and wound healing. N Engl J Med. 1986;315:1650–9.
12. Tsuchiya N, Sawada Y, Endo I, Saito K, Uemura Y, et al. Biomarkers for the early diagnosis of hepatocellular carcinoma. World J Gastroenterol. 2015;21:10573–83.
13. Tatarinov I, Tatarinov S. Detection of embryo-specific alpha-globulin in the blood serum of a patient with primary liver cancer. Vopr Med Khim. 1964;10:90–1.
14. Khattab M, Fouad M, Ahmed E. Role of biomarkers in the prediction and diagnosis of hepatocellular carcinoma. World J Hepatol. 2015;7:2474–81.
15. Hu B, Tian X, Sun J, Meng X. Evaluation of individual and combined applications of serum biomarkers for diagnosis of hepatocellular carcinoma: a meta-analysis. Int J Mol Sci. 2013;14:23559–80.
16. Trevisani F, DÍntino PE, Morselli-Labate AM, Mazzella G, Accogli G, et al. Serum alpha-fetoprotein for diagnosis of hepatocellular carcinoma in patients with chronic liver disease: influence of HBsAg and anti-HCV status. J Hepatol. 2001;34:570–5.
17. Tsuchiya N, Sawada Y, Endo I, Saito K, Uemura Y, et al. Biomarkers for the early diagnosis of hepatocellular carcinoma. World J Gastroenterol. 2015;37:10573–83.
18. Naraki T, Kohno N, Saito H, Fujimoto Y, Ohhira M, et al. Gamma-Carboxyglutamic acid content of hepatocellular carcinoma-associated des-gamma-carboxy prothrombin. Biochim Biophys Acta. 2002;1586:287–98.
19. Libbrecht L, Severi T, Cassiman D, Vander Borght S, Pirenne J, et al. Glypican-3 expression distinguishes small hepatocellular carcinomas from cirrhosis, dysplastic nodules, and focal nodular hyperplasia-like nodules. Am J Surg Pathol. 2006;30:1405–11.
20. Shafizadeh N, Ferrell LD, Kakar S. Utility and limitations of glypican-3 expression for the diagnosis of hepatocellular carcinoma at both ends of the differentiation spectrum. Mod Pathol. 2006;21:1011–8.
21. Motomura Y, Senju S, Nakatsura T, Matsuyoshi H, Hirata S, et al. Embryonic stem cell-derived dendritic cells expressing glypican-3, a recently identified oncofetal antigen, induce protective immunity against highly metastatic mouse melanoma. Cancer Res. 2006;66:2414–22.
22. Shevede LA, Das S, Clark DW, Samant RS. Osteopontin: an effector and an effect of tumor metastasis. Curr Mol Med. 2010;10:71–81.
23. Kawashima R, Mochida S, Matsui A, YouLu TuZ Y, Ishikawa K, et al. Expression of osteopontin in Kupffer cells and hepatic macrophages and Stellate cells in rat liver after carbon tetrachloride intoxication: a possible factor for macrophage migration into hepatic necrotic areas. Biochem Biophys Res Commun. 1999;256:527–31.
24. Shand S, Plymoth A, Ge S, Feng Z, Rosen HR, et al. Identification of osteopontin as a novel marker for early hepatocellular carcinoma. Hepatology. 2012;55:483–90.
25. Kladney RD, Cui X, Bulla GA, Brunt EM, Fimmel CJ, et al. Expression of GP73, a resident Golgi membrane protein, in viral and nonviral liver disease. Hepatology. 2002;35:1431–40.
26. Mao Y, Yang H, Xu H, Lu X, Sang X, et al. Golgi protein 73 (GOLPH2) is a valuable serum marker for hepatocellular carcinoma. Gut. 2010;59:1687–93.
27. Hu JS, Wu DW, Liang S, Miao XY. GP73, a resident Golgi glycoprotein, is sensibility and specificity for hepatocellular carcinoma of diagnosis in a hepatitis B-endemic Asian population. Med Oncol. 2010;27:339–45.
28. Hussein MM, Ibrahim AA, Abdella HM, Montasser IF, Hassan MI. Evaluation of serum squamous cell carcinoma antigen as a novel biomarker for diagnosis of hepatocellular carcinoma in Egyptian patients. Indian J Cancer. 2008;45:167–72.
29. Giannelli G, MArinosci F, Trerotoli P, Volpe A, Quaranta M, et al. SCCA antigen combined with alpha-fetoprotein as serologic marker of HCC. Int J Cancer. 2005;117:506–9.
30. Pontisso P, Quarta S, Caberlotto C, Beneduce L, Marino M, et al. Progressive increase of SCCA-IgM immune complexes in cirrhotic patients is associated with development of hepatocellular carcinoma. Int J Cancer. 2006;119:735–40.

31. Sharma MC, Koltowski L, Ownbey RT, Tuszynski P, Sharam MC. Angiogenesis-associated protein annexin II in breast cancer: selective expression in invasive breast cancer and contribution to tumor invasion and progression. Exp Mol Pathol. 2006;81:146–56.
32. Shiozawa Y, Havens AM, Jung Y, Ziegler AM, Pedersen EA, et al. Annexin II/annexin II receptor axis regulates adhesion, migration, homing, and growth of prostate cancer. J Cell Biochem. 2008;105:370–80.
33. Hollas H, Aukrust I, Grimmer S, Strand E, Flatmark T, et al. Annexin A2 recognises a specific region in the 3′-UTR of its cognate messenger RNA. Biochim Biophys Acta. 2006;1763:1325–34.
34. Muramatsu T. Midkine and pleiotrophin: two related proteins involved in development, survival, inflammation and tumorigenesis. J Biochem. 2002;132:359–71.
35. Shaheen KY, Abdel-Mageed AI, Safwat E, AlBreedy AM. The value of serum midkine level in diagnosis of hepatocellular carcinoma. Int J Hepatol. 2015;2015:146389.
36. Zhu WW, Guo L, Guo JJ, Jia HL, Zhu M, et al. Evaluation of midkine as a diagnostic serum biomarker in hepatocellular carcinoma. Clin Cancer Res. 2013;19:3944–54.
37. Rankin EB, Fuh KC, Taylor TE, Krieg AJ, Musser M, et al. AXL is an essential factor and therapeutic target for metastatic ovarian cancer. Cancer Res. 2010;70:7570–9.
38. Reichl P, Fang M, Starlinger P, Staufer K, Nenutil R, et al. Multicenter analysis of soluble Axl reveals diagnostic value for very early stage hepatocellular carcinoma. Int J Cancer. 2015;137:385–94.
39. Nordberg J, Arnér ES. Reactive oxygen species, antioxidants, and the mammalian thioredoxin system. Free Radic Biol Med. 2001;31:1287–312.
40. Kakolyris S, Giatromanolaki A, Koukourakis M, Powis G, Souglakos J, et al. Thioredoxin expression is associated with lymph node status and prognosis in early operable non-small cell lung cancer. Clin Cancer Res. 2001;7:3087–91.
41. Ferracin M, Veronese A, Negrini M. Micromarkers: miRNA in cancer diagnosis and prognosis. Expert Rev Mol Diagn. 2010;10:297–308.
42. Schütte K, Schulz C, Link A, Malfertheiner P. Current biomarkers for hepatocellular carcinoma: surveillance, diagnosis and prediction of prognosis. World J Hepatol. 2015;7:139–49.
43. Murakami Y, Yasuda T, Saigo K, Urashima T, Toyoda H, et al. Comprehensive analysis of microRNA expression patterns in hepatocellular carcinoma and non-tumorous tissues. Oncogene. 2006;25:2537–45.
44. Zhou J, Yu L, Gao X, Hu J, Wang J, et al. Plasma microRNA panel to diagnose hepatitis B virus-related hepatocellular carcinoma. J Clin Oncol. 2011;29:4781–8.
45. Abdalla MA, Haj-Ahmad Y. Promising candidate urinary microRNA Biomarkers for early detection of hepatocellular carcinoma among high-risk hepatitis C virus Egyptian patients. J Cancer. 2012;3:19–31.
46. He L, Tian DA, Li PY, He XX. Mouse models of liver cancer: progress and recommendations. Oncotarget. 2015;6(27):23306–22. doi:10.18632/oncotarget.4202.
47. Newell P, Villanueva A, Friedman SL, Koike K, Llovet JM. Experimental models of hepatocellular carcinoma. J Hepatol. 2008;48(5):858–79. doi:10.1016/j.jhep.2008.01.008.
48. Bakiri L, Wagner EF. Mouse models for liver cancer. Mol Oncol. 2013;7(2):206–23. doi:10.1016/j.molonc.2013.01.005.
49. Li Y, Tang Z-Y, Hou J-X. Hepatocellular carcinoma: insight from animal models. Nat Rev Gastroenterol Hepatol. 2012;9(1):32–43.
50. Kim SM, Im GH, Lee DG, Lee JH, Lee WJ, Lee IS. Mn(2+)-doped silica nanoparticles for hepatocyte-targeted detection of liver cancer in T1-weighted MRI. Biomaterials. 2013;34(35):8941–8. doi:10.1016/j.biomaterials.2013.08.009.
51. Baek S, Mueller A, Lim Y-S, Lee HC, Lee Y-J, Gong G, Kim JS, Ryu J-S, Oh SJ, Lee SJ, Bacher-Stier C, Fels L, Koglin N, Schatz CA, Dinkelborg LM, Moon DH. (4S)-4-(3-18F-fluoropropyl)-l-glutamate for imaging of xC transporter activity in hepatocellular carcinoma using PET: preclinical and exploratory clinical studies. J Nucl Med. 2013;54(1):117–23. doi:10.2967/jnumed.112.108704.

52. Matouk IJ, DeGroot N, Mezan S, Ayesh S, Abu-Lail R, Hochberg A, Galun E. The H19 non-coding RNA is essential for human tumor growth. PLoS One. 2007;2(9)
53. Zheng T, Yin D, Lu Z, Wang J, Li Y, Chen X, Liang Y, Song X, Qi S, Sun B, Xie C, Meng X, Pan S, Liu J, Jiang H, Liu L. Nutlin-3 overcomes arsenic trioxide resistance and tumor metastasis mediated by mutant p53 in Hepatocellular Carcinoma. Mol Cancer. 2014;13:133. doi:10.1186/1476-4598-13-133.
54. Liu L, Cao Y, Chen C, Zhang X, McNabola A, Wilkie D, Wilhelm S, Lynch M, Carter C. Sorafenib blocks the RAF/MEK/ERK pathway, inhibits tumor angiogenesis, and induces tumor cell apoptosis in hepatocellular carcinoma model PLC/PRF/5. Cancer Res. 2006;66(24):11851–8. doi:10.1158/0008-5472.CAN-06-1377.
55. Fuchs BC, Fujii T, Dorfman JD, Goodwin JM, Zhu AX, Lanuti M, Tanabe KK. Epithelial-to-mesenchymal transition and integrin-linked kinase mediate sensitivity to epidermal growth factor receptor inhibition in human hepatoma cells. Cancer Res. 2008;68(7):2391–9. doi:10.1158/0008-5472.can-07-2460.
56. Li Y, Yu DC, Chen Y, Amin P, Zhang H, Nguyen N, Henderson DR. A hepatocellular carcinoma-specific adenovirus variant, CV890, eliminates distant human liver tumors in combination with doxorubicin. Cancer Res. 2001;61(17):6428–36.
57. Pei Z, Chu L, Zou W, Zhang Z, Qiu S, Qi R, Gu J, Qian C, Liu X. An oncolytic adenoviral vector of Smac increases antitumor activity of TRAIL against HCC in human cells and in mice. Hepatology. 2004;39(5):1371–81. doi:10.1002/hep.20203.
58. Yoo BK, Emdad L, Su ZZ, Villanueva A, Chiang DY, Mukhopadhyay ND, Mills AS, Waxman S, Fisher RA, Llovet JM, Fisher PB, Sarkar D. Astrocyte elevated gene-1 regulates hepatocellular carcinoma development and progression. J Clin Invest. 2009;119(3):465–77. doi:10.1172/JCI36460.
59. Chen L-D, Liu J, Yu X-F, He M, Pei X-F, Tang Z-Y, Wang Q-Q, Pang D-W, Li Y. The biocompatibility of quantum dot probes used for the targeted imaging of hepatocellular carcinoma metastasis. Biomaterials. 2008;29(31):4170–6.
60. doi:http://dx.doi.org/10.1016/j.biomaterials.2008.07.025
61. Chen K-F, Yeh P-Y, Yeh K-H, Lu Y-S, Huang S-Y, Cheng A-L. Down-regulation of phospho-Akt is a major molecular determinant of bortezomib-induced apoptosis in hepatocellular carcinoma cells. Cancer Res. 2008;68(16):6698–707. doi:10.1158/0008-5472.can-08-0257.
62. Hu J, Dong A, Fernandez-Ruiz V, Shan J, Kawa M, Martinez-Anso E, Prieto J, Qian C. Blockade of Wnt signaling inhibits angiogenesis and tumor growth in hepatocellular carcinoma. Cancer Res. 2009;69(17):6951–9. doi:10.1158/0008-5472.CAN-09-0541.
63. Zhu LM, Shi DM, Dai Q, Cheng XJ, Yao WY, Sun PH, Ding Y, Qiao MM, Wu YL, Jiang SH, Tu SP. Tumor suppressor XAF1 induces apoptosis, inhibits angiogenesis and inhibits tumor growth in hepatocellular carcinoma. Oncotarget. 2014;5(14):5403–15.
64. Peng H, Dara L, Li TW, Zheng Y, Yang H, Tomasi ML, Tomasi I, Giordano P, Mato JM, Lu SC. MAT2B-GIT1 interplay activates MEK1/ERK 1 and 2 to induce growth in human liver and colon cancer. Hepatology. 2013;57(6):2299–313. doi:10.1002/hep.26258.
65. Yao X, Hu JF, Daniels M, Yien H, Lu H, Sharan H, Zhou X, Zeng Z, Li T, Yang Y, Hoffman AR. A novel orthotopic tumor model to study growth factors and oncogenes in hepatocarcinogenesis. Clin Cancer Res. 2003;9(7):2719–26.
66. Yang XZ, Dou S, Sun TM, Mao CQ, Wang HX, Wang J. Systemic delivery of siRNA with cationic lipid assisted PEG-PLA nanoparticles for cancer therapy. J Control Release. 2011;156(2):203–11. doi:10.1016/j.jconrel.2011.07.035.
67. Wang R, Zhao N, Li S, Fang JH, Chen MX, Yang J, Jia WH, Yuan Y, Zhuang SM. MicroRNA-195 suppresses angiogenesis and metastasis of hepatocellular carcinoma by inhibiting the expression of VEGF, VAV2, and CDC42. Hepatology. 2013;58(2):642–53. doi:10.1002/hep.26373.
68. Yan XL, Jia YL, Chen L, Zeng Q, Zhou JN, Fu CJ, Chen HX, Yuan HF, Li ZW, Shi L, Xu YC, Wang JX, Zhang XM, He LJ, Zhai C, Yue W, Pei XT. Hepatocellular carcinoma-associated

mesenchymal stem cells promote hepatocarcinoma progression: role of the S100A4-miR155-SOCS1-MMP9 axis. Hepatology. 2013;57(6):2274–86. doi:10.1002/hep.26257.
69. Gauttier V, Judor JP, Le Guen V, Cany J, Ferry N, Conchon S. Agonistic anti-CD137 antibody treatment leads to antitumor response in mice with liver cancer. Int J Cancer. 2014;135(12):2857–67. doi:10.1002/ijc.28943.
70. McIlroy D, Barteau B, Cany J, Richard P, Gourden C, Conchon S, Pitard B. DNA/amphiphilic block copolymer nanospheres promote low-dose DNA vaccination. Mol Ther. 2009;17(8):1473–81. doi:10.1038/mt.2009.84.
71. Avella DM, Li G, Schell TD, Liu D, Zhang SSM, Lou X, Berg A, Kimchi ET, Tagaram HRS, Yang Q, Shereef S, Garcia LS, Kester M, Isom HC, Rountree CB, Staveley-O'Carroll KF. Regression of established hepatocellular carcinoma is induced by chemoimmunotherapy in an orthotopic murine model. Hepatology. 2012;55(1):141–52. doi:10.1002/hep.24652.
72. Zhao W, Zhang L, Xu Y, Zhang Z, Ren G, Tang K, Kuang P, Zhao B, Yin Z, Wang X. Hepatic stellate cells promote tumor progression by enhancement of immunosuppressive cells in an orthotopic liver tumor mouse model. Lab Invest. 2014;94(2):182–91. doi:10.1038/labinvest.2013.139.
73. Kornek M, Lukacs-Kornek V, Limmer A, Raskopf E, Becker U, Klockner M, Sauerbruch T, Schmitz V. 1,2-dioleoyl-3-trimethylammonium-propane (DOTAP)-formulated, immune-stimulatory vascular endothelial growth factor a small interfering RNA (siRNA) increases antitumoral efficacy in murine orthotopic hepatocellular carcinoma with liver fibrosis. Mol Med. 2008;14(7–8):365–73. doi:10.2119/2008-00003.Kornek.
74. Pitot HC, Dragan YP. Facts and theories concerning the mechanisms of carcinogenesis. FASEB J. 1991;5(9):2280–6.
75. Rajewsky MF, Dauber W, Frankenberg H. Liver carcinogenesis by diethylnitrosamine in the rat. Science. 1966;152(3718):83–5.
76. Solt DB, Medline A, Farber E. Rapid emergence of carcinogen-induced hyperplastic lesions in a new model for the sequential analysis of liver carcinogenesis. Am J Pathol. 1977;88(3):595–618.
77. Farber E, Sarma DS. Hepatocarcinogenesis: a dynamic cellular perspective. Lab Invest. 1987;56(1):4–22.
78. Diwan BA, Rice JM, Ohshima M, Ward JM. Interstrain differences in susceptibility to liver carcinogenesis initiated by N-nitrosodiethylamine and its promotion by phenobarbital in C57BL/6NCr, C3H/HeNCrMTV- and DBA/2NCr mice. Carcinogenesis. 1986;7(2):215–20.
79. Vesselinovitch SD, Mihailovich N. Kinetics of diethylnitrosamine hepatocarcinogenesis in the infant mouse. Cancer Res. 1983;43(9):4253–9.
80. McCay PB, Lai EK, Poyer JL, DuBose CM, Janzen EG. Oxygen- and carbon-centered free radical formation during carbon tetrachloride metabolism. Observation of lipid radicals in vivo and in vitro. J Biol Chem. 1984;259(4):2135–43.
81. Ghoshal AK, Ahluwalia M, Farber E. The rapid induction of liver cell death in rats fed a choline-deficient methionine-low diet. Am J Pathol. 1983;113(3):309–14.
82. Li X, Benjamin IS, Alexander B. Reproducible production of thioacetamide-induced macronodular cirrhosis in the rat with no mortality. J Hepatol. 2002;36(4):488–93.
83. Rushmore TH, Ghazarian DM, Subrahmanyan V, Farber E, Ghoshal AK. Probable free radical effects on rat liver nuclei during early hepatocarcinogenesis with a choline-devoid low methionine diet. Cancer Res. 1987;47(24 Pt 1):6731–40.
84. Coulouarn C, Gomez-Quiroz LE, Lee J-S, Kaposi-Novak P, Conner EA, Goldina TA, Onishchenko GE, Factor VM, Thorgeirsson SS. Oncogene-specific gene expression signatures at preneoplastic stage in mice define distinct mechanisms of hepatocarcinogenesis. Hepatology. 2006;44(4):1003–11. doi:10.1002/hep.21293.
85. Groos J, Bannasch P, Schwarz M, Kopp-Schneider A. Comparison of mode of action of four hepatocarcinogens: a model-based approach. Toxicol Sci. 2007;99(2):446–54. doi:10.1093/toxsci/kfm183.

86. Lee JS, Chu IS, Mikaelyan A, Calvisi DF, Heo J, Reddy JK, Thorgeirsson SS. Application of comparative functional genomics to identify best-fit mouse models to study human cancer. Nat Genet. 2004;36(12):1306–11. doi:10.1038/ng1481.
87. Poirier LA. Hepatocarcinogenesis by diethylnitrosamine in rats fed high dietary levels of lipotropes. J Natl Cancer Inst. 1975;54(1):137–40.
88. Tamano S, Merlino GT, Ward JM. Rapid development of hepatic tumors in transforming growth factor alpha transgenic mice associated with increased cell proliferation in precancerous hepatocellular lesions initiated by N-nitrosodiethylamine and promoted by phenobarbital. Carcinogenesis. 1994;15(9):1791–8.
89. Yaswen P, Goyette M, Shank PR, Fausto N. Expression of c-Ki-ras, c-Ha-ras, and c-myc in specific cell types during hepatocarcinogenesis. Mol Cell Biol. 1985;5(4):780–6.
90. Nagy P, Evarts RP, Marsden E, Roach J, Thorgeirsson SS. Cellular distribution of c-myc transcripts during chemical hepatocarcinogenesis in rats. Cancer Res. 1988;48(19):5522–7.
91. Rao MS, Lalwani ND, Watanabe TK, Reddy JK. Inhibitory effect of antioxidants ethoxyquin and 2(3)-tert-butyl-4-hydroxyanisole on hepatic tumorigenesis in rats fed ciprofibrate, a peroxisome proliferator. Cancer Res. 1984;44(3):1072–6.
92. Sakurai T, Kudo M, Umemura A, He G, Elsharkawy AM, Seki E, Karin M. p38α inhibits liver fibrogenesis and consequent hepatocarcinogenesis by curtailing accumulation of reactive oxygen species. Cancer Res. 2013;73(1):215–24. doi:10.1158/0008-5472.can-12-1602.
93. Fujii T, Fuchs BC, Yamada S, Lauwers GY, Kulu Y, Goodwin JM, Lanuti M, Tanabe KK. Mouse model of carbon tetrachloride induced liver fibrosis: histopathological changes and expression of CD133 and epidermal growth factor. BMC Gastroenterol. 2010;10(1):1–11. doi:10.1186/1471-230x-10-79.
94. Ghebranious N, Sell S. The mouse equivalent of the human p53ser249 mutation p53ser246 enhances aflatoxin hepatocarcinogenesis in hepatitis B surface antigen transgenic and p53 heterozygous null mice. Hepatology. 1998;27(4):967–73. doi:10.1002/hep.510270411.
95. Tuveson DA, Jacks T. Technologically advanced cancer modeling in mice. Curr Opin Genet Dev. 2002;12(1):105–10. doi:10.1016/S0959-437X(01)00272-6.
96. Frese KK, Tuveson DA. Maximizing mouse cancer models. Nat Rev Cancer. 2007;7(9):645–58. doi:10.1038/nrc2192.
97. Kamegaya Y, Hiasa Y, Zukerberg L, Fowler N, Blackard JT, Lin W, Choe WH, Schmidt EV, Chung RT. Hepatitis C virus acts as a tumor accelerator by blocking apoptosis in a mouse model of hepatocarcinogenesis. Hepatology. 2005;41(3):660–7. doi:10.1002/hep.20621.
98. Lerat H, Honda M, Beard MR, Loesch K, Sun J, Yang Y, Okuda M, Gosert R, Xiao S, Weinman SA, Lemon SM. Steatosis and liver cancer in transgenic mice expressing the structural and nonstructural proteins of hepatitis C virus. Gastroenterology. 2002;122(2):352–65.
99. Jhappan C, Stahle C, Harkins RN, Fausto N, Smith GH, Merlino GT. TGFα overexpression in transgenic mice induces liver neoplasia and abnormal development of the mammary gland and pancreas. Cell. 1990;61(6):1137–46. doi:10.1016/0092-8674(90)90076-Q.
100. Lou DQ, Molina T, Bennoun M, Porteu A, Briand P, Joulin V, Vasseur-Cognet M, Cavard C. Conditional hepatocarcinogenesis in mice expressing SV 40 early sequences. Cancer Lett. 2005;229(1):107–14. doi:10.1016/j.canlet.2004.12.032.
101. Shachaf CM, Kopelman AM, Arvanitis C, Karlsson Å, Beer S, Mandl S, Bachmann MH, Borowsky AD, Ruebner B, Cardiff RD, Yang Q, Bishop JM, Contag CH, Felsher DW. MYC inactivation uncovers pluripotent differentiation and tumour dormancy in hepatocellular cancer. Nature. 2004;431(7012):1112–7. doi:10.1038/nature03043.
102. Chisari FV, Filippi P, Buras J, McLachlan A, Popper H, Pinkert CA, Palmiter RD, Brinster RL. Structural and pathological effects of synthesis of hepatitis B virus large envelope polypeptide in transgenic mice. Proc Natl Acad Sci U S A. 1987;84(19):6909–13.
103. Chisari FV, Klopchin K, Moriyama T, Pasquinelli C, Dunsford HA, Sell S, Pinkert CA, Brinster RL, Palmiter RD. Molecular pathogenesis of hepatocellular carcinoma in hepatitis B virus transgenic mice. Cell. 1989;59(6):1145.

104. Chisari FV, Pinkert CA, Milich DR, Filippi P, McLachlan A, Palmiter RD, Brinster RL. A transgenic mouse model of the chronic hepatitis B surface antigen carrier state. Science. 1985;230(4730):1157–60.
105. Koike K, Moriya K, Iino S, Yotsuyanagi H, Endo Y, Miyamura T, Kurokawa K. High-level expression of hepatitis B virus HBx gene and hepatocarcinogenesis in transgenic mice. Hepatology. 1994;19(4):810–9.
106. Toshkov I, Chisari FV, Bannasch P. Hepatic preneoplasia in hepatitis B virus transgenic mice. Hepatology. 1994;20(5):1162–72. doi:10.1016/0270-9139(94)90752-8.
107. Koike K, Moriya K, Kimura S. Role of hepatitis C virus in the development of hepatocellular carcinoma: transgenic approach to viral hepatocarcinogenesis. J Gastroenterol Hepatol. 2002;17(4):394–400. doi:10.1046/j.1440-1746.2002.02763.x.
108. Fan CY, Pan J, Usuda N, Yeldandi AV, Rao MS, Reddy JK. Steatohepatitis, spontaneous peroxisome proliferation and liver tumors in mice lacking peroxisomal fatty acyl-Coa oxidase: implications for peroxisome proliferator-activated receptor α natural ligand metabolism. J Biol Chem. 1998;273(25):15639–45. doi:10.1074/jbc.273.25.15639.
109. Geller SA, Nichols WS, Kim S, Tolmachoff T, Lee S, Dycaico MJ, Felts K, Sorge JA. Hepatocarcinogenesis is the sequel to hepatitis in Z#2 α1-antitrypsin transgenic mice: histopathological and DNA ploidy studies. Hepatology. 1994;19(2):389–97.
110. Colnot S, Decaens T, Niwa-Kawakita M, Godard C, Hamard G, Kahn A, Giovannini M, Perret C. Liver-targeted disruption of Apc in mice activates β-catenin signaling and leads to hepatocellular carcinomas. Proc Natl Acad Sci U S A. 2004;101(49):17216–21. doi:10.1073/pnas.0404761101.
111. Harada N, Oshima H, Katoh M, Tamai Y, Oshima M, Taketo MM. Hepatocarcinogenesis in mice with β-catenin and Ha-ras gene mutations. Cancer Res. 2004;64(1):48–54. doi:10.1158/0008-5472.CAN-03-2123.
112. Conner EA, Lemmer ER, Sánchez A, Factor VM, Thorgeirsson SS. E2F1 blocks and c-Myc accelerates hepatic ploidy in transgenic mouse models. Biochem Biophys Res Commun. 2003;302(1):114–20. doi:10.1016/S0006-291X(03)00125-6.
113. Dalemans W, Perraud F, Le Meur M, Gerlinger P, Courthey M, Pavirani A. Heterologous protein expression by transimmortalized differentiated liver cell lines derived from transgenic mice (hepatomas/α1 antitrypsin/ONC mouse). Biologicals. 1990;18(3):191–8. doi:10.1016/1045-1056(90)90006-L.
114. Perraud F, Dalemans W, Gendrault JL, Dreyer D, Ali-Hadji D, Faure T, Pavirani A. Characterization of trans-immortalized hepatic cell lines established from transgenic mice. Exp Cell Res. 1991;195(1):59–65. doi:10.1016/0014-4827(91)90500-T.
115. Sandgren EP, Quaife CJ, Pinkert CA, Palmiter RD, Brinster RL. Oncogene-induced liver neoplasia in transgenic mice. Oncogene. 1989;4(6):715–24.
116. Santoni-Rugiu E, Nagy P, Jensen MR, Factor VM, Thorgeirsson SS. Evolution of neoplastic development in the liver of transgenic mice co- expressing c-myc and transforming growth factor-α. Am J Pathol. 1996;149(2):407–28.
117. Santoni-Rugiu E, Jensen MR, Thorgeirsson SS. Disruption of the pRb/E2F pathway and inhibition of apoptosis are major oncogenic events in liver constitutively expressing c-myc and transforming growth factor α. Cancer Res. 1998;58(1):123–34.
118. Murakami H, Sanderson ND, Nagy P, Marino PA, Merlino G, Thorgeirsson SS. Transgenic mouse model for synergistic effects of nuclear oncogenes and growth factors in tumorigenesis: interaction of c-myc and transforming growth factor α in hepatic oncogenesis. Cancer Res. 1993;53(8):1719–23.
119. Conner EA, Lemmer ER, Omori M, Wirth PJ, Factor VM, Thorgeirsson SS. Dual functions of E2F-1 in a transgenic mouse model of liver carcinogenesis. Oncogene. 2000;19(44):5054–62.
120. Tonjes RR, Lohler J, O'Sullivan JF, Kay GF, Schmidt GH, Dalemans W, Pavirani A, Paul D. Autocrine mitogen IgEGF cooperates with c-myc or with the Hcs locus during hepatocarcinogenesis in transgenic mice. Oncogene. 1995;10(4):765–8.

121. Kitisin K, Ganesan N, Tang Y, Jogunoori W, Volpe EA, Kim SS, Katuri V, Kallakury B, Pishvaian M, Albanese C, Mendelson J, Zasloff M, Rashid A, Fishbein T, Evans SRT, Sidawy A, Reddy EP, Mishra B, Johnson LB, Shetty K, Mishra L. Disruption of transforming growth factor-β signaling through β-spectrin ELF leads to hepatocellular cancer through cyclin D1 activation. Oncogene. 2007;26(50):7103–10. doi:10.1038/sj.onc.1210513.
122. Tang Y, Katuri V, Dillner A, Mishra B, Deng CX, Mishra L. Disruption of transforming growth factor-β signaling in ELF β-spectrin-deficient mice. Science. 2003;299(5606):574–7. doi:10.1126/science.1075994.
123. Sakata H, Takayama H, Sharp R, Rubin JS, Merlino G, LaRochelle WJ. Hepatocyte growth factor/scatter factor overexpression induces growth, abnormal development, and tumor formation in transgenic mouse livers. Cell Growth Differ. 1996;7(11):1513–23.
124. Shiota G, Wang TC, Nakamura T, Schmidt EV. Hepatocyte growth factor in transgenic mice: effects on hepatocyte growth, liver regeneration and gene expression. Hepatology. 1994;19(4):962–72.
125. Santoni-Rugiu E, Preisegger KH, Kiss A, Audolfsson T, Shiota G, Schmidt EV, Thorgeirsson SS. Inhibition of neoplastic development in the liver by hepatocyte growth factor in a transgenic mouse model. Proc Natl Acad Sci U S A. 1996;93(18):9577–82. doi:10.1073/pnas.93.18.9577.
126. Shiota G, Kawasaki H, Nakamura T, Schmidt EV. Characterization of double transgenic mice expressing hepatocyte growth factor and transforming growth factor α. Res Commun Mol Pathol Pharmacol. 1995;90(1):17–24.
127. Rogler CE, Yang D, Rossetti L, Donohoe J, Alt E, Chang CJ, Rosenfeld R, Neely K, Hintz R. Altered body composition and increased frequency of diverse malignancies in insulin-like growth factor-II transgenic mice. J Biol Chem. 1994;269(19):13779–84.
128. Harris TM, Rogler LE, Rogler CE. Reactivation of the maternally imprinted IGF2 allele in TGFα induced hepatocellular carcinomas in mice. Oncogene. 1998;16(2):203–9.
129. Haybaeck J, Zeller N, Wolf MJ, Weber A, Wagner U, Kurrer MO, Bremer J, Iezzi G, Graf R, Clavien P-A, Thimme R, Blum H, Nedospasov SA, Zatloukal K, Ramzan M, Ciesek S, Pietschmann T, Marche PN, Karin M, Kopf M, Browning JL, Aguzzi A, Heikenwalder M. A lymphotoxin-driven pathway to hepatocellular carcinoma. Cancer Cell. 2009;16(4):295.
130. Katzenellenbogen M, Pappo O, Barash H, Klopstock N, Mizrahi L, Olam D, Jacob-Hirsch J, Amariglio N, Rechavi G, Mitchell LA, Kohen R, Domany E, Galun E, Goldenberg D. Multiple adaptive mechanisms to chronic liver disease revealed at early stages of liver carcinogenesis in the Mdr2-knockout mice. Cancer Res. 2006;66(8):4001–10. doi:10.1158/0008-5472.CAN-05-2937.
131. Mauad TH, CMJ VN, Dingemans KP, JJM S, Schinkel AH, RGE N, Van Den Bergh Weerman MA, Verkruisen RP, Groen AK, RPJ OE, Van Der Valk MA, Borst P, GJA O. Mice with homozygous disruption of the mdr2 P-glycoprotein gene a novel animal model for studies of nonsuppurative inflammatory cholangitis and hepatocarcinogenesis. Am J Pathol. 1994;145(5):1237–45.
132. Tward AD, Jones KD, Yant S, Siu TC, Sheung TF, Chen X, Kay MA, Wang R, Bishop JM. Distinct pathways of genomic progression to benign and malignant tumors of the liver. Proc Natl Acad Sci U S A. 2007;104(37):14771–6. doi:10.1073/pnas.0706578104.
133. Wang R, Ferrell LD, Faouzi S, Maher JJ, Michael Bishop J. Activation of the Met receptor by cell attachment induces and sustains hepatocellular carcinomas in transgenic mice. J Cell Biol. 2001;153(5):1023–33. doi:10.1083/jcb.153.5.1023.
134. Takami T, Kaposi-Novak P, Uchida K, Gomez-Quiroz LE, Conner EA, Factor VM, Thorgeirsson SS. Loss of hepatocyte growth factor/c-Met signaling pathway accelerates early stages of N-nitrosodiethylamine-induced hepatocarcinogenesis. Cancer Res. 2007;67(20):9844–51. doi:10.1158/0008-5472.CAN-07-1905.
135. Jacks T, Remington L, Williams BO, Schmitt EM, Halachmi S, Bronson RT, Weinberg RA. Tumor spectrum analysis in p53-mutant mice. Curr Biol. 1994;4(1):1–7. doi:10.1016/S0960-9822(00)00002-6.

136. Lewis BC, Klimstra DS, Socci ND, Xu S, Koutcher JA, Varmus HE. The absence of p53 promotes metastasis in a novel somatic mouse model for hepatocellular carcinoma. Mol Cell Biol. 2005;25(4):1228–37. doi:10.1128/MCB.25.4.1228-1237.2005.
137. Chen YW, Klimstra DS, Mongeau ME, Tatem JL, Boyartchuk V, Lewis BC. Loss of p53 and Ink4a/Arf cooperate in a cell autonomous fashion to induce metastasis of hepatocellular carcinoma cells. Cancer Res. 2007;67(16):7589–96. doi:10.1158/0008-5472.CAN-07-0381.
138. Xue W, Zender L, Miething C, Dickins RA, Hernando E, Krizhanovsky V, Cordon-Cardo C, Lowe SW. Senescence and tumour clearance is triggered by p53 restoration in murine liver carcinomas. Nature. 2007;445(7128):656–60. doi:10.1038/nature05529.
139. Campbell JS, Hughes SD, Gilbertson DG, Palmer TE, Holdren MS, Haran AC, Odell MM, Bauer RL, Ren HP, Haugen HS, Yeh MM, Fausto N. Platelet-derived growth factor C induces liver fibrosis, steatosis, and hepatocellular carcinoma. Proc Natl Acad Sci U S A. 2005;102(9):3389–94. doi:10.1073/pnas.0409722102.
140. Horie Y, Suzuki A, Kataoka E, Sasaki T, Hamada K, Sasaki J, Mizuno K, Hasegawa G, Kishimoto H, Iizuka M, Naito M, Enomoto K, Watanabe S, Mak TW, Nakano T. Hepatocyte-specific Pten deficiency results in steatohepatitis and hepatocellular carcinomas. J Clin Invest. 2004;113(12):1774.
141. Watanabe S, Horie Y, Kataoka E, Sato W, Dohmen T, Ohshima S, Goto T, Suzuki A. Non-alcoholic steatohepatitis and hepatocellular carcinoma: lessons from hepatocyte-specific phosphatase and tensin homolog (PTEN)-deficient mice. J Gastroenterol Hepatol. 2007;22(Suppl. 1):S96–S100. doi:10.1111/j.1440-1746.2006.04665.x.
142. Dubois N, Bennoun M, Allemand I, Molina T, Grimber G, Daudet-Monsac M, Abelanet R, Briand P. Time-course development of differentiated hepatocarcinoma and lung metastasis in transgenic mice. J Hepatol. 1991;13(2):227–39. doi:10.1016/0168-8278(91)90819-W.
143. Manickan E, Satoi J, Wang TC, Liang TJ. Conditional liver-specific expression of simian virus 40 T antigen leads to regulatable development of hepatic neoplasm in transgenic mice. J Biol Chem. 2001;276(17):13989–94.
144. Messing A, Chen HY, Palmiter RD, Brinster RL. Peripheral neuropathies, hepatocellular carcinomas and islet cell adenomas in transgenic mice. Nature. 1985;316(6027):461–3. doi:10.1038/316461a0.
145. Schirmacher P, Held WA, Yang D, Biempica L, Rogler CE. Selective amplification of periportal transitional cells precedes formation of hepatocellular carcinoma in SV40 large tag transgenic mice. Am J Pathol. 1991;139(1):231–41.
146. Sepulveda AR, Finegold MJ, Smith B, Slagle BL, DeMayo JL, Shen RF, Woo SLC, Butel JS. Development of a transgenic mouse system for the analysis of stages in liver carcinogenesis using tissue-specific expression of SV40 large T-antigen controlled by regulatory elements of the human α-1-antitrypsin gene. Cancer Res. 1989;49(21):6108–17.
147. Farazi PA, Glickman J, Horner J, DePinho RA. Cooperative interactions of p53 mutation, telomere dysfunction, and chronic liver damage in hepatocellular carcinoma progression. Cancer Res. 2006;66(9):4766–73. doi:10.1158/0008-5472.CAN-05-4608.
148. Lee GH, Merlino G, Fausto N. Development of liver tumors in transforming growth factor α transgenic mice. Cancer Res. 1992;52(19):5162–70.
149. Sandgren EP, Luetteke NC, Palmiter RD, Brinster RL, Lee DC. Overexpression of TGFα in transgenic mice: induction of epithelial hyperplasia, pancreatic metaplasia, and carcinoma of the breast. Cell. 1990;61(6):1121–35. doi:10.1016/0092-8674(90)90075-.
150. Sanderson N, Factor V, Nagy P, Kopp J, Kondaiah P, Wakefield L, Roberts AB, Sporn MB, Thorgeirsson SS. Hepatic expression of mature transforming growth factor β1 in transgenic mice results in multiple tissue lesions. Proc Natl Acad Sci U S A. 1995;92(7):2572–6. doi:10.1073/pnas.92.7.2572.
151. Schnur J, Nagy P, Sebestyén A, Schaff Z, Thorgeirsson SS. Chemical hepatocarcinogenesis in transgenic mice overexpressing mature TGFβ-1 in liver. Eur J Cancer. 1999;35(13):1842–5. doi:10.1016/S0959-8049(99)00224-5.

152. Schnur J, Oláh J, Szepesi Á, Nagy P, Thorgeirsson SS. Thioacetamide-induced hepatic fibrosis in transforming growth factor beta-1 transgenic mice. Eur J Gastroenterol Hepatol. 2004;16(2):127–33. doi:10.1097/00042737-200402000-00002.
153. Soga M, Kishimoto Y, Kawamura Y, Inagaki S, Makino S, Saibara T. Spontaneous development of hepatocellular carcinomas in the FLS mice with hereditary fatty liver. Cancer Lett. 2003;196(1):43–8.

Neuroendocrine Tumorigenesis

7

Nadine Gantenbein and Johannes Haybaeck

Contents

Before the year 2000, gastroenteropancreatic neuroendocrine tumors (GEP NETs) were not well characterized [3]. The GEP NET incidence has increased worldwide over the last decades [2, 4]. Considering the constantly evolving imaging technology, small asymptomatic lesions in the gut can be identified [5]. NETs are a heterogeneous group found in different locations of the body, e.g., pancreas, foregut, midgut, hindgut, and lung [5, 6]. The regional distribution of NETs over the entire body is schematically displayed in Fig. 7.1. Gastroenteropancreatic (GEP) NETs are with two thirds the most common primary NETs [4, 7]. With one quarter of NETs, they occur in the lung as the second most location [2].

NETs arise from neuroendocrine-programmed cells, which are found throughout the body and are known to excessively produce and secrete molecules like neuropeptides and biologically active neuramines, such as insulin, serotonin, and somatostatin [7–9]. An overview of neuroendocrine-programmed cells is displayed in Fig. 7.2.

N. Gantenbein, Msc
Department of Pathology, Medical University of Graz, Auenbruggerplatz 25, 8036 Graz, Austria

J. Haybaeck, M.D., Ph.D. (✉)
Department of Pathology, Medical Faculty, Otto von Guericke University Magdeburg, 39120 Magdeburg, Germany

Institute of Pathology, Medical University of Graz, 8036 Graz, Austria
e-mail: johannes.haybaeck@med.ovgu.de

J. Haybaeck (ed.), *Mechanisms of Molecular Carcinogenesis – Volume 2*,
DOI 10.1007/978-3-319-53661-3_7

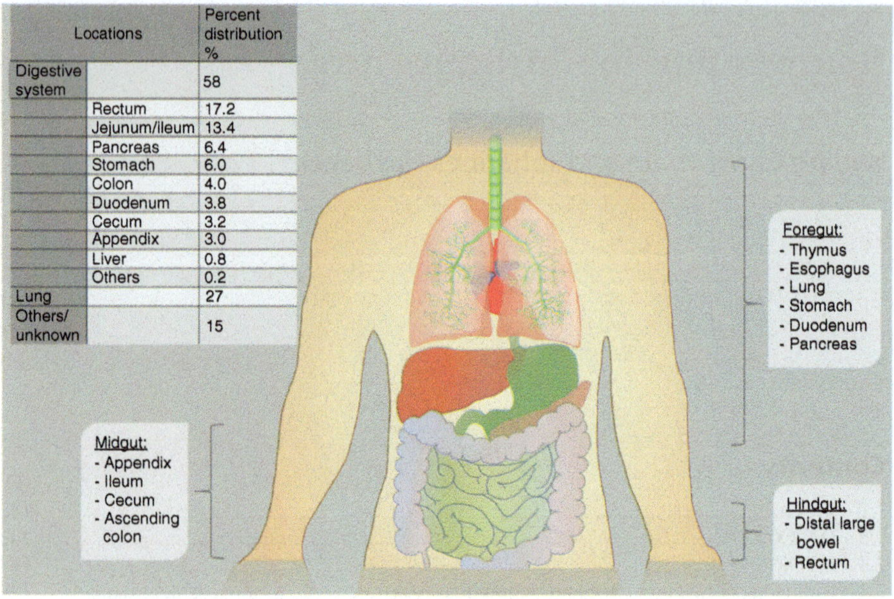

Locations		Percent distribution %
Digestive system		58
	Rectum	17.2
	Jejunum/ileum	13.4
	Pancreas	6.4
	Stomach	6.0
	Colon	4.0
	Duodenum	3.8
	Cecum	3.2
	Appendix	3.0
	Liver	0.8
	Others	0.2
Lung		27
Others/ unknown		15

Fig. 7.1 Overview of NET occurrence in the human body (Adapted of Yao) [4]. NETs are found all over the body; the gastropancreatic system is with 58% the most frequent region, followed by 27% NETs in the lung

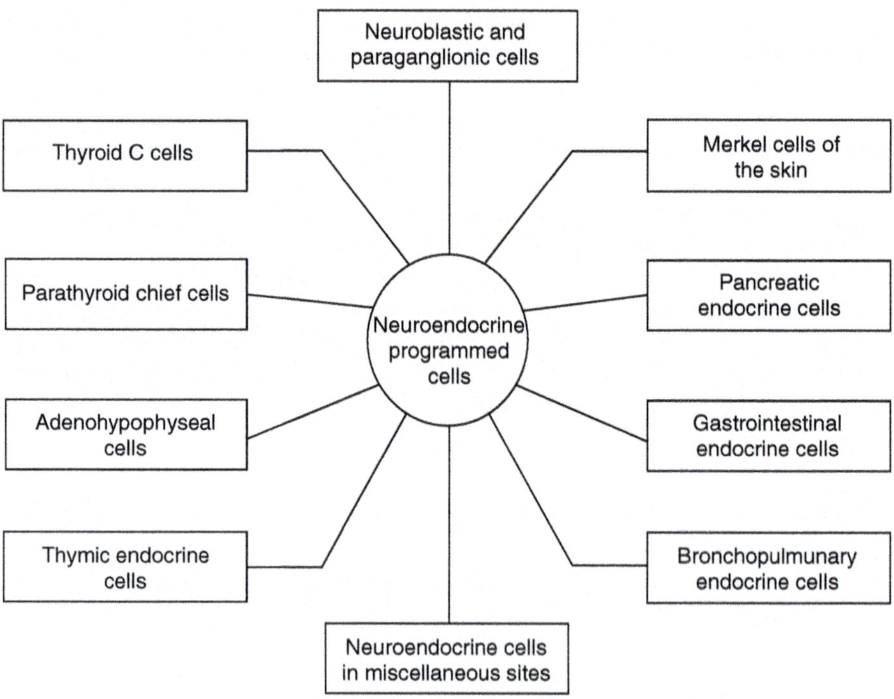

Fig. 7.2 Schematic overview of neuroendocrine-programmed cells (Adapted from Tischler and DeLellis) [10]

The classification of neuroendocrine tumors is based on size, tissue invasion, Ki67 index, and mitotic activity, according to the current WHO classification [5]. KI-67 is a proliferation marker and helps to determine tumor grade and prognosis [11].

There is still no balanced therapy for NETs [5]. Total resection in early stages is unchallenged in curative treatment compared to therapies, such as those with somatostatin analogues, radiotherapy, and chemotherapy, because they are still insufficient [4, 7, 12]. Knowledge on how to suppress hypersecretion or neoplastic growth could lead to a new therapeutic and palliative approach [13]. Due to the lack of mechanistic insights regarding this disease, many whole-genome sequencing approaches on NET patient tissues were initiated in order to identify mutations, which correlate with the development, prediction, or diagnosis of NETs [14]. The most frequent gene alterations in NET patients were found in the following genes: *MEN-1* (encodes menin), *DAXX* (death domain-associated protein), *ATRX* (alpha thalassemia/mental retardation syndrome X linked), and mTOR (mammalian target of rapamycin) with the related pathway members [14–17].

7.1 MEN-1

NETs occur either sporadically or as manifestation of a syndrome, like the multiple endocrine neoplasia type 1 (*MEN-1*) syndrome [16]. A germline mutation in the *MEN-1* tumor suppressor gene, located on the chromosome 11q13, causes this autosomal dominantly inherited condition [15, 16]. This gene encodes the 610 amino acid nuclear protein menin, which is associated with regulation of transcription, genomic stability, cell division, and cell cycle control [10, 18–20]. Over 450 different germline mutations have been identified to date. About two thirds of these mutations are predicted to lead to truncations on the protein [18]. Either truncations or missense in *Men-1* leads to lower protein levels because of proteolytic degradation via the ubiquitin pathway [16, 21]. Mutations in *MEN-1* are associated with a prolonged survival compared to patients without *MEN-1* mutation [14].

7.2 DAXX/ATRX

Likewise, NET patients with mutations in *DAXX/ATRX* have a better survival rate [17, 22]. These mutations affect incorporation of the histone H3.3 complex into telomeres by inducing alternative lengthening of telomeres and chromosomal instability [17, 22].

7.3 mTOR Signaling

Some NET patients were reported to have mutations in the *PTEN*, *PI3K*, and *TSC2*, genes of the mTOR pathway [14]. It seems that these mutations are relevant only for few NET patients because alterations in expression of mTOR pathway members are found in most patients [14, 23, 24]. Therefore, whole-genome sequencing of NETs can help to identify patients which would benefit from therapy with mTOR inhibitors [14].

Chromosomal instability in NET patients is associated with tumor progression. As the extent of genomic changes seems to correlate with disease stage, indicating alterations accumulate during tumor progression [10, 19].

7.4 Biomarkers of Neuroendocrine Neoplasms

At the beginning of their formation, NETs usually do not show specific symptoms over a long time period. The low proliferation rate of most NETs might be an explanation for this phenomenon [25]. Due to their origin, NETs secrete different molecules. This might be a way to look for a tumor marker. Four biomarkers for NETs have been established: chromogranin A (CgA), synaptophysin (SYP), neuron-specific enolase (NSE), and urinary 5-hydroxyindole-3-acetic acid (5-HIAA) [26].

Neuroendocrine cells secrete their products via large dense-core or small synaptic-like vesicles. Those vesicles store proteins like CgA and synaptophysin and therefore serve as markers for neuroendocrine cells [11]. CgA is a member of the chromogranin family and is often observed to be elevated in serum of patients [27]. Immunohistochemistry for CgA can confirm the origin in the tissue [11]. It also seems that CgA is a prognostic marker because it positively correlates with disease progression, liver metastases, and treatment efficiency [8, 11].

For the histopathological diagnosis of NETs, CgA and synaptophysin have to be present [28]. SYP is a calcium-binding integral membrane glycoprotein [11]. It is present in epithelial and neuronal types [10]. SYP is expressed independently from other NET biomarkers [28].

Neuron-specific enolase (NSE) plays a role in glucose metabolism. This enzyme was shown to be present in thyroid and prostatic carcinoma, neuroblastoma, small cell lung carcinoma, carcinoid, gastropancreatic tumor, and neoplasms with a neuroendocrine differentiation [26, 29]. Based on its lacking sensitivity and specificity as biomarker, it is mostly used to confirm the diagnosis or to control the treatment efficacy during follow-up [29].

Serotonin is one of the most hypersecreted hormones in NETs. 5-Hydroxyindoleacetic acid (5-HIAA) has serotonin as substrate and is excreted via the urine, where high levels of 5-HIAA are detected in patients with NETs [30, 31]. Although tryptophan- or serotonin-rich food can elevate 5-HIAA levels, the specificity of this marker is about 88% in NETs [31].

References

1. Modlin IM, Shapiro MD, Kidd M. Siegfried oberndorfer: origins and perspectives of carcinoid tumors. Hum Pathol. 2004;35(12):1440–51.
2. Modlin IM, Lye KD, Kidd M. A 5-decade analysis of 13,715 carcinoid tumors. Cancer. 2003;97(4):934–59.

3. Solcia E, Klöppel G, Sobin LH. Histological typing of endocrine tumours. 2nd ed. WHO, World Health Organization, International Histological Classification of Tumours; 2000.
4. Yao JC, Hassan M, Phan A, Dagohoy C, Leary C, Mares JE, Abdalla EK, Fleming JB, Vauthey JN, Rashid A, Evans DB. One hundred years after 'carcinoid': epidemiology of and prognostic factors for neuroendocrine tumors in 35,825 cases in the United States. J Clin Oncol. 2008;26(18):3063–72.
5. Niederle MB, Hackl M, Kaserer K, Niederle B. GEPNET: the current incidence and staging based on the WHO and ENETS classification: an analysis based on prospectively collected parameters. Endocr Relat Cancer. 2010;17(4):909–18.
6. Klöppel G. Tumour biology and histopathology of neuroendocrine tumours. Best Pract Res Clin Endocrinol Metab. 2007;21(1):15–31.
7. Adler DG. Neuroendocrine tumors: review and clinical update. Hosp Physician. 2007:12–21.
8. Nikou GC, Marinou K, Thomakos P, Papageorgiou D, Sanzanidis V, Nikolaou P, Kosmidis C, Moulakakis A, Mallas E. Chromogranin a levels in diagnosis, treatment and follow-up of 42 patients with non-functioning pancreatic endocrine tumours. Pancreatology. 2008;8(4–5): 510–9.
9. Marx SJ. Molecular genetics of multiple endocrine neoplasia types 1 and 2. Nat Rev Cancer. 2005;5(5):367–75.
10. DeLellis RA. The neuroendocrine system and its tumors: an overview. Am J Clin Pathol. 2001;115(Suppl. 1):5–16.
11. Modlin IM, Oberg K, Chung DC, Jensen RT, de Herder WW, Thakker RV, Caplin M, Delle Fave G, Kaltsas GA, Krenning EP, Moss SF, Nilsson O, Rindi G, Salazar R, Ruszniewski P, Sundin A. Gastroenteropancreatic neuroendocrine tumours. Lancet Oncol. 2008;9(1):61–72.
12. Pfragner R, Behmel A, Hoger H, Beham A, Ingolic E, Stelzer I, Svejda B, Moser VA, Obenauf AC, Siegl V, Haas O, Niederle B. Establishment and characterization of three novel cell lines - P-STS, L-STS, H-STS - derived from a human metastatic midgut carcinoid. Anticancer Res. 2009;29(6):1951–61.
13. Bajetta E, Ferrari L, Martinetti A, Celio L, Procopio G, Artale S, Zilembo N, Di Bartolomeo M, Seregni E, Bombardieri E. Chromogranin A, neuron specific enolase, carcinoembryonic antigen, and hydroxyindole acetic acid evaluation in patients with neuroendocrine tumors. Cancer. 1999;86:858–65.
14. Jiao Y, Shi C, Edil BH, De Wilde RF, Klimstra DS, Maitra A, Schulick RD, Tang LH, Wolfgang CL, Choti A, Velculescu VE, Jr LAD, Vogelstein B, Kinzler KW, Hruban RH, Papadopoulos N. DAXX/ATRX, MEN1, and mTOR pathway genes are frequently altered in pancreatic neuroendocrine tumors. Science. 2011;331(6021):1199–203.
15. Agarwal SK, Kester MB, Debelenko LV, Heppner C, Emmert-Buck MR, Skarulis MC, Doppman JL, Kim YS, Lubensky IA, Zhuang Z, Green JS, Guru SC, Manickam P, Olufemi SE, Liotta LA, Chandrasekharappa SC, Collins FS, Spiegel AM, Burns AL, Marx SJ. Germline mutations of the MEN1 gene in familial multiple endocrine neoplasia type 1 and related states. Hum Mol Genet. 1997;6(7):1169–75.
16. Pieterman CRC, Conemans EB, Dreijerink KMA, De Laat JM, Timmers HTM. Thoracic and duodenopancreatic neuroendocrine tumors in multiple endocrine neoplasia type 1: natural history and function of menin in tumorigenesis. Endocr Relat Cancer. 2014;21(3):R121–42.
17. Marinoni I, Kurrer AS, Vassella E, Dettmer M, Rudolph T, Banz V, Hunger F, Pasquinelli S, Speel EJ, Perren A. Loss of DAXX and ATRX are associated with chromosome instability and reduced survival of patients with pancreatic neuroendocrine tumors. Gastroenterology. 2014;146(2):453–460.e5.
18. Lemos MC, Thakker RV. Multiple endocrine neoplasia type 1 (MEN1): analysis of 1336 mutations reported in the first decade following identification of the gene. Hum Mutat. 2008;29(1):22–32.
19. Asa SL. Pancreatic endocrine tumors. Mod Pathol. 2011;24(Suppl. 2):S66–77.
20. Larsson C, Skogseid B, Oberg K, Nakamura Y, Nordenskjöld M. Multiple endocrine neoplasia type 1 gene maps to chromosome 11 and is lost in insulinoma. Nature. 1988;332(6159):85–7.

21. Yaguchi H, Ohkura N, Takahashi M, Nagamura Y, Kitabayashi I, Tsukada T. Menin missense mutants associated with multiple endocrine neoplasia type 1 are rapidly degraded via the ubiquitin-proteasome pathway. Mol Cell Biol. 2004;24(15):6569–80.
22. Cives M, Kunz PL, Morse B, Coppola D, Schell MJ, Campos T, Nguyen PT, Nandoskar P, Khandelwal V, Strosberg JR. Phase II clinical trial of pasireotide long-acting repeatable in patients with metastatic neuroendocrine tumors. Endocr Relat Cancer. 2015;22(1):1–9.
23. Missiaglia E, Dalai I, Barbi S, Beghelli S, Falconi M, Della Peruta M, Piemonti L, Capurso G, Di Florio A, Delle Fave G, Pederzoli P, Croce CM, Scarpa A. Pancreatic endocrine tumors: expression profiling evidences a role for AKT-mTOR pathway. J Clin Oncol. 2010;28(2):245–55.
24. Perren A, Anlauf M, Henopp T, Rudolph T, Schmitt A, Raffel A, Gimm O, Weihe E, Knoefel WT, Dralle H, Heitz PU, Komminoth P, Klöppel G. Multiple endocrine neoplasia type 1 (MEN1): loss of one MEN1 allele in tumors and monohormonal endocrine cell clusters but not in islet hyperplasia of the pancreas. J Clin Endocrinol Metab. 2007;92(3):1118–28.
25. Neary PC, Redmond PH, Houghton T, Watson GR, Bouchier-Hayes D. Carcinoid disease: review of the literature. Dis Colon Rectum. 1997;40(3):349–62.
26. Modlin IM, Moss SF, Gustafsson BI, Lawrence B, Schimmack S, Kidd M. The archaic distinction between functioning and nonfunctioning neuroendocrine neoplasms is no longer clinically relevant. Langenbeck's Arch Surg. 2011;396:1145–56.
27. Panzuto F, Severi C, Cannizzaro R, Falconi M, Angeletti S, Pasquali A, Corleto VD, Annibale B, Buonadonna A, Pederzoli P, Delle Fave G. Utility of combined use of plasma levels of chromogranin A and pancreatic polypeptide in the diagnosis of gastrointestinal and pancreatic endocrine tumors. J Endocrinol Investig. 2004;27(1):6–11.
28. Bumming P, Nilsson O, Ahlman H, Welbencer A, Andersson MK, Sjolund K, Nilsson B, Bümming P, Sjölund K. Gastrointestinal stromal tumors regularly express synaptic vesicle proteins: evidence of a neuroendocrine phenotype. Endocr Relat Cancer. 2007;14(3):853–63.
29. Andrén-Sandberg A. Pancreatic endocrine tumors. N Am J Med Sci. 2011;3(4):164–6.
30. Oberg K, Knigge U, Kwekkeboom D, Perren A. Neuroendocrine gastro-entero-pancreatic tumors: ESMO Clinical Practice Guidelines for diagnosis, treatment and follow-up. Ann Oncol. 2012;23(Suppl. 7):vii124–30.
31. Modlin IM, Moss SF, Oberg K, Padbury R, Hicks RJ, Gustafsson BI, Wright NA, Kidd M. Gastrointestinal neuroendocrine (carcinoid) tumours: Current diagnosis and management. Med J Aust. 2010;193(1):46–52.

Molecular Etiopathogenesis of Extranodal Marginal Zone B-Cell Lymphoma of Mucosa-Associated Lymphoid Tissue

8

Alexander JA Deutsch, Katharina Troppan,
Karoline Fechter, and Peter Neumeister

Contents

Abstract

Approximately 8% of all non-Hodgkin lymphomas are extranodal marginal zone B-cell lymphomas of mucosa-associated lymphoid tissue, also known as MALT lymphomas. MALT lymphomas arise at several different extranodal sites, with the highest frequency in the stomach, followed by the lung, ocular adnexa and thyroid, and with a low percentage in the small intestine. Interestingly, at least three different, apparently site-specific, chromosomal translocations and mis-

A.JA. Deutsch (✉) • K. Troppan • K. Fechter • P. Neumeister
Division of Hematology, Department of Internal Medicine, Medical University Graz, Graz, Austria
e-mail: alexander.deutsch@medunigraz.at

J. Haybaeck (ed.), *Mechanisms of Molecular Carcinogenesis – Volume 2*,
DOI 10.1007/978-3-319-53661-3_8

sense and frameshift mutations, all affecting the NF-κB signal pathway-related genes, have been implicated in the development and progression of MALT lymphoma. However, these genetic abnormalities alone are not sufficient for malignant transformation. There is now increasing evidence suggesting that the oncogenic product of translocation cooperates with immunological stimulation in oncogenesis, i.e. the association with chronic bacterial or autoaggressive infections. This chapter mainly discusses the genetic aberration and the association with chronic infections of MALT lymphomas and summarizes recent advances in the molecular pathogenesis and therapeutic advances of MALT lymphoma.

8.1 Introduction

Approximately 8% of all non-Hodgkin lymphomas (NHL) are extranodal marginal zone B-cell lymphomas of mucosa-associated lymphoid tissue (MALT), also known as MALT lymphomas, which were first described in 1983 by Isaacson and Wright [1, 2]. They found out that compared to peripheral lymph nodes, primary low-grade gastric B-cell lymphomas and immunoproliferative small intestinal disease had more histological features in common with mucosa-associated lymphoid tissue [1]. Other extranodal low-grade lymphomas arising in mucosal organs, including the salivary gland, lung and thyroid, showed the same histological and clinical features [3–6]. This led to the introduction of the term "MALT lymphoma". MALT lymphomas arise at a wide range of different extranodal sides, including the stomach (70% of cases), lung (14%), ocular adnexa (12%), thyroid (4%) and small intestine (including immunoproliferative small intestinal disease (IPSID), 1%) [7].

The histological feature of MALT lymphoma is an infiltration around B-cell follicles in the region corresponding to the Peyer's patch marginal zone, spreading diffusely into the surrounding tissue. MALT lymphoma cells share the same cytological features and immunophenotype (cluster of differentiation (CD) 20^+, $CD21^+$, $CD35^+$, IgM^+, IgD^-) like marginal zone B cells. This is the reason why the World Health Organisation lymphoma classification of 2001 designates this lymphoma as "extranodal marginal zone B cell lymphoma of mucosa associated lymphoid tissue (MALT lymphoma)" [8]. The lymphoma cells can resemble either follicle centre centrocytes, small lymphocytes or so-called monocytoid B cells. Another important histological feature is the presence of lymphoepithelial lesions formed by invasion of individual mucosal glands or other epithelial structures by lymphoma cells. Transformed blasts and plasma cells are present and scattered beneath the surface epithelium, indicating that MALT lymphoma might participate in the immune response. The lymphoma cells also enter the germinal centres of non-neoplastic B-cell follicles—so-called follicular colonization [9].

In the case of gastric MALT lymphoma, the disease is remarkably indolent and tends to remain localized to the stomach for long periods. The 10-year survival rate for gastric MALT lymphoma is close to 90% with a disease-free survival of approximately 70% [10, 11]. However, MALT lymphoma can progress and transform into aggressive high-grade tumours—extranodal diffuse large B-cell lymphoma (eDLBCL)—whereby the 10-year survival rate drops to approximately 42% [10]. eDLBCLs express B-cell lymphoma protein (BCL)-6 more frequently than do nodal cases, having also a better overall survival rate than the latter [12]. eDLCBL shows different features with respect to MALT lymphoma components: foci of extranodal DLBCL may be seen in MALT lymphoma, suggesting transformation from one to the other. Identification of identically rearranged *Ig* gene between the low- and the high-grade components of the same cases confirmed that transformation processes take place [13]. Some cases of extranodal DLBCL, in which the MALT lymphoma component cannot be detected, are transformed MALT lymphomas that have been completely overgrown by the extranodal DLBCLs. Others are primary extranodal DLBCLs with a germinal centred-like phenotype (CD10 and BCL6 positive) [10]. Transformed MALT lymphomas are CD10 and BCL2 negative [14], but, in contrast to MALT lymphomas, usually express BCL6. Notably, there is no difference in clinical behaviour between transformed MALT lymphoma and extranodal DLBCL [10].

8.2 Clinical Diagnosis and Staging

The initial staging procedure to assess dissemination of lymphomas must include an esophagogastroduodenoscopy (EGD) with multiple biopsies (at least ten from macroscopic visible lesions) but also from regions with normal appearance, i.e. each region of the stomach, duodenum and gastroesophageal junction, respectively [15, 16]. Following diagnosis, further work-up should include physical examination (including Waldeyer's ring and peripheral lymph nodes), lab parameters (complete blood count, LDH, beta-2 macroglobulin levels, serum immunofixation, HIV, hepatitis B and C serologically) and CT scans involving the abdominal, pelvic and thoracic region. Endoscopic ultrasound (EUS) is also recommended for the initial evaluation of regional lymph nodes and the depth of the gastric wall infiltration [15–17], parameters which are of highly prognostic value for the response to *H. pylori* eradication [18]. EUS is the only investigation able to differentiate between the stages I1E, I2E and II1E according to the modified Ann Arbor classification system. A bone marrow biopsy is also recommended (but no longer judged mandatory in current guidelines), particularly in case of failure to eradication therapy or before initiating systemic hemato-oncological treatment [15], although the percentage of patients with bone marrow involvement may be as little as 2% in some series. Because of frequent multifocal involvement [17], an ileocolonoscopy should also be considered in patients with gastric MALT lymphoma.

The question on the optimal staging system in gastrointestinal MALT lymphomas remains controversial. The most widely used system is the Ann Arbor staging system modified according to Musshoff [19] and Radaszkiewicz [20].

This staging system differentiates between the prognostically relevant dissemination extent, i.e. involvement of adjacent (II1E) and distant lymph (II2E) nodes, and the depth of infiltration of the gastric wall, i.e. mucosa and submucosa (I1E) versus the muscularis propria and submucosa (I2E). The Paris staging system is based on the TNM classification but has not been validated in prospective studies so far [21].

8.3 Genetic Aberrations

8.3.1 Molecular Mechanisms to Generate Antibody Diversity Are Responsible for Genetic Aberrations

As indispensable components of adaptive immunity, B cells are responsible for the humoral immune response by producing antibodies in response to antigens. B cells express special surface markers termed B-cell receptors (BCRs). B cells are produced in the bone marrow as immature B cells, which further migrate to secondary lymphoid tissues where they develop into mature B cells. The maturation process of B cells is accompanied by the changing presence of differentiation markers on the cell surface [22, 23].

During different stages of B-cell maturation, BCRs undergo high levels of genetic rearrangement in V (variable), D (diversity) and J (joining) segments of variable regions of heavy and light chains of immunoglobulins called VDJ recombination, thereby generating a rich repertoire of diverse antibodies providing a broad-spectrum cover against the majority of antigens [24, 25].

The next steps of B-cell maturation take place in secondary lymphoid organs where the B-cell response is dependent on T cells. The secondary lymphoid organs consist of a germinal centre (GC), including dark and light zones. B-cell maturation in secondary lymphoid organs initiates in the dark zone, and B cells mature during their migration towards the light zone. Movement of B cells through the GC is thought to be managed by chemotaxis [26]. B cells first enter the centroblast (CB) of the dark zone of GC, where the variable regions of BCRs undergo somatic hypermutation (SHM) to increase the diversity of immunoglobulins. Cells then migrate to centrocytes (CC) of the GC light zone to experience a process called class switching recombination (CSR) in order to create different classes of immunoglobulins (IgG, IgM, IgE, IgA). Afterwards, B cells can be activated in a T-cell-dependent manner for effective response to foreign antigens. However, not all B cells require T-cell-dependent mechanisms to respond to antigens [27, 28].

The VDJ recombination occurs at early B-cell development stages in the bone marrow. During this process, three gene segments (V, D and J) of the variable region of the heavy chains and two (V and J) of the variable region of the light chains are assembled. The DNA located between the rearranging gene segments is deleted by two endonucleases called recombination-activating genes (RAG) 1 and 2 causing double-strand breaks. The heavy chain assembly occurs in two steps. In the first

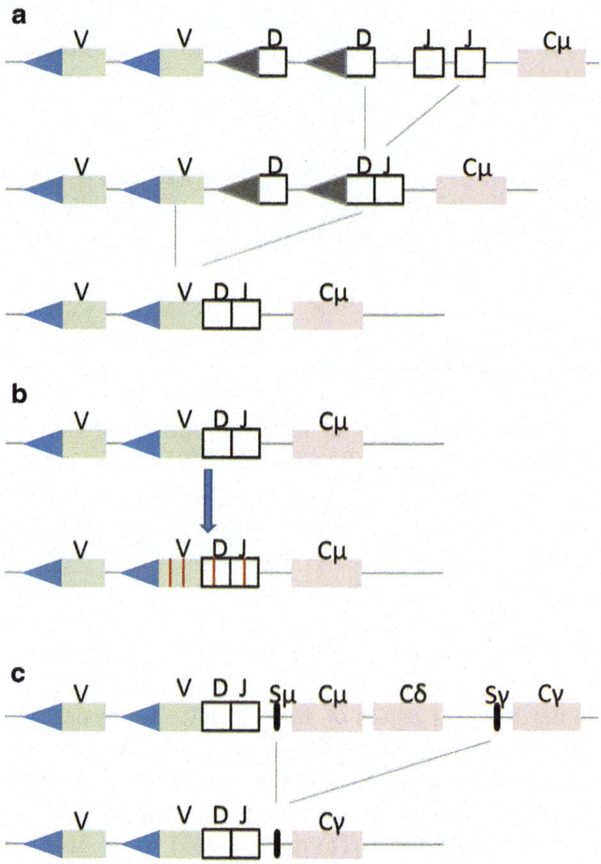

Fig. 8.1 Molecular mechanisms being involved in the diversity of antibodies. (**a**) VDJ recombination, (**b**) somatic hypermutation, and (**c**) class switch recombination

step, a DH gene segment is rearranged to a JH segment, and in the second step, VH segment is rearranged to a DH joint (Fig. 8.1) [27, 28].

The SHM introduces point mutations, deletions and duplications in the rearranged variable regions of heavy and light chains and their flanking sequences (Fig. 8.1) [25]. In the CSR, the expressed heavy chain constant region (C_H) gene (usually Cμ and Cδ) is replaced by a downstream C_H gene. The recombination involves deletion of the DNA between repetitive regions (switch regions sμ, sγ and sα) upstream of the recombining C_H genes (Fig. 8.1) [29]. The enzyme activation-induced cytidine deaminase (AID) is required for SHM and CSR. In the first phase of both molecular processes, AID deaminates cytidine to uracil to generate a U-G mismatch. U-G mismatches normally are corrected by base excision repair (BER) and mismatch repair (MMR) pathways. In the second phase of SHM, U-G mismatches are repaired by error-prone BER and MMR pathway in order to generate mutations [30]. BER and MMR are facilitated by the action of uracil-DNA glycosylase (UNG) and mutS homolog (MSH)2/MSH6, respectively. CSR is initiated by

AID-mediated cytidine deamination on the opposing DNA strands within the switch region. The action of UNG and apurinic/apyrimidinic endonuclease 1 or MSH2/MSH6 activity cause double-strand breaks. Further processing and joining of broken switch regions are facilitated by factors involved in classic and alternative non-homologous end joining [31].

The fact that VDJ recombination, SHM and CSR cause double-strand breaks and that the majority of the translocations in lymphoid malignancies involve the Ig gene locus clearly demonstrates the crucial role of these molecular mechanisms in lymphomagenesis.

8.3.2 Translocations

There are four main recurrent chromosomal translocations associated with the pathogenesis of MALT lymphomas: t(1;14)(p22;q32), t(11;18)(q21;q21), t(14;18)(q32;q21) and t(3;14)(p14.1;q32) [32–35]. The frequency of genetic aberrations is also dependent on the primary site of disease. Translocation t(11;18)(q21;q21) was mainly found in pulmonary and gastric tumours, whereas t(14;18)(q32;q21) was the most detected one in the ocular adnexa, orbit, skin and salivary glands MALT lymphoma [36].

The t(3;14)(p14.1;q32) translocation is the most recently described and established juxtaposition of the transcription factor forkhead box protein (FOX)P1 next to the enhancer region of the Ig heavy chain genes [34]. Overexpression of FOXP1 followed by chromatin immunoprecipitation in lymphoma cells demonstrated that FOXP1 acts as transcriptional repressor of multiple proapoptotic genes and that it suppresses caspase-dependent apoptosis [37].

The t(11;18)(q21;q21) translocation is the most common translocation, occurring in 15–40% of all MALT lymphomas [32, 36]. It is restricted to MALT lymphomas and has not been found in nodal or splenic marginal zone lymphomas. In most of these translocation-positive cases, it is the sole chromosomal aberration, and only exceptionally has it been detected in de novo DLBCL arising at mucosal sites [38–40]. The t(11;18)(q21;q21) has been found in MALT lymphomas at a number of anatomic sites, including the lung, stomach and intestine and, less commonly, the skin, orbit and salivary gland [36, 41]. It has also been associated with cases that do not response to *H. pylori* eradication [42, 43], and it is rarely seen in transformed MALT lymphomas [40]. The t(11;18) translocation represents the fusion of the *apoptosis inhibitor 2* (*API2*) gene on chromosome 11 and the *MALT lymphoma-associated translocation 1* (*MALT1*) gene on chromosome 18 [32]. Breakpoints observed in this translocation are clustered in the region of intron 7 and exon 8 of the *API2* gene and intron 4, 6, 7 and 8 of the *MALT1* gene. High frequencies of deletions and duplications in both genes are also found, implying that multiple double-strand DNA breaks (DSBs) must have occurred during the translocation process, being apparently the product from illegitimate non-homologous end joining after DSBs [44]. The resulting fusion transcript always comprises the N-terminal API2 with three intact baculovirus inhibitors of apoptosis repeat (BIR) domains and the

C-terminal *MALT1* region containing an intact caspase-like domain [32, 41, 45]. t(11;18) cases show a nuclear overexpression of BCL10 protein (Fig. 8.2) [46].

The t(1;14)(p22;q32) translocation occurs in 1–2% of MALT lymphomas and has been reported in the stomach, lung and skin [36]. The entire coding sequence of the *BCL10* gene on chromosome 1 is relocated to the immunoglobulin heavy chain (*IgH*) enhancer region on chromosome 14. This results in the nuclear overexpression of the BCL10 protein. The t(1;14) translocation has exclusively been reported in MALT lymphomas, whereby these cases typically display additional genomic alterations. Patients with advanced stage show this translocation, and they do not respond to *H. pylori* eradication (Fig. 8.2) [46].

The t(14;18)(q32;q21) translocation occurring in 15–20% of MALT lymphomas brings the *MALT 1* gene under the control of the *IGH* enhancer region on chromosome 14 [33]. This translocation occurs more frequently in non-gastrointestinal MALT lymphomas. In contrast to t(11;18), the t(14;18) is associated with other cytogenetic abnormalities [36]. t(14;18)-positive cases also show an overexpression of BCL10 protein, but with a cytoplasmic localisation in contrast to t(1;14) and t(11;18) [47, 48].

t(11;18), t(14;18) and t(1;14) indicate an association of these three translocations and BCL10 and MALT1 in activating the NF-κB pathway in lymphocytes as an oncogenic event [49, 50]. Physiologically, BCL10 binds to the Ig-like domain of MALT1, thereby inducing the MALT1 oligomerization. The BCL10-MALT1 complex promotes ubiquitylation of IκB kinase-γ, release of NF-κB and subsequent translocation to the nucleus, where it transactivates genes, such as those encoding factors for cytokines and growth factors for cellular activation, proliferation and survival [51]. In MALT lymphoma with t(1;14)(p22;q32), BCL10 is believed to form oligomers through its caspase recruitment domain (CARD), triggering the MALT1 oligomerization and aberrant *NF-κB* activation without the need for upstream signalling. In lymphoma cases with t(14;18)(q32;q21), MALT1 is overexpressed. MALT1 does not have a structural domain so that it can mediate self-oligomerization, and it does not activate NF-κB in vitro [49, 50]. It is likely that MALT1 interacts with and stabilizes BCL10, causing an accumulation in the cytoplasm of t(14;18)-positive tumour cells. The resulting oligomerization of MALT1 activates NF-κB [52]. In MALT lymphomas with the t(11;18)(q21;q21), the baculovirus inhibitor of apoptosis repeat (BIR) domain of the API2-MALT1 mediates self-oligomerization, which leads to activation of NF-κB (Fig. 8.2) [53, 54].

8.3.3 Numeric Chromosomal Aberrations: Trisomies and Deletions

Other cytogenetic alterations are trisomies 3, 12 and/or 18, which are present as a sole abnormality in 22% of the cases, but they are often associated with one of the four translocations [36].

Taji et al. [55] detected trisomy 3 as the most frequent aberration in gastrointestinal MALT lymphomas with a frequency of up to 35%. Also partial trisomies of

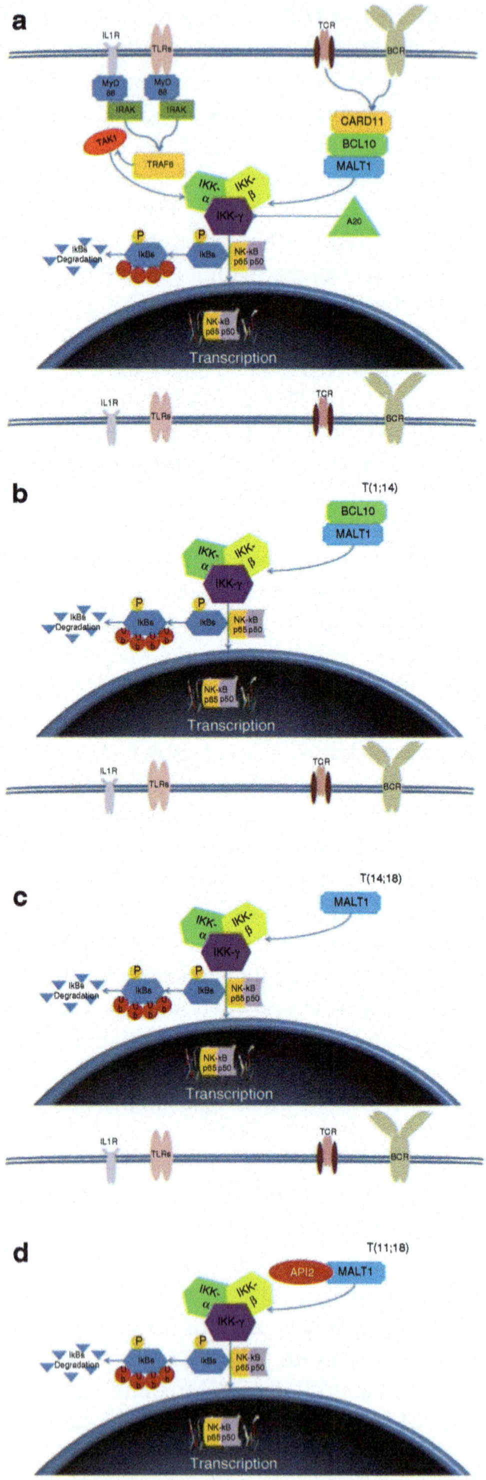

Fig. 8.2 Translocations affecting the NF-κB activation pathway: (**a**) Signaling from the TLR, IL-1R and antigen receptor activates the canonical NF-κB pathway, which is characterized by activation of the IKK complex, phosphorylation and degradation of IκB. A20 is a negative regulator. (**b**) t(1;14)(p22;q32) results in the nuclear overexpression of the BCL10 protein. It is believed to form oligomers through its CARD domain thereby triggering the MALT1 oligomerization and aberrant NF-κB activation. (**c**) t(14;18)(q32;q21) causes overexpression of MALT1. It is thought that it oligimerizes through interaction with BCL10 causing NF-κB activation. (**d**) t(11;18)(q21;q21) the BIR domain of the API2-MALT1 mediates self-oligomerization leading to an activation of NF-κB. *TLR* Toll-like receptor, *IL-1R* interleukin 1 receptor, *BCR* B-cell receptor, *TCR* T-cell receptor, *RIP1* receptor-interacting protein 1, *TRAF* TNF-associated factor, *TAK1* transforming growth factor b-activating kinase, *TAB* TAK-binding protein, *IKK* inhibitor of NF-κB kinase, *IκB* inhibitor of NF-κB

chromosomes 3 and 18 have to be taken into account, which was proven by Krugmann et al. [56]. In contrast, Ott et al. reported an incidence of only 20% of trisomy 3 in low-grade MALT lymphoma and an even lower rate in high-grade ones [41]. The genetic mechanism by which trisomy 3 may contribute to lymphomagenesis has not yet been experimentally addressed. However, an increased gene dosage effect resulting from higher-copy numbers of genes relevant to lymphoma development is likely to explain the biological consequences underlying chromosomal trisomies. Several promising candidate genes are located on chromosome 3 which have been implicated in lymphomagenesis, such as the proto-oncogene BCL6 and the transcription factor FOXP1 [38]. One of our previous studies describes that CCR4—a chemokine receptor genomically located on chromosome 3 (3p24)—was highly expressed in trisomy 3 + MALT lymphomas, whereas transcripts for this chemokine receptors were missing in trisomy 3 − MALT lymphomas [57].

Apart from the typical chromosomal translocation tumour necrosis factor, alpha-induced protein 3 (A20, TNFAIP3) has been identified to be frequently deleted in ocular adnexal MALT lymphoma detected by array comparative genomic hybridization [58–60]. As a key player in the NF-κB pathway, A20 acts as a tumour suppressor gene in lymphomas. In ocular adnexal MALT lymphomas, complete A20 inactivation is associated with poor lymphoma-free survival [58, 61]. A20 deletion occurred in MALT lymphoma of the ocular adnexa (19%), salivary gland (8%), thyroid (11%) and liver (1/2), but not or with rather undetectable frequencies in the lung, stomach and skin [58, 59, 61]. However, A20 inactivation alone is not sufficient for malignant transformation, but it may represent a promising therapeutical target for future investigations [62].

8.3.4 Somatic Mutations

To our knowledge, the number of studies investigating somatic mutations in MALT lymphomas is low, and whole-genome sequencing approach has not been performed until now. Our group reported somatic missense mutations in PIM1 and cMyc (v-Myc avian myelocytomatosis viral oncogene homolog) in 46 and 30% of MALT lymphomas (gastric and extragastric origin), 30 and 41% of transformed MALT lymphomas and 9.1% of 8 of 11 primary cutaneous marginal zone B-cell lymphomas (PCMZL) [63]—considered as part of the group of MALT lymphomas [64, 65]. Du et al. [66] detected missense and frameshift mutations in p53 in 20.8% of MALT lymphomas and 30% of transformed MALT lymphoma (both mainly gastric origin).

Mutation analysis of NF-κB signalling pathway-related genes—A20, Card11, CD79B and Myd88, known to be frequently mutated in aggressive lymphomas [67–70]—demonstrated that 6% of MALT lymphoma cases exhibited missense or frameshift mutations in the Myd88 locus, and 28.6% of ocular adnexal MALT lymphomas were mutated in the A20 locus [61, 71, 72]. Liu et al. [73] reported that Card11 and CD79B were not affected in their cohort of ocular adnexal MALT lymphomas.

8.3.5 Epigenetics

Min et al. [74] demonstrated CpG hypermethylation of the tumour suppressor genes p16 and p57 in 41.7% and 29.2% of low-grade MALT lymphoma cases, respectively.

Additionally, CpG hypermethylation of A20 is detected in 26% of MALT lymphomas investigated, including ocular adnexal cases and lymphomas located in the salivary and thyroid glands [61].

8.4 The Connection to Long-Lasting Chronic Infection

MALT lymphomas and splenic marginal zone (MZ) lymphomas, two lymphomas originating from MZ B cells, are often associated with long-lasting infection of microbial species that do not infect or transform lymphoid cells. These microbial pathogens trigger a sustained lymphoid proliferation, giving a selective advantage to lymphoid clones that still remain dependent on antigen stimulation. MZ cells are anatomically positioned in the lymphoid organs and in the mucosa-associated lymphoid tissue to function as first defence mechanism against invading pathogens. MZ B cells participate in T-cell-independent "innate-like" immune responses to microbial pathogens and can rapidly proliferate and differentiate into IgM or even class switch to other isotype-secreting plasma cells, producing the bulk of primary antibodies. The low activation threshold of these cells may predispose them to neoplastic transformation [75].

Gastric MALT lymphomas show a strong association with chronic *H. pylori* infection, which is an association that satisfies Koch's postulates for an etiologic agent [76]. Other infectious associations, though not fulfilling these criteria, have been reported for *Borrelia burgdorferi* (skin) [77], *Campylobacter jejuni* (intestine) [78] and hepatitis C virus (splenic marginal zone lymphoma) [79]. Other chronic inflammatory reactions or autoimmune diseases have been further associated with MALT lymphomas, including Sjogren's disease [80]. Especially in ocular adnexa MALT lymphoma, representing 5–15% of all extranodal lymphomas, the occurrence of *Chlamydia psittaci* is of special interest. Ferreri et al. [81] demonstrated an association between ocular adnexa MALT lymphomas and infection with *Chlamydia psittaci* in Italian patients. Presence of *Chlamydia psittaci* DNA was detected in 80% of lymphoma samples. Moreover, bacterial DNA was found in 43% of peripheral blood mononuclear cells of patients, but not in healthy donors. More than 80% of these patients went into lymphoma regression when *Chlamydia psittaci* was successfully eradicated by doxycycline administration [82]. In a large study of 142 cases, Chanudet et al. [83] described an overall prevalence of *Chlamydia psittaci* infection in ocular adnexa MALT lymphoma in 22%, but marked geographic variation, with highest incidences in Germany (47%), the East Coast of the United States (35%) and the Netherlands (29%). In our small Austrian study, we detected *Chlamydia psittaci* in 7 out of 13 samples of ocular adnexal MALT lymphomas; in contrast, only one of 17 gastrointestinal specimens was tested positive [84]. Additionally, we detected an almost 100% frequency, dependent on different sites of lymphoma manifestation, of *Chlamydia psittaci* in non-gastrointestinal MALT lymphomas, suggesting a possible involvement of this pathogen [85].

A role for antigen-driven clonal expansion of the lymphoma cells is shown by the evidence of the ongoing somatic hypermutation in the immunoglobulin V (*IgV*) genes [86]. The involvement of an antigen is further supported by evidence of clonal evolution within the tumour, suggesting pressure to increase affinity of the immunoglobulin for the antigen [87]. The early stages of gastric lymphoma development may be facilitated by antigen-driven T cells, specific for the *H. pylori* organism [88]. Eradication of the infection causing a cure rate up to 75% is consistent with this postulate [89] . However, less is known about the role of the host immune response, as demonstrated by the fact that only a minority of infected patients develop gastric MALT lymphomas, [90] as MALT lymphomagenesis may correlate with different cytokines and HLA polymorphisms [91, 92].

8.5 Pathogenesis of MALT Lymphomas

The evolution of gastric MALT lymphomas is a multistage process starting with the infection of *H. pylori* resulting in the recruitment of B, T cells and other inflammatory cells to the gastric mucosa. The infiltrated B cells are stimulated by the *H. pylori*-specific T cells and undergo malignant transformation due to acquisition of genetic abnormalities. One example is the association between *H. pylori* infection and gastric MALT lymphomas. *H. pylori* has been shown to stimulate tumour cell growth with the help of T cells [88]. Epithelial cells are activated by chronic infectious stimuli, expressing high levels of HLA-DR and co-stimulatory molecules, including CD80, on their surface. These cells may be able to present antigens along with HLA molecules to T cells. CD40 ligand molecules expressed on the activated T cells can react with CD40 molecules on B cells, upregulating the expression of CD20 on B cells. CD20, in turn, reacts with the CD28 molecule on CD4 T cells, strongly activating the latter. Activated CD4 T cells can stimulate B cells through CD40 ligand (CD40L)-CD40 interaction, in conjunction with the action of various cytokines and chemokines. In this interaction among epithelial cells, T cells and B cells may allow these cells to live cooperatively in lymphoepithelial lesions (LEL) and not to undergo apoptosis [93]. Lymphoepithelial lesions are thought to be the origin of lymphomas [94]. The transition from a polyclonal to a monoclonal lesion is thought to be facilitated by chronic stimulation with exogenous or autoantigens, thus increasing the frequency of transformation [95–97]. MALT lymphomas with *H. pylori*-dependent alterations at least in the early phases like trisomies 3, 12 and 18 can progress and become *H. pylori* independent, transforming into high-grade tumours following the mechanism already described above. Complete inactivation of the tumour suppressor gene *p53*, homologous deletion of the *p16* gene and chromosomal translocation of *cMYC* and *BCL6* are associated with the transformation of the MALT lymphomas [66, 98–101]. MALT lymphomas which do not show a t(11;18) and have an amplification at 3q27 may be a source of high-grade transformation [102]. MALT lymphomas with t(11;18) are *H. pylori* independent like cases in contrast to t(1;14), which are able to transform [7].

8.6 MALT Lymphomas Are Targeted by the Aberrant Somatic Hypermutation

The aberrant somatic hypermutation (ASHM), which was first described in DLBCL, has been identified to crucially contribute to the development of lymphoid neoplasms. In DLBCL the physiological process of the somatic hypermutation, occurring in the rearranged *V* genes to generate antibody diversity of germinal centre B cells and of all germinal centre-derived B-cell tumours [103, 104], aberrantly targets the 5′ sequences of several proto-oncogenes relevant to lymphomagenesis, including *PIM1*, *PAX5*, *RhoH/TTF* and *cMYC*. This phenomenon occurs in >50% of DLBCL, but is rare in indolent lymphomas like MALT lymphomas [64, 104–108]. The pathogenesis of most B-NHLs is associated with distinct genetic lesions, including chromosomal translocations and aberrant somatic hypermutation, which arise from mistakes during CSR and SHM occurring in the germinal centre [103, 104, 109, 110]. AID is an enzyme required for SHM and CSR, and mistargeting of AID to known proto-oncogenes, linked to B-cell tumorigenesis in germinal centre B cells, combined with a breakdown of protective high fidelity repair mechanism, has been shown to be a principal contributor to the pathogenesis of B-NHL [109, 110]. Our group described that 13 (76.5%) of 17 cases of MALT lymphomas and all of 17 (100%) cases of extranodal DLBCL—still exhibiting a low-grade MALT lymphoma component (transformed MALT lymphoma)—were targeted by the ASHM. Expression levels of AID were associated with the mutational load caused by the ASHM [64]. Additionally, 8 of 11 PCMZL (72.7%)—considered as part of the group of MALT lymphomas [65]—displayed the molecular features typical for the ASHM [63]. Interestingly, *H. pylori* infection upregulates AID expression via NF-κB in gastric cells in vitro and in vivo. The *H. pylori*-mediated AID upregulation causes an accumulation of p53 mutation in vitro [111]. Thus, it might be hypothesised that *H. pylori* infection causes an upregulation of AID in B cells and that it has an impact on MALT lymphomagenesis by causing genetic alterations.

8.7 B-Cell Receptor Signalling in MALT Lymphoma

The B-cell receptor (BCR) signalling pathway, physiologically involved in development and differentiation of normal B cells, has been identified to play a crucial role in lymphomagenesis and acts as an important target for therapeutical interventions [112]. Activation of this pathway is driven by multiple factors, including chronic exposure to antigens, for instance, *H. pylori*. Together with the chronical inflammatory status caused by *H. pylori*, antigens may drive MALT lymphomagenesis; however, a direct connection between the BCR pathway and *H. pylori* has not been identified [113]. Nonetheless, early stage *H. pylori*-positive MALT lymphomas can be cured by eradication of *H. pylori* infection alone, supporting a causative role [78].

The downstream target of the BCR signalling, NF-κB, can be activated independently of BCR signalling by the MALT1 fusion protein and BCL10 overexpression [32]. The MALT1 fusion protein is a result of t(11;18)(q21;q21), occurring in more advanced cases of MALT lymphomas. Many MALT lymphomas require MALT1 for NF-κB activation. The importance of MALT1 protease activity was shown recently by the dependency of NF-κB-restricted B-cell lymphomas on this proteolytic activity. Consequently, therapeutic targeting of MALT1 protease activity might therefore become a promising approach for the treatment of MALT lymphomas and other B-cell lymphomas associated with deregulated NF-κB signalling [114] as MALT lymphomas, harbouring these translocations, show impaired response to antibiotic eradication therapy [115].

8.8 Chemokine Receptors in MALT Lymphomas

Chemokines, also known as proinflammatory, chemotactic cytokines, represent a huge superfamily of peptides with diverse biological functions. Chemokines interact with a target cell by binding to chemokine receptors. There exists numerous chemokines and chemokine receptors, but no single chemokine is assigned to a single receptor. Chemokine signalling can coordinate cell movement during inflammation, as well as the homeostatic transport of haematopoietic stem cell, lymphocytes and dendritic cells [116–118]. The homeostatic transport of the precursor B cell to secondary lymphoid tissue is essential for B-cell development. CCR6, CCR7, chemokine (C-X-C motif) receptor (CXCR)3, CXCR4 and CXCR5 play a crucial role in this homing process; therefore, the group of these five chemokine receptors is called B-cell homeostatic chemokine receptors [119–121]. The group of activation-dependent chemokine receptors, which are expressed on effector leukocytes (including activated effector/memory T cells), play an essential role in inflammatory processes responsible for migration towards chemokines produced by inflamed cells [117]. Our expression analysis of 19 well-characterized chemokine receptors in MALT lymphomas demonstrated a distinct signature of the chemokine receptor expression pattern of extragastric MALT lymphomas compared to gastric MALT lymphomas. Comparing gastric with extragastric MALT lymphomas, upregulation of CXCR1 and CXCR2 accompanied by downregulation of CCR8 and chemokine (C-X3-C motif) receptor (CX3CR)1 and loss of chemokine (C motif) receptor (XCR)1 expression in extragastric MALT lymphomas appear to be key determinants for the site of origin of MALT lymphomagenesis [57]. In our second study on the chemokine receptors in MALT lymphomas, we found that CXCR4 expression was missing in gastric MALT lymphomas or gastric eDLBCL compared to nodal lymphomas, nodal marginal B-cell lymphomas and nodal DLBCL, which exhibited a strong expression [122], indicating that CXCR4 expression is associated with nodal manifestation. Additionally, we found that CXCL12 and CXCR7—a CXCL12 receptor—were upregulated during progression of gastric MALT lymphomas to gastric eDLBCL [122], suggesting at least in part an implication of this signalling pathway in high-grade transformation of gastric MALT lymphomas.

8.9 Treatment

Independent of stage, *H. pylori* eradication by antibiotic therapy is the established first-line therapy of gastric MALT lymphomas [15, 16]. Eradication therapy in patients with localized disease leads to complete lymphoma remission in 60–90% and a 10-year overall survival of almost 90%, respectively [123, 124]. In case of *H. pylori* reinfection/recrudescence with or without a relapsing lymphoma component, further eradication is indicated [15]. Negative predictive factors for the regression of gastric MALT lymphomas after *H. pylori* eradication are t(18;11) positivity, *H. pylori* negativity and lymph node involvement [18]. In addition, the presence of an underlying autoimmune condition [125, 126] has also been suggested as a negative prognostic parameter for lymphoma regression after *H. pylori* eradication. However, patients with gastric MALT lymphomas refractory to *H. pylori* eradication, persistent lymphoma manifestations or patients with disseminated disease are candidates for further systemic or local oncological treatment:

Radiotherapy is a curative option for patients with stage I and II disease not achieving lymphoma regression after successful *H. pylori* eradication [15]. Radiotherapy showed a higher remission rate compared to chemotherapy, being similar to surgery. They conclude that radiotherapy seems to be the most suitable treatment in these patients [127] as MALT lymphomas are highly susceptible to radiation [128]. Risk factors for treatment failure were a large-cell component and an exophytic growth pattern [129].

In the presence of disseminated or advanced disease, chemotherapy and/or immunotherapy with the anti-CD 20 antibody rituximab is the treatment of choice, but no standard chemotherapy has been defined so far. Recently, chemotherapy is also increasingly being used in localized disease as already discussed. Only a few compounds have been tested specifically in gastric MALT lymphomas. Oral alkylating agents such as chlorambucil or cyclophosphamide administered for a median treatment duration of 1 year can result in a high rate of disease control [130], but they are less effective in t(11;18)-positive gastric MALT lymphomas [131]. The compound with the best documented long-term follow-up is the purine analogue cladribine (2-CdA), which exerts excellent antitumour activity [132]. However, concerns have been raised because of the (rare) development of a treatment-related myelodysplastic syndrome in a patient cohort [133]. Aggressive anthracycline-containing chemotherapy (R-CHOP (cyclophosphamide, doxorubicine, Oncovin, prednisolone)) is discouraged in this indolent disease and should be reserved for patients with a high-tumour burden or "transformed" MALT lymphomas with an additional large-cell component. The activity of rituximab has also been demonstrated in several phase II studies with an overall response rate up to 77%, and the t(11;18)(q21;q21) translocation seems not to be a predictive marker of response or subsequent relapse [134].

Conclusion

MALT lymphomas represent a heterogeneous group of lymphoid neoplasms with a large number of different genetic alterations depending on the site of origin [32–36]. Interestingly, most of the genetic alterations affect NF-κB signalling pathway-related genes causing constitutive activation of the NF-κB pathway [49, 50, 53, 114]. This, together with the fact that treatment of MALT lymphoma patients with bortezomib [135, 136]—a proteasome inhibitor with inhibitory effects on the NF-κB signal pathway [137]—is sucessful demonstrates the crucial role of NF-κB in MALT lymphomagenesis. To our knowledge, activated NF-κB is also found in MALT lymphoma patients without any translocation or mutation in one of the NF-κB signalling pathway-related genes. Hence, more studies on genetic alterations with a whole-genome/transcriptome approach are needed to clarify the molecular mechanism of NF-κB activation.

The development of MALT lymphomas is strongly associated with chronic infection by pathogens or autoantigens [76, 77, 79, 80, 138]. Eradication of the causing bacterial pathogen by antibiotics causes remission in the majority of MALT lymphoma patients [82, 89]. However, from our point of view, more studies on bacterial and viral pathogens screening by using a next-generation sequencing approach and additionally analysing the potentially restricted usage of variable gene of the immunoglobulin genes will elucidate the connection of MALT lymphomagenesis and chronic infections especially in the setting of extragastric origin.

References

1. Isaacson P, Wright DH. Malignant lymphoma of mucosa-associated lymphoid tissue. A distinctive type of B-cell lymphoma. Cancer. 1983;52:1410–6.
2. The Non-Hodgkin's Lymphoma Classification. A clinical evaluation of the International Lymphoma Study Group classification of non-Hodgkin's lymphoma. Blood. 1997; 89:3909–18.
3. Addis BJ, Hyjek E, Isaacson PG. Primary pulmonary lymphoma: a re-appraisal of its histogenesis and its relationship to pseudolymphoma and lymphoid interstitial pneumonia. Histopathology. 1988;13:1–17.
4. Hyjek E, Isaacson PG. Primary B cell lymphoma of the thyroid and its relationship to Hashimoto's thyroiditis. Hum Pathol. 1988;19:1315–26.
5. Hyjek E, Smith WJ, Isaacson PG. Primary B-cell lymphoma of salivary glands and its relationship to myoepithelial sialadenitis. Hum Pathol. 1988;19:766–76.
6. Isaacson P, Wright DH. Extranodal malignant lymphoma arising from mucosa-associated lymphoid tissue. Cancer. 1984;53:2515–24.
7. Isaacson PG, Du M-Q. MALT lymphoma: from morphology to molecules. Nat Rev Cancer. 2004;4:644.
8. Isaacson PG. WHO classification of tumours: pathology and genetics tumours of haematopoietic and lymphoid tissues. Lyon: IARC Press; 2001.

9. Isaacson PG, Wotherspoon AC, Diss T, Pan LX. Follicular colonization in B-cell lymphoma of mucosa-associated lymphoid tissue. Am J Surg Pathol. 1991;15:819.
10. Cogliatti SB. Primary B-cell gastric lymphoma: a clinicopathological study of 145 patients. Gastroenterology. 1991;101:1159.
11. Thieblemont C, Berger F, Dumontet C, Moullet I, Bouafia F, Felman P, Salles G, Coiffier B. Mucosa-associated lymphoid tissue lymphoma is a disseminated disease in one third of 158 patients analyzed. Blood. 2000;95:802–6.
12. Lopez-Guillermo A, Colomo L, Jimenez M, Bosch F, Villamor N, Arenillas L, Muntanola A, Montoto S, Gine E, Colomer D, Bea S, Campo E, Montserrat E. Diffuse large B-cell lymphoma: clinical and biological characterization and outcome according to the nodal or extranodal primary origin. J Clin Oncol. 2005;23:2797–804.
13. Chan JK, Ng CS, Isaacson PG. Relationship between high-grade lymphoma and low-grade B-cell mucosa-associated lymphoid tissue lymphoma (MALToma) of the stomach. Am J Pathol. 1990;136:1153.
14. Villuendas R, Piris MA, Orradre JL, Mollejo M, Rodriguez R, Morente M. Different bcl-2 protein expression in high-grade B-cell lymphomas derived from lymph node or mucosa-associated lymphoid tissue. Am J Pathol. 1991;139:989–93.
15. Ruskone-Fourmestraux A, Fischbach W, Aleman BM, Boot H, Du MQ, Megraud F, Montalban C, Raderer M, Savio A, Wotherspoon A, Group E. EGILS consensus report. Gastric extranodal marginal zone B-cell lymphoma of MALT. Gut. 2011;60:747–58.
16. Zucca E, Copie-Bergman C, Ricardi U, Thieblemont C, Raderer M, Ladetto M, GROUP, E. G. W. Gastric marginal zone lymphoma of MALT type: ESMO Clinical Practice Guidelines for diagnosis, treatment and follow-up. Ann Oncol. 2013;24(Suppl. 6):vi144–8.
17. Raderer M, Wohrer S, Streubel B, Troch M, Turetschek K, Jager U, Skrabs C, Gaiger A, Drach J, Puespoek A, Formanek M, Hoffmann M, Hauff W, Chott A. Assessment of disease dissemination in gastric compared with extragastric mucosa-associated lymphoid tissue lymphoma using extensive staging: a single-center experience. J Clin Oncol. 2006;24:3136–41.
18. Ruskone-Fourmestraux A, Lavergne A, Aegerter PH, Megraud F, Palazzo L, De Mascarel A, Molina T, Rambaud JL. Predictive factors for regression of gastric MALT lymphoma after anti-Helicobacter pylori treatment. Gut. 2001;48:297–303.
19. Musshoff K. Clinical staging classification of non-Hodgkin's lymphomas (author's transl). Strahlentherapie. 1977;153:218–21.
20. Radaszkiewicz T, Dragosics B, Bauer P. Gastrointestinal malignant lymphomas of the mucosa-associated lymphoid tissue: factors relevant to prognosis. Gastroenterology. 1992;102:1628–38.
21. Ruskone-Fourmestraux A, Dragosics B, Morgner A, Wotherspoon A, De Jong D. Paris staging system for primary gastrointestinal lymphomas. Gut. 2003;52:912–3.
22. Lebien TW, Tedder TF. B lymphocytes: how they develop and function. Blood. 2008;112:1570–80.
23. Nagasawa T. Microenvironmental niches in the bone marrow required for B-cell development. Nat Rev Immunol. 2006;6:107–16.
24. Alberts B. Molecular biology of the cell. New York: Garland Science; 2002.
25. Kuppers R, Zhao M, Hansmann ML, Rajewsky K. Tracing B cell development in human germinal centres by molecular analysis of single cells picked from histological sections. EMBO J. 1993;12:4955–67.
26. Perez-Andres M, Paiva B, Nieto WG, Caraux A, Schmitz A, Almeida J, Vogt Jr RF, Marti GE, Rawstron AC, Van Zelm MC, Van Dongen JJ, Johnsen HE, Klein B, Orfao A, Primary Health Care Group of Salamanca for the Study of MBL. Human peripheral blood B-cell compartments: a crossroad in B-cell traffic. Cytometry B Clin Cytom. 2010;78(Suppl. 1):S47–60.
27. Soulas-Sprauel P, Rivera-Munoz P, Malivert L, Le Guyader G, Abramowski V, Revy P, De Villartay JP. V(D)J and immunoglobulin class switch recombinations: a paradigm to study the regulation of DNA end-joining. Oncogene. 2007;26:7780–91.
28. Xu Z, Fulop Z, Zhong Y, Evinger III AJ, Zan H, Casali P. DNA lesions and repair in immunoglobulin class switch recombination and somatic hypermutation. Ann N Y Acad Sci. 2005;1050:146–62.

29. Liu YJ, Malisan F, De Bouteiller O, Guret C, Lebecque S, Banchereau J, Mills FC, Max EE, Martinez-Valdez H. Within germinal centers, isotype switching of immunoglobulin genes occurs after the onset of somatic mutation. Immunity. 1996;4:241–50.
30. Stavnezer J. Complex regulation and function of activation-induced cytidine deaminase. Trends Immunol. 2011;32:194–201.
31. Bothmer A, Robbiani DF, Feldhahn N, Gazumyan A, Nussenzweig A, Nussenzweig MC. 53BP1 regulates DNA resection and the choice between classical and alternative end joining during class switch recombination. J Exp Med. 2010;207:855–65.
32. Dierlamm J, Baens M, Wlodarska I, Stefanova-Ouzounova M, Hernandez JM, Hossfeld DK, de Wolf-Peeters C, Hagemeijer A, Van Den Berghe H, Marynen P. The apoptosis inhibitor gene API2 and a novel 18q gene, MLT, are recurrently rearranged in the t(11;18)(q21;q21) associated with mucosa-associated lymphoid tissue lymphomas. Blood. 1999;93:3601–9.
33. Streubel B, Lamprecht A, Dierlamm J, Cerroni L, Stolte M, Ott G, Raderer M, Chott A. T(14;18)(q32;q21) involving IGH and MALT1 is a frequent chromosomal aberration in MALT lymphoma. Blood. 2003;101:2335–9.
34. Streubel B, Vinatzer U, Lamprecht A, Raderer M, Chott A. T(3;14)(p14.1;q32) involving IGH and FOXP1 is a novel recurrent chromosomal aberration in MALT lymphoma. Leukemia. 2005;19(4):652–8.
35. Willis TG, Jadayel DM, Du M-Q, Peng H, Perry AR, Abdul-Rauf M, Price H, Karran L, Majekodunmi O, Wlodarska I. Bcl10 is involved in t(1;14)(p22;q32) of MALT B cell lymphoma and mutated in multiple tumor types. Cell. 1999;96:35.
36. Streubel B, Simonitsch-Klupp I, Mullauer L, Lamprecht A, Huber D, Siebert R, Stolte M, Trautinger F, Lukas J, Puspok A, Formanek M, Assanasen T, Muller-Hermelink HK, Cerroni L, Raderer M, Chott A. Variable frequencies of MALT lymphoma-associated genetic aberrations in MALT lymphomas of different sites. Leukemia. 2004;18:1722–6.
37. Van Keimpema M, Gruneberg LJ, Mokry M, Van Boxtel R, Koster J, Coffer PJ, Pals ST, Spaargaren M. FOXP1 directly represses transcription of pro-apoptotic genes and cooperates with NF-kappaB to promote survival of human B-cells. Blood. 2014;124(23):3431–40.
38. Dierlamm J, Wlodarska I, Michaux L, Stefanova M, Hinz K, Van Den Berghe H, Hagemeijer A, Hossfeld DK. Genetic abnormalities in marginal zone B-cell lymphoma. Hematol Oncol. 2000a;18:1–13.
39. Rosenwald A, Ott G, Stilgenbauer S, Kalla J, Bredt M, Katzenberger T, Greiner A, Ott MM, Gawin B, Dohner H, Muller-Hermelink HK. Exclusive detection of the t(11;18)(q21;q21) in extranodal marginal zone B cell lymphomas (MZBL) of MALT type in contrast to other MZBL and extranodal large B cell lymphomas. Am J Pathol. 1999;155:1817–21.
40. Takada S, Yoshino T, Taniwaki M, Nakamura N, Nakamine H, Oshima K, Sadahira Y, Inagaki H, Oshima K, Tadaatsu A. Involvement of the chromosomal translocation t(11;18) in some mucosa-associated lymphoid tissue lymphomas and diffuse large B-cell lymphomas of the ocular adnexa: evidence from multiplex reverse transcriptase-polymerase chain reaction and fluorescence in situ hybridization on using formalin-fixed, paraffin-embedded specimens. Mod Pathol. 2003;16:445–52.
41. Ott G, Katzenberger T, Greiner A, Kalla J, Rosenwald A, Heinrich U, Ott MM, Muller-Hermelink HK. The t(11;18)(q21;q21) chromosome translocation is a frequent and specific aberration in low-grade but not high-grade malignant non-Hodgkin's lymphomas of the mucosa-associated lymphoid tissue (MALT-) type. Cancer Res. 1997;57:3944–8.
42. Liu H. T(11;18)(q21;q21) is associated with advanced mucosa-associated lymphoid tissue lymphoma that expresses nuclear BCL10. Blood. 2001;98:1182.
43. Ye H. Variable frequencies of t(11;18)(q21;q21) in MALT lymphomas of different sites: significant association with CagA strains of H pylori in gastric MALT lymphoma. Blood. 2003;102:1012.
44. Liu H, Hamoudi RA, Ye H, Ruskone-Fourmestraux A, Dogan A, Isaacson PG, Du M-Q. t(11;18)(q21;q21) of mucosa-associated lymphoid tissue lymphoma results from illegitimate non-homologous end joining following double strand breaks. Br J Haematol. 2004;125:318–29.

45. Remstein ED, James CD, Kurtin PJ. Incidence and subtype specificity of API2-MALT1 fusion translocations in extranodal, nodal, and splenic marginal zone lymphomas. Am J Pathol. 2000;156:1183.
46. Ye H. BCL10 expression in normal and neoplastic lymphoid tissue: nuclear localization in MALT lymphoma. Am J Pathol. 2000;157:1147.
47. Ye H, Gong L, Liu H, Hamoudi RA, Shirali S, Ho L, Chott A, Streubel B, Siebert R, Gesk S, Martin-Subero JI, Radford JA, Banerjee S, Nicholson AG, Ranaldi R, Remstein ED, Gao Z, Zheng J, Isaacson PG, Dogan A, Du MQ. MALT lymphoma with t(14;18)(q32;q21)/IGH-MALT1 is characterized by strong cytoplasmic MALT1 and BCL10 expression. J Pathol. 2005;205:293–301.
48. Nakagawa M, Hosokawa Y, Yonezumi M, Izumiyama K, Suzuki R, Tsuzuki S, Asaka M, Seto M. MALT1 contains nuclear export signals and regulates cytoplasmic localization of BCL10. Blood. 2005;106:4210–6.
49. Lucas PC. Bcl10 and MALT1, independent targets of chromosomal translocation in malt lymphoma, cooperate in a novel NF-[kappa]B signaling pathway. J Biol Chem. 2001;276:19012.
50. Uren GA. Identification of paracaspases and metacaspases: two ancient families of caspase-like proteins, one of which plays a key role in MALT lymphoma. Mol Cell. 2000;6:961.
51. Zhou H. Bcl10 activates the NF-[kappa]B pathway through ubiquitination of NEMO. Nature. 2004;427:167.
52. Sanchez-Izquierdo D. MALT1 is deregulated by both chromosomal translocation and amplification in B-cell non-Hodgkin lymphoma. Blood. 2003;101:4539.
53. Hozak RR, Manji GA, Friesen PD. The BIR motifs mediate dominant interference and oligomerization of inhibitor of apoptosis Op-IAP. Mol Cell Biol. 2000;20:1877.
54. Mcallister-Lucas LM. Bimp1, a MAGUK family member linking protein kinase C activation to Bcl10-mediated NF-[kappa]B induction. J Biol Chem. 2001;276:30589.
55. Taji S, Nomura K, Matsumoto Y, Sakabe H, Yoshida N, Mitsufuji S, Nishida K, Horiike S, Nakamura S, Morita M, Taniwaki M. Trisomy 3 may predict a poor response of gastric MALT lymphoma to Helicobacter pylori eradication therapy. World J Gastroenterol. 2005;11:89–93.
56. Krugmann J, Tzankov A, Dirnhofer S, Fend F, Wolf D, Siebert R, Probst P, Erdel M. Complete or partial trisomy 3 in gastro-intestinal MALT lymphomas co-occurs with aberrations at 18q21 and correlates with advanced disease stage: a study on 25 cases. World J Gastroenterol. 2005;11:7384–5.
57. Deutsch AJ, Aigelsreiter A, Steinbauer E, Fruhwirth M, Kerl H, Beham-Schmid C, Schaider H, Neumeister P. Distinct signatures of B-cell homeostatic and activation-dependent chemokine receptors in the development and progression of extragastric MALT lymphomas. J Pathol. 2008;215:431–44.
58. Chanudet E, Ye H, Ferry J, Bacon CM, Adam P, Muller-Hermelink HK, Radford J, Pileri SA, Ichimura K, Collins VP, Hamoudi RA, Nicholson AG, Wotherspoon AC, Isaacson PG, Du MQ. A20 deletion is associated with copy number gain at the TNFA/B/C locus and occurs preferentially in translocation-negative MALT lymphoma of the ocular adnexa and salivary glands. J Pathol. 2009;217:420–30.
59. Honma K, Tsuzuki S, Nakagawa M, Karnan S, Aizawa Y, Kim WS, Kim YD, Ko YH, Seto M. TNFAIP3 is the target gene of chromosome band 6q23.3-q24.1 loss in ocular adnexal marginal zone B cell lymphoma. Genes Chromosomes Cancer. 2008;47:1–7.
60. Kim WS, Honma K, Karnan S, Tagawa H, Kim YD, Oh YL, Seto M, Ko YH. Genome-wide array-based comparative genomic hybridization of ocular marginal zone B cell lymphoma: comparison with pulmonary and nodal marginal zone B cell lymphoma. Genes Chromosomes Cancer. 2007;46:776–83.
61. Chanudet E, Huang Y, Ichimura K, Dong G, Hamoudi RA, Radford J, Wotherspoon AC, Isaacson PG, Ferry J, Du MQ. A20 is targeted by promoter methylation, deletion and inactivating mutation in MALT lymphoma. Leukemia. 2010;24:483–7.

62. Du MQ. MALT lymphoma: many roads lead to nuclear factor-kappab activation. Histopathology. 2011;58:26–38.
63. Deutsch AJ, Fruhwirth M, Aigelsreiter A, Cerroni L, Neumeister P. Primary cutaneous marginal zone B-cell lymphomas are targeted by aberrant somatic hypermutation. J Invest Dermatol. 2009;129:476–9.
64. Deutsch AJ, Aigelsreiter A, Staber PB, Beham A, Linkesch W, Guelly C, Brezinschek RI, Fruhwirth M, Emberger W, Buettner M, Beham-Schmid C, Neumeister P. MALT lymphoma and extranodal diffuse large B-cell lymphoma are targeted by aberrant somatic hypermutation. Blood. 2007;109:3500–4.
65. Willemze R, Jaffe ES, Burg G, Cerroni L, Berti E, Swerdlow SH, Ralfkiaer E, Chimenti S, Diaz-Perez JL, Duncan LM, Grange F, Harris NL, Kempf W, Kerl H, Kurrer M, Knobler R, Pimpinelli N, Sander C, Santucci M, Sterry W, Vermeer MH, Wechsler J, Whittaker S, Meijer CJ. WHO-EORTC classification for cutaneous lymphomas. Blood. 2005;105:3768–85.
66. Du M, Peng H, Singh N, Isaacson PG, Pan L. The accumulation of p53 abnormalities is associated with progression of mucosa-associated lymphoid tissue lymphoma. Blood. 1995;86:4587–93.
67. Davis RE, Ngo VN, Lenz G, Tolar P, Young RM, Romesser PB, Kohlhammer H, Lamy L, Zhao H, Yang Y, Xu W, Shaffer AL, Wright G, Xiao W, Powell J, Jiang JK, Thomas CJ, Rosenwald A, Ott G, Muller-Hermelink HK, Gascoyne RD, Connors JM, Johnson NA, Rimsza LM, Campo E, Jaffe ES, Wilson WH, Delabie J, Smeland EB, Fisher RI, Braziel RM, Tubbs RR, Cook JR, Weisenburger DD, Chan WC, Pierce SK, Staudt LM. Chronic active B-cell-receptor signalling in diffuse large B-cell lymphoma. Nature. 2010;463:88–92.
68. Kato M, Sanada M, Kato I, Sato Y, Takita J, Takeuchi K, Niwa A, Chen Y, Nakazaki K, Nomoto J, Asakura Y, Muto S, Tamura A, Iio M, Akatsuka Y, Hayashi Y, Mori H, Igarashi T, Kurokawa M, Chiba S, Mori S, Ishikawa Y, Okamoto K, Tobinai K, Nakagama H, Nakahata T, Yoshino T, Kobayashi Y, Ogawa S. Frequent inactivation of A20 in B-cell lymphomas. Nature. 2009;459:712–6.
69. Lenz G, Davis RE, Ngo VN, Lam L, George TC, Wright GW, Dave SS, Zhao H, Xu W, Rosenwald A, Ott G, Muller-Hermelink HK, Gascoyne RD, Connors JM, Rimsza LM, Campo E, Jaffe ES, Delabie J, Smeland EB, Fisher RI, Chan WC, Staudt LM. Oncogenic CARD11 mutations in human diffuse large B cell lymphoma. Science. 2008;319:1676–9.
70. Ngo VN, Young RM, Schmitz R, Jhavar S, Xiao W, Lim KH, Kohlhammer H, Xu W, Yang Y, Zhao H, Shaffer AL, Romesser P, Wright G, Powell J, Rosenwald A, Muller-Hermelink HK, Ott G, Gascoyne RD, Connors JM, Rimsza LM, Campo E, Jaffe ES, Delabie J, Smeland EB, Fisher RI, Braziel RM, Tubbs RR, Cook JR, Weisenburger DD, Chan WC, Staudt LM. Oncogenically active MYD88 mutations in human lymphoma. Nature. 2011; 470:115–9.
71. Bi Y, Zeng N, Chanudet E, Huang Y, Hamoudi RA, Liu H, Dong G, Watkins AJ, Ley SC, Zou L, Chen R, Zhu X, Du MQ. A20 inactivation in ocular adnexal MALT lymphoma. Haematologica. 2012;97:926–30.
72. Li ZM, Rinaldi A, Cavalli A, Mensah AA, Ponzoni M, Gascoyne RD, Bhagat G, Zucca E, Bertoni F. MYD88 somatic mutations in MALT lymphomas. Br J Haematol. 2012;158:662–4.
73. Liu F, Karube K, Kato H, Arita K, Yoshida N, Yamamoto K, Tsuzuki S, Kim W, Ko YH, Seto M. Mutation analysis of NF-kappaB signal pathway-related genes in ocular MALT lymphoma. Int J Clin Exp Pathol. 2012;5:436–41.
74. Min KO, Seo EJ, Kwon HJ, Lee EJ, Kim WI, Kang CS, Kim KM. Methylation of p16(INK4A) and p57(KIP2) are involved in the development and progression of gastric MALT lymphomas. Mod Pathol. 2006;19:141–8.
75. Suarez F, Lortholary O, Hermine O, Lecuit M. Infection-associated lymphomas derived from marginal zone B cells: a model of antigen-driven lymphoproliferation. Blood. 2006;107: 3034–44.
76. Isaacson PG, Spencer J. The biology of low grade MALT lymphoma. J Clin Pathol. 1995;48:395–7.

77. Cerroni L, Zochling N, Putz B, Kerl H. Infection by Borrelia burgdorferi and cutaneous B-cell lymphoma. J Cutan Pathol. 1997;24:457–61.
78. Parsonnet J, Isaacson PG. Bacterial infection and MALT lymphoma. N Engl J Med. 2004;350:213–5.
79. Negri E, Little D, Boiocchi M, La Vecchia C, Franceschi S. B-cell non-Hodgkin's lymphoma and hepatitis C virus infection: a systematic review. Int J Cancer. 2004;111:1–8.
80. Ferraccioli GF, Sorrentino D, De Vita S, Casatta L, Labombarda A, Avellini C, Dolcetti R, Di Luca D, Beltrami CA, Boiocchi M, Bartoli E. B cell clonality in gastric lymphoid tissues of patients with Sjogren's syndrome. Ann Rheum Dis. 1996;55:311–6.
81. Ferreri AJ, Campo E, Ambrosetti A, Ilariucci F, Seymour JF, Willemze R, Arrigoni G, Rossi G, Lopez-Guillermo A, Berti E, Eriksson M, Federico M, Cortelazzo S, Govi S, Frungillo N, Dell'oro S, Lestani M, Asioli S, Pedrinis E, Ungari M, Motta T, Rossi R, Artusi T, Iuzzolino P, Zucca E, Cavalli F, Ponzoni M. Anthracycline-based chemotherapy as primary treatment for intravascular lymphoma. Ann Oncol. 2004a;15:1215–21.
82. Ferreri AJ, Govi S, Pasini E, Mappa S, Bertoni F, Zaja F, Montalban C, Stelitano C, Cabrera ME, Giordano Resti A, Politi LS, Doglioni C, Cavalli F, Zucca E, Ponzoni M, Dolcetti R. Chlamydophila psittaci eradication with doxycycline as first-line targeted therapy for ocular adnexae lymphoma: final results of an international phase II trial. J Clin Oncol. 2012;30:2988–94.
83. Chanudet E, Adam P, Nicholson AG, Wotherspoon AC, Ranaldi R, Goteri G, Pileri SA, Ye H, Muller-Hermelink HK, Du MQ. Chlamydiae and mycoplasma infections in pulmonary MALT lymphoma. Br J Cancer. 2007;97:949–51.
84. Aigelsreiter A, Leitner E, Deutsch AJ, Kessler HH, Stelzl E, Beham-Schmid C, Beham A, Krugmann J, Dinges HP, Linkesch W, Neumeister P. Chlamydia psittaci in MALT lymphomas of ocular adnexals: the Austrian experience. Leuk Res. 2008;32:1292–4.
85. Aigelsreiter A, Gerlza T, Deutsch AJ, Leitner E, Beham-Schmid C, Beham A, Popper H, Borel N, Pospischil A, Raderer M, Kessler HH, Neumeister P. Chlamydia psittaci infection in nongastrointestinal extranodal MALT lymphomas and their precursor lesions. Am J Clin Pathol. 2011;135:70–5.
86. Du MQ. Intestinal dissemination of gastric mucosa-associated lymphoid tissue lymphoma. Blood. 1996b;88:4445.
87. Du M. Ongoing mutation in MALT lymphoma immunoglobulin gene suggests that antigen stimulation plays a role in the clonal expansion. Leukemia. 1996a;10:1190.
88. Hussell T, Isaacson PG, Crabtree JE, Spencer J. The response of cells from low-grade B-cell gastric lymphomas of mucosa-associated lymphoid tissue to Helicobacter pylori. Lancet. 1993;342:571.
89. Wotherspoon AC. Regression of primary low-grade B-cell gastric lymphoma of mucosa-associated lymphoid tissue type after eradication of Helicobacter pylori. Lancet. 1993;342:575.
90. Suerbaum S, Michetti P. Helicobacter pylori infection. N Engl J Med. 2002;347:1175–86.
91. Reimer P, Fischbach W, Goebeler M-E, Kraus M, Goldmann S, Müller C, Wilhelm M. Decreased frequency of HLA-B35 in patients with gastric MALT lymphoma. Ann Hematol. 2004;83:232.
92. Rollinson S, Levene AP, Mensah FK, Roddam PL, Allan JM, Diss TC, Roman E, Jack A, Maclennan K, Dixon MF, Morgan GJ. Gastric marginal zone lymphoma is associated with polymorphisms in genes involved in inflammatory response and antioxidative capacity. Blood. 2003;102:1007–11.
93. Ogawa N, Ping L, Zhenjun L, Takada Y, Sugai S. Involvement of the interferon-gamma-induced T cell-attracting chemokines, interferon-gamma-inducible 10-kd protein (CXCL10) and monokine induced by interferon-gamma (CXCL9), in the salivary gland lesions of patients with Sjögren's syndrome. Arthritis Rheum. 2002;46:2730–41.
94. Schmid U, Helbron D, Lennert K. Development of malignant lymphoma in myoepithelial sialadenitis (Sjogren's syndrome). Virchows Arch A Pathol Anat Histol. 1982;395:11–43.

95. Bahler DW, Miklos JA, Swerdlow SH. Ongoing Ig gene hypermutation in salivary gland mucosa-associated lymphoid tissue-type lymphomas. Blood. 1997;89:3335–44.
96. Berard CW, Greene MH, Jaffe ES, Magrath I, Ziegler J. NIH conference. A multidisciplinary approach to non-hodgkin's lymphomas. Ann Intern Med. 1981;94:218–35.
97. Kipps TJ, Tomhave E, Chen PP, Fox RI. Molecular characterization of a major autoantibody-associated cross-reactive idiotype in Sjogren's syndrome. J Immunol. 1989;142:4261–8.
98. Chen YW, Liang AC, Au WY, Chu KM, Wong KY, Hu X, Lu L, Tang JC, Chan KW, Beh SL, Kwong YL, Liang RH, Srivastava G. Multiple BCL6 translocation partners in individual cases of gastric lymphoma. Blood. 2003;102:1931–2. author reply 1932
99. Liang R, Chan WP, Kwong YL, Xu WS, Srivastava G, Ho FC. High incidence of BCL-6 gene rearrangement in diffuse large B-cell lymphoma of primary gastric origin. Cancer Genet Cytogenet. 1997;97:114–8.
100. Neumeister P, Hoefler G, Beham-Schmid C, Schmidt H, Apfelbeck U, Schaider H, Linkesch W, Sill H. Deletion analysis of the p16 tumor suppressor gene in gastrointestinal mucosa-associated lymphoid tissue lymphomas. Gastroenterology. 1997;112:1871–5.
101. Pan LX, Ramani P, Diss TC, Liang LN, Isaacson PG. Epstein-Barr virus associated lymphoproliferative disorder with fatal involvement of the gastrointestinal tract in an infant. J Clin Pathol. 1995;48:390–2.
102. Starostik P, Patzner J, Greiner A, Schwarz S, Kalla J, Ott G, Muller-Hermelink HK. Gastric marginal zone B-cell lymphomas of MALT type develop along 2 distinct pathogenetic pathways. Blood. 2002;99:3–9.
103. Kuppers R, Dalla-Favera R. Mechanisms of chromosomal translocations in B cell lymphomas. Oncogene. 2001;20:5580–94.
104. Pasqualucci L, Neumeister P, Goossens T, Nanjangud G, Chaganti RS, Kuppers R, Dalla-Favera R. Hypermutation of multiple proto-oncogenes in B-cell diffuse large-cell lymphomas. Nature. 2001a;412:341–6.
105. Bodor C, Bognar A, Reiniger L, Szepesi A, Toth E, Kopper L, Matolcsy A. Aberrant somatic hypermutation and expression of activation-induced cytidine deaminase mRNA in mediastinal large B-cell lymphoma. Br J Haematol. 2005;129:373–6.
106. Dijkman R, Tensen CP, Buettner M, Niedobitek G, Willemze R, Vermeer MH. Primary cutaneous follicle center lymphoma and primary cutaneous large B-cell lymphoma, leg type, are both targeted by aberrant somatic hypermutation but demonstrate differential expression of AID. Blood. 2006;107:4926–9.
107. Gaidano G, Pasqualucci L, Capello D, Berra E, Deambrogi C, Rossi D, Maria Larocca L, Gloghini A, Carbone A, Dalla-Favera R. Aberrant somatic hypermutation in multiple subtypes of AIDS-associated non-Hodgkin lymphoma. Blood. 2003;102:1833–41.
108. Halldorsdottir AM, Fruhwirth M, Deutsch A, Aigelsreiter A, Beham-Schmid C, Agnarsson BA, Neumeister P, Richard Burack W. Quantifying the role of aberrant somatic hypermutation in transformation of follicular lymphoma. Leuk Res. 2008;32(7):1015–21.
109. Liu M, Duke JL, Richter DJ, Vinuesa CG, Goodnow CC, Kleinstein SH, Schatz DG. Two levels of protection for the B cell genome during somatic hypermutation. Nature. 2008;451:841–5.
110. Pasqualucci L, Bhagat G, Jankovic M, Compagno M, Smith P, Muramatsu M, Honjo T, Morse III HC, Nussenzweig MC, Dalla-Favera R. AID is required for germinal center-derived lymphomagenesis. Nat Genet. 2008;40:108–12.
111. Matsumoto Y, Marusawa H, Kinoshita K, Endo Y, Kou T, Morisawa T, Azuma T, Okazaki IM, Honjo T, Chiba T. Helicobacter pylori infection triggers aberrant expression of activation-induced cytidine deaminase in gastric epithelium. Nat Med. 2007;13:470–6.
112. Young RM, Staudt LM. Targeting pathological B cell receptor signalling in lymphoid malignancies. Nat Rev Drug Discov. 2013;12:229–43.
113. Niemann CU, Wiestner A. B-cell receptor signaling as a driver of lymphoma development and evolution. Semin Cancer Biol. 2013;23:410–21.

114. Mcallister-Lucas LM, Baens M, Lucas PC. MALT1 protease: a new therapeutic target in B lymphoma and beyond? Clin Cancer Res. 2011;17:6623–31.
115. Liu H, Ruskon-Fourmestraux A, Lavergne-Slove A, Ye H, Molina T, Bouhnik Y, Hamoudi RA, Diss TC, Dogan A, Megraud F, Rambaud JC, Du MQ, Isaacson PG. Resistance of t(11;18) positive gastric mucosa-associated lymphoid tissue lymphoma to Helicobacter pylori eradication therapy. Lancet. 2001;357:39–40.
116. Balkwill F. Cancer and the chemokine network. Nat Rev Cancer. 2004;4:540–50.
117. Campbell DJ, Kim CH, Butcher EC. Chemokines in the systemic organization of immunity. Immunol Rev. 2003;195:58–71.
118. Laurence AD. Location, movement and survival: the role of chemokines in haematopoiesis and malignancy. Br J Haematol. 2006;132:255–67.
119. Bowman EP, Campbell JJ, Soler D, Dong Z, Manlongat N, Picarella D, Hardy RR, Butcher EC. Developmental switches in chemokine response profiles during B cell differentiation and maturation. J Exp Med. 2000;191:1303–18.
120. Cyster JG. Chemokines, sphingosine-1-phosphate, and cell migration in secondary lymphoid organs. Annu Rev Immunol. 2005;23:127–59.
121. Muller G, Hopken UE, Lipp M. The impact of CCR7 and CXCR5 on lymphoid organ development and systemic immunity. Immunol Rev. 2003;195:117–35.
122. Deutsch AJ, Steinbauer E, Hofmann NA, Strunk D, Gerlza T, Beham-Schmid C, Schaider H, Neumeister P. Chemokine receptors in gastric MALT lymphoma: loss of CXCR4 and upregulation of CXCR7 is associated with progression to diffuse large B-cell lymphoma. Mod Pathol. 2013;26:182–94.
123. Fischbach W, Goebeler-Kolve ME, Dragosics B, Greiner A, Stolte M. Long term outcome of patients with gastric marginal zone B cell lymphoma of mucosa associated lymphoid tissue (MALT) following exclusive Helicobacter pylori eradication therapy: experience from a large prospective series. Gut. 2004;53:34–7.
124. Wundisch T, Thiede C, Morgner A, Dempfle A, Gunther A, Liu H, Ye H, Du MQ, Kim TD, Bayerdorffer E, Stolte M, Neubauer A. Long-term follow-up of gastric MALT lymphoma after Helicobacter pylori eradication. J Clin Oncol. 2005;23:8018–24.
125. Troch M, Woehrer S, Streubel B, Weissel M, Hoffmann M, Mullauer L, Chott A, Raderer M. Chronic autoimmune thyroiditis (Hashimoto's thyroiditis) in patients with MALT lymphoma. Ann Oncol. 2008;19:1336–9.
126. Wohrer S, Troch M, Streubel B, Zwerina J, Skrabs C, Formanek M, Hauff W, Hoffmann M, Mullauer L, Chott A, Raderer M. MALT lymphoma in patients with autoimmune diseases: a comparative analysis of characteristics and clinical course. Leukemia. 2007;21:1812–8.
127. Zullo A, Hassan C, Andriani A, Cristofari F, Bassanelli C, Spinelli GP, Tomao S, Morini S. Treatment of low-grade gastric MALT-lymphoma unresponsive to Helicobacter pylori therapy: a pooled-data analysis. Med Oncol. 2010;27:291–5.
128. Wirth A, Gospodarowicz M, Aleman BM, Bressel M, Ng A, Chao M, Hoppe RT, Thieblemont C, Tsang R, Moser L, Specht L, Szpytma T, Lennard A, Seymour JF, Zucca E. Long-term outcome for gastric marginal zone lymphoma treated with radiotherapy: a retrospective, multi-centre, International Extranodal Lymphoma Study Group study. Ann Oncol. 2013;24:1344–51.
129. Tsang RW, Gospodarowicz MK, Pintilie M, Wells W, Hodgson DC, Sun A, Crump M, Patterson BJ. Localized mucosa-associated lymphoid tissue lymphoma treated with radiation therapy has excellent clinical outcome. J Clin Oncol. 2003;21:4157–64.
130. Hammel P, Haioun C, Chaumette MT, Gaulard P, Divine M, Reyes F, Delchier JC. Efficacy of single-agent chemotherapy in low-grade B-cell mucosa-associated lymphoid tissue lymphoma with prominent gastric expression. J Clin Oncol. 1995;13:2524–9.
131. Levy M, Copie-Bergman C, Gameiro C, Chaumette MT, Delfau-Larue MH, Haioun C, Charachon A, Hemery F, Gaulard P, Leroy K, Delchier JC. Prognostic value of translocation t(11;18) in tumoral response of low-grade gastric lymphoma of mucosa-associated lymphoid tissue type to oral chemotherapy. J Clin Oncol. 2005;23:5061–6.

132. Jager G, Neumeister P, Quehenberger F, Wohrer S, Linkesch W, Raderer M. Prolonged clinical remission in patients with extranodal marginal zone B-cell lymphoma of the mucosa-associated lymphoid tissue type treated with cladribine: 6 year follow-up of a phase II trial. Ann Oncol. 2006;17:1722–3.
133. Jager G, Hofler G, Linkesch W, Neumeister P. Occurrence of a myelodysplastic syndrome (MDS) during first-line 2-chloro-deoxyadenosine (2-CDA) treatment of a low-grade gastrointestinal MALT lymphoma. Case report and review of the literature. Haematologica. 2004;89(4):ECR01.
134. Martinelli G, Laszlo D, Ferreri AJ, Pruneri G, Ponzoni M, Conconi A, Crosta C, Pedrinis E, Bertoni F, Calabrese L, Zucca E. Clinical activity of rituximab in gastric marginal zone non-Hodgkin's lymphoma resistant to or not eligible for anti-Helicobacter pylori therapy. J Clin Oncol. 2005;23:1979–83.
135. Conconi A, Martinelli G, Lopez-Guillermo A, Zinzani PL, Ferreri AJ, Rigacci L, Devizzi L, Vitolo U, Luminari S, Cavalli F, Zucca E, International Extranodal Lymphoma Study Group. Clinical activity of bortezomib in relapsed/refractory MALT lymphomas: results of a phase II study of the International Extranodal Lymphoma Study Group (IELSG). Ann Oncol. 2011;22:689–95.
136. Troch M, Jonak C, Mullauer L, Puspok A, Formanek M, Hauff W, Zielinski CC, Chott A, Raderer M. A phase II study of bortezomib in patients with MALT lymphoma. Haematologica. 2009;94:738–42.
137. Panwalkar A, Verstovsek S, Giles F. Nuclear factor-kappaB modulation as a therapeutic approach in hematologic malignancies. Cancer. 2004;100:1578–89.
138. Ferreri AJ, Guidoboni M, Ponzoni M, De Conciliis C, Dell'oro S, Fleischhauer K, Caggiari L, Lettini AA, Dal Cin E, Ieri R, Freschi M, Villa E, Boiocchi M, Dolcetti R. Evidence for an association between Chlamydia psittaci and ocular adnexal lymphomas. J Natl Cancer Inst. 2004b;96:586–94.

Molecular Pathogenesis of Prostate Cancer

9

Maria A. Smolle, Johannes Haybaeck, and Martin Pichler

Contents

M.A. Smolle (✉)
Institute of Pathology, Medical University of Graz, Graz, Austria

Department of Orthopaedics and Trauma, Medical University of Graz, Graz, Austria
e-mail: maria.smolle@stud.medunigraz.at

J. Haybaeck, M.D., Ph.D.
Department of Pathology, Medical Faculty, Otto von Guericke University Magdeburg, 39120 Magdeburg, Germany

Institute of Pathology, Medical University of Graz, 8036 Graz, Austria
e-mail: johannes.haybaeck@med.ovgu.de

M. Pichler
Division of Clinical Oncology, Medical University of Graz, Graz, Austria
e-mail: martin.pichler@medunigraz.at

J. Haybaeck (ed.), *Mechanisms of Molecular Carcinogenesis – Volume 2*,
DOI 10.1007/978-3-319-53661-3_9

Abstract

With a lifetime risk of 17%, prostate cancer is the second most common malignancy in men. Although the vast majority of patients will have an excellent prognosis, a rapid progression of disease with metastatic spread is seen in some. Therefore, therapy may vary from active surveillance over hormone therapy and chemotherapy to radical prostatectomy. Thus, the molecular heterogeneity of prostate cancer is supposed to alter the tumour's biological behaviour.

This section starts with a brief introduction of prostate cancer in order to give readers a quick overview. Main signalling pathways, genes and proteins involved in the molecular pathogenesis of prostate cancer will be discussed in the second part. One emphasis is placed on the interactions between the different proteins, genes and signalling pathways. Another focus is on the characterisation of the impact of aberrant molecular profiles on the prognosis of patients with prostate cancer.

9.1 Introduction

Following lung cancer, prostate cancer (PC) is the second most common malignancy in men. The lifetime risk for being diagnosed with PC is 17%, with a steadily increasing incidence during life. An estimated 180,890 new cases are going to be diagnosed in the USA in 2016 [1]. Among these patients, 84% are predicted to survive the next 10 years, indicating that prostate cancer has a quite differing clinical course [2]. Epidemiology indicates that the major risk factors for PC are advanced age, black race and a family history of prostate cancer.

Since the initiation of screening men for PC by measuring prostate-specific antigen (PSA) levels in 1985, the lifetime risk for men being diagnosed with PC has nearly doubled. This may be caused by the fact that PC is not a homogenous disease. Slow-growing tumours that would not cause any problems during the man's life are often detected via PSA screening. However, also highly aggressive carcinomas significantly impairing the patient's prognosis are found [3]. As a consequence, treatment may vary from active surveillance, over radical prostatectomy and radiotherapy, to hormonal therapy and chemotherapy in late-stage patients. Interestingly, in certain subpopulations, large clinical trials were unable to show a better prognosis for PC patients following radical prostatectomy in comparison to active surveillance [4].

9.1.1 Histological Subtypes

The World Health Organization (WHO) differentiates between epithelial, neuroendocrine, mesenchymal, haematolymphoid and prostatic stromal tumours. The most common subtype is the "classic" acinar adenocarcinoma, accounting for more than 90% of histologically confirmed PCs. Rare subtypes include the ductal adenocarcinoma, urothelial carcinoma, squamous cell carcinoma and small cell carcinoma [5]. The Gleason scoring system is generally used to determine the differentiation of the tumour based on the glandular architecture, which is also a histopathological prognosticator of aggressiveness and commonly used in risk stratification tools [6].

9.1.2 Principles of Therapy

PCs limited to the prostate itself are subdivided into three risk groups, depending on tumour stage, PSA levels and the Gleason score. The treatment options in this group are chosen with regard to patient's comorbidities, patient's age, biological behaviour of the tumour and the patient's individual preferences. Radical prostatectomy may be suitable for patients without significant comorbidities, young age and low-to intermediate-risk disease [7]. However, surgical complications, including impotence and urinary incontinence, are common. Radical radiotherapy is an option for patients with older age, multi-morbid PC patients and can be either achieved by external beam therapy or brachytherapy [8]. Results are comparable to surgery, yet a direct comparison in a prospective clinical controlled trial is still missing. Hormonal therapy by androgen deprivation used in localised disease has a temporary effect, as hormone-refractory disease may develop [8]. Active surveillance by routinely performed blood tests for PSA, clinical examination and biopsies of the prostate every 2 years may be suitable for low-risk PC patients [9].

In patients with metastatic disease, hormonal therapy with LHRH agonists or antagonists may be chosen as first-line therapy. However, this long-standing gold standard was revised recently when results of two large prospective clinical trials (STAMPEDE and CHAARTED trial) were published. Based on these novel results, first-line therapy in metastatic PC should be hormonal therapy plus cytotoxic chemotherapy with Taxotere 75 mg/m^2 for six cycles until disease progression [10, 11]. Although initial tumour response rates are often good, most patients will develop castration resistance, first commonly noticed by increasing PSA levels, followed by clinical or radiological features of progressive disease [8]. In this clinical scenario, treatment with the CYP17A1 inhibitor abiraterone or the selective androgen-receptor blocker enzalutamide may be effective [12, 13].

Though surgical castration for metastatic PC decreases the levels of circulating androgens, side effects, such as fatigue, loss of libido and osteoporosis as a long-term toxicities, may not be acceptable to all patients [8]. External beam radiotherapy is chosen for local symptom palliation, whereas intravenous application of radioactive radium-223 chloride is the treatment of choice for patients with multiple bone metastases, as median overall survival can be prolonged significantly [14]. Other therapeutic options include second-line chemotherapy with cabazitaxel, a novel taxane-based agent which was associated with an overall survival benefit in patients previously treated with Taxotere [15, 16].

9.2 Molecular Pathogenesis

Various proteins, genes and signalling pathways are involved in the pathogenesis of PC. Table 9.1 gives an overview of genes and gene products involved in tumour development. In Fig. 9.1, the interactions between proteins and pathways are outlined.

9.2.1 Cancer-Testis Antigens (CTAs)

The cancer-testis antigens (CTAs) constitute a family of tumour-associated antigens that are normally only expressed in testicular germ cells. However, upregulation of these specific antigens is also frequently seen in PC [17]. Due to few antigen-presenting cells within the prostate, lack of MHC molecules on the surface of testicular cells and the natural blood-testis barrier, CTAs are thought to be privileged regarding the body's immune response [18]. The CTA superfamily includes members from the MAGE and NY-ESO-1 family of antigens that are both found in metastatic malignant melanoma [19]. SSX proteins constitute another subgroup of CTAs

Table 9.1 Genes involved in the pathogenesis of prostate cancer

Gene	Protein	Function	Protein expression
Akt	Akt	Oncogene	Overexpression
AR	Androgen receptor	Oncogene	Overexpression
AURKA	Aurora kinase A	Oncogene	Overexpression
CCND1	Cyclin D1	Oncogene	Overexpression
CCND2	Cyclin D2	Oncogene	Overexpression
CDKN1B	p27	Tumour suppressor gene	Underexpression
ERG	ERG	Oncogene	Overexpression
ESR1	ER-alpha	Oncogene	Overexpression
ESR2	ER-beta	Tumour suppressor gene	Underexpression
HOXB13	HOXB13	Function unknown	Overexpression
HOXC6	HOXC6	Oncogene	Overexpression
MAGI2	MAGI2	Function unknown	
MAK2K	MEK	Oncogene	Overexpression
mTOR	mTOR	Oncogene	Overexpression
MYCN	N-Myc	Oncogene	Overexpression
p53	p53	Tumour suppressor gene	Underexpression
PTEN	PTEN	Tumour suppressor gene	Underexpression
RAF-1	Raf	Oncogene	Overexpression
RB1	pRB	Tumour suppressor gene	Underexpression
SIX1	SIX1	Oncogene	Overexpression
SSX2	SSX2	Function unknown	Overexpression
TGF-beta	TGF-beta	Oncogenic potential	Overexpression

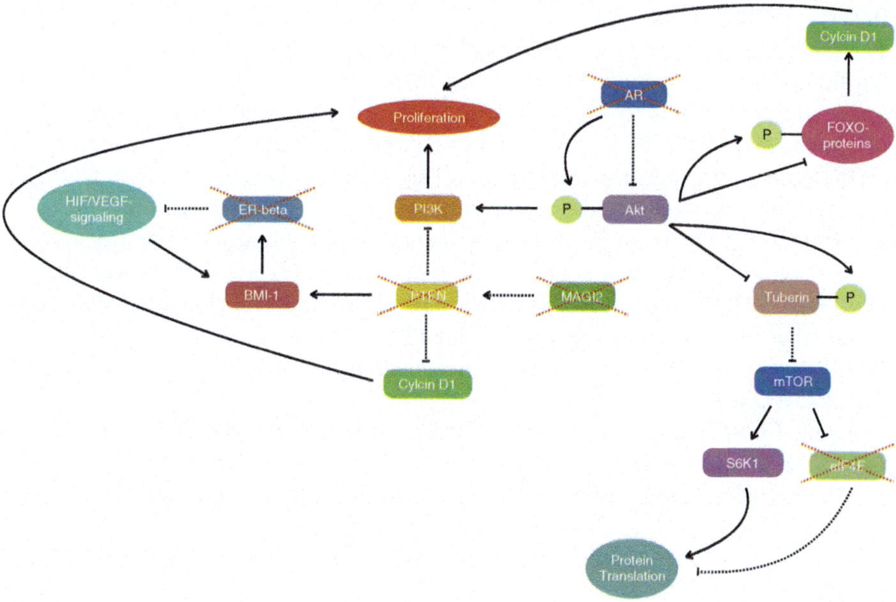

Fig. 9.1 Proteins involved in the pathogenesis of prostate cancer

and are aberrantly expressed in colon cancer, lymphoma, breast cancer, ovarian carcinoma and prostate cancer [20, 21]. All of these proteins are possible targets for cancer immunotherapy [19].

9.2.1.1 SSX Family

Synovial sarcoma X chromosome breakpoint (SSX) proteins belong to the CTA superfamily [22]. Their detailed function is still unknown. Altered expression of SSX family members is present in different tumour types, including PC [21, 23].

Not only *SSX2*, but also *SSX1* and *SSX5*, can be upregulated in PC cell lines under treatment with 5-aza-2′-deoxycytidine (=azacitidine) [24]. In human PC tissue, SSX2 is almost exclusively expressed in metastatic lesions, whereas it seems to be absent in primary tumours [25]. This is concordant with the observation that SSX2 p103-111 peptide-specific T-cell levels are higher in blood samples of PC patients with metastases [26]. Therefore, SSX2 might serve as an immunotherapeutic target in advanced stages of PC for future clinical trials.

9.2.2 HOX Family Members

The human genome comprises four *HOX* clusters, located on chromosome 2 (*HOXD*), 12 (*HOXC*), 7 (*HOXA*) and 17 (*HOXB*) [27]. Genes of the *HOX* clusters are expressed at different stages during human development.

9.2.2.1 HOXB13

The *HOXB13* is a homeobox transcription factor gene located on chromosome 17. All homeobox superfamily members are characterised by the so-called homeodomain, a highly conserved DNA-binding domain. HOX13 gene members, for example, are necessary during the embryonal development of the prostate. In adult tissue, HOXB13 is still expressed at a high level in healthy prostate cells [28]. HOXB13 actively interacts with the androgen receptor (AR), which is why a potential role of this HOX family member in the development of PC is supposed. Experiments have shown that PC cell growth is both mediated by an overexpression and suppression of HOXB13 [29, 30]. However, it seems more likely that *HOXB13* acts as an oncogene rather than a tumour suppressor gene [31].

9.2.2.2 HOXC6

The gene for the transcription factor HOXC6 is located on chromosome 12 together with *HOXC4* and *HOXC5*, all encoding two isoforms. Comparing all *HOXC* genes, the *HOXC6* gene seems to be upregulated most in PC [32]. Suppression of *HOXC6* leads to a decreased cell proliferation rate, thus exerting oncogenic functions. However, altered expression levels of HOXC6 are not associated with the patient's prognosis [32].

9.2.3 SIX1

SIX1, a member of the Six family of homeodomain transcription factors, is upregulated in cancers of different origin, including rhabdomyosarcoma, hepatocellular carcinoma and breast cancer [33, 34].

In PC, relative overexpression of SIX1 is present in comparison to surrounding healthy prostate tissue [35]. As the expression levels increase with ongoing tumour dedifferentiation, high levels of SIX1 are associated with a poorer patient's prognosis [35]. Although the exact mechanism of SIX1 in the pathogenesis of PC is still unclear, a possible oncogenic role of this transcription factor is plausible.

9.2.4 Cell Cycle

9.2.4.1 Cyclins and Cyclin-Dependent Kinases (CDKs)

Cyclin-dependent kinases (CDKs) constitute a family of serine-threonine protein kinases. With their regulatory subunits, the cyclins, CDKs play a crucial role in the regulation of the cell cycle [36]. The G1 phase is controlled by the three members of D cyclins by binding to the retinoblastoma protein (pRB) [37].

In PC pathogenesis, overexpression of cyclin D1 is commonly found [38]. PTEN usually targets cyclin D1, both impairing its protein stability and inhibiting translocation of cyclin D1 into the nucleus [39]. In absence of PTEN, the consecutive overexpression of cyclin D1 leads to increased proliferation of PC cells [40].

Cyclin D2 is known for its oncogenic potential in gastric and colon cancer, as overexpression of this specific cyclin correlates with cancer progression [41, 42]. On the other hand, depletion of cyclin D2 induced by promoter hypermethylation is associated with a poor prognosis in breast, pancreatic and lung cancer [43].

In the AR-positive PC cell line LnCaP, high cyclin D2-levels seem to have a protective effect, whilst no inhibitory effect emerges for PC3 cells [44]. Therefore, an AR-dependent mechanism is supposed.

9.2.4.2 Cyclin-Dependent Kinase Inhibitors (CKIs)

The two cyclin-dependent kinase inhibitor (CKI) families Cip/Kip and INK4 exert their regulatory function by inactivating cyclin/CDK complexes [45]. Two proteins are encoded by the *INK4a/ARF* locus, namely, p16INK4a and p14ARF (p19INK4d), both members of the INK4-inhibitor group. They are involved in the regulation of the pRB- and p53-pathway and pointedly inhibit the catalytic domains of CDK4 and CDK6 [45].

P21, p27 and p57 all belong to the Cip/Kip family of CKIs. P27 usually prevents the transition from G1 into S-phase. Therefore, p27 facilitates cell cycle exit under normal conditions. Decreased levels of the tumour suppressor p27 are commonly found in PC and correlate with aggressive potential, high Gleason grades and a poor prognosis [46].

9.2.4.3 Retinoblastoma Protein (pRB)

The retinoblastoma protein (pRB) is involved in the transition from G1 into S-phase of cell cycle. It normally binds members of the E2F family of transcription factors, thus preventing cell cycle progression [47]. Inactivation of pRB by its phosphorylation leads to a release of E2F members. Therefore, genes responsible for cell cycle progression are expressed [47]. Raf-1 directly phosphorylates pRB and induces upregulation of cyclin D1 [48]. Moreover, inactivation of pRB is also an effect of the Ras/PI3K/Akt pathway [49].

9.2.5 Tumour Suppressor Genes

9.2.5.1 MAGI2

The synaptic scaffolding molecule MAGI2 belongs to the membrane-associated guanylate kinase superfamily. It collects and merges specific cellular signalling proteins, including PTEN and atrophin-1 [50]. MAGI2 interacts with the C-terminal end of PTEN, therefore enabling the conversion of PIP3 into PIP2 [51]. MAGI2 positively influences the activity of PTEN by suppressing its protein degradation [50].

However, *MAGI2* downregulation is not associated with PTEN expression levels, indicating that different genomic events are responsible for changes in the two genes [52].

9.2.5.2 P53

The human p53 protein is built up of 393 amino acids, with the first 42 amino acids constituting the domain for transcriptional activation. P53 is both essential for cell cycle regulation and initiation of cellular apoptosis [53]. MDM2 is a negative regulator of p53, as MDM2 has a E3 ubiquitin ligase activity. Therefore, the formation of the MDM2-p53 complex leads to ubiquitination and proteasome degradation of p53 [54].

Mutations of p53 are present in more than half of human cancers, also in PC [55]. Inactivating mutations of p53 result in a dysregulated cell cycle and impaired apoptosis, hence strongly influencing tumour progression [56].

The presence of the $Arg^{72}Pro$ functional polymorphism in the *TP53* gene, however, is associated with a decreased PC risk [57]. When the single-nucleotide polymorphism (SNP) 309 in the *MDM2* gene is present, a protective joint effect together with $Arg^{72}Pro$ polymorphism emerges, resulting in a favourable prognosis for patients with PC [57].This may be explained by the fact that the SNP309 reduces MDM2 protein expression, whilst the $Arg^{72}Pro$ polymorphism leads to an increased transcriptional activity of p53 [58].

9.2.6 Oncogenes

9.2.6.1 AURKA

The *Aurora kinase A* (*AURKA*) gene is located on chromosome 20, a commonly amplified gene region in human solid tumours. It is responsible for centrosome maturation and separation, as well as centrosomal microtubule stabilisation and nucleation in the cell cycle [59]. Thus, upregulation of *AURKA* results in a permanently active kinase during all stages of the cell cycle. The upregulation is then seen both in the nucleus and the cytoplasm [60]. Overexpression of AURKA induces activity of several transcription factors with oncogenic potential [61].

In "conventional" PC (i.e., prostatic adenocarcinoma), amplification of *AURKA* is present in about 5% of cases, most often accompanied by upregulation of *MYCN*. Moreover, in 40% of primary neuroendocrine prostate cancers (NEPCs) or treatment-related neuroendocrine prostate cancers (t-NEPCs), co-amplification of *AURKA* and *MYCN* can be found [62]. Interestingly, *AURKA* and *MYCN* amplification was present in 65% of prostatic adenocarcinoma tissue samples from patients who later developed t-NEPCs [63]. Moreover, these alterations were present in PC tissue independently from Gleason scores, PSA levels and pathological tumour stages. Therefore, *AURKA* and *MYCN* amplifications could serve as prognostic and predictive biomarkers in PC, as patients harbouring this genetic aberration seem to be at high risk of developing extremely aggressive t-NEPCs [63].

9.2.6.2 MYCN

Amplification of the *MYCN* gene on the short arm of chromosome 2 is a frequent event in PC [64]. Overexpression of the N-Myc protein induces glutaminolysis, thus adapting the mitochondrial metabolism in a glutamine-dependent manner [65].

AURKA stabilises N-Myc by inhibiting its Fbxw7-mediated degradation [66]. N-Myc is essential for proper brain development, as it induces proliferation of granule neuron precursors in the hindbrain and forebrain [67]. Amplification of the *MYCN* region is found in nearly 20% of neuroblastoma cases, positively correlating with an aggressive biological pattern, advanced tumour stages and thus a poor patient's prognosis [68].

Not surprisingly, co-amplification of *AURKA* and *MYCN* is a common condition in PC, especially in those tumours harbouring the potential of transdifferentiating into a neuroendocrine subtype [63].

9.2.6.3 ERG

One of the 30 ETS-family genes of transcription factors is *ERG* (ETS-related gene), located on chromosome 21 [69]. *ETS* genes either act as repressors or activators of transcription, therefore being important for embryogenesis and early neuronal development, as well as lifelong haematopoiesis [70]. ERG is not expressed in normal prostate cells [71]. However, consistent upregulation of *ERG* is found in PC, caused by a gene fusion with the androgen-driven promoter of the *TMPRSS2* gene [72]. The *TMPRSS2* promotor contains androgen-sensitive elements, causing the overexpression of *ERG* in presence of androgens [73].

ERG plays a crucial role in the epithelial to mesenchymal transformation (EMT). ERG-induced upregulation of the *FZD4* gene encoding for a receptor of the frizzled family leads to reduction of epithelial markers and concurrent induction of mesenchymal ones [74]. Loss of E-cadherin, activation of metalloproteases and upregulation of the *vimentin* gene are all consequences of ERG overexpression, leading to increased invasive potential of tumour cells [75].

ERG impairs AR-related transcription. It represses two promoters of PSA, namely, trefoil factor 3 and prostein, which are both normally induced by AR [76]. Moreover, upregulation of *MYC* is seen in ERG-positive PC cells, whilst high AR levels block the promoter of *MYC* in ERG-deprived cells [77].

Although data on ERG as a prognostic and predictive biomarker in PC are rather conflicting, most authors conclude that the *ERG-TMPRSS2* fusion constitutes an early event in the development of PC [70].

9.2.7 Oestrogen-Receptor Pathway

Oestrogens affect the pathogenesis of PC by interacting with the oestrogen-receptor alpha (ER-alpha) and ER-beta. ER-alpha is encoded by the *ESR1* gene and expressed mainly in the prostatic stroma. On the other hand, *ESR2* encodes for ER-beta, which is predominantly found in the epithelium of the prostate [78].

9.2.7.1 ESR1

In the healthy prostate, ER-alpha is necessary for proper development and maintenance of fibroblast proliferation in the stroma [79]. Prostatic dysplasia is promoted by ER-alpha through induction of massive proliferation and inflammation [80].

ER-alpha expression is only present in the stroma of benign and malignant prostatic tissue samples. Following hormone therapy with LHRH analogues (e.g., degarelix), in 25% of cases, an overexpression of ER-alpha in the epithelium will develop. Moreover, expression levels of ER-alpha increase constantly during prolonged hormonal castration using degarelix. The overexpression of ER-alpha positively correlates with proliferation of PC cells [81].

9.2.7.2 ESR2

ER-beta expression is constantly lowered with increasing dedifferentiation from prostatic intraepithelial neoplasia over low Gleason scores to poorly differentiated tumours [82]. ER-beta seems to have a potential as suppressor, because the overexpression of the receptor is associated with decreased proliferation rates and invasiveness in PC cell lines [83].

Loss during the pathogenesis of PC is epigenetically regulated by hypermethylation of the ER-beta promoter CpG islands, resulting in transcriptional silencing [84].

Loss of PTEN is associated with decreased levels of ER-beta. PTEN deletion induces the polycomb complex protein BMI-1. High levels of BMI-1 impede ER-beta transcription in PC cells and therefore induce HIF-1/VEGF signalling. In a positive feedback loop, the activation of HIV-1/VEGF signalling sustains BMI-1 expression [85].

9.2.8 Androgen-Receptor Pathway

The AR belongs to the superfamily of steroid hormones. Heat-shock proteins (HSPs) bind ARs when they are inactive, i.e., free from ligands as dihydrotestosterone. After binding a ligand to the ligand-binding domain (LBD), the AR dimerises. Its N-terminal domain is then phosphorylated, resulting in a release of the HSP. Next, the receptor dimer is located into the nucleus and binds to specific elements of the DNA, thus enhancing transcription and activation of signalling pathways [86].

In PC, *AR* gene amplification is usually low at baseline. Especially androgen deprivation therapy with substances, such as leuproline, triptorelin and goserelin, enhances *AR* gene amplification [87]. Moreover, patients with castration-resistant prostate cancer (CRPC) present with consistently high androgen levels within the tumour tissue. Therefore, AR-related signalling pathways are overactive [88]. Both the transformation of "weak" androgens into "potent" ones (e.g., dihydrotestosterone) and the enhanced production of androgens in general within the tumour may cause the increased androgen levels [89].

During hormonal therapy, a conversion from hormone-sensitive tumours into CRPCs is commonly seen. The novel therapeutic drugs enzalutamide and arbiraterone are effective in 60% of these patients. However, about 40% show primary resistance to these drugs, and almost all patients treated with the novel therapeutics will develop a secondary resistance [90]. Enzalutamide directly binds to the AR as an antagonist, thus impeding AR translocation into the nucleus [91]. The production of testosterone precursors, including dehydroepiandrostenedione (DHEA) and androstenedione, is impaired by arbiraterone, as it irreversibly inhibits CYP17A1 [92]. Interestingly, the metabolic product of arbiraterone, D4A (delta-4-abiraterone), seems to antagonise the AR similarly to enzalutamide [93].

Point mutations of the AR are predominantly found in CRPC samples. They may result in loss of function, no change or gain in function [94]. Many AR splice variants have been described lacking the LBD, resulting in an independent and constitutive activation of AR-related signalling pathways [95]. Therefore, PCs harbouring specific AR splice variants, such as ARV7, are resistant to enzalutamide and arbiraterone [96].

9.2.9 PI3K/Akt/mTOR Pathway

Phosphoinositide 3-kinases (PI3Ks) are proteins important for several cellular mechanisms, including growth and proliferation, vesicle trafficking, glucose homeostasis and metabolism [97]. Most PI3K members bind to regulatory subunits in order to exert their specific functions. They are subdivided into three classes: Vps34, the only representative of PI3K class III, is important for trafficking of vesicles and interacts with class I PI3Ks by regulating mTORC1 signalling [98]. Class II PI3Ks are effective without binding to a regulatory subunit. They are activated by receptor tyrosine kinases, cytokine receptors and integrins, thus inducing internalisation of receptors and intensifying membrane trafficking [97].

The functions of class I PI3Ks are best known. Whilst PI3K1A (subtype IA) is activated by receptor tyrosine kinases, PI3K1B (subtype IB) is induced by G-protein-coupled receptors [99].

9.2.9.1 Akt

PI3Ks catalyse the formation of phosphatidylinositol-3,4,5-triphosphate (PIP3) in the membrane of cells. PIP3 itself activates several downstream targets, including Akt. This serine-threonine protein kinase is involved in the regulation of the cell cycle, cell survival, protein synthesis and cellular growth [100]. During the cell cycle, Akt-mediated phosphorylation of FOXO proteins results in cyclin D1 expression, which is why cell cycle inhibition is impaired [101].

9.2.9.2 mTOR

Akt activates the mammalian target of rapamycin (mTOR) through phosphorylation and inhibition of tuberin, a product of the tuberous sclerosis complex-2 (TSC2) [102]. mTOR induces protein translation by activating the ribosomal S6 kinase (S6K1) and concurrently inhibiting the eukaryotic translation initiation factor 4E (eIF4E) [103]. Targeting mTOR with the mTOR-inhibitor rapamycin and its derivatives results in decreased proliferation of tumour cells, both in vitro and in vivo using xenograft models [104]. In PC, however, a chronic inhibition of the PI3K/Akt/mTOR pathway leads to increased AR and PSA levels, resulting in a robust resistance to enzalutamide [105].

9.2.9.3 PTEN

Loss of heterozygosity (LOH) or inactivation of phosphate and tensin homologue (PTEN) is one of the most common genetic alterations in PC [106]. The tumour suppressor gene encoding for PTEN is located on chromosome 10. The main function of PTEN is reformation of PIP2 out of PIP3. Therefore, the PI3K/Akt/mTOR pathway is blocked [107].

High Gleason scores, an aggressive behaviour and advanced tumour stages are associated with this genetic aberration [108]. PTEN deletion leads to overexpression of the polycomb complex protein BMI-1. This protein, in turn, represses oestrogen-receptor (ER)-beta transcription in PC cells. Suppression of ER-beta enables activation of HIF-1/VEGF signalling, causing even higher levels of BMI-1 [85].

Moreover, loss of PTEN leads to an uncontrolled activation of the PI3K/Akt/mTOR pathway. About 40% of primary and up to 70% of metastatic PC are partially driven by this pathological pathway activation [64].

In PTEN-negative PCs, a two-way crosstalk between the PI3K/Akt/mTOR and AR pathway has been observed [109]. Whilst blockage of the AR leads to an activation of PI3K through increased Akt phosphorylation, the inhibition of the PI3K/Akt/mTOR pathway results in enhanced AR protein levels [109]. As a consequence, targeting one or the other signalling pathway is less effective than a simultaneous inhibition of AR and PI3K, both in hormone-sensitive and castration-resistant PCs [105].

9.2.10 Transforming Growth Factor-Beta Pathway

The transforming growth factor-beta (TGF-beta) exerts its function by binding to the transmembrane receptors TbetaR-I and TbetaR-II, thus inducing several downstream signalling pathways [110]. TGF-beta suppresses cellular growth in healthy tissues, whereas it develops oncogenic potential in tumours by overexpression [111]. This sounds contradictory, as high levels of this tumour suppressor should induce growth arrest. However, studies have shown that PC cells did not respond to high TGF-beta levels, whilst concurrent activation of the Akt pathway was observed [112]. This may be due to frequent loss of PTEN in PC, promoting activation of the PI3K/Akt/mTOR pathway [64].

TGF-beta exerts oncogenic functions by promoting EMT [113]. In normal ectodermal epithelium, selected cells, from time to time, undergo EMT by forming stress actin fibres, losing epithelial polarity and removing cellular adhesion markers from the outer membrane, including E-cadherin and integrin beta-1 [114, 115]. EMT is associated with tumour invasion and metastasis, as transformed cells breach the extracellular matrix [112]. Interestingly, both TGF-beta activation and PI3K/Akt/mTOR activation are required for sufficient EMT, thus enhancing invasiveness and metastatic potential of PC cells [112].

9.2.11 Ras/Raf/MAPK Pathway

The activation of the Ras/Raf/MAPK pathway leads to proliferation of PC cells both in an AR-independent and AR-dependent manner [116]. The activation of Raf by specific members of the Ras GTPase family leads to initiation of the MAPK cascade by induction of the mitogen-activated protein kinase (MAPKK, MEK).In a positive feedback loop, the mitogen-activated protein kinase (MAPK) sustains high Raf levels [117]. Both MEK and Raf are frequently overexpressed in localised and advanced stages of PC [117].

MAPK and Raf independently phosphorylate and thus activate the transcription factors AP-1 (c-Jun) and c-myc [118]. Activated AP-1 induces AR-associated genes, therefore bypassing AR-dependent signalling [119]. However, Raf is able to circumvent the MAPK pathway through direct phosphorylation and inactivation of pRB, leading to cell cycle progression [118]. As a consequence, high levels of MAPK and Raf-1 correlate with an aggressive behaviour and progression into CRPC.

Conclusion

Several genes, proteins and signalling pathways are involved in the molecular pathogenesis of PC. Their interactions have an influence on the tumour's biological behaviour and aggressiveness. Moreover, the adaptability of PC to specific therapeutic agents is facilitated by heavily interconnected pathways. Therefore, the tumour's variability should be kept in mind at any time during treatment, and prompt action may be necessary once disease progression is apparent.

References

1. Siegel RL, Miller KD, Jemal A. Cancer statistics, 2016. CA Cancer J Clin. 2016;66(1):7–30.
2. Mitchell T, Neal DE. The genomic evolution of human prostate cancer. Br J Cancer. 2015;113(2):193–8.
3. Heidegger I, Massoner P, Sampson N, Klocker H. The insulin-like growth factor (IGF) axis as an anticancer target in prostate cancer. Cancer Lett. 2015;367(2):113–21.
4. Wilt TJ, Brawer MK, Jones KM, Barry MJ, Aronson WJ, Fox S, et al. Radical prostatectomy versus observation for localized prostate cancer. N Engl J Med. 2012;367(3):203–13.
5. Eble JN, Sauter G, Epstein JI, Sesterhenn IA. In: Kleihues P, Sobin LH, editors. World Health Organization classification of tumours. Lyon, France: IARC Press; 2002.
6. Gleason DF. Classification of prostatic carcinomas. Cancer Chemother Rep. 1966;50(3):125–8.
7. Bill-Axelson A, Holmberg L, Ruutu M, Garmo H, Stark JR, Busch C, et al. Radical prostatectomy versus watchful waiting in early prostate cancer. N Engl J Med. 2011;364(18):1708–17.
8. Attard G, Parker C, Eeles RA, Schroder F, Tomlins SA, Tannock I, et al. Prostate cancer. Lancet. 2016;387(10013):70–82.
9. Parker C. Active surveillance: towards a new paradigm in the management of early prostate cancer. Lancet Oncol. 2004;5(2):101–6.
10. James ND, Sydes MR, Clarke NW, Mason MD, Dearnaley DP, Spears MR, et al. Addition of docetaxel, zoledronic acid, or both to first-line long-term hormone therapy in prostate cancer (STAMPEDE): survival results from an adaptive, multiarm, multistage, platform randomised controlled trial. Lancet. 2016;387(10024):1163–77.
11. Vale CL, Burdett S, Rydzewska LH, Albiges L, Clarke NW, Fisher D, et al. Addition of docetaxel or bisphosphonates to standard of care in men with localised or metastatic, hormone-sensitive prostate cancer: a systematic review and meta-analyses of aggregate data. Lancet Oncol. 2016;17(2):243–56.
12. Beer TM, Armstrong AJ, Rathkopf DE, Loriot Y, Sternberg CN, Higano CS, et al. Enzalutamide in metastatic prostate cancer before chemotherapy. N Engl J Med. 2014;371(5):424–33.

13. Ryan CJ, Smith MR, de Bono JS, Molina A, Logothetis CJ, de Souza P, et al. Abiraterone in metastatic prostate cancer without previous chemotherapy. N Engl J Med. 2013;368(2):138–48.
14. Wissing MD, van Leeuwen FW, van der Pluijm G, Gelderblom H. Radium-223 chloride: Extending life in prostate cancer patients by treating bone metastases. Clin Cancer Res. 2013;19(21):5822–7.
15. Tannock IF, de Wit R, Berry WR, Horti J, Pluzanska A, Chi KN, et al. Docetaxel plus prednisone or mitoxantrone plus prednisone for advanced prostate cancer. N Engl J Med. 2004;351(15):1502–12.
16. de Bono JS, Oudard S, Ozguroglu M, Hansen S, Machiels JP, Kocak I, et al. Prednisone plus cabazitaxel or mitoxantrone for metastatic castration-resistant prostate cancer progressing after docetaxel treatment: a randomised open-label trial. Lancet. 2010;376(9747):1147–54.
17. Bricard G, Bouzourene H, Martinet O, Rimoldi D, Halkic N, Gillet M, et al. Naturally acquired MAGE-A10- and SSX-2-specific CD8+ T cell responses in patients with hepatocellular carcinoma. J Immunol. 2005;174(3):1709–16.
18. Simpson AJ, Caballero OL, Jungbluth A, Chen YT, Old LJ. Cancer/testis antigens, gametogenesis and cancer. Nat Rev Cancer. 2005;5(8):615–25.
19. Velazquez EF, Jungbluth AA, Yancovitz M, Gnjatic S, Adams S, O'Neill D, et al. Expression of the cancer/testis antigen NY-ESO-1 in primary and metastatic malignant melanoma (MM)--correlation with prognostic factors. Cancer Immun. 2007;7:11.
20. Tureci O, Chen YT, Sahin U, Gure AO, Zwick C, Villena C, et al. Expression of SSX genes in human tumors. Int J Cancer. 1998;77(1):19–23.
21. Gure AO, Wei IJ, Old LJ, Chen YT. The SSX gene family: characterization of 9 complete genes. Int J Cancer. 2002;101(5):448–53.
22. Gure AO, Tureci O, Sahin U, Tsang S, Scanlan MJ, Jager E, et al. SSX: a multigene family with several members transcribed in normal testis and human cancer. Int J Cancer. 1997;72(6):965–71.
23. Cronwright G, Le Blanc K, Gotherstrom C, Darcy P, Ehnman M, Brodin B. Cancer/testis antigen expression in human mesenchymal stem cells: down-regulation of SSX impairs cell migration and matrix metalloproteinase 2 expression. Cancer Res. 2005;65(6):2207–15.
24. Dubovsky JA, McNeel DG. Inducible expression of a prostate cancer-testis antigen, SSX-2, following treatment with a DNA methylation inhibitor. Prostate. 2007;67(16):1781–90.
25. Smith HA, Cronk RJ, Lang JM, McNeel DG. Expression and immunotherapeutic targeting of the SSX family of cancer-testis antigens in prostate cancer. Cancer Res. 2011;71(21):6785–95.
26. Smith HA, McNeel DG. Vaccines targeting the cancer-testis antigen SSX-2 elicit HLA-A2 epitope-specific cytolytic T cells. J Immunother. 2011;34(8):569–80.
27. Graham A. Developmental patterning. The Hox code out on a limb. Curr Biol. 1994;4(12):1135–7.
28. Economides KD, Capecchi MR. Hoxb13 is required for normal differentiation and secretory function of the ventral prostate. Development. 2003;130(10):2061–9.
29. Norris JD, Chang CY, Wittmann BM, Kunder RS, Cui H, Fan D, et al. The homeodomain protein HOXB13 regulates the cellular response to androgens. Mol Cell. 2009;36(3):405–16.
30. Jung C, Kim RS, Zhang HJ, Lee SJ, Jeng MH. HOXB13 induces growth suppression of prostate cancer cells as a repressor of hormone-activated androgen receptor signaling. Cancer Res. 2004;64(24):9185–92.
31. Ewing CM, Ray AM, Lange EM, Zuhlke KA, Robbins CM, Tembe WD, et al. Germline mutations in HOXB13 and prostate-cancer risk. N Engl J Med. 2012;366(2):141–9.
32. Hamid AR, Hoogland AM, Smit F, Jannink S, van Rijt-van de Westerlo C, Jansen CF, et al. The role of HOXC6 in prostate cancer development. Prostate. 2015;75(16):1868–76.
33. Coletta RD, Christensen K, Reichenberger KJ, Lamb J, Micomonaco D, Huang L, et al. The Six1 homeoprotein stimulates tumorigenesis by reactivation of cyclin A1. Proc Natl Acad Sci U S A. 2004;101(17):6478–83.

34. Yu Y, Davicioni E, Triche TJ, Merlino G. The homeoprotein Six1 transcriptionally activates multiple protumorigenic genes but requires Ezrin to promote metastasis. Cancer Res. 2006;66(4):1982–9.
35. Zeng J, Shi R, Cai CX, Liu XR, Song YB, Wei M, et al. Increased expression of Six1 correlates with progression and prognosis of prostate cancer. Cancer Cell Int. 2015;15:63.
36. Pestell RG, Albanese C, Reutens AT, Segall JE, Lee RJ, Arnold A. The cyclins and cyclin-dependent kinase inhibitors in hormonal regulation of proliferation and differentiation. Endocr Rev. 1999;20(4):501–34.
37. Weinberg RA. The retinoblastoma protein and cell cycle control. Cell. 1995;81(3):323–30.
38. Han EK, Lim JT, Arber N, Rubin MA, Xing WQ, Weinstein IB. Cyclin D1 expression in human prostate carcinoma cell lines and primary tumors. Prostate. 1998;35(2):95–101.
39. Radu A, Neubauer V, Akagi T, Hanafusa H, Georgescu MM. PTEN induces cell cycle arrest by decreasing the level and nuclear localization of cyclin D1. Mol Cell Biol. 2003;23(17):6139–49.
40. Chen Y, Martinez LA, LaCava M, Coghlan L, Conti CJ. Increased cell growth and tumorigenicity in human prostate LNCaP cells by overexpression to cyclin D1. Oncogene. 1998;16(15):1913–20.
41. Takano Y, Kato Y, van Diest PJ, Masuda M, Mitomi H, Okayasu I. Cyclin D2 overexpression and lack of p27 correlate positively and cyclin E inversely with a poor prognosis in gastric cancer cases. Am J Pathol. 2000;156(2):585–94.
42. Mermelshtein A, Gerson A, Walfisch S, Delgado B, Shechter-Maor G, Delgado J, et al. Expression of D-type cyclins in colon cancer and in cell lines from colon carcinomas. Br J Cancer. 2005;93(3):338–45.
43. Virmani A, Rathi A, Heda S, Sugio K, Lewis C, Tonk V, et al. Aberrant methylation of the cyclin D2 promoter in primary small cell, nonsmall cell lung and breast cancers. Int J Cancer. 2003;107(3):341–5.
44. Kobayashi T, Nakamura E, Shimizu Y, Terada N, Maeno A, Kobori G, et al. Restoration of cyclin D2 has an inhibitory potential on the proliferation of LNCaP cells. Biochem Biophys Res Commun. 2009;387(1):196–201.
45. Sherr CJ, Roberts JM. Inhibitors of mammalian G1 cyclin-dependent kinases. Genes Dev. 1995;9(10):1149–63.
46. Yang RM, Naitoh J, Murphy M, Wang HJ, Phillipson J, deKernion JB, et al. Low p27 expression predicts poor disease-free survival in patients with prostate cancer. J Urol. 1998;159(3):941–5.
47. Bracken AP, Pasini D, Capra M, Prosperini E, Colli E, Helin K. EZH2 is downstream of the pRB-E2F pathway, essential for proliferation and amplified in cancer. EMBO J. 2003;22(20):5323–35.
48. Malumbres M, Pellicer A. RAS pathways to cell cycle control and cell transformation. Front Biosci. 1998;3:d887–912.
49. Diehl JA, Cheng M, Roussel MF, Sherr CJ. Glycogen synthase kinase-3beta regulates cyclin D1 proteolysis and subcellular localization. Genes Dev. 1998;12(22):3499–511.
50. Hu Y, Li Z, Guo L, Wang L, Zhang L, Cai X, et al. MAGI-2 inhibits cell migration and proliferation via PTEN in human hepatocarcinoma cells. Arch Biochem Biophys. 2007;467(1):1–9.
51. Engelman JA. Targeting PI3K signalling in cancer: opportunities, challenges and limitations. Nat Rev Cancer. 2009;9(8):550–62.
52. Mahdian R, Nodouzi V, Asgari M, Rezaie M, Alizadeh J, Yousefi B, et al. Expression profile of MAGI2 gene as a novel biomarker in combination with major deregulated genes in prostate cancer. Mol Biol Rep. 2014;41(9):6125–31.
53. Levine AJ. p53, the cellular gatekeeper for growth and division. Cell. 1997;88(3):323–31.
54. Oliner JD, Pietenpol JA, Thiagalingam S, Gyuris J, Kinzler KW, Vogelstein B. Oncoprotein MDM2 conceals the activation domain of tumour suppressor p53. Nature. 1993;362(6423):857–60.

55. Aas T, Borresen AL, Geisler S, Smith-Sorensen B, Johnsen H, Varhaug JE, et al. Specific P53 mutations are associated with de novo resistance to doxorubicin in breast cancer patients. Nat Med. 1996;2(7):811–4.
56. Petitjean A, Mathe E, Kato S, Ishioka C, Tavtigian SV, Hainaut P, et al. Impact of mutant p53 functional properties on TP53 mutation patterns and tumor phenotype: lessons from recent developments in the IARC TP53 database. Hum Mutat. 2007;28(6):622–9.
57. Xue L, Han X, Liu R, Wang Z, Li H, Chen Q, et al. MDM2 and P53 polymorphisms contribute together to the risk and survival of prostate cancer. Oncotarget. 2015;7(22):31825–31.
58. Pim D, Banks L. p53 polymorphic variants at codon 72 exert different effects on cell cycle progression. Int J Cancer. 2004;108(2):196–9.
59. Katayama H, Sasai K, Kloc M, Brinkley BR, Sen S. Aurora kinase-A regulates kinetochore/chromatin associated microtubule assembly in human cells. Cell Cycle. 2008;7(17):2691–704.
60. Nikonova AS, Astsaturov I, Serebriiskii IG, Dunbrack Jr RL, Golemis EA. Aurora A kinase (AURKA) in normal and pathological cell division. Cell Mol Life Sci. 2013;70(4):661–87.
61. Karthigeyan D, Prasad SB, Shandilya J, Agrawal S, Kundu TK. Biology of Aurora A kinase: implications in cancer manifestation and therapy. Med Res Rev. 2011;31(5):757–93.
62. Beltran H, Rickman DS, Park K, Chae SS, Sboner A, MacDonald TY, et al. Molecular characterization of neuroendocrine prostate cancer and identification of new drug targets. Cancer Discov. 2011;1(6):487–95.
63. Mosquera JM, Beltran H, Park K, MacDonald TY, Robinson BD, Tagawa ST, et al. Concurrent AURKA and MYCN gene amplifications are harbingers of lethal treatment-related neuroendocrine prostate cancer. Neoplasia. 2013;15(1):1–10.
64. Taylor BS, Schultz N, Hieronymus H, Gopalan A, Xiao Y, Carver BS, et al. Integrative genomic profiling of human prostate cancer. Cancer Cell. 2010;18(1):11–22.
65. Wise DR, DeBerardinis RJ, Mancuso A, Sayed N, Zhang XY, Pfeiffer HK, et al. Myc regulates a transcriptional program that stimulates mitochondrial glutaminolysis and leads to glutamine addiction. Proc Natl Acad Sci U S A. 2008;105(48):18782–7.
66. Otto T, Horn S, Brockmann M, Eilers U, Schuttrumpf L, Popov N, et al. Stabilization of N-Myc is a critical function of Aurora A in human neuroblastoma. Cancer Cell. 2009;15(1):67–78.
67. Knoepfler PS, Cheng PF, Eisenman RN. N-myc is essential during neurogenesis for the rapid expansion of progenitor cell populations and the inhibition of neuronal differentiation. Genes Dev. 2002;16(20):2699–712.
68. Seeger RC, Brodeur GM, Sather H, Dalton A, Siegel SE, Wong KY, et al. Association of multiple copies of the N-myc oncogene with rapid progression of neuroblastomas. N Engl J Med. 1985;313(18):1111–6.
69. Leprince D, Gegonne A, Coll J, de Taisne C, Schneeberger A, Lagrou C, et al. A putative second cell-derived oncogene of the avian leukaemia retrovirus E26. Nature. 1983; 306(5941):395–7.
70. Adamo P, Ladomery MR. The oncogene ERG: a key factor in prostate cancer. Oncogene. 2016;35(4):403–14.
71. Deramaudt TB, Remy P, Stiegler P. Identification of interaction partners for two closely-related members of the ETS protein family, FLI and ERG. Gene. 2001;274(1-2):169–77.
72. Vanaja DK, Cheville JC, Iturria SJ, Young CY. Transcriptional silencing of zinc finger protein 185 identified by expression profiling is associated with prostate cancer progression. Cancer Res. 2003;63(14):3877–82.
73. Tomlins SA, Rhodes DR, Perner S, Dhanasekaran SM, Mehra R, Sun XW, et al. Recurrent fusion of TMPRSS2 and ETS transcription factor genes in prostate cancer. Science. 2005;310(5748):644–8.
74. Leshem O, Madar S, Kogan-Sakin I, Kamer I, Goldstein I, Brosh R, et al. TMPRSS2/ERG promotes epithelial to mesenchymal transition through the ZEB1/ZEB2 axis in a prostate cancer model. PLoS One. 2011;6(7):e21650.
75. Becker-Santos DD, Guo Y, Ghaffari M, Vickers ED, Lehman M, Altamirano-Dimas M, et al. Integrin-linked kinase as a target for ERG-mediated invasive properties in prostate cancer models. Carcinogenesis. 2012;33(12):2558–67.

76. Wright ME, Tsai MJ, Aebersold R. Androgen receptor represses the neuroendocrine transdifferentiation process in prostate cancer cells. Mol Endocrinol. 2003;17(9):1726–37.
77. Bernard D, Pourtier-Manzanedo A, Gil J, Beach DH. Myc confers androgen-independent prostate cancer cell growth. J Clin Invest. 2003;112(11):1724–31.
78. Makela S, Strauss L, Kuiper G, Valve E, Salmi S, Santti R, et al. Differential expression of estrogen receptors alpha and beta in adult rat accessory sex glands and lower urinary tract. Mol Cell Endocrinol. 2000;170(1–2):219–29.
79. Chen M, Hsu I, Wolfe A, Radovick S, Huang K, Yu S, et al. Defects of prostate development and reproductive system in the estrogen receptor-alpha null male mice. Endocrinology. 2009;150(1):251–9.
80. Prins GS, Birch L, Couse JF, Choi I, Katzenellenbogen B, Korach KS. Estrogen imprinting of the developing prostate gland is mediated through stromal estrogen receptor alpha: studies with alphaERKO and betaERKO mice. Cancer Res. 2001;61(16):6089–97.
81. Shaw GL, Whitaker H, Corcoran M, Dunning MJ, Luxton H, Kay J, et al. The early effects of rapid androgen deprivation on human prostate cancer. Eur Urol. 2015;70(2):214–8.
82. Leav I, Lau KM, Adams JY, McNeal JE, Taplin ME, Wang J, et al. Comparative studies of the estrogen receptors beta and alpha and the androgen receptor in normal human prostate glands, dysplasia, and in primary and metastatic carcinoma. Am J Pathol. 2001;159(1):79–92.
83. Cheng J, Lee EJ, Madison LD, Lazennec G. Expression of estrogen receptor beta in prostate carcinoma cells inhibits invasion and proliferation and triggers apoptosis. FEBS Lett. 2004;566(1–3):169–72.
84. Zhu X, Leav I, Leung YK, Wu M, Liu Q, Gao Y, et al. Dynamic regulation of estrogen receptor-beta expression by DNA methylation during prostate cancer development and metastasis. Am J Pathol. 2004;164(6):2003–12.
85. Mak P, Li J, Samanta S, Chang C, Jerry DJ, Davis RJ, et al. Prostate tumorigenesis induced by PTEN deletion involves estrogen receptor beta repression. Cell Rep. 2015;10(12):1982–91.
86. Darshan MS, Loftus MS, Thadani-Mulero M, Levy BP, Escuin D, Zhou XK, et al. Taxane-induced blockade to nuclear accumulation of the androgen receptor predicts clinical responses in metastatic prostate cancer. Cancer Res. 2011;71(18):6019–29.
87. Edwards J, Krishna NS, Mukherjee R, Watters AD, Underwood MA, Bartlett JM. Amplification of the androgen receptor may not explain the development of androgen-independent prostate cancer. BJU Int. 2001;88(6):633–7.
88. Montgomery RB, Mostaghel EA, Vessella R, Hess DL, Kalhorn TF, Higano CS, et al. Maintenance of intratumoral androgens in metastatic prostate cancer: a mechanism for castration-resistant tumor growth. Cancer Res. 2008;68(11):4447–54.
89. Mohler JL, Gregory CW, Ford 3rd OH, Kim D, Weaver CM, Petrusz P, et al. The androgen axis in recurrent prostate cancer. Clin Cancer Res. 2004;10(2):440–8.
90. Antonarakis ES, Lu C, Wang H, Luber B, Nakazawa M, Roeser JC, et al. AR-V7 and resistance to enzalutamide and abiraterone in prostate cancer. N Engl J Med. 2014;371(11):1028–38.
91. Scher HI, Beer TM, Higano CS, Anand A, Taplin ME, Efstathiou E, et al. Antitumour activity of MDV3100 in castration-resistant prostate cancer: a phase 1–2 study. Lancet. 2010;375(9724):1437–46.
92. Attard G, Reid AH, Yap TA, Raynaud F, Dowsett M, Settatree S, et al. Phase I clinical trial of a selective inhibitor of CYP17, abiraterone acetate, confirms that castration-resistant prostate cancer commonly remains hormone driven. J Clin Oncol. 2008;26(28):4563–71.
93. Li Z, Bishop AC, Alyamani M, Garcia JA, Dreicer R, Bunch D, et al. Conversion of abiraterone to D4A drives anti-tumour activity in prostate cancer. Nature. 2015;523(7560):347–51.
94. Koochekpour S. Androgen receptor signaling and mutations in prostate cancer. Asian J Androl. 2010;12(5):639–57.
95. Hu R, Dunn TA, Wei S, Isharwal S, Veltri RW, Humphreys E, et al. Ligand-independent androgen receptor variants derived from splicing of cryptic exons signify hormone-refractory prostate cancer. Cancer Res. 2009;69(1):16–22.

96. Nakazawa M, Antonarakis ES, Luo J. Androgen receptor splice variants in the era of enzalutamide and abiraterone. Horm Cancer. 2014;5(5):265–73.
97. Engelman JA, Luo J, Cantley LC. The evolution of phosphatidylinositol 3-kinases as regulators of growth and metabolism. Nat Rev Genet. 2006;7(8):606–19.
98. Backer JM. The regulation and function of Class III PI3Ks: novel roles for Vps34. Biochem J. 2008;410(1):1–17.
99. Katso R, Okkenhaug K, Ahmadi K, White S, Timms J, Waterfield MD. Cellular function of phosphoinositide 3-kinases: implications for development, homeostasis, and cancer. Annu Rev Cell Dev Biol. 2001;17:615–75.
100. Vivanco I, Sawyers CL. The phosphatidylinositol 3-Kinase AKT pathway in human cancer. Nat Rev Cancer. 2002;2(7):489–501.
101. Luo J, Manning BD, Cantley LC. Targeting the PI3K-Akt pathway in human cancer: rationale and promise. Cancer Cell. 2003;4(4):257–62.
102. Manning BD, Cantley LC. United at last: the tuberous sclerosis complex gene products connect the phosphoinositide 3-kinase/Akt pathway to mammalian target of rapamycin (mTOR) signalling. Biochem Soc Trans. 2003;31(Pt 3):573–8.
103. Fingar DC, Salama S, Tsou C, Harlow E, Blenis J. Mammalian cell size is controlled by mTOR and its downstream targets S6K1 and 4EBP1/eIF4E. Genes Dev. 2002;16(12):1472–87.
104. Neshat MS, Mellinghoff IK, Tran C, Stiles B, Thomas G, Petersen R, et al. Enhanced sensitivity of PTEN-deficient tumors to inhibition of FRAP/mTOR. Proc Natl Acad Sci U S A. 2001;98(18):10314–9.
105. Qi W, Morales C, Cooke LS, Johnson B, Somer B, Mahadevan D. Reciprocal feedback inhibition of the androgen receptor and PI3K as a novel therapy for castrate-sensitive and -resistant prostate cancer. Oncotarget. 2015;6(39):41976–87.
106. Li J, Yen C, Liaw D, Podsypanina K, Bose S, Wang SI, et al. PTEN, a putative protein tyrosine phosphatase gene mutated in human brain, breast, and prostate cancer. Science. 1997;275(5308):1943–7.
107. Cantley LC, Neel BG. New insights into tumor suppression: PTEN suppresses tumor formation by restraining the phosphoinositide 3-kinase/AKT pathway. Proc Natl Acad Sci U S A. 1999;96(8):4240–5.
108. Cairns P, Okami K, Halachmi S, Halachmi N, Esteller M, Herman JG, et al. Frequent inactivation of PTEN/MMAC1 in primary prostate cancer. Cancer Res. 1997;57(22):4997–5000.
109. Carver BS, Chapinski C, Wongvipat J, Hieronymus H, Chen Y, Chandarlapaty S, et al. Reciprocal feedback regulation of PI3K and androgen receptor signaling in PTEN-deficient prostate cancer. Cancer Cell. 2011;19(5):575–86.
110. Massague J. TGF-beta signal transduction. Annu Rev Biochem. 1998;67:753–91.
111. Wakefield LM, Roberts AB. TGF-beta signaling: positive and negative effects on tumorigenesis. Curr Opin Genet Dev. 2002;12(1):22–9.
112. Ao M, Williams K, Bhowmick NA, Hayward SW. Transforming growth factor-beta promotes invasion in tumorigenic but not in nontumorigenic human prostatic epithelial cells. Cancer Res. 2006;66(16):8007–16.
113. Ellenrieder V, Hendler SF, Boeck W, Seufferlein T, Menke A, Ruhland C, et al. Transforming growth factor beta1 treatment leads to an epithelial-mesenchymal transdifferentiation of pancreatic cancer cells requiring extracellular signal-regulated kinase 2 activation. Cancer Res. 2001;61(10):4222–8.
114. Guo Y, Kyprianou N. Overexpression of transforming growth factor (TGF) beta1 type II receptor restores TGF-beta1 sensitivity and signaling in human prostate cancer cells. Cell Growth Differ. 1998;9(2):185–93.
115. Yue J, Mulder KM. Activation of the mitogen-activated protein kinase pathway by transforming growth factor-beta. Methods Mol Biol. 2000;142:125–31.
116. Bakin RE, Gioeli D, Sikes RA, Bissonette EA, Weber MJ. Constitutive activation of the Ras/mitogen-activated protein kinase signaling pathway promotes androgen hypersensitivity in LNCaP prostate cancer cells. Cancer Res. 2003;63(8):1981–9.

117. Weinstein-Oppenheimer CR, Blalock WL, Steelman LS, Chang F, McCubrey JA. The Raf signal transduction cascade as a target for chemotherapeutic intervention in growth factor-responsive tumors. Pharmacol Ther. 2000;88(3):229–79.
118. Mukherjee R, McGuinness DH, McCall P, Underwood MA, Seywright M, Orange C, et al. Upregulation of MAPK pathway is associated with survival in castrate-resistant prostate cancer. Br J Cancer. 2011;104(12):1920–8.
119. Sato N, Sadar MD, Bruchovsky N, Saatcioglu F, Rennie PS, Sato S, et al. Androgenic induction of prostate-specific antigen gene is repressed by protein-protein interaction between the androgen receptor and AP-1/c-Jun in the human prostate cancer cell line LNCaP. J Biol Chem. 1997;272(28):17485–94.

Molecular Carcinogenesis of Urinary Bladder Cancer

10

Rita Seeböck and Johannes Haybaeck

Contents

Abstract

Since the 1970s, research on molecular carcinogenesis of urinary bladder cancer (UBC) has concentrated on the dual-track mode, which describes the genesis of papillary tumors on the one hand and invasive, non-papillary carcinomas on the other hand [1, 2]. This model was for a long term the basic meshwork of molecular UBC categorization and research design. Sjödahl et al. [3] recently proposed a novel approach describing a molecular taxonomy for bladder cancer based on gene expression profiling data obtained from tumors of all grades and stages.

R. Seeböck (✉)
Department Life Sciences, IMC University of Applied Sciences, Krems, Austria
e-mail: rita.seeboeck@fh-krems.ac.at

J. Haybaeck, M.D., Ph.D.
Department of Pathology, Medical Faculty, Otto von Guericke University Magdeburg, 39120 Magdeburg, Germany

Institute of Pathology, Medical University of Graz, 8036 Graz, Austria
e-mail: johannes.haybaeck@med.ovgu.de

J. Haybaeck (ed.), *Mechanisms of Molecular Carcinogenesis – Volume 2*,
DOI 10.1007/978-3-319-53661-3_10

This milestone work introduced five new categories into which UBC cases can be divided. Based on these insights, it was possible to catch up with other tumor entities, behind which the knowledge about molecular carcinogenesis in UBC was lacking for a long time. This chapter introduces the classic dual-track model of papillary versus non-papillary urinary bladder carcinogenesis followed by the molecular taxonomy developed in the last few years, highlighting the five novel subcategories of UBC introduced recently: urobasal A (UroA), UroB, genomically unstable (GU), squamous cell carcinoma-like (SCCL), and "infiltrated."

10.1 Incidence and Morphology

Almost half a million people worldwide are diagnosed with UBC every year, and more than 165,000 individuals succumb to the disease. The gender ratio of male to female UBC patients is 3 to 1 [4]. After prostate, lung, and colorectal cancers, UBC is the fourth most common cancer in males worldwide. Correspondingly, UBC is the ninth most common cancer in world's females [5]. The incidence rate, as well as the mortality of UBC, has remained relatively stable within the last two decades. Nevertheless, with an increasing expectancy of life, the number of incidence cases and deaths is also on the rise. The rate of UBC development increases sharply with age, with nine out of ten new cases diagnosed in patients older than 65 years [6].

The urinary bladder is a hollow organ lined with urothelium. The urothelium, earlier also known as transitional epithelium, is a multilayered epithelium lining the inner wall of the urinary bladder. The special organization within the urothelium correlates with the cells' level of differentiation. The basal cells top the basement membrane and are separated from the umbrella cells facing the bladder lumen by four to six layers of intermediate cells [7, 8]. UBC patients are stratified by pathologic stage and grade, the basis of clinical decision-making. The staging differentiates between non-muscle-invasive (Tis, Ta, and T1) and muscle-invasive (T2, T3, and T4) tumors according to their invasion depth, where Ta tumors are restricted to the urothelium, T1 tumors invade the lamina propria, and T2, T3, and T4 tumors invade the superficial muscle, perivesical fat layer, and surrounding organs, respectively (Fig. 10.1). The carcinomas in situ, stage Tis, are poorly understood and believed to be a precursor of muscle-invasive tumors [2].

The urothelium as inner bladder lining, being a reservoir for urine, is inevitably exposed to carcinogens. The highest risk factor identified is tobacco, which can be associated with ~50% of all UBC cases [1, 9]. The vast majority of UBC cases consist of the transitional cell carcinoma (TCC) or urothelial carcinoma (UC) type. Approximately 1% each can be classified as squamous cell carcinomas, adenocarcinomas, small cell carcinomas, or sarcomas. The tumors of the squamous cell carcinoma category show a different morphology, resembling flat cells as found on the surface of the skin. These tumors are commonly associated with an invasive growth pattern. Adenocarcinomas of the urinary bladder are also invasive tumors that resemble gland-forming cells as involved in colon cancer. Small cell carcinomas

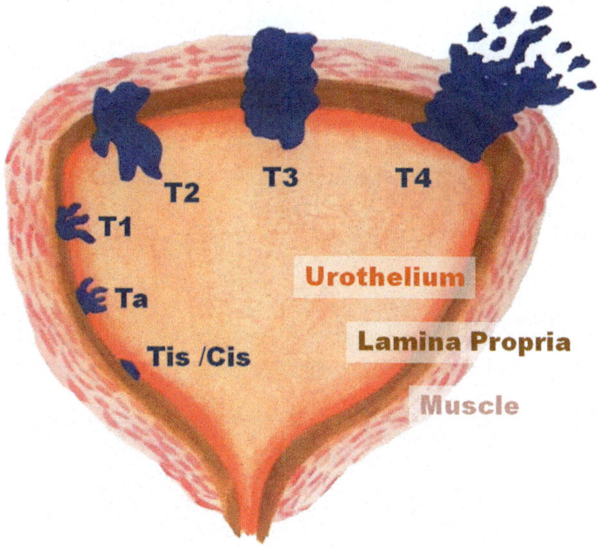

Fig. 10.1 UBC tumors are staged by their invasiveness. Tis/CIS betoken superficial tumors and also Ta tumors are restricted to the urothelium. T1 tumors invade the lamina propria, T2 tumors invade the superficial muscle layer, T3 tumors go completely beyond the muscle layer into the perivesical fat, and T4 tumors reach from inside the bladder through all denoted layers and also affect other organs

originate from nerve-like neuroendocrine cells. These tumors are also characterized by fast growth and early disease progression, which requires harsh chemotherapeutic treatment. Sarcomas are a rare group of tumors that progress from muscle cells of the urinary bladder. In general, all kinds of tumors occurring in the urinary bladder undergo similar treatments in their early stages and only require a different, specific chemotherapeutic regimen in advanced stages [9].

10.2 Dual-Track Model of Carcinogenesis

Eighty percent of all UBC cases follow a developmental pattern of low-grade papillary urothelial carcinomas. These are characterized by urothelium of increased thickness (>7 cell layers) and papillary growth, where slender, fingerlike extensions grow toward the hollow bladder center [9]. The other 20% of UBC cases generally arise from a carcinoma in situ (CIS), which is a flat lesion that does not show any growth toward the hollow part of the bladder. Indeed, these tumors tend to grow into deeper layers of the urothelium. Therefore, the growth pattern is referred to as invasive non-papillary (Fig. 10.1). This morphological differentiation into invasive non-papillary urothelial carcinomas and papillary tumors provides the basis for the categorization of single UBC cases. Research conducted in the field of UBC aimed at a molecular description and discrimination of exactly these two subgroups.

The tumorigenic papillary growth of urothelial cells is triggered by abnormal and excessive activation of signals that lead to an increased or accelerated cell proliferation. In this sense, the fibroblast growth factor receptor (FGFR)3-rat sarcoma (RAS) signaling network ([10, 11]) (Fig. 10.2) is of high importance for papillary carcinomas.

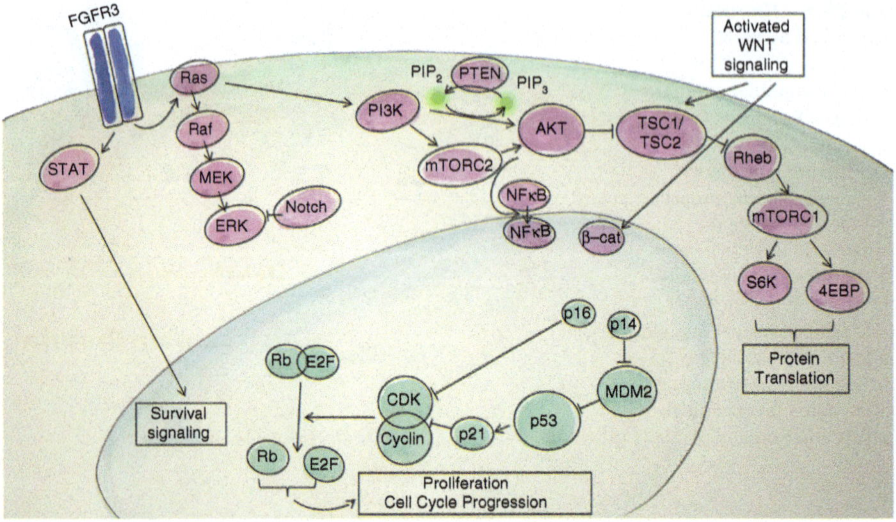

Fig. 10.2 Molecular pathways orchestrating urinary bladder carcinogenesis and their interrelations. FGFR3 activates downstream signaling pathways upon extracellular growth factor triggering. Among the effectors are STAT proteins that will initiate gene expression to promote cellular survival. More important for UBC is activation of the RAS/RAF/MEK/ERK signaling cascade, which drives cellular proliferation. ERK can be inhibited by activated Notch. A second downstream target of RAS important for UBC is PI3K, which activates AKT and therewith protein translation, proliferation, and tumor expansion via the mTOR signaling network. AKT also supports the translocation of NF-κB to the nucleus where it can act as a transcription factor. The TSC1/TSC2 complex is also activated by WNT signaling, which drives nuclear translocation of β-catenin, especially in muscle-invasive tumors. In the nucleus, the cell cycle is regulated by p53 and Rb. p53 is ubiquitinated by MDM2. MDM2 is inhibited by p14 expression, so that p53 is released to activate p21. p21 and p16 act as CDK inhibitors and set cyclins free to further release Rb from E2F and drive the cell cycle into progression

FGFR3 is a receptor tyrosine kinase that is mutated in 30–40% of low-grade papillary bladder tumors. Mutations most frequently (50–80%) affect exon 5 and thereby the receptor's extracellular domain [12, 13]. FGFR3 phosphorylation initiates a wide range of signaling cascades, the most predominant being the RAS/RAF/MEK/ERK pathway, controlling cell growth and proliferation. The RAS protein naturally occurs in three different isotypes: HRAS, KRAS, and NRAS. All three are expressed and can be mutated in UBC, but mutations are rather rare (<10%) and show no correlation to tumor subgroups, stage, or grade [11, 14]. *HRAS* mutations were reported to be more frequent than mutations in *KRAS* and *NRAS* and, furthermore, could be correlated to the development of superficial tumors of low histopathological grade [15–17]. As is shown in Fig. 10.2, FGFR3 signaling also activates PI3K and STAT pathways that can crucially influence cell proliferation, differentiation, and growth. Noninvasive papillary tumors are relatively harmless compared to their invasive counterpart and can be commonly treated with transurethral resection. Nevertheless, in about 15% of cases, tumors that have shown a noninvasive papillary growth pattern, probably for years, suddenly alter their behavior and proliferate in an aggressive manner comparable to muscle-invasive non-papillary tumors [18]. A key question in research was, and still is, to determine a predictor to early detect such a "switch" scenario and eventually provide protection.

In the dual-track model of urinary carcinogenesis, the non-papillary branch was recently introduced, characterized by *FGRF3* and *HRAS* mutations and involving these signaling cascades [2]. The second branch, representing tumors that grow in a non-papillary but highly aggressive and invasive manner, is associated with two different proteins: tumor suppressor p53 and retinoblastoma (Rb) protein [1, 2].

An urothelial CIS evolving into an invasive non-papillary bladder tumor frequently shows mutations targeting a potent tumor suppressor, such as *p53* or the retinoblastoma susceptibility gene *Rb*. These two genes are well known for their control of a normal cell cycle and their wide spectrum of target genes, regulating cellular pathways ranging from development, proliferation, differentiation, and DNA repair to apoptosis and controlled cell death (Fig. 10.2). Tumors that have lost the activity of p53 and/or Rb have no chance to protect the genome from further damage and thereby trigger this highly aggressive, muscle-invasive cancer progression.

10.3 Pathways Orchestrating Urinary Bladder Carcinogenesis

The two-stranded model or urinary bladder carcinogenesis was the basic scheme of classification for many years. The best described marker genes are *FGFR3* and *HRAS* for the papillary strand and *p53* and *Rb* for the non-papillary strand. Of course, the literature contains many more markers and molecular pathways involved in the fine-tuning of cancer development and progression. The most important pathways and marker genes/proteins are summarized in the following section and graphically represented in Fig. 10.2.

10.3.1 FGFR3: Transducer of Extracellular Signals

FGFR3 belongs to the family of receptor tyrosine kinases and is the most frequently mutated oncogene in UBC, with mutations occurring in more than 70% of noninvasive and up to 20% of invasive tumors [19]. FGFR3 activation is strongly associated with low-grade noninvasive, papillary tumors. The constitutive receptor activation is established most commonly by mutations in FGFR3s extracellular domain, leading to dimerization and stimulation of tyrosine kinase activity. Overexpression of a wild-type FGFR3 is another frequent feature of UBC but tends to appear rather in invasive tumors [20].

FGFR3 is physiologically activated by the extracellular binding of FGFs, initiating an intracellular downstream cascade that can ultimately influence cell growth, migration of tumor cells, differentiation, and angiogenesis. The most important downstream pathways are the RAS followed by PI3K and STAT signaling cascades. RAS and PI3K pathways are discussed below, STAT pathways are only beginning to be investigated in UBC, but there is no clear evidence of an involvement in tumor development or progression [21, 22]. In this respect, STAT1 activation and STAT3 expression, with emphasis on cancer stem cell fate, are subject to ongoing research [7]. The future of UBC research will be influenced by breast cancer-related knowledge, as explained at the end of this chapter. STAT signaling is the key to understanding molecular carcinogenesis of breast cancer [23, 24] and therefore presents an interesting future field also in UBC research.

As mentioned above, FGFR3 is important in the regulatory network orchestrating angiogenesis. The formation of tumor microvessels is an essential step in the development of invasive and metastatic tumors. Matrix metalloproteinases (MMP) 2 and 9, which are secreted to disintegrate basement membranes as a preparative step of invasion, are overexpressed in high-grade and high-stage UBC cases. They can activate FGFs, which increase FGFR3 signaling and consequently the angiogenic potential of tumor cells. In parallel, aberrant p53, present in most high-grade UBC cases, is associated with lacking regulation of the antiangiogenic regulator thrombospondin-1 (TSP-1). These two strands are the major triggers of tumor angiogenesis in UBC [25].

The FGFR3 isoform expressed also influences its activity. Normal urothelium usually harbors the IIIb isoform, which has a preference for FGF1. A typical splice variant (Δ8–10) is secreted and lacks the transmembrane domain, therefore acting as a negative regulator capturing FGFs or blocking full-length receptors [26]. A SNP in the genetic vicinity of FGFR3 is implicated to increase the risk of FGFR3-driven bladder cancer development, although the underlying mechanisms could not have been explained to date [27].

As a gate between extracellular signals and the many facets of intracellular response, FGFR3 is a prominent target for novel tumor therapeutic approaches [19, 28].

10.3.2 RAS Signaling

RAS proteins function in cellular signaling as small GTPases. This family comprises 39 proteins encoded in the human genome. The spectrum of cellular mechanisms co-orchestrated by RAS signaling is wide and ranges from growth, proliferation, and differentiation over cytoskeletal rearrangements to adhesion and motility [29]. RAS proteins localize to the plasma membrane, where they are activated upon extracellular stimuli by GTPase-activating proteins (GAPs), of which 14 are known to be encoded in our genome [30].

It has been known for three decades that in bladder cancer, the *HRAS* subtype of *RAS* oncogenes is frequently mutated and that UBC is therewith among the rare diseases primarily associated with *HRAS* mutations and not *NRAS* or *KRAS* mutations, which frequently occur in lymphoid malignancies and melanomes or in lung, pancreas, or colon carcinomas, respectively [29, 31]. Most mutations driving carcinogenesis by a RAS protein affect their ability to hydrolyze GTP, such as the codon 12 and 13 substitutions. In *HRAS*, important for bladder cancer, mutation patterns are a bit different to *KRAS* or *NRAS*, with the highest incidence in codon 12 (more than 50%) followed by codon 61, and less than 10% of mutations occur in codon 13 [32].

In general, *RAS* mutations are detected at lower rates than in other entities, with approximately 15% of cases presenting *HRAS* mutations, of which 60% show the predominant G12V substitution [32]. In UBC, next to the mutation of *RAS*, their protein overexpression seems to play a crucial role, as in ~80% of cases, one of the three subtypes is expressed at an increased level [32]. The oncogenic transformations of *RAS* provoke a constant activation of signaling and therefore an overreaction in downstream effects.

Among RAS effectors, the Ser/Thr kinase RAF was the first to be identified in 1993 [33, 34]. RAF is able to signal trough the mitogen-activated protein kinase (MAPK)/extracellular signal-regulated kinase (ERK1/2) and MAPK kinase (MEK), which is a kind of prototype for plasma membrane to nucleus signaling. A second effector of RAS proteins is phosphoinositide 3-kinases (PI3K), which can further signal to the Ser/Thr kinase AKT, protein kinase B, and the transcription factor nuclear factor-κB (NF-κB), thereby rearranging gene expression profiles on the translation and transcription level. Karnoub and Weinberg summarized the RAS signaling network and gave a comprehensive overview of RAS' "split personalities" [29].

A recent insight revealed an interesting role of Notch signaling in bladder cancer, where activated Notch reduces ERK phosphorylation and therefore RAS downstream signaling. Therewith, loss of function in Notch signaling, respective alterations of which were identified in up to 60% of UBC cases, could be identified as driving event in UBC. Impaired Notch signaling leads to upregulation of epithelial-mesenchymal transition (EMT) effectors, such as SNAIL, SLUG, or ZEB2, parallel to a concomitant suppression of E-cadherin [35–37].

10.3.3 PI3K Signaling

PI3K is one target of RAS signaling that can contribute to the molecular carcinogenesis of UBC. Upon activation of PI3K, its p110 catalytic subunit executes phosphorylation of phosphatidylinositol-4,5-bisphosphate (PIP_2) to phosphatidylinositol-3,4,5-trisphosphate (PIP_3), which then recruits AKT to the plasma membrane, where it can be activated by mammalian target of rapamycin complex 2 (TORC2). Downstream of AKT, the PI3K pathway branches significantly and may regulate a multitude of cellular processes. One of these is the negative regulation of tuberous sclerosis complex1 (TSC1)/TSC2, triggering mTOR activation [38]. Via mTOR and eukaryotic initation factors (eIFs) protein translation is regulated. Of note, eIF3a was suggested to play a role in UBC also beyond its translation initiating activity [39].

The mutational spectrum of *PIK3CA* alterations in UBC is different to other tumor entities, featuring mutations in the helical domain E542K and E545K, which require RAS binding [40].

Phosphatase and tensin homolog (PTEN) is a tumor suppressor, essential for PI3K signaling, catalyzing the PIP_3 to PIP_2 reaction. In UBC, PTEN expression levels are commonly found to be reduced [41]. This is, on the one hand, due to low expression levels per se in urothelium, and, on the other hand, associations were made to p53 alterations which might lead to a decrease in PTEN levels [38].

Further downstream in the PI3K pathway, we find AKT, which is mutated only in rare cases of UBC and has not yet been associated with the development or progression of these tumors [38]. AKT phosphorylates TSC2 in the TSC1/TSC2 complex, thereby releasing the suppression of mTOR activation. TSC1, encoded on the chromosomal arm 9q, is frequently lost in UBC cases. This loss of TSC1 function, due to chromosomal loss of heterozygosity (LOH), has been reported for no other human malignancies than UBC [38, 42].

A common expression pattern in UBC is a reduction in the PTEN levels and increased phosphorylated AKT with a strong nuclear expression of β-catenin, revealing a putative link of PI3K and WNT signaling in UBC genesis. In general, WNT pathway components are altered in papillary and muscle-invasive UBC cases, with the most frequent event being increased nuclear localization of β-catenin in muscle-invasive tumors [43–45].

10.3.4 Gross DNA Integrity

Chromosomal aberrations occur frequently in UBC and range from abnormal numbers of chromosomes to deletions and amplifications that can affect all chromosomes. Especially with expression of altered p53 and/or Rb proteins, it is not easy to define if chromosomal alterations are causes or consequences, but there are some scenarios where research revealed the details of the associated molecular carcinogenesis.

Low-grade non-muscle-invasive UBC cases commonly have near-diploid karyotypes and low rates of genomic rearrangements. On the contrary, high-grade muscle-invasive tumors are frequently aneuploid, harboring a multitude of genetic alterations. In this second case, many DNA repair and damage checkpoint, as well as chromosome maintenance genes, are reported to be frequently altered in muscle-invasive tumors.

The earliest detected and most common chromosomal aberration in UBC is LOH of chromosome 9. LOH of 9q is a characteristic feature of low-grade and noninvasive tumors but is rare in CIS or invasive tumors [13, 46]. Localized on chromosome 9 are several potent cell-cycle regulating genes, like cyclin-dependent kinase inhibitor 2A (*CDKN2A*), encoding for p16 and p14^{ARF}, patched 1 (*PTCH1*), *TSC1*, and several others [47]. As for CDKN2A, heterozygous and homozygous deletions are found in up to 50% of UBC cases [48]. Other genetic loci that are altered in UBC include LOH of 15q, which is present in about 40% of UBC cases, and loss of 8p in up to 30% of primarily high-grade cases [49, 50]. Majewski et al. analyzed UBC using a genomic mapping approach and related their findings to the dual-track model of urinary bladder carcinogenesis. Besides the above introduced genetic alterations, they highlighted a set of six genetic regions (3q22-q24, 5q22-q31, 9q21-q22, 10q26, 13q14, 17p13) and their gene products which also include novel target genes referred to as forerunner genes. They found similar frequencies in the loss of these genetic regions for both papillary and non-papillary tumors, suggesting that both branches derive from contrastable low-grade intraurothelial lesions [18].

Tumors with lost Rb activity and resulting increased E2F-driven transcription have induced MAD2 gene activation, which plays a critical role in controlling mitotic spindles and chromosome segregation, running the cells to aneuploidy. Another mechanism frequently affecting chromosome segregation and sister chromatid cohesion is mutation of STAG2, encoding a subunit of cohesin [51, 52]. Loss of STAG2 expression occurs rather in chromosomally stable tumors and is suggested not to provoke aneuploidy [53].

10.3.5 p53, Rb, and Cell-Cycle Regulation

p53 is probably the most famous tumor suppressor and well-studied cellular regulator of DNA repair and apoptosis, as well as a major cell-cycle gatekeeper. Genetic changes in the p53 gene are found in almost every kind of human cancer. In UBC, *p53* mutations are strongly associated with high tumor grade, invasive behavior, increased risk of recurrence, and adverse clinical outcome. Structural and/or functional defects are recorded in more than half of UBC cases [13, 47]. As already mentioned, among others, the gene loci of *p53*, on chromosome 17p13, as well as *Rb*, on chromosome 13q14, are frequently deleted in high-grade muscle-invasive tumors. Therefore, alterations in the p53/Rb cell-cycle regulatory network are described as a characteristic feature especially of the group of invasive UBC cases [54]. To regulate the cell cycle, p53 activates transcription of the cyclin-dependent kinase (CDK) inhibitor p21 that blocks cyclin/CDK interaction. Release of CDKs enables Rb phosphorylation and the release of E2F. E2F proteins are transcription factors which, on the one hand, drive cell cycle to progression and, on the other hand, reduce p53 activity. A loss-of-function mutation of p53 is detectable in approximately 60% of invasive bladder tumors; inactive p53 is reported in up to 80% of cases [55]. Most of the mutations in the p53 gene affect its DNA-binding domain, hampering correct transcriptional activation of target genes. Mutated p53 proteins may also disable functional execution of wild-type p53 proteins by dimerization with them [12]. Besides the genetic or mutational inactivation of p53, its function can be inactivated by overexpression of MDM2. MDM2 binds p53, inducing its ubiquitination and degradation. This physiological autoregulatory loop fails upon MDM2 overexpression in about 30% of UBC cases, predominantly in high-grade tumors [56].

Rb is in its phosphorylated state an essential regulator of cell-cycle progression. Active, dephosphorylated Rb binds and blocks the transcription factor E2F. Upon phosphorylation, E2F is released to induce transcription of genes necessary for S-phase progression. In UBC, Rb mutations and concomitant hyperphosphorylation are reported, the latter also being due to lost CDKI p16 expression or cyclin D1 overexpression [57]. With the cell-cycle control of Rb missing, cells accumulate genetic aberrations and are associated with genetic instability.

The three pillars of genetic instability, p53, and Rb alterations greatly interact and push each other toward more aggressive and adverse tumor phenotypes.

10.4 New Approaches to UBC Classification

The two-pathway model described at the beginning of this section has dominated the literature on bladder cancer, but cannot provide explanations for the distinct heterogeneity within the two categories in respect to their clinical behavior. In an attempt to answer the long-pending question for a better subgrouping, Sjödahl et al. [3] introduced five major subtypes of bladder cancer stages, based on mRNA expression analysis of 308 tumor cases. The groups are termed urobasal A (UroA), UroB, genomically unstable (GU), squamous cell carcinoma-like (SCCL), and "infiltrated," which refers to a tumor infiltration by non-tumor cells and the corresponding increased presence of immune cells. The molecular subtypes are also prognostic features, as UroA tumors were associated with longer survival. GU and the infiltrated group showed intermediate prognosis, whereas that of UroB and SCCL cases was the worst. This differential behavior of molecular subgroups keeps its evidence across all tumor stages and grades, so that even high-grade, G3 tumors of immanent adverse prognosis could be split into groups of better (UroA), intermediate (GU, infiltrated), and worse (UroB, SCCL) survival [3]. Not only with respect to survival should the molecular subtypes be seen as a characterization independent of pathological stratification. Nevertheless, it can be reported that Ta tumors are frequently of the UroA subtype, T1 tumors of UroA and GU subtype, and T2–T4 tumors of any subtype. As for tumor grading, low-grade G1 and G2 tumors are commonly of the UroA subtype. For high-grade tumors, no predominance was analyzed. Consequently, the molecular subtypes can aid in a finer differentiation of tumors within every pathologically predefined category. For the installment of the five molecular subtypes, a wide range of mRNA expression profiles was evaluated, and protein levels were compared. Key features that best describe the single subtype are listed below.

UroA tumors are characterized by increased expression of FGFR3, TP63, CCND1, and KRT5. *FGFR3* is frequently mutated; the *FGFR3* gene signature is strongly expressed, including *FGFR3*, *TP63*, *IRS1*, *SEMA4B*, *PTPN13*, and *TMPRSS4*. UroA tumors express *CCND1*, *RBL2*, and *ID* genes. KRT5 can be overexpressed, in addition to the fact that UroA tumors generally express keratins KRT5, KRT13, KRT15, and KRT17 in a pattern resembling normal urothelium. These tumors show a very good prognosis [3].

GU tumors are characterized by frequent *TP53* mutation and expression of CCNE, CDH1, and ERBB2 on reduced cytokeratin expression. Genes frequently overexpressed in GU tumors include *KPNA2*, *HMOX1*, *CTSL1*, and *CTSL2*. PTEN is characteristically downregulated. From a cell-cycle point of view, genes important for later stages, including *CCNA*, *CCNB*, and *CDC20*, are expressed. GU tumors have lost physiological keratin expression patterns and commonly present only with KRT20, a marker of umbrella cells. This supports the hypothesis that the tumor cells lost stromal contact. GU tumors are predominantly of high tumor grade. With 40% of GU tumors being muscle-invasive, this subtype represents a high-risk group [3, 47].

SCCL tumors are characterized by an aberrant expression of keratins (KRT4, KRT6A, KRT6B, KRT6C, KRT14), not typical for the urothelium. These keratins are associated with basal cells and squamous cell differentiation. Other proteins expressed comprise EGFR and P-cadherin [47]. SCCL tumors have a poor prognosis and seem to affect women more than men [3] (Sjödahl et al. 2013).

UroB tumors show similarities to UroA tumors, including high FGFR3 expression and mutation frequency, as well as high expression of *CCND1* and the *FGFR3* gene signature. A vast disparity is the frequent TP53 mutation in UroB tumors along with expression of SCCL-specific keratins. Nonetheless, it is supposed that UroB progressed out of UroA tumors [3].

The infiltrated tumors are per se characterized by the infiltrated non-tumorous cells. The gene expression profiles of these tumors are compromised by the infiltrates of T lymphocytes, myofibroblasts, and, to a lesser extent, endothelial cells [3].

Other groups working on gene expression profiling of bladder cancer introduced a different nomenclature for the single subtypes, discussed and summarized by McConkey et al. [58], pointing out that urinary bladder cancer can be generally divided into two more basal and two rather luminal intrinsic subtypes. The luminal subtypes occur with different labels, including "papillary," "p53-like," "genetically unstable," or according to the Cancer Genome Atlas research network "cluster I" and "cluster II." The basal subtypes may be referred to as "urobasal," "squamous," or "cluster III" and "cluster IV" [55, 58, 59].

The discovery of the distinct molecular subtypes by gene expression profiling was found to strikingly resemble molecular subtypes established for breast cancer. In 2000, Perou et al. [60] presented with their "molecular portraits" the first whole-genome approach to molecular profiling of a cancer entity. Analyzing the whole-genome messenger RNA expression profile, they discovered the intrinsic subtypes of breast cancer based on shared gene expression patterns: claudin-low, human epidermal growth factor receptor 2 (HER2)-enriched, luminal A, and luminal B. Many parallels between the intrinsic subtypes of bladder cancer and breast cancer were drawn [58, 59, 61]. The claudin-low subtype of breast cancer is characterized by low expression of claudin tight junction proteins claudin 3, 4, and 7, as well as induced EMT markers. Damrauer et al. [61] discovered a subset of basal UC cases that display the expression pattern indicative of claudin-low breast cancer. This adds another feature to the great similarity of especially high-grade UC tumors with breast cancer. The claudin-low subtype does not alter subtype characteristics in respect to overall survival to the remaining basal-type UC cases [61].

As stated by Choi et al. [59], who compared the clinical behavior of different UC subtypes with the corresponding breast cancer subtypes, it will be of great interest and relevance to apply the huge knowledge on breast cancer to the respective subtypes in future bladder cancer research.

References

1. Koss LG, Tiamson EM, Robbins MA. Mapping cancerous and precancerous bladder changes. A study of the urothelium in ten surgically removed bladders. JAMA. 1974;227:281–6.
2. Spiess PE, Czerniak B. Dual-track pathway of bladder carcinogenesis: practical implications. Arch Pathol Lab Med. 2006;130:844–52.

3. Sjödahl G, Lauss M, Lovgren K, Chebil G, Gudjonsson S, Veerla S, Patschan O, Aine M, Ferno M, Ringner M, Mansson W, Liedberg F, Lindgren D, Hoglund M. A molecular taxonomy for urothelial carcinoma. Clin Cancer Res. 2012;18:3377–86.
4. Ferlay J, Steliarova-Foucher E, Lortet-Tieulent J, Rosso S, Coebergh JW, Comber H, Forman D, Bray F. Cancer incidence and mortality patterns in Europe: estimates for 40 countries in 2012. Eur J Cancer. 2013;49:1374–403.
5. Jemal A, Bray F, Center MM, Ferlay J, Ward E, Forman D. Global cancer statistics. CA Cancer J Clin. 2011;61:69–90.
6. Yee DS, Ishill NM, Lowrance WT, Herr HW, Elkin EB. Ethnic differences in bladder cancer survival. Urology. 2011;78:544–9.
7. Ho PL, Kurtova A, Chan KS. Normal and neoplastic urothelial stem cells: getting to the root of the problem. Nat Rev Urol. 2012;9:583–94.
8. McConkey DJ, Lee S, Choi W, Tran M, Majewski T, Lee S, Siefker-Radtke A, Dinney C, Czerniak B. Molecular genetics of bladder cancer: emerging mechanisms of tumor initiation and progression. Urol Oncol. 2010;28:429–40.
9. Kucuk U, Pala EE, Cakir E, Sezer O, Bayol U, Divrik RT, Cakmak O. Clinical, demographic and histopathological prognostic factors for urothelial carcinoma of the bladder. Cent European J Urol. 2015;68:30–6.
10. Billerey C, Chopin D, Aubriot-Lorton MH, Ricol D, Gil Diez de Medina S, Van Rhijn B, Bralet MP, Lefrere-Belda MA, Lahaye JB, Abbou CC, Bonaventure J, Zafrani ES, van der Kwast T, Thiery JP, Radvanyi F. Frequent FGFR3 mutations in papillary non-invasive bladder (pTa) tumors. Am J Pathol. 2001;158(6):1955–9.
11. Ouerhani S, Elgaaied AB. The mutational spectrum of HRAS, KRAS, NRAS and FGFR3 genes in bladder cancer. Cancer Biomark. 2011;10:259–66.
12. Al Hussain TO, Akhtar M. Molecular basis of urinary bladder cancer. Adv Anat Pathol. 2013;20:53–60.
13. Wu XR. Urothelial tumorigenesis: a tale of divergent pathways. Nat Rev Cancer. 2005;5:713–25.
14. Ouerhani S, Bougatef K, Soltani I, Elgaaied AB, Abbes S, Menif S. The prevalence and prognostic significance of KRAS mutation in bladder cancer, chronic myeloid leukemia and colorectal cancer. Mol Biol Rep. 2013;40:4109–14.
15. Boulalas I, Zaravinos A, Karyotis I, Delakas D, Spandidos DA. Activation of RAS family genes in urothelial carcinoma. J Urol. 2009;181:2312–9.
16. Pasin E, Josephson DY, Mirta AP, Cote RJ, Stein JP. Superficial bladder cancer: an update on etiology, molecular development, classification, and natural history. Rev Urol. 2008;10:31–43.
17. Przybojewska B, Jagiello A, Jalmuzna P. H-RAS, K-RAS, and N-RAS gene activation in human bladder cancers. Cancer Genet Cytogenet. 2000;121:73–7.
18. Majewski T, Lee S, Jeong J, Yoon DS, Kram A, Kim MS, Tuziak T, Bondaruk J, Lee S, Park WS, Tang KS, Chung W, Shen L, Ahmed SS, Johnston DA, Grossman HB, Dinney CP, Zhou JH, Harris RA, Snyder C, Filipek S, Narod SA, Watson P, Lynch HT, Gazdar A, Bar-Eli M, Wu XF, McConkey DJ, Baggerly K, Issa JP, Benedict WF, Scherer SE, Czerniak B. Understanding the development of human bladder cancer by using a whole-organ genomic mapping strategy. Lab Invest. 2008;88:694–721.
19. Liu X, Zhang W, Geng D, He J, Zhao Y, Yu L. Clinical significance of fibroblast growth factor receptor-3 mutations in bladder cancer: a systematic review and meta-analysis. Genet Mol Res. 2014;13:1109–20.
20. Iyer G, Milowsky MI. Fibroblast growth factor receptor-3 in urothelial tumorigenesis. Urol Oncol. 2013;31:303–11.
21. DiPierro GB, Gulia C, Cristini C, Fraietta G, Marini L, Grande P, Gentile V, Piergentili R. Bladder cancer: a simple model becomes complex. Curr Genomics. 2012;13:395–415.
22. Su WC, Kitagawa M, Xue N, Xie B, Garofalo S, Cho J, Deng C, Horton WA, Fu XY. Activation of Stat1 by mutant fibroblast growth-factor receptor in thanatophoric dysplasia type II dwarfism. Nature. 1997;386:288–92.
23. Furth PA. STAT signaling in different breast cancer sub-types. Mol Cell Endocrinol. 2014;382:612–5.

24. Tymoszuk P, Charoentong P, Hackl H, Spilka R, Muller-Holzner E, Trajanoski Z, Obrist P, Revillion F, Peyrat JP, Fiegl H, Doppler W. High STAT1 mRNA levels but not its tyrosine phosphorylation are associated with macrophage infiltration and bad prognosis in breast cancer. BMC Cancer. 2014;14:257.
25. Grossfeld GD, Ginsberg DA, Stein JP, Bochner BH, Esrig D, Groshen S, Dunn M, Nichols PW, Taylor CR, Skinner DG, Cote RJ. Thrombospondin-1 expression in bladder cancer: association with p53 alterations, tumor angiogenesis, and tumor progression. J Natl Cancer Inst. 1997;89:219–27.
26. Tomlinson DC, L'Hote CG, Kennedy W, Pitt E, Knowles MA. Alternative splicing of fibroblast growth factor receptor 3 produces a secreted isoform that inhibits fibroblast growth factor-induced proliferation and is repressed in urothelial carcinoma cell lines. Cancer Res. 2005;65:10441–9.
27. Kiemeney LA, Sulem P, Besenbacher S, Vermeulen SH, Sigurdsson A, Thorleifsson G, Gudbjartsson DF, Stacey SN, Gudmundsson J, Zanon C, Kostic J, Masson G, Bjarnason H, Palsson ST, Skarphedinsson OB, Gudjonsson SA, Witjes JA, Grotenhuis AJ, Verhaegh GW, Bishop DT, Sak SC, Choudhury A, Elliott F, Barrett JH, Hurst CD, De Verdier PJ, Ryk C, Rudnai P, Gurzau E, Koppova K, Vineis P, Polidoro S, Guarrera S, Sacerdote C, Campagna M, Placidi D, Arici C, Zeegers MP, Kellen E, Gutierrez BS, Sanz-Velez JI, Sanchez-Zakabardo M, Valdivia G, Garcia-Prats MD, Hengsterl JG, Blaszkewicz M, Dietrich H, Ophoff RA, Van der Berg LH, Alexiusdottir K, Kristjansson K, Geirsson G, Nikulasson S, Petursdottir V, Kong A, Thorgeirsson T, Mungan NA, Lindblom A, Van Es MA, Porru S, Buntinx F, Golka K, Mayordomo JI, Kumar R, Matullo G, Steineck G, Kiltie AE, Aben KK, Jonsson E, Thorsteinsdottir U, Knowles MA, Rafnar T, Stefansson K. A sequence variant at 4p16.3 confers susceptibility to urinary bladder cancer. Nat Genet. 2010;42:415–9.
28. Pandith AA, Shah ZA, Siddiqi MA. Oncogenic role of fibroblast growth factor receptor 3 in tumorigenesis of urinary bladder cancer. Urol Oncol. 2013;31:398–406.
29. Karnoub AE, Weinberg RA. Ras oncogenes: split personalities. Nat Rev Mol Cell Biol. 2008;9:517–31.
30. Maertens O, Cichowski K. An expanding role for RAS GTPase activating proteins (RAS GAPs) in cancer. Adv Biol Regul. 2014;55:1–14.
31. Fujita J, Yoshida O, Yuasa Y, Rhim JS, Hatanaka M, Aaronson SA. Ha-ras oncogenes are activated by somatic alterations in human urinary tract tumours. Nature. 1984;309:464–6.
32. Fernandez-Medarde A, Santos E. Ras in cancer and developmental diseases. Genes Cancer. 2011;2:344–58.
33. Moodie SA, Willumsen BM, Weber MJ, Wolfman A. Complexes of Ras.GTP with Raf-1 and mitogen-activated protein kinase kinase. Science. 1993;260:1658–61.
34. Zhang XF, Settleman J, Kyriakis JM, Takeuchi-Suzuki E, Elledge SJ, Marshall MS, Bruder JT, Rapp UR, Avruch J. Normal and oncogenic p21ras proteins bind to the amino-terminal regulatory domain of c-Raf-1. Nature. 1993;364:308–13.
35. Fernandez-Marcos PJ, Serrano M, Maraver A. Bladder cancer and the Notch pathway. Oncotarget. 2015;6:1346–7.
36. Rampias T, Vgenopoulou P, Avgeris M, Polyzos A, Stravodimos K, Alavanis C, Scorilas A, Klinakis A. A new tumor suppressor role for the Notch pathway in bladder cancer. Nat Med. 2014;20:1199–205.
37. Seton-Rogers S. Tumour suppressors: Notch blocks bladder tumorigenesis. Nat Rev Cancer. 2014;14
38. Knowles MA, Platt FM, Ross RL, Hurst CD. Phosphatidylinositol 3-kinase (PI3K) pathway activation in bladder cancer. Cancer Metastasis Rev. 2009;28:305–16.
39. Spilka R, Ernst C, Bergler H, Rainer J, Flechsig S, Vogetseder A, Lederer E, Benesch M, Brunner A, Geley S, Eger A, Bachmann F, Doppler W, Obrist P, Haybaeck J. eIF3a is overexpressed in urinary bladder cancer and influences its phenotype independent of translation initiation. Cell Oncol. 2014;37:253–67.
40. Samuels Y, Velculescu VE. Oncogenic mutations of PIK3CA in human cancers. Cell Cycle. 2004;3:1221–4.

41. Tsuruta H, Kishimoto H, Sasaki T, Horie Y, Natsui M, Shibata Y, Hamada K, Yajima N, Kawahara K, Sasaki M, Tsuchiya N, Enomoto K, Mak TW, Nakano T, Habuchi T, Suzuki A. Hyperplasia and carcinomas in Pten-deficient mice and reduced PTEN protein in human bladder cancer patients. Cancer Res. 2006;66:8389–96.
42. Platt FM, Hurst CD, Taylor CF, Gregory WM, Harnden P, Knowles MA. Spectrum of phosphatidylinositol 3-kinase pathway gene alterations in bladder cancer. Clin Cancer Res. 2009;15:6008–17.
43. Baumgart E, Cohen MS, Silva Neto B, Jacobs MA, Wotkowicz C, Rieger-Christ KM, Biolo A, Zeheb R, Loda M, Libertino JA, Summerhayes IC. Identification and prognostic significance of an epithelial-mesenchymal transition expression profile in human bladder tumors. Clin Cancer Res. 2007;13:1685–94.
44. Kastritis E, Murray S, Kyriakou F, Horti M, Tamvakis N, Kavantzas N, Patsouris ES, Noni A, Legaki S, Dimopoulos MA, Bamias A. Somatic mutations of adenomatous polyposis coli gene and nuclear b-catenin accumulation have prognostic significance in invasive urothelial carcinomas: evidence for Wnt pathway implication. Int J Cancer. 2009;124:103–8.
45. Zhu X, Kanal Y, Saito A, Kondo Y, Hirohashi S. Aberrant expression of beta-catenin and mutation of exon 3 of the beta-catenin gene in renal and urothelial carcinomas. Pathol Int. 2000;50:945–52.
46. Spruck CH, Ohneseit PF, Gonzalez-Zulueta M, Esrig D, Miyao N, Tsai YC, Lerner SP, Schmutte C, Yang AS, Cote R, et al. Two molecular pathways to transitional cell carcinoma of the bladder. Cancer Res. 1994;54:784–8.
47. Knowles MA, Hurst CD. Molecular biology of bladder cancer: new insights into pathogenesis and clinical diversity. Nat Rev Cancer. 2015;15:25–41.
48. Baud E, Catilina P, Bignon YJ. p16 involvement in primary bladder tumors: analysis of deletions and mutations. Int J Oncol. 1999;14:441–5.
49. Knowles MA, Shaw ME, Proctor AJ. Deletion mapping of chromosome 8 in cancers of the urinary bladder using restriction fragment length polymorphisms and microsatellite polymorphisms. Oncogene. 1993;8:1357–64.
50. Natrajan R, Louhelainen J, Williams S, Laye J, Knowles MA. High-resolution deletion mapping of 15q13.2-q21.1 in transitional cell carcinoma of the bladder. Cancer Res. 2003;63:7657–62.
51. Kong X, Ball Jr AR, Pham HX, Zeng W, Chen HY, Schmiesing JA, Kim JS, Berns M, Yokomori K. Distinct functions of human cohesin-SA1 and cohesin-SA2 in double-strand break repair. Mol Cell Biol. 2014;34:685–98.
52. Solomon DA, Kim JS, Bondaruk J, Shariat SF, Wang ZF, Elkahloun AG, Ozawa T, Gerard J, Zhuang D, Zhang S, Navai N, Siefker-Radtke A, Phillips JJ, Robinson BD, Rubin MA, Volkmer B, Hautmann R, Kufer R, Hogendoorn PC, Netto G, Theodorescu D, James CD, Czerniak B, Miettinen M, Waldman T. Frequent truncating mutations of STAG2 in bladder cancer. Nat Genet. 2013;45:1428–30.
53. Balbas-Martinez C, Sagrera A, Carrillo-De-Santa-Pau E, Earl J, Marquez M, Vazquez M, Lapi E, Castro-Giner F, Beltran S, Bayes M, Carrato A, Cigudosa JC, Dominguez O, Gut M, Herranz J, Juanpere N, Kogevinas M, Langa X, Lopez-Knowles E, Lorente JA, Lloreta J, Pisano DG, Richart L, Rico D, Salgado RN, Tardon A, Chanock S, Heath S, Valencia A, Losada A, Gut I, Malats N, Real FX. Recurrent inactivation of STAG2 in bladder cancer is not associated with aneuploidy. Nat Genet. 2013;45:1464–9.
54. Mirta AP, Hansel DE, Cote RJ. Prognostic value of cell-cycle regulation biomarkers in bladder cancer. Semin Oncol. 2012;39:524–33.
55. Cancer Genome Atlas Research. Comprehensive molecular characterization of urothelial bladder carcinoma. Nature. 2014;507:315–22.
56. Simon R, Struckmann K, Schraml P, Wagner U, Forster T, Moch H, Fijan A, Bruderer J, Wilber K, Mihatsch MJ, Gasser T, Sauter G. Amplification pattern of 12q13-q15 genes (MDM2, CDK4, GLI) in urinary bladder cancer. Oncogene. 2002;21:2476–83.

57. Quentin T, Henke C, Korabiowska M, Schlott T, Zimmerman B, Kunze E. Altered mRNA expression of the Rb and p16 tumor suppressor genes and of CDK4 in transitional cell carcinomas of the urinary bladder associated with tumor progression. Anticancer Res. 2004;24:1011–23.
58. McConkey DJ, Choi W, Dinney CP. New insights into subtypes of invasive bladder cancer: considerations of the clinician. Eur Urol. 2014;66:609–10.
59. Choi W, Porten S, Kim S, Willis D, Plimack ER, Hoffman-Censits J, Roth B, Cheng T, Tran M, Lee IL, Melquist J, Bondaruk J, Majewski T, Zhang S, Pretzsch S, Baggerly K, Siefker-Radtke A, Czerniak B, Dinney CP, McConkey DJ. Identification of distinct basal and luminal subtypes of muscle-invasive bladder cancer with different sensitivities to frontline chemotherapy. Cancer Cell. 2014;25:152–65.
60. Perou CM, Sorlie T, Eisen MB, Van de Rijn M, Jeffrey SS, Rees CA, Pollack JR, Ross DT, Johnsen H, Akslen LA, Fluge O, Pergamenschikov A, Williams C, Zhu SX, Lonning PE, Borresen-Dale AL, Brown PO, Botstein D. Molecular portraits of human breast tumours. Nature. 2000;406:747–52.
61. Damrauer JS, Hoadley KA, Chism DD, Fan C, Tiganelli CJ, Wobker SE, Yey JJ, Milowsky MI, Iyer G, Parker JS, Kim WY. Intrinsic subtypes of high-grade bladder cancer reflect the hallmarks of breast cancer biology. Proc Natl Acad Sci U S A. 2014;111:3110–5.

Molecular Carcinogenesis in Gynecologic Neoplasias

11

Elisabeth Smolle

Contents

Abstract

Breast cancer is the most common cancer in women worldwide and accounted for 1.7 million new cases in 2012, which is a quarter of all new cases of cancer. In 2014, the American Cancer Society reported 235,030 new cases of breast cancer and 40,430 deaths related to breast cancer. Among women in the United States, breast cancer is the most common malignancy, the second most common cause of death from cancer and a leading cause of premature mortality from cancer in women. Ovarian cancer occurs with a lifetime risk of 1.4% in the general female population, but with a risk of 15–56% in women carrying a germline mutation of the BRCA1 and BRCA2 genes. Epithelial ovarian cancer (EOC) is the leading cause of death among gynecologic cancers in the western world and the fifth leading cause of

E. Smolle
Department of Internal Medicine, Division of Pulmonology, Medical University Graz, Auenbruggerplatz 15, 8036 Graz, Austria
e-mail: elisabeth.smolle@medunigraz.at

J. Haybaeck (ed.), *Mechanisms of Molecular Carcinogenesis – Volume 2*,
DOI 10.1007/978-3-319-53661-3_11

cancer-related death in women. Worldwide, about 200,000 women are newly diagnosed, with 125,000 disease-related deaths every year. Endometrial cancer is the most common gynecologic cancer in the USA, accounting for 40,100 new cases and 7470 deaths per year. Endometrial cancer shares similar patterns of distribution by age and geography with ovarian cancer.
Overall gynecologic malignancies pose a significant disease burden, and novel therapeutic strategies are needed to decrease morbidity and mortality from gynecological cancer. Understanding the molecular characteristics of gynecological cancer ist crucial to develop new targeted therapies.

11.1 Introduction

Breast cancer is the most common cancer in women worldwide and accounted for 1.7 million new cases in 2012, which is a quarter of all new cases of cancer [1]. In 2014, the American Cancer Society reported 235,030 new cases of breast cancer and 40,430 deaths related to breast cancer [2]. Among women in the United States, breast cancer is the most common malignancy, the second most common cause of death from cancer and a leading cause of premature mortality from cancer in women [3]. As estimated for the year 2016, 246,660 cases of invasive breast cancer will be newly diagnosed in US women, and 40,450 breast cancer-related deaths will occur [4]. Notably, overall breast cancer incidence rates were stable from 2004 to 2012 [3].

Ovarian cancer occurs with a lifetime risk of 1.4% in the general female population, but with a risk of 15–56% in women carrying a germline mutation of the *BRCA1* and *BRCA2* genes [5–7]. Epithelial ovarian cancer (EOC) is the leading cause of death among gynecologic cancers in the western world and the fifth leading cause of cancer-related death in women [8]. Worldwide, about 200,000 women are newly diagnosed, with 125,000 disease-related deaths every year [9]. Although primary cytoreductive surgery and combination chemotherapy with platinum have improved the patients' prognosis, the 5-year survival rate for those with malignant ovarian cancer is still ~40% [10, 11]. The fallopian tube epithelium is one of the likely progenitor cell types for the most common and deadly ovarian carcinoma histotype, namely, high-grade serous carcinoma [12]. Morphological, immunological, and gene expression analysis of high-grade serous carcinoma also suggests a close relationship to fallopian tube epithelium, rather than ovarian surface epithelium [13, 14]. Risk factors for nonheritable ovarian cancer include nulliparity, infertility, the number of lifetime ovulations, and the use of estrogen only-hormone replacement therapy [15].

Endometrial cancer is the most common gynecologic cancer in the USA, accounting for 40,100 new cases and 7470 deaths per year [16]. Endometrial cancer shares similar patterns of distribution by age and geography with ovarian cancer [17]. However, endometrial cancer rates drop sharply in the age group >65, while ovarian cancer rates continue to rise well into a woman's 80s. In industrialized and Northern European populations, much higher rates of endometrial cancer are observed and lower rates in third world countries. Notably, endometrial cancer rates correlate with ovarian cancer rates significantly, and both diseases are associated with per capita fat intake [18–20].

Cervical cancer is the second leading cause of cancer-related death in young women worldwide [21]. Cervical cancer is very frequent in some countries, counting among the top three leading cancers, for example, in India, with an incidence of 22/100,000/year [1, 22]. However, the mortality rate declined by 80% in the twentieth century, mainly because of the implementation of screening programs [2]. Cervical cancer is caused by high-risk human papillomavirus (HPV) infection [23]. Women with early-stage cervical cancer can be potentially cured with radical hysterectomy or a combination of radio- and chemotherapy [24]. Unfortunately, up to 17% of women recur, either locally or distant within the first 2 years after completing treatment [25, 26]. Local recurrence of cervical cancer after primary surgery is problematic, since it occurs oftentimes as central pelvic recurrence, spreading into contiguous tissues [27]. Even after recurrence, treatment can be performed with curative intent. However, both radiation therapy and pelvic exenteration result in suboptimal local tumor control and survival rates [28, 29].

Gynecologic malignancies are a challenge both for clinicians and researchers, with the aim to lower incidences and improve the patients' outcome. Understanding the molecular carcinogenesis of gynecologic malignancies will be essential to outline new prognostic or predictive biomarkers as well as therapeutic targets and to develop novel treatment strategies.

11.2 Molecular Biology of Gynecologic Carcinogenesis

11.2.1 PTEN

The tumor suppressor gene phosphate and tensin homologue (*PTEN*) is mutated in certain types of cancer, also contributing significantly to the pathogenesis of endometrial cancer. *PTEN* mutations influence not only cancer progression but also response to therapy [30–34]. In fact, *PTEN* mutations were observed in 40–80% of endometrial cancer cases [35–37]. *PTEN* is located on chromosome 10q23, a gene region where loss of heterozygosity is often observed in human cancers. Somatic deletions or mutations of *PTEN* were found in many human sporadic neoplasias, among them endometrial cancer, colorectal cancer, and glioblastoma [35, 38, 39]. In endometrial cancer, loss of *PTEN* functionality has been found to happen particularly early [36, 40]. A high intake of ω-3 polyunsaturated fatty acids (PUFAs) has been associated with a reduced risk of endometrial cancer. Thus, Pan et al. have tried to demonstrate the impact of PUFAs in endometrial cancer of *PTEN*-mutant mice. In this study, a mfat-1 transgene was overexpressed in *PTEN* +/− mice, allowing endogenous production of ω-3 PUFAs [40]. A fish oil-enriched diet or expression of mfat-1 transgene significantly inhibited the growth of xenograft tumors derived from RL95-2 cells bearing a *PTEN* null mutation. Moreover, at the cellular level, ω-3 PUFA treatment decreased the viability of RL95-2 cells, as well as protein kinase B (AKT) phosphorylation and cyclin D1 expression [40]. These events were mediated by reduction of cylooxygenase-2 (COX-2) expression and prostaglandin E_2 (PGE_2) production. These findings show that ω-3 PUFAs have a protective effect on endometrial cancer resulting from *PTEN* mutation.

The phosphatidyl 3-kinase (PI3K) pathway is often induced in breast cancer through loss of the tumor suppressor *PTEN* [30, 41, 42]. When *PTEN* loss leads to mammary tumorigenesis, elevated PI3K pathway activity and reduced EGFR activity are observed [43]. Thus, loss of *PTEN* has obviously distinct effects on cancer signaling. *PTEN* can also dephosphorylate certain proteins and can enter and function within the nucleus or be secreted from cells to affect their neighbors [44–49]. Loss of *PTEN*, in addition to p53 mutation and the effect on breast tumorigenesis, was also investigated by Wang et al. [50]. For this purpose, WAP-Cre:Pten(f/f):p53(lox.stop.lox_R270H) composite mice were generated in which *PTEN* is deleted and a p53-R270H mutation is induced. Combined *PTEN* deletion plus p53 mutation facilitated the formation of four distinct mammary tumors, including poorly differentiated adenocarcinoma, and spindle/mesenchymal-like lesions in the mouse model [50].

It is also known that the PI3K/Akt pathway plays a role in chemoresistance to platinum-based neoadjuvant chemotherapy in cervical cancer [51]. In a study by Guo et al., genetic polymorphisms in the PI3K/Akt pathway and chemotherapeutic outcomes following platinum-based therapy were investigated in a subset of 259 patients with squamous cell cervical cancer. In total, 17 single nucleotide polymorphisms in four genes (*PI3KCA*, *Akt1*, *Akt2*, and *PTEN*) were associated with response to chemotherapy. In another study, the role of the PTEN/PI3K/Akt signaling pathway in response to the tyrosine kinase inhibitor gefitinib in cervical cancer was investigated [52]. Gefitinib targets the epidermal growth factor receptor (EGFR), and clinical response in cervical cancer patients is generally poor. microRNA-221 (miR-221) features a significantly increased expression in cervical cancer as compared to adjacent normal tissue [52]. Interestingly, upregulation of miR-221 expression in cervical cancer cells decreased *PTEN* expression, resulting in increased pAkt expression. Gefitinib sensitivity was decreased by the upregulation of miR-221. This study shows that gefitinib sensitivity of cervical cancer cells is reduced by miR-221 through the PTEN/PI3K/Akt signaling pathway; and thus miR-221 represents a potential target to increase the sensitivity to gefitinib in cervical cancer treatment [52]. Also Yang et al. recently showed that the PI3K/Akt signaling pathway is a key regulator in cervical carcinogenesis [53]. microRNA-494 (miR-494) was previously shown to directly target *PTEN*. The authors aimed at analyzing the significance of miR-494 concerning *PTEN* expression in cervical cancer, correlating miR-494 expression with clinicopathological data [53]. Importantly, miR-494 was significantly upregulated in human cervical cancer cell lines and tissues, which was associated with *PTEN* downregulation, adverse clinicopathological characteristics, poor overall and progression-free survival, and a poor prognosis. Inhibition of miR-494 suppressed cell proliferation and cell growth by targeting the 3′-untranslated region (3′-UTR) of *PTEN* mRNA. This investigation highlights that miR-494 has an essential role in the carcinogenesis and progression of cervical cancer by deregulating *PTEN* functionality [53].

Endometrioid ovarian tumors are characterized by specific molecular alterations, namely, mutations of β-catenin [54, 55] and *PTEN* [56] in one third and 20% of cases, respectively. Both mutations are already observed in well-differentiated, stage 1 tumors and are thus thought to happen as an early event [57]. The fact that loss of heterozygosity (LOH) at 10q23 (the location of *PTEN*) and *PTEN* mutation were identified both in endometrioid ovarian carcinoma and endometriosis points out that endometrioid carcinomas arise from endometriosis implanted on the ovary

[56–58]. Also in ovarian clear cell carcinomas, *PTEN* mutations have been reported in 8–40% of cases, based on preliminary studies [58, 59].

Summing up the above-mentioned studies, *PTEN* strongly influences the pathogenesis of different entities of gynecologic malignancies. Therefore, components of the PTEN/PI3K/Akt signaling pathway may serve as therapeutic targets in the future.

11.2.2 Cancer-Associated Fibroblasts

Cancer-associated fibroblasts (CAFs) are the most abundant cell type in breast cancer stroma, producing a plethora of chemokines, growth factors, and proteins of the extracellular matrix (ECM) that may contribute to dissemination and metastasis [60]. Increasing evidence suggests that stromal involvement, including the function of CAFs, is crucial for carcinogenesis. It was found, for instance, that a positive expression of matrix metallopeptidase 13 (MMP13) and lectin, galactoside-binding, soluble 1 (LGALS1) in CAFs was associated with an enhanced odds ratio for regional metastasis in breast cancer [61, 62]. Also in ovarian cancer, CAFs are one of the major components of the tumor stroma and have shown a supportive role for tumor progression [63]. Natriuretic peptide B (NPPB), a CAF-specific secretory protein, was recently identified as a tumor biomarker for ovarian cancer. It was found that NPPB is expressed in 60% of primary ovarian CAF tissues, but not in the healthy ovarian stroma [64]. As the unique signaling crosstalk networks of activated CAFs are identified, this would provide the possibility of reprogramming activated CAFs back to "normal state," which could be a therapeutic strategy against tumor progression [65]. It is known that CAFs secrete CAF-specific proteins, cytokines, and growth factors and form an ECM which is beneficial for tumor progression, facilitating cell growth and angiogenesis [66–69]. Among the commonly known CAF-derived factors, fibroblast activation protein alpha (FAPα) has been demonstrated to promote the proliferation in ovarian cancer, as well as invasion, via α3β1 integrin receptor and the upregulation of the extracellular signal-regulated kinases (ERK) signaling pathway in ovarian cancer cells [70, 71]. CAF-derived SDF-1, also known as chemokine (C-X-C motif) ligand 12 (CXCL12), has been demonstrated to promote tumor growth, motility, and tumor angiogenesis in multiple cancer types, including ovarian cancer, by interaction with the chemokine (C-X-C motif) receptor 4 (CXCR4) receptor on cancer cells [72–76]. In a study by Lau and colleagues, reciprocal tumor–stroma interaction in ovarian cancer was reported [77]. According to their study, CAFs responded to lymphotoxin via lymphotoxin-β receptor and the NF-κB signaling pathway. Ovarian cancer cell-derived lymphotoxin upregulates chemokine (C-X-C motif) ligand 11 (CXCL11) secretion by CAFs, and CXCL11 activates chemokine (C-X-C motif) receptor 3 (CXCR3) on the ovarian cancer cells, facilitating their proliferation and motility [77]. In a study by McAndrews et al., it was shown that stromal fibroblasts are more adherent to invasive ovarian cancer cells, as compared to noninvasive ovarian cancer cells. Such adhesion is mediated by cadherin 11 and 2, which are highly expressed by invasive cancer cells and CAFs [78].

In a recent study the molecular profiles of fibroblasts from normal ovary and high-grade serous ovarian tumors were examined to identify novel potential therapeutic targets [79]. Thereby 2300 genes that are significantly differentially expressed in CAFs were identified. The expression of one of these genes, connective tissue growth factor (*CTGF*), was also confirmed by immunohistochemistry. The CTGF protein is a secretion product of the tumor microenvironment and is currently being pursued as a therapeutic target in pancreatic cancer. It was found that CTGF promotes migration and peritoneal adhesion of ovarian cancer cells. These effects are blocked by FG-3019, a human monoclonal antibody against CTGF, currently under clinical investigation as a therapeutic agent [79]. Immunohistochemistry of CTGF expression in high-grade ovarian serous carcinoma showed that the highest level of tumor stromal CTGF expression was correlated with the poorest prognosis. Thus, CTGF serves as a therapeutic target for the treatment of high-grade serous ovarian cancer.

An investigation by Huang and colleagues found CD117 expression in fibroblast-like stromal cells as being an indicator for an unfavorable clinical outcome in ovarian carcinoma patients [80]. The stem cell factor (SCF) receptor CD117 (c-kit) is widely used for the identification of hematopoietic stem cells and cancer stem cells. In a variety of cancers, it was shown that CD117 expression in carcinoma cells indicated an unfavorable prognosis. The authors of this study evaluated the immunohistochemical expression of CD117 in a serial of 242 epithelial ovarian cancer cases. Indeed, CD117 expression in fibroblast-like stromal cells was closely linked to an advanced FIGO stage, poor differentiation grade, and histological subtype ($p < 0.05$) and was also significantly associated with poor overall survival and progression-free survival. The CD117-positive fibroblast-like stromal cells were positive for mesenchymal stem/stromal cell marker CD73 but negative for the fibroblast marker fibroblast activation protein (FAP) and also negative for α smooth muscle actin (α-SMA), indicating that the CD117+/CD73+ fibroblast-like stromal cells are a subtype of mesenchymal stem cells in the tumor stroma [80]. This study highlights once again the important influence of tumor stroma cells on cancer progression and the patients' prognosis.

Summing up the above-mentioned data, increasing evidence shows that CAFs can modulate cancer phenotypes, cancer cell growth, motility, and invasiveness. Thus it is crucial to determine which protein factors are exclusively produced by CAFs. With the aid of cell type-specific expression profiles, crosstalk signaling networks can be identified, which may contribute to the development of novel targeted therapeutics for the treatment of cancer.

11.2.3 Galectin-1 Expression in Epithelial Ovarian Cancer and in Cervical Cancer

Galectins form a gene family of widely distributed carbohydrate-binding proteins characterized by their affinity for β-galactoside-containing glycans [81]. Currently 14 members of this family are known [82, 83]. Galectin-1 is a 14-kDa laminin-binding galectin and a member of the galectin family of β-galactoside-binding proteins classified as a prototype galectin [82]. Galectin acts via both intracellular sugar-independent interactions with other proteins and extracellular sugar-dependent autocrine or paracrine interactions [84]. In some experimental studies, galectin-1 has been linked to the invasion and metastasis formation of cancer cells [85, 86], as well as to the promotion of angiogenesis [87, 88] and the protection of the tumor from host immune responses [89, 90]. Galectin-1 has been reported to be upregulated in thyroid carcinoma [91, 92]. Similarly, an increased galectin-1 expression has been correlated with the potential to metastasize of tumorigenic cells, most likely by affecting cell motility and invasion, altering extracellular matrices [86, 93]. Moreover, accumulation of galectin-1 in the peritumoral stroma of breast cancer and ovarian cancer regulates both cancer cell proliferation and invasiveness [62, 94]. Galectin-1 is also strongly expressed in ovarian cancer, promoting tumor progression and chemoresistance to cisplatin [95, 96]. According to a recent investigation, galectin-1 can be released from ovarian cancer cells and from CAFs, which was detected in cell culture [81]. In the peripheral circulation of a majority of patients suffering from epithelial ovarian cancer, galectin-1 could be detected. In 140 patients with epithelial ovarian cancer, galectin-1 serum concentrations were investigated and compared to serum levels in 70 healthy individuals. Elevated serum galectin-1 concentrations were associated with tumor progression. Also, serum galectin-1 levels were significantly higher in patients with metastatic disease when compared to patients with localized tumors. The authors conclude that increased galectin-1 serum levels favor metastasis by enhancing the adhesive interaction between tumor cells and proteins of the extracellular matrix and by promoting tumor cell embolization through increased cell adhesion and dissemination of tumor cells into the circulation [81]. Increased expression levels of galectin-1 were observed in ovarian carcinoma samples, as compared to normal ovarian tissue, which was confirmed by qRT-PCR and Western blot [81]. Thus, galectin-1 can serve as a biomarker for the prognosis of ovarian cancer patients, as its dysregulation contributes to tumor progression.

Another study found that the intensity of galectin-1 expression significantly and independently correlated with poor survival in cervical cancer, as observed by immunofluorescent staining on formalin-fixed, paraffin-embedded tissue samples [21]. Disease-specific survival for tumor galectin-1 expression intensity was evaluated by correcting for FIGO stage, lymph node metastasis, and vaso-invasion. Altogether, samples of 155 patients were included in this analysis. Tumor galectin-1 intensity was also an independent predictor of poor disease-free survival. Patients with strong tumor galectin-1 expression had increased tumor invasion and received postoperative radiotherapy treatment more frequently. Thus, galectin-1 expression is associated with a more aggressive tumor behavior, which was also confirmed by a study conducted by Huang et al., who reported on a correlation between tumor expression of galectin-1 and poor survival in cervical cancer patients who were treated with curative-intent radiation therapy [97]. Therefore, also for cervical cancer, tumor galectin-1 expression is a biomarker for poor survival [21].

11.2.4 Cathepsin D and Cathepsin L

Cathepsin D (CathD) is a soluble lysosomal aspartic endopeptidase primarily involved in degrading unfolded or nonfunctional proteins intracellularly [98, 99]. Normally, procathepsin D (pCathD), the precursor of active CathD, is not secreted extracellularly. However, it has been found that in some conditions pCathD or CathD escape the usual pathway and are secreted from cells [100]. Recent data suggests that CathD has a potential role in tumor progression, both in its intracellular and extracellular form. Intracellular CathD, for example, has a role in apoptosis. By inhibition of enzymatically active cytosolic CathD, using the inhibitor pepstatin A (pepA), apoptosis induced by IFN-gamma or oxidative stress was delayed [101–103]. The role of CathD in apoptosis induction has been found to be related to caspases, as the caspase inhibitor Z-VAD-FMK added in combination with pepA induced a reduction in cell death, suggesting an association between caspases and cytosolic CathD [104, 105]. pCathD and CathD were demonstrated to induce proliferation and migration of cancer cells, fibroblasts, and endothelial cells [106]. Interestingly, the acidic pH in tumor microenvironment promotes the conversion of pCathD into the mature and biologically active CathD [99]. Hypersecretion of CathD has been demonstrated in numerous cancer types including ovarian cancer, breast cancer, and endometrial cancer, among others [107–114]. Immunohistochemical studies have shown an enhanced CathD expression to be an indicator of malignancy in serous ovarian carcinoma [115–117]. In an investigation on omental metastasis of ovarian cancer, a higher CathD expression in omental lesions of serous ovarian carcinoma was observed, as compared to the omentum of patients with benign ovarian cystadenoma. Moreover, high omental expression of CathD was linked to a poor disease-specific survival [118].

CathD was extensively studied in human primary breast cancer. In estrogen receptor-positive breast cancer cell lines, upregulation of CathD was observed [119]. This data was also supported by experiments with MCF7 cell lines, revealing that pCathD/CathD were overexpressed and hypersecreted from these cells. Moreover it was found that purified pCathD from MCF-7 breast cancer cells stimulated MCF-7 cell growth [120].

Cathepsin L (CathL) is a ubiquitous cysteine proteinase playing an important role in degrading endocytosed proteins and intracellular proteins as well [121, 122]. CathL has been linked to tumor invasion and metastasis, as it degrades compounds of the ECM such as proteoglycans, elastin, laminin or fibronectin, as well as collagens I, II, IX, and XI [123–128]. An increased level of secreted CathL was observed in the sera of epithelial ovarian cancer patients, as compared to patients with benign ovarian tumors or normal ovaries [129, 130]. Moreover, a significant increase in mRNA levels in the ovarian malignancies was observed. CathL has been suggested to be involved in invasion and metastasis of ovarian cancer and may thus be a marker of advanced disease. This is also supported by a study by Winiarski et al., who demonstrated CathL to be increasingly expressed in the endothelium of blood vessels within omentum hosting metastatic ovarian high-grade serous carcinoma, whereas an increased CathL expression was not found in the omentum of patients with benign ovarian tumors [118]. In SKOV3 ovarian cancer cell lines, downregulation of CathL significantly inhibited the proliferative and invasive capability [131].

Summing up the results above, CathD and CathL both play a role in cancer, especially by breaking down ECM compounds, and thereby facilitating invasion.

11.2.5 Biomarkers in Cervical Cancer

Today there are numerous assays for the detection of nucleic acids of oncogenic and non-oncogenic human papillomaviruses (HPVs) in the cervical tissue. HPV testing has a high sensitivity, with a high negative predictive value, because the absence of carcinogenic HPV indicates a very low risk of cervical intraepithelial neoplasia grade 3 (CIN3) or cancer [132–134]. However, HPV assays do not discriminate between transient and persistent HPV infections [135]. Still, according to long-term prospective cohort studies and randomized clinical trials, HPV testing is highly effective for the detection of CIN2 or worse in women aged 30 years or older and for the diagnosis of adenocarcinoma [136, 137]. There are assays for HPV DNA testing, for HPV RNA testing, as well as for the detection of HPV proteins [135].

The functional inactivation of p53 and pRb onco-suppressors by the oncoproteins E6 and E7 leads to the alteration of several cellular pathways that are relevant for cell transformation and cancer development. E7 expression determines the inactivation of pRb with a consequent increase of free E2F in the cell and an increase of cyclin-dependent kinase inhibitor p16 (p16INK4a) and augmented proliferation (increased Ki-67 expression) [138, 139]. Thus, p16 overexpression, identified by immunohistochemistry or immunosorbent assay (ELISA), can be considered as a marker of HPV infection and of activated expression of viral oncogenes and for virus-induced cell cycle deregulation [140]. The use of p16INK4 immunohistochemistry also improves diagnostic accuracy, reliability, and quality in histopathology of cervical tissue samples [141]. A large meta-analysis of 17 studies showed that the pooled sensitivity of p16INK4a to detect CIN2 or worse was 83.2% and 83.8% in atypical squamous cells of unknown significance (ASCUS) and low squamous intraepithelial lesions (SIL), respectively, and the pooled specificities were 71% and 65.7%, respectively [142].

The proliferation antigen Ki-67 is usually expressed during the G2 and mitotic phases of the cell cycle and was found to be a reliable indicator of the growth fraction of a tumor according to many studies. In a series of 138 cervical cone biopsies, it was shown that p16INK4 and Ki-67 are co-expressed in dysplastic lesions only [143]. Thus, a dual p16/Ki-67 immunocytochemistry assay is now available as an adjunctive test in cervical cancer screening. The sensitivity of this dual assay was found to be 92.2% in ASCUS and 94.2 in low SIL, and specificities were 80.6% and 68%, respectively [144].

The minichromosome maintenance protein 2 (MCM 2) and topoisomerase II alpha (TOP2A) proteins are expressed in cells with aberrant S-phases and in HPV-transformed cells, in association with an elevated expression of the HPV E6/E7 proteins [145]. There is also an assay based on an antibody cocktail recognizing both MCM2 and TOP2A proteins. The use of this assay for the triage of women testing positive for high-risk HPVs was found to increase the specificity and the positive predictive value of the screening as compared to the high-risk HPV test alone [146].

The E6 proteins of oncogenic HPVs promote the transcription of telomerase reverse transcriptase (TERT), which stabilizes and repairs the repeated DNA sequences at the telomere end of the chromosomes [147]. Notably, gain of chromosome 3q, containing the sequence for the telomerase RNA component (TERC), and gain of chromosome 5p, containing the TERT gene, are associated with CIN2 or worse, with a specificity of 97% [148]. Evaluating gains of chromosomes 3q and 5p with fluorescence in situ hybridization (FISH) and multiplex ligation-dependent probe amplification (MLPA) may be a useful marker for the identification of progressing lesions.

microRNAs might also be involved in the pathogenesis of cervical cancer [149]. Specifically, miR-9, miR-127, miR-145, miR-146a, miR-199a, miR-200a, and miR-424 have been found to be dysregulated in cervical carcinoma [150, 151]. In an analysis by Li et al. on women with CIN, it was shown that miR-218 levels were lower in patients with high-risk HPV than in those with low-risk or intermediate-risk HPVs [152]. Wang et al. demonstrated a significantly decreased expression of miR-375 in 170 cervical cancer tissues, compared to samples of normal tissue, suggesting that a downregulation of miR-375 is involved in the progression of cervical cancer [153]. Therefore, the evaluation of specific miRNAs could represent novel candidate markers for cancer screening and the prognostic evaluation of patients with cervical neoplasia.

DNA methylation is one of the epigenetic mechanisms influencing gene transcription, the structure of chromatin, genomic stability, as well as the inactivation of imprinted genes and X chromosome [154]. The analysis of the DNA methylation pattern as a biomarker in clinical oncology seems to be a promising approach [155]. Recently it was found that methylation of viral and cellular DNA is a potential biomarker for improving the accuracy of cervical cancer screening [156]. There seems to be a direct relationship between the methylation status of the *HPV L1* gene and diagnosis of CIN2. However, some studies reported a decreased methylation of CpG sites within the HPV regulatory region, while others showed an increased methylation in this viral region to be associated with CIN2. Two studies have described a whole-genome analysis of methylation patterns of HPV16, HPV31, HPV18, and HPV45 in a large cohort study [157, 158]. According to these studies, elevated DNA methylation on multiple CG sites in the L1, L2, E2, and E4 open reading frames (ORFs) is significantly associated with CIN2 or worse [157, 158]. The methylation status of several human genes was also shown to be a relevant event pointing out cervical carcinogenesis. For instance, the treatment of HPV-positive cervical cancer cell lines with demethylating agents, coupled to expression microarrays, allowed the identification of genes encoding the secreted protein, acidic, cysteine-rich (SPARC) and the tissue factor pathway inhibitor 2 (TFPI2) protein as being highly methylated in invasive cervical cancer [159]. Another approach based on restriction landmark genomic scanning (RLGS) was used to identify genes encoding for nucleolar protein 4 (NOL4) and lipoma HMGIC fusion partner-like 4 (LHFPL4) as being methylated in cervical cancer [160]. Interestingly, aberrant methylation can be detected in cervical cancer smears up to 7 years prior to the diagnosis of cervical cancer, suggesting that gene methylation analysis may be a valuable strategy for the triage of women positive for high-risk HPVs [161]. Moreover, certain genes display specific methylation patterns for cervical adenocarcinoma (in situ), and detecting these in cervical scrapings can therefore be a guide for appropriate therapy [162].

11.2.6 microRNAs in Breast Cancer

microRNAs (miRNAs, miRs) are a novel class of noncoding, single-stranded RNAs, which were first described in 1993 by Lee et al. in *C. elegans* [163]. When first detected it was believed that miRNAs are just "genomic trash." However, miRNAs as small regulatory RNA molecules (with an approximate length of about 22 nucleotides) posttranscriptionally inhibit gene expression by degrading or blocking the translation of messenger RNA (mRNA) targets [164]. miRNAs suppress the translation of their target mRNAs by binding to their 3′-untranslated region, but also other mechanisms have been described [165–167]. miRNA loci are statistically overrepresented at fragile genomic regions commonly amplified or deleted in human cancers, which implies a connection of miRNAs with cancer initiation and progression [168, 169]. Most miRNAs are thought to be direct suppressor miRs or posttranscriptional repressors of known oncogenes, but some may also act as promotors of tumorigenesis (onco-miRs) [170–172]. The relevance of miRNAs in breast cancer has been demonstrated by several groups so far [172–174]. Notably, some miRNAs were found to be upregulated in breast cancer, as compared to normal breast tissue, while others were downregulated, also concordant with the hypothesis that some miRNAs act as onco-miRs and others as tumor suppressor miRs [174]. Loss of several tumor suppressor miRNAs (miR-206, miR-17-5p, miR-125a, miR-125b, miR-200, let-7, miR-34, and miR-31) and overexpression of certain oncogenic miRNAs (miR-21, miR-155, miR-10b, miR-373, and miR-520c) have been observed in breast cancer [173]. Still, the gene network orchestrated by these miRNAs remains largely unknown, although some key targets have been identified, which might influence the tumor phenotype. In a study by Schrauder and colleagues, 153 whole blood samples of early-stage breast cancer patients and healthy control individuals were analyzed [172]. The purpose was to outline some potential disease biomarker miRNAs for early-stage breast cancer. Using the Geniom® Realtime Analyzer microarray platform, 59 deregulated miRNAs were outlined in the whole blood samples, compared to healthy controls. Thirteen of these were significantly upregulated, and 46 were significantly downregulated. Among the most upregulated miRNAs in early-stage breast cancer were miR-4306, miR-202, miR-4257, miR-1323, miR-335, miR-497, miR-106b, miR-922, and miR-516b [172]. Among the most downregulated miRNAs were miR-718, miR-625, miR-1471, miR-193a-3p, miR-182, miR-1915, miR-564, miR-107, miR-2355, and miR-3186-3p [172]. It is known that several miRNAs are not only aberrantly expressed in human breast cancer tissue but also that their expression levels correlate with clinical stage and clinicopathological variables like hormone receptor status and tumor subtype and clinical variables like metastatic potential, progression-free survival, and overall survival [174–179]. Tissue-based miRNA expression profiling of the inflammatory breast cancer subtype outlined some miRNAs to be associated with the difference between inflammatory and noninflammatory breast cancer. Among others, miR-335 was found to be increasingly expressed in inflammatory breast cancer [180].

Cai et al. have performed an analysis where the function of miR-205 in breast cancer cell lines and its influence on docetaxel sensitivity was investigated [181]. Two breast cancer cell lines, MDA-MB-231 and MCF-7, were investigated with and without miR-205 overexpression. In MDA-MB-231 cells, as well as in MCF-7 cells, miR-205 was shown to increase cell sensitivity to docetaxel, as detected with a cell proliferation assay [181]. The authors of this study also performed a colony formation assay to assess whether miR-205 could inhibit the clonogenic survival of the MDA-MB-231 cancer cells. Indeed, the cells treated with miR-205 combined with docetaxel showed a significantly decreased colony formation ability. Similar results were also observed for the breast cancer cell line MCF-7. Thus, miR-205 suppresses breast cancer cell proliferation and has a synergistic effect with docetaxel treatment. The effect of miR-205 together with docetaxel was also investigated in vivo via a mouse xenograft model. MDA-MB-231 cells, and cells stably expressing miR-205, were subcutaneously injected into the flank region of athymic nude mice, and docetaxel was injected directly into the xenografts from day 9 once every 3 days. It was found that miR-205 overexpression and docetaxel treatment inhibited tumor cell growth in vivo. MiR-205 overexpression and docetaxel treatment alone inhibited tumor growth, but an even greater inhibitory effect was observed when combining miR-205 with docetaxel [181].

11.2.7 microRNAs in Ovarian Cancer

It has been demonstrated that miRNAs are involved in various cellular functions in ovarian cancer, ranging from carcinogenesis, cell cycle, apoptosis, proliferation, invasion, and metastasis to the development of chemoresistance [182]. The deregulation of cancer-related miRNAs is due to several factors, including chromosomal rearrangements, aberrations in genomic copy numbers, epigenetics, abnormal maturation pathways and the regulation of miRNAs by transcription factors, as well as miRNA–miRNA interactions [183]. Various miRNAs have been reported to be differentially expressed among different histotypes in ovarian cancer [184]. The abnormally expressed miRNAs promote tumorigenesis by inactivating tumor suppressor genes and/or by activating oncogenes [61, 185]. The transforming growth factor-β (TGF-β) signaling pathway is one of the best characterized pathways which is known to play a role in ovarian carcinogenesis and particularly in epithelial–mesenchymal transition (EMT), modulated by miR-181a [182]. Also the PI3K/AKT pathway, the G-PCR signaling pathway, the Wnt/β-catenin pathway, and the ERK5 pathway are known to be involved in the formation of ovarian cancer, as these pathways are deregulated by miRNAs [186]. The miRNA precursor let-7 targets several oncogenes, such as c-Myc, ras, high-mobility group A (HMGA), Janus protein tyrosine kinase (JAK), signal transducer and activator of transcription 3 (STAT3), and NIRF. It has also been suggested that let-7a promotes tumorigenesis, proliferation, and invasion by regulating the cell cycle through the NIRF/p53/p21/CDK signaling pathway [187]. In the pathogenesis of ovarian cancer, the transformation of epithelial cells to mesenchymal cells marks the inception of cancer development and invasion. This epithelial–mesenchymal transition (EMT) is characterized by cellular molecular reprogramming and phenotypic changes in cells. This reprogramming occurs when the level of E-cadherin protein is reduced [188]. The miR-200 family has been found to play an important role in this transition by targeting ZEB-1 and ZEB-2, the transcriptional repressors of E-cadherin genes [189]. miR-200a, miR-200b, miR-200c, miR-141, and miR-429 belong to the miR-200 family. A strong and positive correlation between the expression of E-cadherin and miR-200c has been reported in ovarian cancer tissues. Overexpression of miR-200a, miR-200b, and miR-200c and/or overexpression of miR-141 downregulates ZEB1/ZEB2 levels, which leads to higher levels of E-cadherin and epithelial phenotype [190]. However, another study reported that ZEB1/ZEB2 can also block E-cadherin expression by inhibition of the transcription of miR-200 family members by binding to clusters of the miR-200 promoter [191]. Notably, the downregulation of the miR-200 family leads to increased expression of β-tubulin III, which results in chemoresistance in ovarian cancer patients on paclitaxel-based treatment [192]. A reduced expression of miR-200c is associated with recurrence in ovarian cancer. Moreover, miR-200 overexpression also significantly inhibits ovarian cancer cell invasiveness and metastasis by downregulating matrix metallopeptidase 3 (MMP3) [193]. miR-34a/miR-34b/miR-34c induced by p53 is downregulated in ovarian cancer, as the p53 mutation promotes epithelial–mesenchymal transition of cancer cells by increasing the expression of Snail1 protein. The miR-34 family of miRNAs suppresses Snail1 activity, binding to highly conserved 3′-untranslated region in Snail1 and its regulatory molecules. Thus, mutated p53 downregulates miR-34a/miR-34b/miR-34c to maintain the level of Snail1 protein [194, 195]. During the progression of ovarian cancer, oxidative stress occurs which also affects cancer proliferation and is mediated by miRNAs. MiR-141 and miR-200a are known to modulate oxidative stress response by targeting p38α. Overexpression of these two miRNAs leads to p38α deficiency and increases tumor growth. However, overexpression of miR-141 and miR-200a also improves the response to chemotherapeutic agents, as demonstrated in mouse models. High-grade human ovarian adenocarcinomas with increased miR-200a expression show decreased p38α levels and associated oxidative stress [196]. There is also a correlation between the miR200a-dependent stress signature and an improved survival of patients, with a better response to treatment [196].

MiR-214 constitutively activates the PTEN/AKT pathway, leading to chemoresistance in different types of tumors, including ovarian cancer [197]. An increased miR-214 expression is involved in resistance to cisplatin therapy by downregulation of PTEN protein and consecutive activation of the PI3K/AKT/mTOR pathway, which enhances tumor cell survival [197]. MiR-21 also blocks apoptosis in cancer cells by targeting PTEN and programmed cell death 4 (PDCD4), activating the AKT pathway [198]. Similarly, downregulation of miR-100, a tumor suppressor miRNA, mediates increased sensitivity to everolimus in ovarian cancer cell lines and tissues, primarily via the repression of mTOR–AKT signaling [199, 200].

Recently it was found that the aberrant expression pattern of miRNAs can be a powerful tool to diagnose ovarian cancer at its primary stage before metastatic spread occurs. Hypomethylation of miRNA genes is an epigenetic mechanism, upregulating miR-21, miR-203, and miR-205 in ovarian cancer, as compared to the normal ovary [201]. MiR-30c, miR-30d, and miR-30e are frequently upregulated, whereas miR-493 is usually downregulated in ovarian carcinomas compared to normal ovarian cell lines [202]. Expression levels of the four miRNAs, miR-30c, miR-30d, miR-30e-3p, and miR-370, were found to be significantly higher in ovarian carcinoma than in benign ovarian tissue, and miR-181d, miR-30a-3p, and miR-532-5p were significantly different between ovarian borderline tumors and ovarian carcinoma. Notably, miR-370, which is highly upregulated in early stages of ovarian cancer, can be used as a biomarker for the early detection of ovarian cancer. Moreover, downregulation of miR-30c, miR-30d, miR-30e-3p, and miR-532-5p is associated with Her2/neu oncogene overexpression [203].

Recently a study showed the potential of miRNAs as stable blood-based noninvasive biomarkers using serum miRNAs from epithelial ovarian cancer patients. Eight miRNAs were differentially expressed between cancer patients and normal controls, five (miR-21, miR-92, miR-93, miR-126, and miR-29a) were significantly overexpressed, and three (miR-155, miR-127, and miR-99b) were significantly downregulated in the serum of cancer patients as compared to their matching controls [204].

miRNAs may also serve as prognostic biomarkers in ovarian cancer. For example, overexpression of miR-23a, miR-27a, miR-21, and miR24-2 is significantly associated with a poor prognosis in patients with progressive disease during first-line chemotherapy. Overexpression of miR-378 is associated with increased chemosensitivity, whereas its under-expression is associated with chemoresistance among patients treated with a platinum-based chemotherapy [205, 206]. Downregulation of miR-22 in epithelial ovarian cancer is linked to overall survival and progression-free survival, serving as an efficient prognostic factor [207]. Three miRNAs (miR-484, miR-642, and miR-217) have been reported to predict chemoresistance in serous epithelial ovarian carcinomas. The response to chemotherapy in association with these three miRNAs was shown in a large multicenter cohort, and miR-484 was also found to be involved in the control of tumor angiogenesis, probably serving as a future therapeutic target as an anti-angiogenetic agent in serous epithelial ovarian cancer [208].

Conclusion

With this chapter, we have tried to give a brief insight into gynecologic carcinogenesis on the molecular level, summing up the most recent and most relevant studies on this topic.

Evidently there are many pathways, molecules, and even microRNAs providing potential therapeutic targets in the future for the treatment of gynecologic cancer.

References

1. Ferlay J, Soerjomataram I, Ervik M, Bray FD. International Agency for Research on Cancer; GLOBOCAN 2012: Cancer Incidence, Mortality and Prevalence Worldwide. http://www.globocan.iarc.fr/pages/fact_sheets_cancer.aspx. Accessed 21 Mar 2014.
2. Siegel R, Ma J, Zou Z, Jemal A. Cancer statistics, 2014. CA Cancer J Clin. 2014;64(1):9–29.
3. Howlander N, Noone A, Krapcho M, et al. SEER cancer statistics review, 1975–2012. Bethesda: National Cancer Institute; 2015.
4. Siegel RL, Miller KD, Jemal A. Cancer statistics, 2016. CA Cancer J Clin. 2016;66(1):7–30.
5. Bougie O, Weberpals JI. Clinical considerations of BRCA1- and BRCA2-mutation carriers: a review. Int J Surg Oncol. 2011;2011:374012.
6. King MC, Marks JH, Mandell JB, New York Breast Cancer Study Group. Breast and ovarian cancer risks due to inherited mutations in BRCA1 and BRCA2. Science. 2003; 302(5645):643–6.
7. National Cancer Institute, DCCPS, Surveillance Research Program, Surveillance Systems Branch. Surveillance, Epidemiology, and End Results (SEER) Program. Research Data (1973–2009). Released April 2012 based on November 2011 submission. www.seer.cancer.gov.
8. DeSantis CE, Lin CC, Mariotto AB, Siegel RL, Stein KD, Kramer JL, et al. Cancer treatment and survivorship statistics, 2014. CA Cancer J Clin. 2014;64(4):252–71.
9. Parkin DM, Bray F, Ferlay J, Pisani P. Global cancer statistics, 2002. CA Cancer J Clin. 2005;55(2):74–108.
10. Johnatty SE, Beesley J, Paul J, Fereday S, Spurdle AB, Webb PM, et al. ABCB1 (MDR 1) polymorphisms and progression-free survival among women with ovarian cancer following paclitaxel/carboplatin chemotherapy. Clin Cancer Res. 2008;14(17):5594–601.
11. Deraco M, Baratti D, Laterza B, Balestra MR, Mingrone E, Macri A, et al. Advanced cytoreduction as surgical standard of care and hyperthermic intraperitoneal chemotherapy as promising treatment in epithelial ovarian cancer. Eur J Surg Oncol. 2011;37(1):4–9.
12. Li J, Fadare O, Xiang L, Kong B, Zheng W. Ovarian serous carcinoma: recent concepts on its origin and carcinogenesis. J Hematol Oncol. 2012;5:8. doi:10.1186/1756-8722-5-8.
13. Marquez RT, Baggerly KA, Patterson AP, Liu J, Broaddus R, Frumovitz M, et al. Patterns of gene expression in different histotypes of epithelial ovarian cancer correlate with those in normal fallopian tube, endometrium, and colon. Clin Cancer Res. 2005;11(17):6116–26.
14. Bell DA. Origins and molecular pathology of ovarian cancer. Mod Pathol. 2005;18(Suppl 2):S19–32.
15. Lacey Jr JV, Mink PJ, Lubin JH, Sherman ME, Troisi R, Hartge P, et al. Menopausal hormone replacement therapy and risk of ovarian cancer. JAMA. 2002;288(3):334–41.
16. American Career Society. Cancer facts and figures 2008. 2008.
17. Merritt MA, Cramer DW. Molecular pathogenesis of endometrial and ovarian cancer. Cancer Biomark. 2010;9(1–6):287–305.
18. Parazzini F, Franceschi S, La Vecchia C, Fasoli M. The epidemiology of ovarian cancer. Gynecol Oncol. 1991;43(1):9–23.
19. Parazzini F, La Vecchia C, Bocciolone L, Franceschi S. The epidemiology of endometrial cancer. Gynecol Oncol. 1991;41(1):1–16.
20. Prentice RL, Thomson CA, Caan B, Hubbell FA, Anderson GL, Beresford SA, et al. Low-fat dietary pattern and cancer incidence in the Women's Health Initiative Dietary Modification Randomized Controlled Trial. J Natl Cancer Inst. 2007;99(20):1534–43.
21. Punt S, Thijssen VL, Vrolijk J, de Kroon CD, Gorter A, Jordanova ES. Galectin-1, -3 and -9 expression and clinical significance in squamous cervical cancer. PLoS One. 2015;10(6):e0129119.
22. Curado MP, Edwards B, Shin HR, Storm H, Ferlay J, Heanue M. Cancer incidence in five continents. IX ed. IARC Scientific Publications No. 160. Lyon: International Agency for Research on Cancer; 2007.

23. Walboomers JM, Jacobs MV, Manos MM, Bosch FX, Kummer JA, Shah KV, et al. Human papillomavirus is a necessary cause of invasive cervical cancer worldwide. J Pathol. 1999;189(1):12–9.
24. Bansal N, Herzog TJ, Shaw RE, Burke WM, Deutsch I, Wright JD. Primary therapy for early-stage cervical cancer: radical hysterectomy vs radiation. Am J Obstet Gynecol. 2009;201(5):485.e1–9.
25. SEER data for 2000–2004. http://seer.cancer.gov/. Accessed 25 April 2011.
26. Ries LAG, Harkins D, Krapcho M, et al. SEER cancer statistics review, 1975 to 2003. Bethesda: National Cancer Institute; 2006.
27. Di Saia PJ, Creasman WT, Mannel RS, McMeekin DS, Mutch DG. Invasive cervical cancer. In: Di Saia PJ, Creasman WT, editors. Clinical gynecologic oncology. 8th ed. Philadelphia: Elsevier; 2012. p. 51–119.
28. Estape R, Angioli R. Surgical management of advanced and recurrent cervical cancer. Semin Surg Oncol. 1999;16(3):236–41.
29. Jobsen JJ, Leer JW, Cleton FJ, Hermans J. Treatment of locoregional recurrence of carcinoma of the cervix by radiotherapy after primary surgery. Gynecol Oncol. 1989;33(3):368–71.
30. Davis NM, Sokolosky M, Stadelman K, Abrams SL, Libra M, Candido S, et al. Deregulation of the EGFR/PI3K/PTEN/Akt/mTORC1 pathway in breast cancer: possibilities for therapeutic intervention. Oncotarget. 2014;5(13):4603–50.
31. Yuan TL, Cantley LC. PI3K pathway alterations in cancer: variations on a theme. Oncogene. 2008;27(41):5497–510.
32. Weigelt B, Warne PH, Downward J. PIK3CA mutation, but not PTEN loss of function, determines the sensitivity of breast cancer cells to mTOR inhibitory drugs. Oncogene. 2011;30(29):3222–33.
33. Wheler JJ, Moulder SL, Naing A, Janku F, Piha-Paul SA, Falchook GS, et al. Anastrozole and everolimus in advanced gynecologic and breast malignancies: activity and molecular alterations in the PI3K/AKT/mTOR pathway. Oncotarget. 2014;5(10):3029–38.
34. Stambolic V. Cancer: precise control of localized signals. Nature. 2015;522(7554):38–40.
35. Tashiro H, Blazes MS, Wu R, Cho KR, Bose S, Wang SI, et al. Mutations in PTEN are frequent in endometrial carcinoma but rare in other common gynecological malignancies. Cancer Res. 1997;57(18):3935–40.
36. Mutter GL, Lin MC, Fitzgerald JT, Kum JB, Baak JP, Lees JA, et al. Altered PTEN expression as a diagnostic marker for the earliest endometrial precancers. J Natl Cancer Inst. 2000;92(11):924–30.
37. Dinkelspiel HE, Wright JD, Lewin SN, Herzog TJ. Contemporary clinical management of endometrial cancer. Obstet Gynecol Int. 2013;2013:583891.
38. Di Cristofano A, Pandolfi PP. The multiple roles of PTEN in tumor suppression. Cell. 2000;100(4):387–90.
39. Hsu CP, Kao TY, Chang WL, Nieh S, Wang HL, Chung YC. Clinical significance of tumor suppressor PTEN in colorectal carcinoma. Eur J Surg Oncol. 2011;37(2):140–7.
40. Pan J, Cheng L, Bi X, Zhang X, Liu S, Bai X, et al. Elevation of omega-3 polyunsaturated fatty acids attenuates PTEN-deficiency induced endometrial cancer development through regulation of COX-2 and PGE2 production. Sci Rep. 2015;5:14958.
41. Cully M, You H, Levine AJ, Mak TW. Beyond PTEN mutations: the PI3K pathway as an integrator of multiple inputs during tumorigenesis. Nat Rev Cancer. 2006;6(3):184–92.
42. Adams JR, Schachter NF, Liu JC, Zacksenhaus E, Egan SE. Elevated PI3K signaling drives multiple breast cancer subtypes. Oncotarget. 2011;2(6):435–47.
43. Liu JC, Wang DY, Egan SE, Zacksenhaus E. Common and distinct features of mammary tumors driven by Pten-deletion or activating Pik3ca mutation. Oncotarget. 2016;7:9060–8.
44. Hopkins BD, Fine B, Steinbach N, Dendy M, Rapp Z, Shaw J, et al. A secreted PTEN phosphatase that enters cells to alter signaling and survival. Science. 2013;341(6144):399–402.

45. Bassi C, Ho J, Srikumar T, Dowling RJ, Gorrini C, Miller SJ, et al. Nuclear PTEN controls DNA repair and sensitivity to genotoxic stress. Science. 2013;341(6144):395–9.
46. Song MS, Carracedo A, Salmena L, Song SJ, Egia A, Malumbres M, et al. Nuclear PTEN regulates the APC-CDH1 tumor-suppressive complex in a phosphatase-independent manner. Cell. 2011;144(2):187–99.
47. Song MS, Salmena L, Pandolfi PP. The functions and regulation of the PTEN tumour suppressor. Nat Rev Mol Cell Biol. 2012;13(5):283–96.
48. Zhang S, Huang WC, Li P, Guo H, Poh SB, Brady SW, et al. Combating trastuzumab resistance by targeting SRC, a common node downstream of multiple resistance pathways. Nat Med. 2011;17(4):461–9.
49. Shi Y, Wang J, Chandarlapaty S, Cross J, Thompson C, Rosen N, et al. PTEN is a protein tyrosine phosphatase for IRS1. Nat Struct Mol Biol. 2014;21(6):522–7.
50. Wang S, Liu JC, Kim D, Datti A, Zacksenhaus E. Targeted Pten deletion plus p53-R270H mutation in mouse mammary epithelium induces aggressive claudin-low and basal-like breast cancer. Breast Cancer Res. 2016;18(1):9. doi:10.1186/s13058-015-0668-y.
51. Guo L, Wu H, Zhu J, Zhang C, Ma J, Lan J, et al. Genetic variations in the PI3K/AKT pathway predict platinum-based neoadjuvant chemotherapeutic sensitivity in squamous cervical cancer. Life Sci. 2015;143:217–24.
52. Du J, Wang L, Li C, Yang H, Li Y, Hu H, et al. MicroRNA-221 targets PTEN to reduce the sensitivity of cervical cancer cells to gefitinib through the PI3K/Akt signaling pathway. Tumour Biol. 2016;37:3939–47.
53. Yang YK, Xi WY, Xi RX, Li JY, Li Q, Gao YE. MicroRNA-494 promotes cervical cancer proliferation through the regulation of PTEN. Oncol Rep. 2015 May;33(5):2393–401.
54. Moreno-Bueno G, Gamallo C, Perez-Gallego L, de Mora JC, Suarez A, Palacios J. beta-Catenin expression pattern, beta-catenin gene mutations, and microsatellite instability in endometrioid ovarian carcinomas and synchronous endometrial carcinomas. Diagn Mol Pathol. 2001;10(2):116–22.
55. Wu R, Zhai Y, Fearon ER, Cho KR. Diverse mechanisms of beta-catenin deregulation in ovarian endometrioid adenocarcinomas. Cancer Res. 2001;61(22):8247–55.
56. Obata K, Morland SJ, Watson RH, Hitchcock A, Chenevix-Trench G, Thomas EJ, et al. Frequent PTEN/MMAC mutations in endometrioid but not serous or mucinous epithelial ovarian tumors. Cancer Res. 1998;58(10):2095–7.
57. Shih I, Kurman RJ. Ovarian tumorigenesis: a proposed model based on morphological and molecular genetic analysis. Am J Pathol. 2004;164(5):1511–8.
58. Sato N, Tsunoda H, Nishida M, Morishita Y, Takimoto Y, Kubo T, et al. Loss of heterozygosity on 10q23.3 and mutation of the tumor suppressor gene PTEN in benign endometrial cyst of the ovary: possible sequence progression from benign endometrial cyst to endometrioid carcinoma and clear cell carcinoma of the ovary. Cancer Res. 2000; 60(24):7052–6.
59. Hashiguchi Y, Tsuda H, Inoue T, Berkowitz RS, Mok SC. PTEN expression in clear cell adenocarcinoma of the ovary. Gynecol Oncol. 2006;101(1):71–5.
60. Folgueira MA, Maistro S, Katayama ML, Roela RA, Mundim FG, Nanogaki S, et al. Markers of breast cancer stromal fibroblasts in the primary tumour site associated with lymph node metastasis: a systematic review including our case series. Biosci Rep. 2013;33(6) doi:10.1042/BSR20130060.
61. Zhang B, Cao X, Liu Y, Cao W, Zhang F, Zhang S, et al. Tumor-derived matrix metalloproteinase-13 (MMP-13) correlates with poor prognoses of invasive breast cancer. BMC Cancer. 2008;8:83. doi:10.1186/1471-2407-8-83.
62. Jung EJ, Moon HG, Cho BI, Jeong CY, Joo YT, Lee YJ, et al. Galectin-1 expression in cancer-associated stromal cells correlates tumor invasiveness and tumor progression in breast cancer. Int J Cancer. 2007;120(11):2331–8.

63. Yeung TL, Leung CS, Li F, Wong SS, Mok SC. Targeting Stromal-Cancer Cell Crosstalk Networks in Ovarian Cancer Treatment. Biomolecules. 2016;6(1) doi:10.3390/biom6010003.
64. Lawrenson K, Grun B, Lee N, Mhawech-Fauceglia P, Kan J, Swenson S, et al. NPPB is a novel candidate biomarker expressed by cancer-associated fibroblasts in epithelial ovarian cancer. Int J Cancer. 2015;136(6):1390–401.
65. Yeung TL, Leung CS, Mok SC. CAF reprogramming inhibits ovarian cancer progression. Cell Cycle. 2014;13(24):3783–4.
66. Yeung TL, Leung CS, Wong KK, Samimi G, Thompson MS, Liu J, et al. TGF-beta modulates ovarian cancer invasion by upregulating CAF-derived versican in the tumor microenvironment. Cancer Res. 2013;73(16):5016–28.
67. Serini G, Gabbiani G. Mechanisms of myofibroblast activity and phenotypic modulation. Exp Cell Res. 1999;250(2):273–83.
68. Erez N, Glanz S, Raz Y, Avivi C, Barshack I. Cancer associated fibroblasts express pro-inflammatory factors in human breast and ovarian tumors. Biochem Biophys Res Commun. 2013;437(3):397–402.
69. Mueller MM, Fusenig NE. Friends or foes—bipolar effects of the tumour stroma in cancer. Nat Rev Cancer. 2004;4(11):839–49.
70. Chen H, Yang WW, Wen QT, Xu L, Chen M. TGF-beta induces fibroblast activation protein expression; fibroblast activation protein expression increases the proliferation, adhesion, and migration of HO-8910PM [corrected]. Exp Mol Pathol. 2009;87(3):189–94.
71. Yang W, Han W, Ye S, Liu D, Wu J, Liu H, et al. Fibroblast activation protein-alpha promotes ovarian cancer cell proliferation and invasion via extracellular and intracellular signaling mechanisms. Exp Mol Pathol. 2013;95(1):105–10.
72. Ohira S, Itatsu K, Sasaki M, Harada K, Sato Y, Zen Y, et al. Local balance of transforming growth factor-beta1 secreted from cholangiocarcinoma cells and stromal-derived factor-1 secreted from stromal fibroblasts is a factor involved in invasion of cholangiocarcinoma. Pathol Int. 2006;56(7):381–9.
73. Ohira S, Sasaki M, Harada K, Sato Y, Zen Y, Isse K, et al. Possible regulation of migration of intrahepatic cholangiocarcinoma cells by interaction of CXCR4 expressed in carcinoma cells with tumor necrosis factor-alpha and stromal-derived factor-1 released in stroma. Am J Pathol. 2006;168(4):1155–68.
74. Chang SC, Lin PC, Yang SH, Wang HS, Li AF, Lin JK. SDF-1alpha G801A polymorphism predicts lymph node metastasis in stage T3 colorectal cancer. Ann Surg Oncol. 2009;16(8):2323–30.
75. Daly AJ, McIlreavey L, Irwin CR. Regulation of HGF and SDF-1 expression by oral fibroblasts—implications for invasion of oral cancer. Oral Oncol. 2008;44(7):646–51.
76. Matsuo Y, Ochi N, Sawai H, Yasuda A, Takahashi H, Funahashi H, et al. CXCL8/IL-8 and CXCL12/SDF-1alpha co-operatively promote invasiveness and angiogenesis in pancreatic cancer. Int J Cancer. 2009;124(4):853–61.
77. Lau TS, Chung TK, Cheung TH, Chan LK, Cheung LW, Yim SF, et al. Cancer cell-derived lymphotoxin mediates reciprocal tumour-stromal interactions in human ovarian cancer by inducing CXCL11 in fibroblasts. J Pathol. 2014;232(1):43–56.
78. McAndrews KM, Yi J, McGrail DJ, Dawson MR. Enhanced adhesion of stromal cells to invasive cancer cells regulated by cadherin 11. ACS Chem Biol. 2015;10(8):1932–8.
79. Moran-Jones K, Gloss BS, Murali R, Chang DK, Colvin EK, Jones MD, et al. Connective tissue growth factor as a novel therapeutic target in high grade serous ovarian cancer. Oncotarget. 2015;6:44551–62.
80. Huang R, Wu D, Yuan Y, Li X, Holm R, Trope CG, et al. CD117 expression in fibroblasts-like stromal cells indicates unfavorable clinical outcomes in ovarian carcinoma patients. PLoS One. 2014;9(11):e112209.
81. Chen L, Yao Y, Sun L, Zhou J, Liu J, Wang J, et al. Clinical implication of the serum galectin-1 expression in epithelial ovarian cancer patients. J Ovarian Res. 2015;8(1):78. doi:10.1186/s13048-015-0206-7.

82. Barondes SH, Castronovo V, Cooper DN, Cummings RD, Drickamer K, Feizi T, et al. Galectins: a family of animal beta-galactoside-binding lectins. Cell. 1994;76(4):597–8.
83. Barondes SH, Cooper DN, Gitt MA, Leffler H. Galectins. Structure and function of a large family of animal lectins. J Biol Chem. 1994;269(33):20807–10.
84. Camby I, Le Mercier M, Lefranc F, Kiss R. Galectin-1: a small protein with major functions. Glycobiology. 2006;16(11):137R–57R.
85. Liu FT, Rabinovich GA. Galectins as modulators of tumour progression. Nat Rev Cancer. 2005;5(1):29–41.
86. Wu MH, Hong TM, Cheng HW, Pan SH, Liang YR, Hong HC, et al. Galectin-1-mediated tumor invasion and metastasis, up-regulated matrix metalloproteinase expression, and reorganized actin cytoskeletons. Mol Cancer Res. 2009;7(3):311–8.
87. Thijssen VL, Poirier F, Baum LG, Griffioen AW. Galectins in the tumor endothelium: opportunities for combined cancer therapy. Blood. 2007;110(8):2819–27.
88. Thijssen VL, Postel R, Brandwijk RJ, Dings RP, Nesmelova I, Satijn S, et al. Galectin-1 is essential in tumor angiogenesis and is a target for antiangiogenesis therapy. Proc Natl Acad Sci U S A. 2006;103(43):15975–80.
89. Rabinovich GA, Ilarregui JM. Conveying glycan information into T-cell homeostatic programs: a challenging role for galectin-1 in inflammatory and tumor microenvironments. Immunol Rev. 2009;230(1):144–59.
90. Kovacs-Solyom F, Blasko A, Fajka-Boja R, Katona RL, Vegh L, Novak J, et al. Mechanism of tumor cell-induced T-cell apoptosis mediated by galectin-1. Immunol Lett. 2010;127(2):108–18.
91. Xu XC, el-Naggar AK, Lotan R. Differential expression of galectin-1 and galectin-3 in thyroid tumors. Potential diagnostic implications. Am J Pathol. 1995;147(3):815–22.
92. Chiariotti L, Berlingieri MT, Battaglia C, Benvenuto G, Martelli ML, Salvatore P, et al. Expression of galectin-1 in normal human thyroid gland and in differentiated and poorly differentiated thyroid tumors. Int J Cancer. 1995;64(3):171–5.
93. Wu MH, Hong HC, Hong TM, Chiang WF, Jin YT, Chen YL. Targeting galectin-1 in carcinoma-associated fibroblasts inhibits oral squamous cell carcinoma metastasis by down-regulating MCP-1/CCL2 expression. Clin Cancer Res. 2011;17(6):1306–16.
94. van den Brule F, Califice S, Garnier F, Fernandez PL, Berchuck A, Castronovo V. Galectin-1 accumulation in the ovary carcinoma peritumoral stroma is induced by ovary carcinoma cells and affects both cancer cell proliferation and adhesion to laminin-1 and fibronectin. Lab Invest. 2003;83(3):377–86.
95. Kim HJ, Jeon HK, Cho YJ, Park YA, Choi JJ, Do IG, et al. High galectin-1 expression correlates with poor prognosis and is involved in epithelial ovarian cancer proliferation and invasion. Eur J Cancer. 2012;48(12):1914–21.
96. Zhang P, Zhang P, Shi B, Zhou M, Jiang H, Zhang H, et al. Galectin-1 overexpression promotes progression and chemoresistance to cisplatin in epithelial ovarian cancer. Cell Death Dis. 2014;5:e991.
97. Huang EY, Chanchien CC, Lin H, Wang CC, Wang CJ, Huang CC. Galectin-1 is an independent prognostic factor for local recurrence and survival after definitive radiation therapy for patients with squamous cell carcinoma of the uterine cervix. Int J Radiat Oncol Biol Phys. 2013;87(5):975–82.
98. Benes P, Vetvicka V, Fusek M. Cathepsin D—many functions of one aspartic protease. Crit Rev Oncol Hematol. 2008;68(1):12–28.
99. Pranjol MZ, Gutowski N, Hannemann M, Whatmore J. The potential role of the proteases cathepsin D and cathepsin L in the progression and metastasis of epithelial ovarian cancer. Biomolecules. 2015;5(4):3260–79.
100. Mathieu M, Vignon F, Capony F, Rochefort H. Estradiol down-regulates the mannose-6-phosphate/insulin-like growth factor-II receptor gene and induces cathepsin-D in breast cancer cells: a receptor saturation mechanism to increase the secretion of lysosomal proenzymes. Mol Endocrinol. 1991;5(6):815–22.

101. Kagedal K, Johansson U, Ollinger K. The lysosomal protease cathepsin D mediates apoptosis induced by oxidative stress. FASEB J. 2001;15(9):1592–4.
102. Heinrich M, Neumeyer J, Jakob M, Hallas C, Tchikov V, Winoto-Morbach S, et al. Cathepsin D links TNF-induced acid sphingomyelinase to Bid-mediated caspase-9 and -3 activation. Cell Death Differ. 2004;11(5):550–63.
103. Blomgran R, Zheng L, Stendahl O. Cathepsin-cleaved Bid promotes apoptosis in human neutrophils via oxidative stress-induced lysosomal membrane permeabilization. J Leukoc Biol. 2007;81(5):1213–23.
104. Zuzarte-Luis V, Montero JA, Kawakami Y, Izpisua-Belmonte JC, Hurle JM. Lysosomal cathepsins in embryonic programmed cell death. Dev Biol. 2007;301(1):205–17.
105. Zuzarte-Luis V, Montero JA, Torre-Perez N, Garcia-Porrero JA, Hurle JM. Cathepsin D gene expression outlines the areas of physiological cell death during embryonic development. Dev Dyn. 2007;236(3):880–5.
106. Ohri SS, Vashishta A, Proctor M, Fusek M, Vetvicka V. The propeptide of cathepsin D increases proliferation, invasion and metastasis of breast cancer cells. Int J Oncol. 2008;32(2):491–8.
107. Winiarski BK, Wolanska KI, Rai S, Ahmed T, Acheson N, Gutowski NJ, et al. Epithelial ovarian cancer-induced angiogenic phenotype of human omental microvascular endothelial cells may occur independently of VEGF signaling. Transl Oncol. 2013;6(6):703–14.
108. Rochefort H. Cathepsin D in breast cancer: a tissue marker associated with metastasis. Eur J Cancer. 1992;28A(11):1780–3.
109. Rochefort H, Garcia M, Glondu M, Laurent V, Liaudet E, Rey JM, et al. Cathepsin D in breast cancer: mechanisms and clinical applications, a 1999 overview. Clin Chim Acta. 2000;291(2):157–70.
110. Ferrandina G, Scambia G, Bardelli F, Benedetti Panici P, Mancuso S, Messori A. Relationship between cathepsin-D content and disease-free survival in node-negative breast cancer patients: a meta-analysis. Br J Cancer. 1997;76(5):661–6.
111. Foekens JA, Look MP, Bolt-de Vries J, Meijer-van Gelder ME, van Putten WL, Klijn JG. Cathepsin-D in primary breast cancer: prognostic evaluation involving 2810 patients. Br J Cancer. 1999;79(2):300–7.
112. Briozzo P, Badet J, Capony F, Pieri I, Montcourrier P, Barritault D, et al. MCF7 mammary cancer cells respond to bFGF and internalize it following its release from extracellular matrix: a permissive role of cathepsin D. Exp Cell Res. 1991;194(2):252–9.
113. Losch A, Kohlberger P, Gitsch G, Kaider A, Breitenecker G, Kainz C. Lysosomal protease cathepsin D is a prognostic marker in endometrial cancer. Br J Cancer. 1996;73(12):1525–8.
114. Nazeer T, Malfetano JH, Rosano TG, Ross JS. Correlation of tumor cytosol cathepsin D with differentiation and invasiveness of endometrial adenocarcinoma. Am J Clin Pathol. 1992;97(6):764–9.
115. Henzen-Logmans SC, Fieret EJ, Berns EM, van der Burg ME, Klijn JG, Foekens JA. Ki-67 staining in benign, borderline, malignant primary and metastatic ovarian tumors: correlation with steroid receptors, epidermal-growth-factor receptor and cathepsin D. Int J Cancer. 1994;57(4):468–72.
116. Losch A, Schindl M, Kohlberger P, Lahodny J, Breitenecker G, Horvat R, et al. Cathepsin D in ovarian cancer: prognostic value and correlation with p53 expression and microvessel density. Gynecol Oncol. 2004;92(2):545–52.
117. Chai Y, Wu W, Zhou C, Zhou J. The potential prognostic value of cathepsin D protein in serous ovarian cancer. Arch Gynecol Obstet. 2012;286(2):465–71.
118. Winiarski BK, Cope N, Alexander M, Pilling LC, Warren S, Acheson N, et al. Clinical relevance of increased endothelial and mesothelial expression of proangiogenic proteases and VEGFA in the omentum of patients with metastatic ovarian high-grade serous carcinoma. Transl Oncol. 2014;7(2):267–276.e4.
119. Rochefort H, Capony F, Garcia M, Cavailles V, Freiss G, Chambon M, et al. Estrogen-induced lysosomal proteases secreted by breast cancer cells: a role in carcinogenesis? J Cell Biochem. 1987;35(1):17–29.

120. Vignon F, Capony F, Chambon M, Freiss G, Garcia M, Rochefort H. Autocrine growth stimulation of the MCF 7 breast cancer cells by the estrogen-regulated 52K protein. Endocrinology. 1986;118(4):1537–45.
121. Kirschke H, Langner J, Wiederanders B, Ansorge S, Bohley P. Cathepsin L. A new proteinase from rat-liver lysosomes. Eur J Biochem. 1977;74(2):293–301.
122. Kominami E, Ueno T, Muno D, Katunuma N. The selective role of cathepsins B and D in the lysosomal degradation of endogenous and exogenous proteins. FEBS Lett. 1991;287(1–2):189–92.
123. Nguyen Q, Mort JS, Roughley PJ. Cartilage proteoglycan aggregate is degraded more extensively by cathepsin L than by cathepsin B. Biochem J. 1990;266(2):569–73.
124. Nosaka AY, Kanaori K, Teno N, Togame H, Inaoka T, Takai M, et al. Conformational studies on the specific cleavage site of Type I collagen (alpha-1) fragment (157-192) by cathepsins K and L by proton NMR spectroscopy. Bioorg Med Chem. 1999;7(2):375–9.
125. Mason RW, Johnson DA, Barrett AJ, Chapman HA. Elastinolytic activity of human cathepsin L. Biochem J. 1986;233(3):925–7.
126. Ishidoh K, Kominami E. Procathepsin L degrades extracellular matrix proteins in the presence of glycosaminoglycans in vitro. Biochem Biophys Res Commun. 1995;217(2):624–31.
127. Maciewicz RA, Wotton SF, Etherington DJ, Duance VC. Susceptibility of the cartilage collagens types II, IX and XI to degradation by the cysteine proteinases, cathepsins B and L. FEBS Lett. 1990;269(1):189–93.
128. Maciewicz RA, Wotton SF. Degradation of cartilage matrix components by the cysteine proteinases, cathepsins B and L. Biomed Biochim Acta. 1991;50(4–6):561–4.
129. Nishida Y, Kohno K, Kawamata T, Morimitsu K, Kuwano M, Miyakawa I. Increased cathepsin L levels in serum in some patients with ovarian cancer: comparison with CA125 and CA72-4. Gynecol Oncol. 1995;56(3):357–61.
130. Zhang W, Wang S, Wang Q, Yang Z, Pan Z, Li L. Overexpression of cysteine cathepsin L is a marker of invasion and metastasis in ovarian cancer. Oncol Rep. 2014;31(3):1334–42.
131. Zhang L, Wei L, Shen G, He B, Gong W, Min N, et al. Cathepsin L is involved in proliferation and invasion of ovarian cancer cells. Mol Med Rep. 2015;11(1):468–74.
132. Arbyn M, Ronco G, Anttila A, Meijer CJ, Poljak M, Ogilvie G, et al. Evidence regarding human papillomavirus testing in secondary prevention of cervical cancer. Vaccine. 2012;30(Suppl 5):F88–99.
133. Carozzi F, Visioli CB, Confortini M, Iossa A, Mantellini P, Burroni E, et al. hr-HPV testing in the follow-up of women with cytological abnormalities and negative colposcopy. Br J Cancer. 2013;109(7):1766–74.
134. Meijer CJ, Berkhof H, Heideman DA, Hesselink AT, Snijders PJ. Validation of high-risk HPV tests for primary cervical screening. J Clin Virol. 2009;46(Suppl 3):S1–4.
135. Tornesello ML, Buonaguro L, Giorgi-Rossi P, Buonaguro FM. Viral and cellular biomarkers in the diagnosis of cervical intraepithelial neoplasia and cancer. Biomed Res Int. 2013;2013:519619.
136. Ronco G, Giorgi-Rossi P, Carozzi F, Confortini M, Dalla Palma P, Del Mistro A, et al. Efficacy of human papillomavirus testing for the detection of invasive cervical cancers and cervical intraepithelial neoplasia: a randomised controlled trial. Lancet Oncol. 2010;11(3):249–57.
137. Castle PE, Fetterman B, Poitras N, Lorey T, Shaber R, Kinney W. Relationship of atypical glandular cell cytology, age, and human papillomavirus detection to cervical and endometrial cancer risks. Obstet Gynecol. 2010;115(2 Pt 1):243–8.
138. Khleif SN, DeGregori J, Yee CL, Otterson GA, Kaye FJ, Nevins JR, et al. Inhibition of cyclin D-CDK4/CDK6 activity is associated with an E2F-mediated induction of cyclin kinase inhibitor activity. Proc Natl Acad Sci U S A. 1996;93(9):4350–4.
139. Nakao Y, Yang X, Yokoyama M, Ferenczy A, Tang SC, Pater MM, et al. Induction of p16 during immortalization by HPV 16 and 18 and not during malignant transformation. Br J Cancer. 1997;75(10):1410–6.
140. Pannone G, Rodolico V, Santoro A, Lo Muzio L, Franco R, Botti G, et al. Evaluation of a combined triple method to detect causative HPV in oral and oropharyngeal squamous cell

carcinomas: p16 immunohistochemistry, consensus PCR HPV-DNA, and in situ hybridization. Infect Agent Cancer. 2012;7:4. doi:10.1186/1750-9378-7-4.
141. Zhang Q, Kuhn L, Denny LA, De Souza M, Taylor S, Wright Jr TC. Impact of utilizing p16INK4A immunohistochemistry on estimated performance of three cervical cancer screening tests. Int J Cancer. 2007;120(2):351–6.
142. Roelens J, Reuschenbach M, von Knebel Doeberitz M, Wentzensen N, Bergeron C, Arbyn M. p16^{INK4a} immunocytochemistry versus human papillomavirus testing for triage of women with minor cytologic abnormalities: a systematic review and meta-analysis. Cancer Cytopathol. 2012;120(5):294–307.
143. Reuschenbach M, Seiz M, von Knebel DC, Vinokurova S, Duwe A, Ridder R, et al. Evaluation of cervical cone biopsies for coexpression of p16INK4a and Ki-67 in epithelial cells. Int J Cancer. 2012;130(2):388–94.
144. Schmidt D, Bergeron C, Denton KJ, Ridder R, European CINtec Cytology Study Group. p16/ki-67 dual-stain cytology in the triage of ASCUS and LSIL papanicolaou cytology: results from the European equivocal or mildly abnormal papanicolaou cytology study. Cancer Cytopathol. 2011;119(3):158–66.
145. Sahasrabuddhe VV, Luhn P, Wentzensen N. Human papillomavirus and cervical cancer: biomarkers for improved prevention efforts. Future Microbiol. 2011;6(9):1083–98.
146. Depuydt CE, Makar AP, Ruymbeke MJ, Benoy IH, Vereecken AJ, Bogers JJ. BD-ProExC as adjunct molecular marker for improved detection of CIN2+ after HPV primary screening. Cancer Epidemiol Biomarkers Prev. 2011;20(4):628–37.
147. Van Doorslaer K, Burk RD. Association between hTERT activation by HPV E6 proteins and oncogenic risk. Virology. 2012;433(1):216–9.
148. Liu H, Liu S, Wang H, Xie X, Chen X, Zhang X, et al. Genomic amplification of the human telomerase gene (hTERC) associated with human papillomavirus is related to the progression of uterine cervical dysplasia to invasive cancer. Diagn Pathol. 2012;7:147. doi:10.1186/1746-1596-7-147.
149. Fiorucci G, Chiantore MV, Mangino G, Percario ZA, Affabris E, Romeo G. Cancer regulator microRNA: potential relevance in diagnosis, prognosis and treatment of cancer. Curr Med Chem. 2012;19(4):461–74.
150. Gadducci A, Guerrieri ME, Greco C. Tissue biomarkers as prognostic variables of cervical cancer. Crit Rev Oncol Hematol. 2013;86(2):104–29.
151. Xu J, Li Y, Wang F, Wang X, Cheng B, Ye F, et al. Suppressed miR-424 expression via upregulation of target gene Chk1 contributes to the progression of cervical cancer. Oncogene. 2013;32(8):976–87.
152. Li Y, Liu J, Yuan C, Cui B, Zou X, Qiao Y. High-risk human papillomavirus reduces the expression of microRNA-218 in women with cervical intraepithelial neoplasia. J Int Med Res. 2010;38(5):1730–6.
153. Wang F, Li Y, Zhou J, Xu J, Peng C, Ye F, et al. miR-375 is down-regulated in squamous cervical cancer and inhibits cell migration and invasion via targeting transcription factor SP1. Am J Pathol. 2011;179(5):2580–8.
154. Bird A. DNA methylation patterns and epigenetic memory. Genes Dev. 2002;16(1):6–21.
155. Sharma S, Kelly TK, Jones PA. Epigenetics in cancer. Carcinogenesis. 2010;31(1):27–36.
156. Cuzick J, Bergeron C, von Knebel DM, Gravitt P, Jeronimo J, Lorincz AT, et al. New technologies and procedures for cervical cancer screening. Vaccine. 2012;30(Suppl 5):F107–16.
157. Wentzensen N, Sun C, Ghosh A, Kinney W, Mirabello L, Wacholder S, et al. Methylation of HPV18, HPV31, and HPV45 genomes and cervical intraepithelial neoplasia grade 3. J Natl Cancer Inst. 2012;104(22):1738–49.
158. Mirabello L, Sun C, Ghosh A, Rodriguez AC, Schiffman M, Wentzensen N, et al. Methylation of human papillomavirus type 16 genome and risk of cervical precancer in a Costa Rican population. J Natl Cancer Inst. 2012;104(7):556–65.
159. Sova P, Feng Q, Geiss G, Wood T, Strauss R, Rudolf V, et al. Discovery of novel methylation biomarkers in cervical carcinoma by global demethylation and microarray analysis. Cancer Epidemiol Biomarkers Prev. 2006;15(1):114–23.

160. Wang SS, Smiraglia DJ, Wu YZ, Ghosh S, Rader JS, Cho KR, et al. Identification of novel methylation markers in cervical cancer using restriction landmark genomic scanning. Cancer Res. 2008;68(7):2489–97.
161. Steenbergen RD, Kramer D, Braakhuis BJ, Stern PL, Verheijen RH, Meijer CJ, et al. TSLC1 gene silencing in cervical cancer cell lines and cervical neoplasia. J Natl Cancer Inst. 2004;96(4):294–305.
162. Wentzensen N, Sherman ME, Schiffman M, Wang SS. Utility of methylation markers in cervical cancer early detection: appraisal of the state-of-the-science. Gynecol Oncol. 2009;112(2):293–9.
163. Lee RC, Feinbaum RL, Ambros V. The *C. elegans* heterochronic gene lin-4 encodes small RNAs with antisense complementarity to lin-14. Cell. 1993;75(5):843–54.
164. He L, Hannon GJ. MicroRNAs: small RNAs with a big role in gene regulation. Nat Rev Genet. 2004;5(7):522–31.
165. Ambros V. microRNAs: tiny regulators with great potential. Cell. 2001;107(7):823–6.
166. Forman JJ, Legesse-Miller A, Coller HA. A search for conserved sequences in coding regions reveals that the let-7 microRNA targets Dicer within its coding sequence. Proc Natl Acad Sci U S A. 2008;105(39):14879–84.
167. Lytle JR, Yario TA, Steitz JA. Target mRNAs are repressed as efficiently by microRNA-binding sites in the 5′ UTR as in the 3′ UTR. Proc Natl Acad Sci U S A. 2007;104(23):9667–72.
168. Croce CM. Causes and consequences of microRNA dysregulation in cancer. Nat Rev Genet. 2009;10(10):704–14.
169. Calin GA, Sevignani C, Dumitru CD, Hyslop T, Noch E, Yendamuri S, et al. Human microRNA genes are frequently located at fragile sites and genomic regions involved in cancers. Proc Natl Acad Sci U S A. 2004;101(9):2999–3004.
170. Melo SA, Moutinho C, Ropero S, Calin GA, Rossi S, Spizzo R, et al. A genetic defect in exportin-5 traps precursor microRNAs in the nucleus of cancer cells. Cancer Cell. 2010;18(4):303–15.
171. Helland A, Anglesio MS, George J, Cowin PA, Johnstone CN, House CM, et al. Deregulation of MYCN, LIN28B and LET7 in a molecular subtype of aggressive high-grade serous ovarian cancers. PLoS One. 2011;6(4):e18064.
172. Schrauder MG, Strick R, Schulz-Wendtland R, Strissel PL, Kahmann L, Loehberg CR, et al. Circulating micro-RNAs as potential blood-based markers for early stage breast cancer detection. PLoS One. 2012;7(1):e29770.
173. O'Day E, Lal A. MicroRNAs and their target gene networks in breast cancer. Breast Cancer Res. 2010;12(2):201.
174. Iorio MV, Ferracin M, Liu CG, Veronese A, Spizzo R, Sabbioni S, et al. MicroRNA gene expression deregulation in human breast cancer. Cancer Res. 2005;65(16):7065–70.
175. Foekens JA, Sieuwerts AM, Smid M, Look MP, de Weerd V, Boersma AW, et al. Four miRNAs associated with aggressiveness of lymph node-negative, estrogen receptor-positive human breast cancer. Proc Natl Acad Sci U S A. 2008;105(35):13021–6.
176. Lowery AJ, Miller N, Devaney A, McNeill RE, Davoren PA, Lemetre C, et al. MicroRNA signatures predict oestrogen receptor, progesterone receptor and HER2/neu receptor status in breast cancer. Breast Cancer Res. 2009;11(3):R27.
177. Kondo N, Toyama T, Sugiura H, Fujii Y, Yamashita H. miR-206 Expression is down-regulated in estrogen receptor alpha-positive human breast cancer. Cancer Res. 2008;68(13):5004–8.
178. Blenkiron C, Goldstein LD, Thorne NP, Spiteri I, Chin SF, Dunning MJ, et al. MicroRNA expression profiling of human breast cancer identifies new markers of tumor subtype. Genome Biol. 2007;8(10):R214.
179. Yan LX, Huang XF, Shao Q, Huang MY, Deng L, Wu QL, et al. MicroRNA miR-21 overexpression in human breast cancer is associated with advanced clinical stage, lymph node metastasis and patient poor prognosis. RNA. 2008;14(11):2348–60.
180. Van der Auwera I, Limame R, van Dam P, Vermeulen PB, Dirix LY, Van Laere SJ. Integrated miRNA and mRNA expression profiling of the inflammatory breast cancer subtype. Br J Cancer. 2010;103(4):532–41.

181. Cai Y, Yan X, Zhang G, Zhao W, Jiao S. MicroRNA-205 increases the sensitivity of docetaxel in breast cancer. Oncol Lett. 2016;11(2):1105–9.
182. Pal MK, Jaiswar SP, Dwivedi VN, Tripathi AK, Dwivedi A, Sankhwar P. MicroRNA: a new and promising potential biomarker for diagnosis and prognosis of ovarian cancer. Cancer Biol Med. 2015;12(4):328–41.
183. Chen PS, Su JL, Hung MC. Dysregulation of microRNAs in cancer. J Biomed Sci. 2012;19:90. doi:10.1186/1423-0127-19-90.
184. Calin GA, Croce CM. MicroRNA signatures in human cancers. Nat Rev Cancer. 2006;6(11):857–66.
185. Zhang W, Dahlberg JE, Tam W. MicroRNAs in tumorigenesis: a primer. Am J Pathol. 2007;171(3):728–38.
186. Parikh A, Lee C, Joseph P, Marchini S, Baccarini A, Kolev V, et al. microRNA-181a has a critical role in ovarian cancer progression through the regulation of the epithelial-mesenchymal transition. Nat Commun. 2014;5:2977.
187. Wang X, Cao L, Wang Y, Wang X, Liu N, You Y. Regulation of let-7 and its target oncogenes (review). Oncol Lett. 2012;3(5):955–60.
188. Gregory PA, Bracken CP, Bert AG, Goodall GJ. MicroRNAs as regulators of epithelial-mesenchymal transition. Cell Cycle. 2008;7(20):3112–8.
189. Gregory PA, Bert AG, Paterson EL, Barry SC, Tsykin A, Farshid G, et al. The miR-200 family and miR-205 regulate epithelial to mesenchymal transition by targeting ZEB1 and SIP1. Nat Cell Biol. 2008;10(5):593–601.
190. Park SM, Gaur AB, Lengyel E, Peter ME. The miR-200 family determines the epithelial phenotype of cancer cells by targeting the E-cadherin repressors ZEB1 and ZEB2. Genes Dev. 2008;22(7):894–907.
191. Korpal M, Lee ES, Hu G, Kang Y. The miR-200 family inhibits epithelial-mesenchymal transition and cancer cell migration by direct targeting of E-cadherin transcriptional repressors ZEB1 and ZEB2. J Biol Chem. 2008;283(22):14910–4.
192. Leskela S, Leandro-Garcia LJ, Mendiola M, Barriuso J, Inglada-Perez L, Munoz I, et al. The miR-200 family controls beta-tubulin III expression and is associated with paclitaxel-based treatment response and progression-free survival in ovarian cancer patients. Endocr Relat Cancer. 2010;18(1):85–95.
193. Sun N, Zhang Q, Xu C, Zhao Q, Ma Y, Lu X, et al. Molecular regulation of ovarian cancer cell invasion. Tumour Biol. 2014;35(11):11359–66.
194. Kim NH, Kim HS, Li XY, Lee I, Choi HS, Kang SE, et al. A p53/miRNA-34 axis regulates Snail1-dependent cancer cell epithelial-mesenchymal transition. J Cell Biol. 2011;195(3):417–33.
195. Corney DC, Hwang CI, Matoso A, Vogt M, Flesken-Nikitin A, Godwin AK, et al. Frequent downregulation of miR-34 family in human ovarian cancers. Clin Cancer Res. 2010;16(4):1119–28.
196. Mateescu B, Batista L, Cardon M, Gruosso T, de Feraudy Y, Mariani O, et al. miR-141 and miR-200a act on ovarian tumorigenesis by controlling oxidative stress response. Nat Med. 2011;17(12):1627–35.
197. Yang H, Kong W, He L, Zhao JJ, O'Donnell JD, Wang J, et al. MicroRNA expression profiling in human ovarian cancer: miR-214 induces cell survival and cisplatin resistance by targeting PTEN. Cancer Res. 2008;68(2):425–33.
198. Lou Y, Yang X, Wang F, Cui Z, Huang Y. MicroRNA-21 promotes the cell proliferation, invasion and migration abilities in ovarian epithelial carcinomas through inhibiting the expression of PTEN protein. Int J Mol Med. 2010;26(6):819–27.
199. Nagaraja AK, Creighton CJ, Yu Z, Zhu H, Gunaratne PH, Reid JG, et al. A link between mir-100 and FRAP1/mTOR in clear cell ovarian cancer. Mol Endocrinol. 2010;24(2):447–63.
200. Peng DX, Luo M, Qiu LW, He YL, Wang XF. Prognostic implications of microRNA-100 and its functional roles in human epithelial ovarian cancer. Oncol Rep. 2012;27(4):1238–44.
201. Iorio MV, Visone R, Di Leva G, Donati V, Petrocca F, Casalini P, et al. MicroRNA signatures in human ovarian cancer. Cancer Res. 2007;67(18):8699–707.

202. Wyman SK, Parkin RK, Mitchell PS, Fritz BR, O'Briant K, Godwin AK, et al. Repertoire of microRNAs in epithelial ovarian cancer as determined by next generation sequencing of small RNA cDNA libraries. PLoS One. 2009;4(4):e5311.
203. Lee H, Park CS, Deftereos G, Morihara J, Stern JE, Hawes SE, et al. MicroRNA expression in ovarian carcinoma and its correlation with clinicopathological features. World J Surg Oncol. 2012;10:174.
204. Resnick KE, Alder H, Hagan JP, Richardson DL, Croce CM, Cohn DE. The detection of differentially expressed microRNAs from the serum of ovarian cancer patients using a novel real-time PCR platform. Gynecol Oncol. 2009;112(1):55–9.
205. Eitan R, Kushnir M, Lithwick-Yanai G, David MB, Hoshen M, Glezerman M, et al. Tumor microRNA expression patterns associated with resistance to platinum based chemotherapy and survival in ovarian cancer patients. Gynecol Oncol. 2009;114(2):253–9.
206. Xu YZ, Xi QH, Ge WL, Zhang XQ. Identification of serum microRNA-21 as a biomarker for early detection and prognosis in human epithelial ovarian cancer. Asian Pac J Cancer Prev. 2013;14(2):1057–60.
207. Wan WN, Zhang YQ, Wang XM, Liu YJ, Zhang YX, Que YH, et al. Down-regulated miR-22 as predictive biomarkers for prognosis of epithelial ovarian cancer. Diagn Pathol. 2014;9:178. doi:10.1186/s13000-014-0178-8.
208. Vecchione A, Belletti B, Lovat F, Volinia S, Chiappetta G, Giglio S, et al. A microRNA signature defines chemoresistance in ovarian cancer through modulation of angiogenesis. Proc Natl Acad Sci U S A. 2013;110(24):9845–50.

Potential and Challenges of Liquid Biopsies

12

Amin El-Heliebi, Ellen Heitzer, Thomas Kroneis, Shukun Chen, Christoph Haudum, and Julia Fuchs

Contents

A. El-Heliebi (✉) • T. Kroneis • S. Chen • C. Haudum • J. Fuchs
Institute of Cell Biology, Histology and Embryology, Medical University Graz, Graz, Austria
e-mail: Amin.elheliebi@medunigraz.at; thomas.kroneis@gu.se; shukun.chen@medunigraz.at; christoph.haudum@medunigraz.at; fuchs-julia@outlook.com

E. Heitzer
Institute of Human Genetics, Medical University Graz, Graz, Austria
e-mail: ellen.heitzer@medunigraz.at

J. Haybaeck (ed.), *Mechanisms of Molecular Carcinogenesis – Volume 2*,
DOI 10.1007/978-3-319-53661-3_12

Abstract

Although tumor genotyping is still the most currently used method for categorizing tumors for clinical decisions, tumor tissues provide only a snapshot or are often difficult to obtain. To overcome these issues, methods are needed for a rapid, cost-effective, and noninvasive identification of biomarkers at various time points during the course of disease. The analysis of circulating tumor cells (CTCs), cell-free circulating tumor DNA (ctDNA), circulating RNAs, and exosomes, frequently referred to as liquid biopsy, has recently gained enormous momentum. Due to technological advances, novel circulating tumor biomarkers were shown to have a great potential to improve patient treatment in terms of estimation of prognosis, monitoring treatment response, early detection of resistance mechanisms, identification of actionable targets, and detection of minimal residual disease. However, despite all efforts, liquid biopsies are not yet routinely used mainly due to technological hurdles, lack of analytical and pre-analytical standards and conclusive evidence that patients indeed benefit from such analyses. In this chapter, the different entities with respect to state-of-the-art technologies, potential clinical applications, and their limitations are discussed.

12.1 Introduction

Liquid biopsies are noninvasive blood tests that detect and analyze circulating biomarkers, such as circulating tumor cells (CTCs), cell-free circulating tumor DNA (ctDNA), circulating RNAs, and exosomes, released from primary tumors and their metastatic deposits. In general, liquid biopsies are highly beneficial compared to tissue biopsy since repeated sampling during the entire disease course is easily achievable. Here we discuss the current impact and future directions of liquid biopsies, including the biology of the different entities, pros and cons, current state-of-the-art technologies, and their potential clinical applications.

12.2 Circulating Tumor Cell

12.2.1 Biology of Circulating Tumor Cells (CTCs)

The hematogenous dissemination of tumor cells, either as single cell or as tumor cell cluster, is an important step in metastasis formation. Cancer cells can leave their primary site already at early stage of disease [1]. As a first step, tumor cells intravasate into the vascular system followed by physical arrest at capillary walls of a distant organ and eventually extravasate from the vascular system into the parenchyma of target organs [2]. These steps of metastasis formation are shown to be very inefficient, as the number of CTCs greatly exceed the number of formed metastases [3]. Generally, the passage in blood is a very stressful event for CTCs which most of them do not survive [2]. Experimental data in mice have shown that CTCs which travel in clusters are more protected and more proliferative at the target site, eventually bringing their own tumor stroma [4]. Additionally, platelets which are coating the cell surface of CTCs have been shown to promote CTC survival [5]. To progress from a

CTC to a full-blown metastasis, several other factors are necessary. Those factors include the stemness-like features of CTCs [6], suppression of immune defense [7], and a supportive niche by the adjacent stroma to sustain tumor survival and growth [8]. In the metastatic cascade, a biological process called "epithelial to mesenchymal transition" (EMT) has been described as an important factor [9]. The EMT is linked to an upregulation of genes, such as vimentin, N-cadherin, Twist, Snail, and others, which are associated with mesenchymal cells [10]. Vice versa, epithelial markers, such as cytokeratins, are downregulated [10]. Cells which undergo EMT show a stemness-like behavior, being more aggressive in forming a successful metastatic lesion [11]. To address the different subtypes of CTCs, such as mesenchymal- or epithelial-like phenotypes, a broad array of technologies is in place. In the following section, we focus on a few promising technologies, which are described below.

12.2.2 Methodological Aspects for the Isolation and Analysis of CTCs

CTCs are a group of extremely rare cells, with 1–10 cells in 1 mL of peripheral blood which contains few millions of leukocytes and billions of erythrocytes [12]. Therefore, technologies for CTC enrichment are required to be highly sensitive and specific, which remains technically challenging. Up to now, more than 50 different platforms have been described, which are reviewed by Alix-Panabiers et al. [13]. In general, two different approaches are used for the detection and isolation of CTCs. The first approach is making use of biological properties of CTCs, such as expressed epithelial-derived antigens on the cell surface. The second approach is exploiting the physical properties that distinguish CTCs from most peripheral blood cells, including cell density, cell size, electrical charges on the surface of the cell membrane, and deformability of cells. However, since an efficient isolation of CTCs in a viable and intact state is preferable, isolation methods based on physical properties provide a compelling advantage over those relying on fixation and extra- or intracellular staining. A representative overview of currently available methods is shown in Table 1 and discussed in the next paragraph:

1. Exploiting biological properties

   Since CTCs express tumor-associated antigens on their surfaces—that are usually not found on blood cells—these biological features can be targeted by antibodies. Typically, a CTC is defined as a cell with an intact nucleus, being positive for the expression of epithelial cell adhesion molecule (EpCAM) and other epithelial markers, such as cytokeratin (CK), and negative for CD45 [27]. Therefore, EpCAM is a widely used epithelial marker for positive selection of CTCs. In order to obtain a higher specificity, many platforms use combined antibody staining of epithelial markers and the leukocyte-specific surface marker CD45. To date, the gold standard for CTC detection, the Veridex CellSearch system, which was first introduced in 2004, is the only Food and Drug Administration (FDA)-cleared CTC detection device for the enumeration of CTC in 7.5 mL of blood. Using the CellSearch system, numerous studies indicated the prognostic value of CTCs in metastatic breast, colon, prostate, and lung

Table 1 Representative overview of currently available methods

Enrichment device/method	Technique	Features and advantages	Limitations	Reference
Based on biological features				
CellSearch	Magnetic-activated cell sorting	FDA approved	Low flexibility for research applications	Allard et al. [14]
AdnaTest, AdnaGen	Immunomagnetic based	CTC detection and further transcript analysis	Only EpCAM-positive CTC is being detected	Antonarakis et al. [15]
CTC-chip HB-chip CTC-iChip Graphene oxide-Chip	Microfluidics	Gentle CTC enrichment → high viability	Slow, only low volume of blood can be processed	Nagrath et al. [16], Stott et al. [17], Yoon et al. [18], Yu et al. [19]
CellCollector	In vivo EpCAM-based capture	In vivo detection → large volume of blood	Imaging of CTCs on the CellCollector is challenging	Saucedo-Zeni et al. [20]
EPISPOT	Assay to detect secreted proteins	Viable CTCs can be detected	Indirect detection, as only the secreted proteins remain	Alix-Panabieres [21]
Based on physical features (size, density)				
Ficoll, Oncoquick™	Density gradient centrifugation	Density based	Nonspecific cell loss	Gertler et al. [22], He et al. [23]
RosetteSep™	Negative enrichment	Simple and can capture viable CTCs	Nonspecific cell loss	He et al. [23]
ScreenCell	Filtration-based size exclusion	Simple, no additional equipment needed	Size bias, small CTCs may be lost	El-Heliebi et al. [24]
ISET	Filtration-based size exclusion	Simple and fast	Size bias, small CTCs may be lost. Additional equipment needed	Hou et al. [25], Vona et al. [26]

cancer [28–31]. Another promising technology branch is represented by microfluidic chips which allow isolation of CTCs with high viability. One of the first microfluidic chips for CTC isolation was the CTC-chip, based on EpCAM expression [16]. Several generations of new microfluidic chips followed [17, 32,

33]. Using the CTC-iChip, Yu et al. succeeded in culturing CTCs isolated from blood samples of patients with metastatic estrogen receptor (ER)-positive breast cancer [19]. Furthermore, making use of established CTC lines, tumorigenic tests were performed in mice, which provided encouraging strategies for functional characterization of CTCs in addition to a simple enumeration and one-time genetic analyses [19]. Another promising EpCAM-based enrichment system is the AdnaGen system. The AdnaGen system is an optimized combination of antibodies for cell selection and subsequent RT-qPCR for tumor-associated expression patterns. In the first step, CTCs are enriched by magnetic beads coupled to EpCAM antibodies. In the following steps, the cells are lysed, the RNA is reversely transcribed, and specific targets are analyzed using quantitative PCR. For example, in prostate cancer, it was shown that the AdnaGen test could reliably detect the androgen receptor splice variant 7 (AR-V7) in CTCs, which is associated with resistance to antihormonal therapy [15]. Although this technology offers specific molecular characterization of CTCs, it does not support the quantification of the original CTC numbers.

Due to the low abundance of CTCs in the circulation, one of the major limitations for their detection is the low volume that can be obtained from one blood draw. Therefore, the most effective way to increase the CTC detection rate would be to increase the sample volume within the clinically allowable range without burdening the patient. In a study of Lalmahomed et al., it was shown that the analysis of 30 mL instead of 7.5 mL of blood resulted in 20% more patients having detectable CTCs [34]. Another possibility of further increasing the analyzable blood volume is an in vivo enrichment of CTCs. The CellCollector is an in vivo device, based on a medical wire which is coated with anti-EpCAM antibodies enabling the harvest of CTCs expressing EpCAM on their cell surface. The CellCollector is applied by inserting the wire into a cubital vein for 30 min. Within 30 min, it is estimated that 1.5–3 L of blood pass the CellCollector, capturing CTCs as they pass by [20, 35]. Recent data show higher detection rates of CTCs compared to other technologies [36].

An indirect approach to isolate CTCs is based on a negative enrichment strategy. Negative enrichment involves a red blood cell lysis, followed by depletion of CD45+ leukocytes using a magnetic bead separation method [37]. This allows an efficient enrichment of the CTC fraction to a maximum of 1% purity (e.g., 1 CTC to 99 leukocytes) [37]. A similar approach is the RosetteSep system, which combines Ficoll density gradient to remove cells from whole blood with a subsequent depletion of unwanted blood cells [23].

Another method for the detection of viable CTCs is the EPISPOT technology that detects proteins secreted/released/shed from single epithelial cancer cells. After leukocyte depletion, the enriched samples are put into plates that are coated with specific antibodies, directed against specific proteins expressed on the CTC surface. EPISPOT is targeting secreted proteins of CTCs rather than the CTCs themselves [38]. By this technique, viable CTCs can be indirectly counted as they leave a “footprint” of their secreted proteins on a membrane, which can be visualized.

2. Exploiting the physical properties

   There exist several technologies which exploit the physical properties of CTCs for their isolation. Filtration-based size exclusion technologies have been developed, such as ISET (isolation by size of epithelial tumor cells) or ScreenCell, which allow for antigen-independent isolation of CTCs from blood based on their larger size in comparison to hematological cells [24, 25]. CTCs can be isolated from diluted blood using a polycarbonate membrane with 8 μm-sized pores. Blood cells can pass the membrane, while CTCs are captured on the filter and can then be analyzed by light microscopy and immunocytochemistry [24]. Although a promising technology, size filtration-based methods have their limitation in the lack of specificity for CTCs as many non-CTCs are isolated by the filtration devices and may lead to false positivity. Downstream analysis of DNA or RNA can be performed, but needs labor-intensive technologies, such as laser capture microdissection [24, 39–41]. A novel microfluidic platform, called Parsortix, utilizes the size and deformability of cells to enrich CTCs from blood [42, 43]. The technology is based on a chip with physical "steps" in which tumor cells move upward. The CTCs get arrested at the top steps as they get stuck between the most upper step and the top lid of the microfluidic chip [43]. Blood cells which are smaller and more deformable will pass through [43]. Cells can then be forwarded to mRNA and DNA downstream analysis [43], and thereby Parsortix represents an attractive EpCAM-independent solution.

12.2.3 Clinical Use of CTCs

Established routine procedures to investigate a tumor site include imaging technologies and biopsies. As biopsies are an invasive procedure, and usually not conducted in a metastatic setting, liquid biopsies, such as CTCs, can represent a promising alternative. Potential clinical applications of CTCs include the monitoring of cancer progression, the prediction of relapses or drug resistances, and the evaluation of treatment efficiency [44–47].

The most widely used CTC technology currently in clinical testing is still the CellSearch platform since it is the only technology to have received FDA approval for the enumeration of CTC in whole blood in specific cohorts of cancer patients. A landmark study for the clinical application of CTCs that actually led to FDA clearance was published by the group of Cristofanilli in 2004 [48]. The authors showed for the first time that CTC counts in metastatic breast cancer before treatment were an independent predictor of progression-free survival (PFS) and overall survival (OS) [48]. Investigation with several other tumor entities followed these tracks. In castration-resistant prostate cancer (CRPC), CTC enumeration is the most accurate and independent predictor of OS [29]. Furthermore, in metastatic colorectal cancer, the number of CTCs before and during treatment was reported as an independent predictor of PFS and OS in patients [28]. Similar results were obtained from metastatic non-small cell lung cancer (NSCLC), where the CTC numbers were shown to be the strongest predictor of OS [30].

Besides the prognostic utility of CTCs, an important question is whether CTCs can also be used as a predictive marker, i.e., to what extent CTCs can lead to a treatment decision which in the end improves health outcomes. In the SWOG S0500 clinical trial, a total of 595 patients with metastatic breast cancer were recruited and stratified based on repeated CTC counts during the treatment [49]. Patients with consistently high levels of CTCs before, during, and after the first cycle of chemotherapy were switched to a different treatment, while patients with decreasing CTC counts remained on the initial treatment [49]. However, the SWOG S0500 trial failed to show an improved outcome, based on CTC numbers and their resulting change of therapy [49]. Nevertheless, it is not clear yet whether the CTC counts failed as a predictive marker or if the available drugs for metastatic breast cancer failed to improve patient outcome. On the other hand, a recent study showed promising results in metastatic castration-resistant prostate cancer if CTCs were analyzed for androgen receptor splice variants [15]. Detection of the splice variant 7 (AR-V7) of the androgen receptor in CTCs was associated with a resistance to antihormonal therapy with enzalutamide or abiraterone [15]. Moreover, the authors stated that AR-V7 status may serve as a treatment selection marker in metastatic castration-resistant prostate cancer [50]. This study shows that for a clinical utility of CTCs, downstream analysis may pose an important factor.

12.2.4 Limitations and Challenges of CTCs

The greatest limitation to translate CTC research into a clinical application is the scarcity of CTCs in blood. Especially in patients with early cancer disease, the low number of patients with detectable CTCs remains an obstacle. Furthermore, most platforms for CTC isolation need qualified personal and usually come along with high costs. In addition, the molecular analysis of single cells is a challenging and expensive endeavor. These are major contributing factors for its infrequent use in the clinical routine workflow. Nevertheless, from a research point of view, technologies, such as next-generation sequencing, have evolved dramatically in the past years, which will potentially lead to a better and affordable molecular analysis of CTCs which finally could lead to improvements of therapeutic strategies.

12.2.5 Summary

Better insight into the biology of CTCs may help to understand the metastatic cascade. The aforementioned methods show a wide range of different properties of CTCs, such as physical or biological ones. Each technology has its advantages and disadvantages, and maybe each tumor entity will need a specific CTC technology. The clinical application of CTCs was first proven in 2004, and 10 years later, analysis of androgen receptor splice variants in prostate cancer CTCs was shown to be of relevance for a treatment selection. Improvements of CTC isolation and novel molecular analysis tools led to novel diagnostic applications for patient stratification. Especially functional assays of CTCs, such as CTC culture or CTC-derived xenografts, will stimulate CTC research for the coming years.

12.3 Circulating Tumor DNA (ctDNA)

12.3.1 Biology of Cell-Free DNA

The presence of cell-free DNA (cfDNA) in blood of healthy individuals was already described by Mandel and Metais in 1948 [51]. Despite their pioneering work, it took several decades until its clinical utility as a potential biomarker was recognized. Only in the 1970s, the occurrence of higher concentrations of cfDNA in patients with benign conditions, such as systemic lupus erythematosus (SLE) or rheumatoid arthritis, compared to healthy individuals was observed [52, 53]. Leon et al. reported elevated levels of cfDNA in the circulation of cancer patients, and in some patients, even a decrease of cfDNA after successful anticancer therapy could be observed [52]. Another 10 years later, Stroun et al. demonstrated the presence of tumor-specific aberrations in the circulation and therefore provided evidence that certain circulating DNA fragments originate from tumor tissues [54]. These findings were then confirmed by several other groups [55–58]. In the following years, other tumor-specific aberrations, including mutations in tumor suppressors and oncogenes [59], LOH [56], MSI [60], and DNA methylation [61], were identified and provided concrete evidence that cfDNA is released into the circulation by tumors, which is referred to as circulating tumor DNA (ctDNA).

There are now numerous studies about ctDNA available, but little is known about the origin, mechanisms and kinetics of release or clearance of cfDNA and ctDNA. Although it is thought that necrosis and active secretion contribute to cfDNA, the driving force for the release may be apoptosis. Cell-free DNA is highly fragmented, and assessment of the size distribution of cfDNA reveals an enrichment of fragments in the size of nucleoprotein complexes or multiples of them [62, 63]. More specifically, peaks corresponding to nucleosomes (147 bp) and chromatosomes (nucleosome + linker histone 167 bp) have been noted. This was also confirmed in mice experiments where the predominant fragments in plasma from xenografted animals were mononucleosome derived, indicating that apoptosis is the major source of cfDNA [64]. The authors demonstrated that ctDNA features vary during colorectal cancer (CRC) tumor development in nude mice that were xenografted with the human colorectal carcinoma cell lines HT29 or SW620 [64]. In a study of Heitzer et al., total plasma DNA concentrations and tumor-specific KRAS mutations in CRC patients were analyzed, and these data showed that a higher amount of tumor-specific fragments and a higher number of CTCs were linked to biphasic size distributions of plasma DNA fragments. However, despite advanced tumor stage, not all patients had detectable levels of ctDNA in their circulation [62].

It is now clear that cfDNA constitutes of a mixture of DNA released from cells from different tissues of the body. Studies of pregnant women have shown that the placenta is the origin of the cell-free fetal DNA detectable in the maternal circulation [65, 66]. Moreover, the investigation of circulating DNA pools after organ and bone marrow transplantations shed light on the different origins of cfDNA [67, 68]. These studies suggested that in healthy individuals, cfDNA is primarily derived from apoptosis of normal cells of the hematopoietic lineage, and material from other solid tissues contributes only to a small part of cfDNA [67, 68]. These data were confirmed by the Lo group, which used organ-specific DNA methylation signatures established by whole-genome bisulfite sequencing in order to trace back the origin of cfDNA fragments in pregnant women, patients with hepatocellular carcinoma, and subjects following bone marrow and liver transplantation [69]. Consistent with previous reports, the most abundant signature could be attributed to hematopoietic cells. The placental contributions in plasma of pregnant women range from 12.1 to 41.0% [69]. The graft-derived contributions to the plasma in the transplant recipients correlated with those determined using donor-specific genetic markers [69]. In cancer patients, a large part of the circulating DNA fragments could be associated with the primary tumor tissue [69].

In a recent study conducted by Snyder et al., it was shown that cfDNA is the detritus of cell death and that nucleosome phasing is reflected in the fragmentation pattern of cfDNA [70]. Since the boundaries of cfDNA fragments are biased by their association with nucleosomes, the fragmentation patterns of cfDNA might contain evidence of the epigenetic landscape of their tissue(s) of origin [70]. This might be a useful approach for the identification of cfDNA-releasing cells independent of genotypic differences between contributing cell types.

The fact that cfDNA is also released from normal cells is one of the major limitations of cfDNA as a diagnostic marker in cancer patients. The amount of ctDNA is highly variable and can range from less than 1% to more than 90% of total cfDNA [62, 71–73]. Despite continuous improvements in precision and accuracy of sequencing technologies, the fraction of ctDNA can be below the detection limit of these methods, especially in early-stage cancer, where in most cases insufficient levels of ctDNA for comprehensive analyses are present.

Although more and more studies deal with cfDNA origins, the clearance mechanisms of cfDNA are still poorly understood. The short half-life of cfDNA in the circulation suggests a model of ongoing release from apoptotic cells and rapid degradation or filtration [74]. A more recent study of Dennis Lo's group revealed a biphasic clearance with half-lives of about 1 h for the rapid phase and a second phase of 13 h [75]. It is of note that these data come from studies in pregnant women, and it is not clear yet whether these findings can be transferred to cancer patients. In contrast to fetal DNA, the massive accumulation of ctDNA in some patients might be a consequence of massive cell death due to a fast turnover of cancer cells, an inefficient degradation, or a combination of both. Moreover, it is not known how other factors, such as circadian rhythms, inflammation, or particular therapies, influence release and clearance mechanisms.

12.3.2 Methodological Aspects for the Analysis of ctDNA

Due to its high fragmentation and the low abundance in the circulation, the analysis of plasma DNA is challenging. However, recent advances in molecular technologies now offer the necessary sensitivity and specificity to detect small amounts of ctDNA in circulation. Although significantly higher cfDNA concentrations are present in serum than in plasma, plasma turned out to be the better source for ctDNA analyses [76]. In serum, cfDNA is "contaminated" by high molecular weight genomic DNA due to the clotting of white blood cells in the collection tube, leading to their lysis [77]. To minimize cellular degradation, even plasma should be immediately processed after blood collection in standard EDTA tubes. By the addition of cell-preserving reagents, which prevent white blood cell degradation, and inhibit nuclease-mediated DNA degradation, tubes can be stored for up to 14 days at ambient temperature [78]. A variety of methods have been used for the quantification of cfDNA [79, 80]. Although most studies dealing with extraction and quantification methods come from fetal cfDNA, the same issues apply for cfDNA from tumor patients. These are several technical confounders, including storage conditions or processing delay before plasma separation [77, 81], DNA extraction method [82], amplicon size, and target gene choice [83] that can influence the quantitation of cfDNA and, therefore, complicate data analysis, comparability, and reproducibility of the tests. The lack of generally accepted units of measure for cfDNA quantification further aggravates the situation. However, in recent years, efforts were made in order to establish benchmarks for standardization of the extraction and quantification of cfDNA.

Turning to the analysis of ctDNA, there are two different approaches used for the analysis, i.e., targeted and untargeted methods. Targeted methods are limited to the analysis of single or few known mutation or hotspots with clinical implications for therapy decisions, e.g., mutations in KRAS or EGFR. Since the first targeted mutation analyses in plasma or serum in the 1990s, technological progress has brought a number of highly sensitive methods, such as ARMS [84], digital PCR [85, 86], or BEAMing [87], which allow for the identification of mutant alleles at very low frequencies. A particularly sensitive and specific approach is the so-called personalized analysis of rearranged ends (PARE) [72]. This method involves the identification of tumor-specific translocations from the primary tumor that are monitored in plasma by the use of dPCR. This method can be used to detect tumor-specific changes at very low levels, i.e., in early stages, or to identify minimal residual disease; however, the availability of tumor tissue is required [88]. In a study by Heitzer et al., they were able to identify structural rearrangements directly from plasma after targeted enrichment of chromosomal regions that are frequently involved in translocations [89].

Although these methods achieve a high resolution, most of them interrogate only few loci. Novel occurring mutations or mutations in genes that lack mutational hotspots, such as tumor suppressors, are missing. One possibility of including driver genes without hotspots is targeted resequencing of selected genes that are known to be associated with tumorigenesis and progression. To this end there is striving for the establishment of targeted enrichment of larger number of genes or chromosomal regions. CAPP-Seq, an ultrasensitive method for quantifying ctDNA, was introduced by Newman et al. [90]. This method combines optimized library preparation methods for low-input DNA with a multiphase bioinformatics approach to design a "selector" consisting of biotinylated oligonucleotides that target recurrently mutated regions in the cancer of interest [90]. The researchers have shown to detect ctDNA in 100% of patients with stage II–IV nonsmall cell lung cancer and in 50% of patients with stage I and with 96% specificity for mutant allele fractions down to approximately 0.02%. Another approach was developed by Forshew et al., the so-called tagged-amplicon deep sequencing (TAm-Seq) [91], including 5995 genomic bases for low-frequency mutations. Using this approach, they identified cancer-specific mutations present at allele frequencies as low as 2%, with a sensitivity and specificity of >97%.

In contrast, genome-wide, untargeted approaches offer several advantages compared to targeted methods, including the fact that no a priori knowledge about the genetic makeup of the tumor is necessary and that they are not limited to recurrent changes [44, 89, 92]. Such comprehensively designed studies are of particular interest in late-stage cancer since tumors evolve rapidly due to progression and the selective pressure of therapies. The establishment of genome-wide copy number profiles from plasma DNA can now be performed very quickly and cost-effective [89]. Several studies have shown that the evolution and the plasticity of tumors can be effectively tracked using such genome-wide approaches [62, 89, 92–94]. However, these analyses are still lacking sensitivity and require a certain amount of tumor-specific DNA (about 5–10%), which does not apply to many samples from patients in earlier stages. Recent studies demonstrated that even in highly metastasized patients, there are clinical situations where ctDNA is present below optimal levels for the detection of mutations [62, 71, 95]. An untargeted pre-screening methods called mFAST-SeqS can identify samples with sufficient ctDNA levels that are suitable for subsequent analyses with genome-wide methods [96].

12.3.3 Clinical Use of ctDNA

The first effort to use cfDNA as a biomarker focused on the simple quantification of DNA. Several studies reported significant differences in the amount of cfDNA isolated from healthy individuals, patients with benign disease, and cancer patients [57, 97]. In lung cancer patients, higher levels of cfDNA have been observed compared to disease-free heavy smokers, suggesting a new, noninvasive approach for early detection of lung cancer [57]. Kim et al. recently reported that changes in the levels of cfDNA can act as reliable biomarkers to detect cancer early, predict tumor burden, and estimate curative resection and even prognosis in gastric cancer [98]. In contrast, other studies demonstrated no association between cfDNA concentrations and clinical, biological, or histological characteristics [56, 99, 100]. Soon it became clear that the amount of cfDNA alone is not a suitable marker for cancer patients due to the highly variable amounts of circulating DNA fragments that partially overlap with those of healthy individuals [56, 71, 97, 100]. However, the use of tumor-specific alterations in the circulation has already been applied in different scenarios of the therapy management of cancer patients.

12.3.3.1 Monitoring Tumor Burden and Minimal Residual Disease

Studies of the University of Cambridge and the Johns Hopkins University showed that the analysis of ctDNA is a better marker for the detection of recurrence of breast and colorectal cancer than conventional protein tumor markers [71, 101]. Dawson et al. used a personalized assay for a minimally invasive monitoring of treatment response in breast cancer patients. Changes in ctDNA levels showed a greater dynamic range and greater correlation with changes in tumor burden than CA 15-3 or CTCs. Another group from Lund University made use of a combined approach, including whole-genome sequencing of the primary tumor for the identification of rearrangements and digital PCR for monitoring purposes [88]. They retrospectively analyzed a set of breast cancer patients with localized tumors. Patients with detectable amounts of ctDNA after curative surgery developed metastases within a median time frame of 11 months, whereas those patients with no detectable ctDNA showed long-term progression-free survival [88]. A similar study was already reported in 2008 by Bert Vogelstein's group. The authors showed a significant association of decreasing levels of ctDNA with progression-free survival (PFS). On the other hand, in patients in whom tumor-specific mutations could be detected after surgery, the tumors recurred [87].

12.3.3.2 Detection of Resistance Mechanisms

Another paradigm for the clinical use of ctDNA is the early detection of resistance mechanisms, which can only be ensured by a tight monitoring. One of the first clinically used applications is the minimally invasive monitoring of patients with non-small cell lung cancers (NSCLC) that are treated with specific tyrosine kinase inhibitors (TKI). Approximately 10–15% of NSCLC harbor activating mutations in the EGFR gene, which codes for the epidermal growth factor receptor (EGFR), a protein involved in cell proliferation and division. Patients with an activated EGF pathway benefit from an intracellular blockade of the receptor. However, 50% of patients that initially respond well to the treatment develop resistance within several months. The underlying mechanism is in most cases a secondary mutation in EGFR that hinders the TKI from binding its target. A close monitoring of known resistance mechanisms can guide treatment decision and lead to an early adaption of further lines of therapies before the progression becomes clinically obvious. Sorensen et al. were able to monitor decreasing levels of the activating EGFR mutations and occurrence of the resistance-conferring mutation at the same time. The resistance mutation was detected up to 344 days before a clinically evident progression [102]. Other studies using different analysis methods achieved similar results [103–105].

Achievements in minimally invasive tumor monitoring could also be shown in patients with colorectal tumors. Similar to lung carcinoma, the EGF receptor is an important therapeutic target in the treatment of metastatic colorectal cancer. However, patients harboring activating mutations in the KRAS gene in their tumors do not benefit from EGFR antibodies, such as cetuximab and panitumumab. Since KRAS is a downstream component of the EGFR signaling network, activating mutations lead to intrinsic activation of the signal transduction pathway. KRAS mutations are thus a negative predictor of response to EGFR-directed therapy. Although the majority of KRAS wild-type patients benefit from the EGFR blockade, resistance occurs within 3–6 months in almost all patients. Known resistance mechanisms include—in addition to KRAS mutations as the predominant mechanism of resistance—mutations of BRAF and activation of alternate signaling pathways or increased EGFR numbers [106]. All these mechanisms have already been identified in ctDNA. In 2012, Diaz and colleagues determined whether mutant KRAS DNA could be detected in the circulation of 28 CRC patients receiving the anti-EGFR antibody panitumumab [107]. They showed that the appearance of these mutations was very consistent, generally occurring between 5 and 6 months following treatment. In three out of nine cases, mutant KRAS could be identified before radiographic evidence of disease progression. The mean time interval from detectable ctDNA to radiographic evidence of progression was 21 weeks [107]. Other studies showed that a comprehensive genome-wide analysis based on copy number status of ctDNA can further contribute to early detection of resistance mechanisms. The development of resistance to anti-EGFR therapies was associated with acquired gains of KRAS which occurred either as novel focal amplifications or as high-level polysomy of chromosome 12p. Again, in some cases, the resistant clones were detectable in the circulation months before progression was clinically obvious [106, 108]. In addition, focal amplifications of other genes recently shown to be involved in acquired resistance to anti-EGFR therapies, such as MET and ERBB2 [106–109], were minimally and invasively identified in the plasma DNA [106].

Similar approaches are used for prostate cancer patients. A variety of novel agents targeting the androgen receptor (AR) have altered the treatment paradigm of metastatic prostate cancer. Nevertheless, all patients develop inevitable therapeutic resistance. Resistance-conferring aberrations, such as mutations or gene amplification of the AR gene, can be monitored in plasma and may help to quickly adapt treatment based on the molecular nature of the tumor [108, 109]. These data highlight the benefit of moving beyond specific mutations and toward the full spectrum of genomic alterations, i.e., aneuploidy, amplifications, deletions, and translocations, since these aberrations represent some of the most clinically useful genomic targets in cancer (e.g., ERBB2, AR, KRAS amplifications) [110].

The relevance of ctDNA was proven not only by the monitoring of targeted therapies but also in patients under chemotherapy. Sequencing of 15 clinically relevant genes demonstrated the benefits of ctDNA as a marker for treatment response to cytotoxic chemotherapy [111]. In more than 98% of patients, candidate mutations were detected in the tumor, which were then screened with the high resolution using the Safe-SeqS method in plasma [112]. Patients under chemotherapy, who had a significant reduction in the ctDNA levels, had a significantly better response and a better progression-free survival [111]. Using exome sequencing, the group led by Nitzan Rosenfeld was able to identify resistance mechanisms in more than 80% of patients [113]. By the application of genome-wide methods, i.e., low-coverage genome sequencing for the detection of somatic copy number alterations (SCNA), even clonal shifts and the occurrence of focal alteration that contain driver genes can be detected at the chromosome level [89, 95].

12.3.4 Limitations and Challenges of ctDNA

Although the analysis of ctDNA has a great potential for improving therapy management in a cancer patient, there are several issues that have prevented the widespread implementation of ctDNA in clinical routine. The lack of pre-analytical and analytical standards is still a big issue, especially if it comes to comparability and reproducibility of results and the integration of ctDNA analysis in large clinical trials. In this respect, more and more efforts are being made to establish standard operation procedures by international consortiums, e.g., the IMI CANCER-ID project, which includes a number of experts in the field of ctDNA and CTCs (http://www.cancer-id.eu/). It is of utmost interest to find a consensus in which methods will find practical application and how to report results. Furthermore, it is still not clear yet whether the focus should be on specific targets that can be analyzed with high resolution or broader approaches should be used. In this respect, of course, also costs and time play a major role, as well as the question of who will bear the costs for such investigations.

Moreover, the actual clinical long-term benefit of ctDNA analyses for patients needs to be confirmed in large-scale studies with sufficient sample sizes. The discovery of resistance-conferring mutations neither saves a patient's life nor does it increase the quality of life, if there is no drug that bypasses the resistance or may be administered subsequently.

In addition, many of the questions regarding the biology and release of ctDNA are still unanswered: Does ctDNA represent a true portrait of the cancer? Do all tumor locations or all clones of a tumor release the same amounts of ctDNA, or is it just the most dominant and proliferative clone that can be found in the circulation? Is the prevalence of ctDNA in all tumor entities the same? Many of these issues could only be resolved by a comparison of ctDNA and all existing tumor sites by the use of "warm autopsies" which can be quite challenging from an ethical perspective.

12.3.5 Summary

All these studies and many more that could not be discussed in this chapter suggest that the analysis of ctDNA resents a very promising tool in the treatment management of cancer patients. The hitherto most comprehensive study of the group of Diaz showed that in at least 75% of cancer patients with advanced solid tumors, ctDNA can be detected [71]. In later stages, the analysis of ctDNA allows for a comprehensive therapy monitoring, which allows the physicians to respond as quickly as possible to changes in the tumor. Thus, treatments can be adjusted rapidly, and patients can be spared from expensive treatments with very toxic drugs and side effects from the moment the drug is no longer effective. Moreover, novel therapeutic targets that occur in the course of the disease can be identified and offer new treatment options. ctDNA also provides a unique opportunity to learn more about metastasis processes and the related signaling pathways.

Furthermore, ctDNA levels can be used as prognostic markers in order to evaluate whether a patient needs adjuvant treatment after curative surgery in order to eliminate residual cancer cells. Especially in earlier stages, only a subset of patients relapse and therefore profit from adjuvant therapy after a curative resection of the tumor. ctDNA analyses could spare some patients a burdensome and costly therapy. In the near future, ctDNA might also be used as a diagnostic biomarker enabling early detection of cancer. Illumina, the current market leader in the field of NGS, has just founded a 100 million dollar start-up company called GRAIL with the aim to develop a test for early detection of cancer, which would be launched in 2019. Although the detection of cancer in its earliest stage is the "the Holy Grail" in oncology, one has to consider that the detection of specific mutations in the circulation in individuals who do not yet have visible tumor can be problematic.

12.4 Circulating RNA and Exosomes

12.4.1 Biology of Exosomes and Circulating RNA

Besides the two major topics in liquid biopsies, CTCs and ctDNA, circulating RNA is a small but promising field for clinical applications. The term circulating RNA refers primarily to microRNA (miRNA) which either travels as small fragments bound to proteins or encapsulated in exosomes [114, 115]. The number of publications increased tremendously in the past 10 years, of only 85 papers published in the year 2005 to 3680 in 2015, with a peak in 2014 of 4190 papers published with the topic "microRNA in cancer." Longer stretches of RNA transcripts outside of cells are considered to derive from dying cells and having no functional role. Furthermore, RNA molecules can originate from viruses and therefore can pose a serious threat to the whole organism. Therefore, it is not surprising that RNase activity is high in serum, as it was shown that >99% of mRNA added to serum is degraded within 15 s [116]. This is an explanation why long RNAs, like mRNAs, are usually not detectable in blood. However, smaller RNAs with a size below 25 nucleotides, like miRNAs which play an important role in gene silencing and post-transcriptional regulation of gene expression, are very stable in plasma and serum. This is due to the fact that these molecules are too small to be degraded by RNases. They are often bound to subcellular particles like Argonaute proteins or captured in microvesicles preventing them from degradation [117, 118]. Exosomes belong to these microvesicles that are thought to be actively released from cells. They are 30–100 nm in diameter and can be found in blood of cancer patients [119–124]. Exosomes contain proteins and nucleic acids, such as fragmented DNA, RNA, and miRNA [115, 125]. Being encapsulated by lipid bilayers, the molecules are stably preserved, making exosome a good resource for the study of tumor-associated biomarkers. In the following paragraphs, we will focus on miRNA as blood-based biomarker. In this context, they are also discussed as a "liquid biopsy," although miRNAs are not necessarily derived from the tumor tissue, but can be a regulatory change of expression in response to cancer disease.

12.4.2 Methodological Aspects for the Isolation and Analysis of miRNA

miRNAs are exposed to lots of physical and chemical stresses prior to their isolation which affect their stability and quantity in plasma and serum. It is essential for a clinically reliable and robust biomarker to minimize the influence of pre-analytical parameters prior to their analysis. The following paragraphs will give more information about the most important steps in miRNA isolation and detection and some insights into current research.

Recent publications showed that in many pathologic conditions, miRNAs are expeditiously released from tissues into blood circulation. It was also demonstrated that circulating miRNA in peripheral blood was highly stable and protected from degradation conditions, such as extreme pH values or endogenous RNase activity. The protection is given as miRNAs are embedded in microvesicles or exosomes, thereby being inaccessible by degrading enzymes [114, 126]. This feature allows miRNAs to be used as noninvasive biomarkers [127]. MiRNA extraction can be applied to nearly every fluid or tissue of the body. Besides the commonly used sample types like serum and plasma, there are also kits available for miRNA extraction from cerebrospinal fluid, saliva, or urine [127]. As the composition of these body fluids is pretty dissimilar, the isolation methods of miRNA need to be adapted from one tissue/fluid to another [128]. Thus, depending on the specific miRNA and the corresponding sample material, the corresponding extraction kit should be used. Most studies use serum or plasma as sample starting material for miRNA extraction. Comparison of extracellular miRNA stability shows little to no difference in these biological fluids [127], though higher concentrations were steadily found in serum [129].

12.4.2.1 Storage of miRNA Samples

Sample storage conditions can seriously affect the accuracy and reliability of analytical results. Sourvinou et al. investigated the stability of circulating miRNAs in identical plasma samples under different temperature and time conditions [130]. The results showed that for the accurate quantification of cell-free miRNAs, the isolation process should be performed within 48 h after sample collection if plasma samples are kept at −20 °C or −70 °C [130]. Storage at 4 °C leads to a significant decrease in circulating miRNA levels within 24 h [130]. If long-term storage of plasma samples is needed, temperatures of −70 °C rather than −20 °C should be preferred to avoid extensive miRNA degradation [130]. Grasedieck et al. reported similar results with respect to the impact of serum storage conditions on miRNA stability [131].

Experiments indicate that extracted miRNAs are stable for more than 1 year at a temperature of −70 °C [130, 132]. However, results showed differences in the stability of stored miRNAs depending on the extraction buffer [130]. There are also studies that report significant degradation of miRNAs 3 days after isolation [133]. A study of Sourvinou et al. indicated that the use of different elution buffers for miRNA storage might be responsible for differential stability of miRNA [130]. Therefore, they suggested the use of mirVana PARIS kit for long-term storage [130].

12.4.2.2 Extraction of miRNA

Several extraction kits for miRNA are commercially available. As mentioned above for each source of material, the optimal miRNA extraction method needs to be experimentally determined. The following paragraph refers to protocol considerations for isolating miRNAs from blood plasma or serum.

12.4.2.3 Phenol/Chloroform Extraction for miRNA

This procedure relies on the different solubility of cellular components in organic solvents, such as phenol, chloroform, or ethanol. The main components of the phenol/chloroform protocol are phenol and guanidinium thiocyanate, usually marketed as TRIzol [134]. TRIzol denatures proteins, including RNases, which permits long-term storage of samples [135]. After phase separation, RNA gets precipitated with isopropyl alcohol. Due to the fact that miRNAs are small, ample time is needed for recovery. Recently, authors realized that there is selective loss of small RNA molecules with low GC content using TRIzol, especially by analyzing low numbers of cells [136]. Though this publication refers to miRNAs isolated from cells rather than body fluids, nevertheless, it is unknown if these factors also alter RNA extraction from blood plasma or serum [130, 134].

12.4.2.4 Silica-Based miRNA Recovery Methods

There are manifold kits available, of which the most frequently used are the miRVana PARIS and miRNeasy Mini kits. A direct comparison of these methods is difficult as the majority of publications do not report the actual yield and quality of miRNA. The miRVana PARIS kit is a commercially available method of separating nucleic acids and proteins. miRVana PARIS method is a two-part, sequential filtration with increasing ethanol concentrations, used for collection of a highly enriched fractions of RNA molecules shorter than 200 nucleotides. This method works very well for the isolation of miRNA from tissues and body fluids as well, as it requires fluid volumes from 100 μL up to 625 μL [134]. The miRNeasy Mini kit uses a silica-based column technique to recover miRNAs. Some groups report that this kit leads to a two- to threefold better yield than miRVana PARIS kit; however, the available literature of groups that use this kit for miRNA extraction from plasma or serum is very limited. The kit also uses a phenol/guanidine-based lysis to isolate the miRNA from other plasma components, by adsorption on a silica mini-column in the presence of ethanol. Remarkably, using the Qiagen QIAcube, the binding, washing, and elution step can be operated automatically. Consequently, this development can decrease working time and variability [134].

12.4.2.5 Analysis and Measurement of miRNA

Moldovan et al. give a good insight into the three most common miRNA profiling methods, (1) RT-qPCR, (2) microarrays which are a hybridization-based method to detect miRNA expression, and (3) next-generation sequencing and their advantages and disadvantages [134]. Most of the scientific interest is focused on the detection of circulating miRNA in plasma or serum, and the commonly used techniques, such as conventional RT-qPCR, offer both high sensitivity and specificity. The measured miRNA levels can vary depending on the extraction method and the body fluid used. This complicates the comparison of results from different methods as well as between fluids. Therefore, a normalization control needs to be implemented to minimize these variations [134]. There are two main normalization controls: One is the absolute quantification which is done by analyzing a series of probes with known, increasing concentrations [134]. Another approach involves relative quantification, where a small constantly expressed miRNA or a spiked-in miRNA is used for normalization of input amounts [134]. Microarrays offer high-throughput analysis of a large number of miRNAs and can be customized for high flexibility. Nevertheless, microarrays are less specific and sensitive than RT-qPCR, and results obtained by microarrays need to be confirmed by RT-qPCR. The third and probably most promising technology for miRNA analysis is next-generation sequencing. With its extreme sensitivity of one miRNA copy per cell, it has the ability to detect expressions over 6–7 log fold ranges. It is also the only one of the three technologies which is able to detect both known and novel miRNAs. Moreover, it can detect small RNAs like noncoding RNA (ncRNA), small interfering RNA (siRNA), etc. [134]. Even though these advantages are tremendous, the downside of RNA-Seq is its high cost as well as the tremendous amount of computational infrastructure and bioinformatics know-how needed [137].

12.4.3 Clinical Use of Circulating miRNA

At present, there are miRNA panels helping clinicians in determining the origins of cancer in disseminated tumors, as reviewed by Hydbring et al. [138]. These days, approximately 1600 human miRNAs have been placed into miRNA databases based on analyses of RNA deep sequencing data [139]. The majority of miRNA publications refer to the usage of solid tissues even though miRNAs can be readily detected in human serum, plasma, or total blood because of their small size and high stability as described above. The high potential of circulating miRNAs as biomarkers in serum was demonstrated by studies testing patients with diffuse large B-cell lymphoma, highlighting miR-21 as a potential biomarker [140, 141]. In another study, expression level of miR-143 allowed the discrimination between prostate cancer patients and healthy controls [127]. Subsequent studies reported on miRNA detection in patients suffering from breast cancer, colorectal cancer, or squamous cell lung cancer by using whole blood, plasma, or sputum samples [138]. It has been shown that circulating miRNAs may also be used for prognostic purposes. The group of Boeri et al. detected miRNAs with strong prognostic value in lung cancer patients years before the onset of disease by analyzing expression in samples taken before diagnosis, at the time of disease detection, and in disease-free smokers [142].

By analyzing 863 miRNAs from 454 human blood samples, Keller et al. could show that in each disease an average of more than 100 miRNAs were deregulated [143]. The samples were taken from patients suffering from 14 different diseases, including lung cancer, prostate cancer, multiple sclerosis, pancreatic ductal adenocarcinoma, ovarian cancer, melanoma, gastric tumors, pancreatic tumors, chronic obstructive pulmonary disease, pancreatitis, sarcoidosis, periodontitis, and myocardial infarction [143]. By using this data and developing mathematical algorithms, the authors could precisely predict the disease in more than two-thirds of people involved in the study [143].

12.4.4 Limitations and Challenges of Circulating RNA

Even though the field of miRNA is very promising, there are also limitations and challenges. The main issue in miRNA quantification is the lack of validation and standardization. There are huge differences in detectable miRNA expression depending on pre-analytical conditions, such as temperature (e.g., freezing) and additives (e.g., anticoagulants), and the condition of the patient with respect to diet, lifestyle, or drug usage. Standard operating procedures (SOPs) are not in place yet, but need to be defined for sample preparation, extraction, and analysis. As already small differences in sample handling can result in big changes in outcome, published miRNA data which correlate with a specific disease should be critically questioned prior to using their miRNA signatures in clinically meaningful tests [144].

12.4.5 Summary

miRNA detection is an increasingly important field. Although still facing challenges mainly due to the lack of standardization, miRNA detection, not only from serum and plasma but also from other noninvasive collected fluids like saliva or urine, is a promising biomarker for a daily routine clinical approach. Importantly, miRNA diagnostic is not just a further approach to detect disease and its progression; it could also have the potential to initially identify the unknown origins of tumors and metastases.

References

1. Lugo TG, Braun S, Cote RJ, Pantel K, Rusch V. Detection and measurement of occult disease for the prognosis of solid tumors. J Clin Oncol. 2003;21:2609–15. doi:10.1200/JCO.2003.01.153.
2. Massagué J, Obenauf AC. Metastatic colonization by circulating tumour cells. Nature. 2016;529:298–306. doi:10.1038/nature17038.
3. Aceto N, Toner M, Maheswaran S, Haber DA. En route to metastasis: circulating tumor cell clusters and epithelial-to-mesenchymal transition. Trends Cancer. 2015;1:44–52. doi:10.1016/j.trecan.2015.07.006.

4. Duda DG, Duyverman AM, Kohno M, Snuderl M, Steller EJ, Fukumura D, Jain RK. Malignant cells facilitate lung metastasis by bringing their own soil. Proc Natl Acad Sci U S A. 2010;107:21677–82. doi:10.1073/pnas.1016234107.
5. Labelle M, Begum S, Hynes RO. Direct signaling between platelets and cancer cells induces an epithelial-mesenchymal-like transition and promotes metastasis. Cancer Cell. 2011;20:576–90.
6. Barrière G, Riouallon A, Renaudie J, Tartary M, Rigaud M. Mesenchymal and stemness circulating tumor cells in early breast cancer diagnosis. BMC Cancer. 2012;12:114. doi:10.1186/1471-2407-12-114.
7. Quail DF, Joyce JA. Microenvironmental regulation of tumor progression and metastasis. Nat Med. 2013;19:1423–37. doi:10.1038/nm.3394.
8. Cao Z, Ding B-S, Guo P, Lee SB, Butler JM, Casey SC, Simons M, Tam W, Felsher DW, Shido K, Rafii A, Scandura JM, Rafii S. Angiocrine factors deployed by tumor vascular niche induce B cell lymphoma invasiveness and chemoresistance. Cancer Cell. 2014;25:350–65. doi:10.1016/j.ccr.2014.02.005.
9. Tsai JH, Yang J. Epithelial–mesenchymal plasticity in carcinoma metastasis. Genes Dev. 2013;27:2192–206. doi:10.1101/gad.225334.113.
10. Lamouille S, Xu J, Derynck R. Molecular mechanisms of epithelial–mesenchymal transition. Nat Rev Mol Cell Biol. 2014;15:178–96. doi:10.1038/nrm3758.
11. Thiery JP, Acloque H, Huang RYJ, Nieto MA. Epithelial-Mesenchymal Transitions in Development and Disease. Cell. 2009;139:871–90. doi:10.1016/j.cell.2009.11.007.
12. Paterlini-Brechot P, Benali NL. Circulating tumor cells (CTC) detection: clinical impact and future directions. Cancer Lett. 2007;253:180–204. doi:10.1016/j.canlet.2006.12.014.
13. Alix-Panabières C, Pantel K. Challenges in circulating tumour cell research. Nat Rev Cancer. 2014;14:623–31. doi:10.1038/nrc3820.
14. Allard WJ, Matera J, Miller MC, Repollet M, Connelly MC, Rao C, Tibbe AG, Uhr JW, Terstappen LW. Tumor cells circulate in the peripheral blood of all major carcinomas but not in healthy subjects or patients with nonmalignant diseases. Clin. Cancer Res. 2004;10:6897–904. doi:10.1158/1078-0432.CCR-04-0378.
15. Antonarakis ES, Lu C, Wang H, Luber B, Nakazawa M, Roeser JC, Chen Y, Mohammad TA, Chen Y, Fedor HL, Lotan TL, Zheng Q, De Marzo AM, Isaacs JT, Isaacs WB, Nadal R, Paller CJ, Denmeade SR, Carducci MA, Eisenberger MA, Luo J. AR-V7 and resistance to enzalutamide and abiraterone in prostate cancer. N Engl J Med. 2014;371:1028–38. doi:10.1056/NEJMoa1315815.
16. Nagrath S, Sequist LV, Maheswaran S, Bell DW, Irimia D, Ulkus L, Smith MR, Kwak EL, Digumarthy S, Muzikansky A, Ryan P, Balis UJ, Tompkins RG, Haber DA, Toner M. Isolation of rare circulating tumour cells in cancer patients by microchip technology. Nature. 2007;450:1235–9. doi:10.1038/nature06385.
17. Stott SL, Hsu CH, Tsukrov DI, Yu M, Miyamoto DT, Waltman BA, Rothenberg SM, Shah AM, Smas ME, Korir GK, Floyd Jr FP, Gilman AJ, Lord JB, Winokur D, Springer S, Irimia D, Nagrath S, Sequist LV, Lee RJ, Isselbacher KJ, Maheswaran S, Haber DA, Toner M. Isolation of circulating tumor cells using a microvortex-generating herringbone-chip. Proc Natl Acad Sci U S A. 2010;107:18392–7. doi:10.1073/pnas.1012539107.
18. Yoon HJ, Kim TH, Zhang Z, Azizi E, Pham TM, Paoletti C, Lin J, Ramnath N, Wicha MS, Hayes DF, Simeone DM, Nagrath S. Sensitive capture of circulating tumour cells by functionalized graphene oxide nanosheets. Nat Nanotechnol. 2013;8:735–41. doi:10.1038/nnano.2013.194.
19. Yu M, Bardia A, Aceto N, Bersani F, Madden MW, Donaldson MC, Desai R, Zhu H, Comaills V, Zheng Z, Wittner BS, Stojanov P, Brachtel E, Sgroi D, Kapur R, Shioda T, Ting DT, Ramaswamy S, Getz G, Iafrate AJ, Benes C, Toner M, Maheswaran S, Haber DA. Cancer therapy. Ex vivo culture of circulating breast tumor cells for individualized testing of drug susceptibility. Science. 2014;345:216–20. doi:10.1126/science.1253533.

20. Saucedo-Zeni N, Mewes S, Niestroj R, Gasiorowski L, Murawa D, Nowaczyk P, Tomasi T, Weber E, Dworacki G, Morgenthaler NG, Jansen H, Propping C, Sterzynska K, Dyszkiewicz W, Zabel M, Kiechle M, Reuning U, Schmitt M, Lucke K. A novel method for the in vivo isolation of circulating tumor cells from peripheral blood of cancer patients using a functionalized and structured medical wire. Int J Oncol. 2012;41:1241–50. doi:10.3892/ijo.2012.1557.
21. Alix-Panabieres C. EPISPOT assay: detection of viable DTCs/CTCs in solid tumor patients. Recent Results Cancer Res. 2012;195:69–76. doi:10.1007/978-3-642-28160-0_6.
22. Gertler R, Rosenberg R, Fuehrer K, Dahm M, Nekarda H, Siewert JR. Detection of circulating tumor cells in blood using an optimized density gradient centrifugation. Recent Results Cancer Res. 2003;162:149–55.
23. He W, Kularatne SA, Kalli KR, Prendergast FG, Amato RJ, Klee GG, Hartmann LC, Low PS. Quantitation of circulating tumor cells in blood samples from ovarian and prostate cancer patients using tumor-specific fluorescent ligands. Int J Cancer. 2008;123:1968–73. doi:10.1002/ijc.23717.
24. El-Heliebi A, Kroneis T, Zohrer E, Haybaeck J, Fischereder K, Kampel-Kettner K, Zigeuner R, Pock H, Riedl R, Stauber R, Geigl JB, Huppertz B, Sedlmayr P, Lackner C. Are morphological criteria sufficient for the identification of circulating tumor cells in renal cancer? J Transl Med. 2013;11:214. doi:10.1186/1479–5876–11-214.
25. Hou JM, Krebs M, Ward T, Sloane R, Priest L, Hughes A, Clack G, Ranson M, Blackhall F, Dive C. Circulating tumor cells as a window on metastasis biology in lung cancer. Am J Pathol. 2011;178:989–96. doi:10.1016/j.ajpath.2010.12.003.
26. Vona G, Sabile A, Louha M, Sitruk V, Romana S, Schutze K, Capron F, Franco D, Pazzagli M, Vekemans M, Lacour B, Brechot C, Paterlini-Brechot P. Isolation by size of epithelial tumor cells: a new method for the immunomorphological and molecular characterization of circulating tumor cells. Am J Pathol. 2000;156:57–63. doi:10.1016/S0002-9440(10)64706-2.
27. Armstrong AJ, Marengo MS, Oltean S, Kemeny G, Bitting RL, Turnbull JD, Herold CI, Marcom PK, George DJ, Garcia-Blanco MA. Circulating tumor cells from patients with advanced prostate and breast cancer display both epithelial and mesenchymal markers. Mol Cancer Res. 2011;9:997–1007. doi:10.1158/1541-7786.MCR-10-0490.
28. Cohen SJ, Punt CJA, Iannotti N, Saidman BH, Sabbath KD, Gabrail NY, Picus J, Morse MA, Mitchell E, Miller MC, Doyle GV, Tissing H, Terstappen LWMM, Meropol NJ. Prognostic significance of circulating tumor cells in patients with metastatic colorectal cancer. Ann Oncol. 2009;20:1223–9. doi:10.1093/annonc/mdn786.
29. de Bono JS, Scher HI, Montgomery RB, Parker C, Miller MC, Tissing H, Doyle GV, Terstappen LW, Pienta KJ, Raghavan D. Circulating tumor cells predict survival benefit from treatment in metastatic castration-resistant prostate cancer. Clin Cancer Res. 2008;14:6302–9. doi:10.1158/1078-0432.CCR-08-0872.
30. Krebs MG, Sloane R, Priest L, Lancashire L, Hou J-M, Greystoke A, Ward TH, Ferraldeschi R, Hughes A, Clack G, Ranson M, Dive C, Blackhall FH. Evaluation and prognostic significance of circulating tumor cells in patients with non-small-cell lung cancer. J Clin Oncol. 2011;29:1556–63. doi:10.1200/JCO.2010.28.7045.
31. Zhang L, Riethdorf S, Wu G, Wang T, Yang K, Peng G, Liu J, Pantel K. Meta-analysis of the prognostic value of circulating tumor cells in breast cancer. Clin Cancer Res. 2012;18:5701–10. doi:10.1158/1078-0432.CCR-12-1587.
32. Karabacak NM, Spuhler PS, Fachin F, Lim EJ, Pai V, Ozkumur E, Martel JM, Kojic N, Smith K, Chen P, Yang J, Hwang H, Morgan B, Trautwein J, Barber TA, Stott SL, Maheswaran S, Kapur R, Haber DA, Toner M. Microfluidic, marker-free isolation of circulating tumor cells from blood samples. Nat Protoc. 2014;9:694–710. doi:10.1038/nprot.2014.044.
33. Ozkumur E, Shah AM, Ciciliano JC, Emmink BL, Miyamoto DT, Brachtel E, Yu M, Chen P, Morgan B, Trautwein J, Kimura A, Sengupta S, Stott SL, Karabacak NM, Barber TA, Walsh JR, Smith K, Spuhler PS, Sullivan JP, Lee RJ, Ting DT, Luo X, Shaw AT, Bardia A, Sequist LV, Louis DN, Maheswaran S, Kapur R, Haber DA, Toner M. Inertial focusing for tumor

antigen-dependent and -independent sorting of rare circulating tumor cells. Sci Transl Med. 2013;5:179ra47. doi:10.1126/scitranslmed.3005616.
34. Lalmahomed ZS, Kraan J, Gratama JW, Mostert B, Sleijfer S, Verhoef C. Circulating tumor cells and sample size: the more, the better. J Clin Oncol. 2010;28:e288–e289; author reply e290. doi:10.1200/JCO.2010.28.2764.
35. Fortune JB, Feustel P. Effect of patient position on size and location of the subclavian vein for percutaneous puncture. Arch Surg. 2003;1960(138):996–1000; discussion 1001. doi:10.1001/archsurg.138.9.996.
36. Gorges TM, Penkalla N, Schalk T, Joosse SA, Riethdorf S, Tucholski J, Lücke K, Wikman H, Jackson S, Brychta N, von Ahsen O, Schumann C, Krahn T, Pantel K. Enumeration and molecular characterization of tumor cells in lung cancer patients using a novel in vivo device for capturing circulating tumor cells. Clin Cancer Res. 2015; doi:10.1158/1078-0432.CCR-15-1416.
37. Liu Z, Fusi A, Klopocki E, Schmittel A, Tinhofer I, Nonnenmacher A, Keilholz U. Negative enrichment by immunomagnetic nanobeads for unbiased characterization of circulating tumor cells from peripheral blood of cancer patients. J Transl Med. 2011;9:70. doi:10.1186/1479–5876–9-70.
38. Alix-Panabières C, Vendrell J-P, Pellé O, Rebillard X, Riethdorf S, Müller V, Fabbro M, Pantel K. Detection and characterization of putative metastatic precursor cells in cancer patients. Clin Chem. 2007;53:537–9. doi:10.1373/clinchem.2006.079509.
39. El-Heliebi A, Chen S, Kroneis T. Heat-induced fragmentation and adapter-assisted whole genome amplification using GenomePlex® single-cell whole genome amplification kit (WGA4). Methods Mol Biol. 2015a;1347:101–9. doi:10.1007/978-1-4939-2990-0_7.
40. El-Heliebi A, Chen S, Kroneis T. Using multiplex PCR for assessing the quality of whole genome amplified DNA. Methods Mol Biol. 2015b;1347:119–28. doi:10.1007/978-1-4939-2990-0_9.
41. Ma YC, Wang L, Yu FL. Recent advances and prospects in the isolation by size of epithelial tumor cells (ISET) methodology. Technol Cancer Res Treat. 2013;12:295–309. doi:10.7785/tcrt.2012.500328.
42. Chudziak J, Burt DJ, Mohan S, Rothwell DG, Mesquita B, Antonello J, Dalby S, Ayub M, Priest L, Carter L, Krebs MG, Blackhall F, Dive C, Brady G. Clinical evaluation of a novel microfluidic device for epitope-independent enrichment of circulating tumour cells in patients with small cell lung cancer. Analyst. 2016;141:669–78. doi:10.1039/c5an02156a.
43. Hvichia GE, Parveen Z, Wagner C, Janning M, Quidde J, Stein A, Müller V, Loges S, Neves RPL, Stoecklein NH, Wikman H, Riethdorf S, Pantel K, Gorges TM. A novel microfluidic platform for size and deformability based separation and the subsequent molecular characterization of viable circulating tumor cells. Int J Cancer. 2016;138:2894–904. doi:10.1002/ijc.30007.
44. Heitzer E, Auer M, Gasch C, Pichler M, Ulz P, Hoffmann EM, Lax S, Waldispuehl-Geigl J, Mauermann O, Lackner C, Höfler G, Eisner F, Sill H, Samonigg H, Pantel K, Riethdorf S, Bauernhofer T, Geigl JB, Speicher MR. Complex tumor genomes inferred from single circulating tumor cells by array-CGH and next-generation sequencing. Cancer Res. 2013;73:2965–75. doi:10.1158/0008-5472.CAN-12-4140.
45. Pierga J-Y, Hajage D, Bachelot T, Delaloge S, Brain E, Campone M, Diéras V, Rolland E, Mignot L, Mathiot C, Bidard F-C. High independent prognostic and predictive value of circulating tumor cells compared with serum tumor markers in a large prospective trial in first-line chemotherapy for metastatic breast cancer patients. Ann Oncol. 2012;23:618–24. doi:10.1093/annonc/mdr263.
46. Scher HI, Jia X, de Bono JS, Fleisher M, Pienta KJ, Raghavan D, Heller G. Circulating tumour cells as prognostic markers in progressive, castration-resistant prostate cancer: a reanalysis of IMMC38 trial data. Lancet Oncol. 2009;10:233–9. doi:10.1016/S1470-2045(08)70340-1.
47. Tewes M, Aktas B, Welt A, Mueller S, Hauch S, Kimmig R, Kasimir-Bauer S. Molecular profiling and predictive value of circulating tumor cells in patients with metastatic breast cancer: an option for monitoring response to breast cancer related therapies. Breast Cancer Res Treat. 2009;115:581–90. doi:10.1007/s10549-008-0143-x.

48. Cristofanilli M, Budd GT, Ellis MJ, Stopeck A, Matera J, Miller MC, Reuben JM, Doyle GV, Allard WJ, Terstappen LW, Hayes DF. Circulating tumor cells, disease progression, and survival in metastatic breast cancer. N Engl J Med. 2004;351:781–91. doi:10.1056/NEJMoa040766.
49. Smerage JB, Barlow WE, Hortobagyi GN, Winer EP, Leyland-Jones B, Srkalovic G, Tejwani S, Schott AF, O'Rourke MA, Lew DL, Doyle GV, Gralow JR, Livingston RB, Hayes DF. Circulating tumor cells and response to chemotherapy in metastatic breast cancer: SWOG S0500. J Clin Oncol. 2014;32:3483–9. doi:10.1200/JCO.2014.56.2561.
50. Antonarakis ES, Lu C, Luber B, Wang H, Chen Y, Nakazawa M, Nadal R, Paller CJ, Denmeade SR, Carducci MA, Eisenberger MA, Luo J. Androgen receptor splice variant 7 and efficacy of taxane chemotherapy in patients with metastatic castration-resistant prostate cancer. JAMA Oncol. 2015;1:582–91. doi:10.1001/jamaoncol.2015.1341.
51. Mandel P, Metais P. Les acides nucléiques du plasma sanguin chez l'homme. C R Séances Soc Biol Fil. 1948;142:241–3.
52. Leon SA, Ehrlich GE, Shapiro B, Labbate VA. Free DNA in the serum of rheumatoid arthritis patients. J Rheumatol. 1977a;4:139–43.
53. Leon SA, Green A, Ehrlich GE, Poland M, Shapiro B. Avidity of antibodies in SLE: relation to severity of renal involvement. Arthritis Rheum. 1977b;20:23–9.
54. Stroun M, Anker P, Maurice P, Lyautey J, Lederrey C, Beljanski M. Neoplastic characteristics of the DNA found in the plasma of cancer patients. Oncology. 1989;46:318–22.
55. Castells A, Puig P, Móra J, Boadas J, Boix L, Urgell E, Solé M, Capellà G, Lluís F, Fernández-Cruz L, Navarro S, Farré A. K-ras mutations in DNA extracted from the plasma of patients with pancreatic carcinoma: diagnostic utility and prognostic significance. J Clin Oncol. 1999;17:578–84.
56. Chen X, Bonnefoi H, Diebold-Berger S, Lyautey J, Lederrey C, Faltin-Traub E, Stroun M, Anker P. Detecting tumor-related alterations in plasma or serum DNA of patients diagnosed with breast cancer. Clin Cancer Res. 1999;5:2297–303.
57. Sorenson GD, Pribish DM, Valone FH, Memoli VA, Bzik DJ, Yao SL. Soluble normal and mutated DNA sequences from single-copy genes in human blood. Cancer Epidemiol Biomarkers Prev. 1994;3:67–71.
58. Vasioukhin V, Anker P, Maurice P, Lyautey J, Lederrey C, Stroun M. Point mutations of the N-ras gene in the blood plasma DNA of patients with myelodysplastic syndrome or acute myelogenous leukaemia. Br J Haematol. 1994;86:774–9.
59. Powell SM, Zilz N, Beazer-Barclay Y, Bryan TM, Hamilton SR, Thibodeau SN, Vogelstein B, Kinzler KW. APC mutations occur early during colorectal tumorigenesis. Nature. 1992;359:235–7. doi:10.1038/359235a0.
60. Shaw JA, Smith BM, Walsh T, Johnson S, Primrose L, Slade MJ, Walker RA, Coombes RC. Microsatellite alterations plasma DNA of primary breast cancer patients. Clin Cancer Res. 2000;6:1119–24.
61. Esteller M, Sanchez-Cespedes M, Rosell R, Sidransky D, Baylin SB, Herman JG. Detection of aberrant promoter hypermethylation of tumor suppressor genes in serum DNA from non-small cell lung cancer patients. Cancer Res. 1999;59:67–70.
62. Heitzer E, Auer M, Hoffmann EM, Pichler M, Gasch C, Ulz P, Lax S, Waldispuehl-Geigl J, Mauermann O, Mohan S, Pristauz G, Lackner C, Hofler G, Eisner F, Petru E, Sill H, Samonigg H, Pantel K, Riethdorf S, Bauernhofer T, Geigl JB, Speicher MR. Establishment of tumor-specific copy number alterations from plasma DNA of patients with cancer. Int J Cancer. 2013a;133:346–56. doi:10.1002/ijc.28030.
63. Mouliere F, Robert B, Arnau Peyrotte E, Del Rio M, Ychou M, Molina F, Gongora C, Thierry AR. High fragmentation characterizes tumour-derived circulating DNA. PLoS One. 2011;6:e23418. doi:10.1371/journal.pone.0023418.
64. Thierry AR, Mouliere F, Gongora C, Ollier J, Robert B, Ychou M, Del Rio M, Molina F. Origin and quantification of circulating DNA in mice with human colorectal cancer xenografts. Nucleic Acids Res. 2010;38:6159–75. doi:10.1093/nar/gkq421.
65. Chen C, Cram DS, Xie F, Wang P, Xu X, Li H, Song Z, Chen D, Zhang J, Tang S. A pregnancy with discordant fetal and placental chromosome 18 aneuploidies revealed by invasive

and noninvasive prenatal diagnosis. Reprod Biomed Online. 2014;29:136–9. doi:10.1016/j.rbmo.2014.03.017.
66. Masuzaki H, Miura K, Yoshiura K, Yoshimura S, Niikawa N, Ishimaru T. Detection of cell free placental DNA in maternal plasma: direct evidence from three cases of confined placental mosaicism. J Med Genet. 2004;41:289–92.
67. Lo YM, Tein MS, Pang CC, Yeung CK, Tong KL, Hjelm NM. Presence of donor-specific DNA in plasma of kidney and liver-transplant recipients. Lancet. 1998;351:1329–30.
68. Snyder TM, Khush KK, Valantine HA, Quake SR. Universal noninvasive detection of solid organ transplant rejection. Proc Natl Acad Sci U S A. 2011;108:6229–34. doi:10.1073/pnas.1013924108.
69. Sun K, Jiang P, Chan KCA, Wong J, Cheng YKY, Liang RHS, Chan W, Ma ESK, Chan SL, Cheng SH, Chan RWY, Tong YK, Ng SSM, Wong RSM, Hui DSC, Leung TN, Leung TY, Lai PBS, Chiu RWK, Lo YMD. Plasma DNA tissue mapping by genome-wide methylation sequencing for noninvasive prenatal, cancer, and transplantation assessments. Proc Natl Acad Sci U S A. 2015;112:E5503–12. doi:10.1073/pnas.1508736112.
70. Snyder MW, Kircher M, Hill AJ, Daza RM, Shendure J. Cell-free DNA comprises an in vivo nucleosome footprint that informs its tissues-of-origin. Cell. 2016;164:57–68. doi:10.1016/j.cell.2015.11.050.
71. Bettegowda C, Sausen M, Leary RJ, Kinde I, Wang Y, Agrawal N, Bartlett BR, Wang H, Luber B, Alani RM, Antonarakis ES, Azad NS, Bardelli A, Brem H, Cameron JL, Lee CC, Fecher LA, Gallia GL, Gibbs P, Le D, Giuntoli RL, Goggins M, Hogarty MD, Holdhoff M, Hong S-M, Jiao Y, Juhl HH, Kim JJ, Siravegna G, Laheru DA, Lauricella C, Lim M, Lipson EJ, Marie SKN, Netto GJ, Oliner KS, Olivi A, Olsson L, Riggins GJ, Sartore-Bianchi A, Schmidt K, Shih l-M, Oba-Shinjo SM, Siena S, Theodorescu D, Tie J, Harkins TT, Veronese S, Wang T-L, Weingart JD, Wolfgang CL, Wood LD, Xing D, Hruban RH, Wu J, Allen PJ, Schmidt CM, Choti MA, Velculescu VE, Kinzler KW, Vogelstein B, Papadopoulos N, Diaz LA. Detection of circulating tumor DNA in early- and late-stage human malignancies. Sci Transl Med. 2014;6:224ra24. doi:10.1126/scitranslmed.3007094.
72. Leary RJ, Sausen M, Kinde I, Papadopoulos N, Carpten JD, Craig D, O'Shaughnessy J, Kinzler KW, Parmigiani G, Vogelstein B, Diaz LA, Velculescu VE. Detection of chromosomal alterations in the circulation of cancer patients with whole-genome sequencing. Sci Transl Med. 2012;4:162ra154. doi:10.1126/scitranslmed.3004742.
73. Thierry AR, Mouliere F, El Messaoudi S, Mollevi C, Lopez-Crapez E, Rolet F, Gillet B, Gongora C, Dechelotte P, Robert B, Del Rio M, Lamy P-J, Bibeau F, Nouaille M, Loriot V, Jarrousse A-S, Molina F, Mathonnet M, Pezet D, Ychou M. Clinical validation of the detection of KRAS and BRAF mutations from circulating tumor DNA. Nat Med. 2014;20:430–5. doi:10.1038/nm.3511.
74. Lo YM, Zhang J, Leung TN, Lau TK, Chang AM, Hjelm NM. Rapid clearance of fetal DNA from maternal plasma. Am J Hum Genet. 1999;64:218–24. doi:10.1086/302205.
75. Yu SCY, Lee SWY, Jiang P, Leung TY, Chan KCA, Chiu RWK, Lo YMD. High-resolution profiling of fetal DNA clearance from maternal plasma by massively parallel sequencing. Clin Chem. 2013;59:1228–37. doi:10.1373/clinchem.2013.203679.
76. Lee TH, Montalvo L, Chrebtow V, Busch MP. Quantitation of genomic DNA in plasma and serum samples: higher concentrations of genomic DNA found in serum than in plasma. Transfusion. 2001;41:276–82.
77. Jung M, Klotzek S, Lewandowski M, Fleischhacker M, Jung K. Changes in concentration of DNA in serum and plasma during storage of blood samples. Clin Chem. 2003;49:1028–9.
78. Wong D, Moturi S, Angkachatchai V, Mueller R, DeSantis G, van den Boom D, Ehrich M. Optimizing blood collection, transport and storage conditions for cell free DNA increases access to prenatal testing. Clin Biochem. 2013;46:1099–104. doi:10.1016/j.clinbiochem.2013.04.023.
79. Breitbach S, Tug S, Helmig S, Zahn D, Kubiak T, Michal M, Gori T, Ehlert T, Beiter T, Simon P. Direct quantification of cell-free, circulating DNA from unpurified plasma. PLoS One. 2014;9:e87838. doi:10.1371/journal.pone.0087838.

80. Manokhina I, Singh TK, Peñaherrera MS, Robinson WP. Quantification of cell-free DNA in normal and complicated pregnancies: overcoming biological and technical issues. PLoS One. 2014;9:e101500. doi:10.1371/journal.pone.0101500.
81. Hidestrand M, Stokowski R, Song K, Oliphant A, Deavers J, Goetsch M, Simpson P, Kuhlman R, Ames M, Mitchell M, Tomita-Mitchell A. Influence of temperature during transportation on cell-free DNA analysis. Fetal Diagn Ther. 2012;31:122–8. doi:10.1159/000335020.
82. Repiská G, Sedláčková T, Szemes T, Celec P, Minárik G. Selection of the optimal manual method of cell free fetal DNA isolation from maternal plasma. Clin Chem Lab Med. 2013;51:1185–9. doi:10.1515/cclm-2012-0418.
83. Hromadnikova I, Zejskova L, Kotlabova K, Jancuskova T, Doucha J, Dlouha K, Krofta L, Jirasek JE, Vlk R. Quantification of extracellular DNA using hypermethylated RASSF1A, SRY, and GLO sequences—evaluation of diagnostic possibilities for predicting placental insufficiency. DNA Cell Biol. 2010;29:295–301. doi:10.1089/dna.2009.0971.
84. Spindler K-LG, Pallisgaard N, Vogelius I, Jakobsen A. Quantitative cell-free DNA, KRAS, and BRAF mutations in plasma from patients with metastatic colorectal cancer during treatment with cetuximab and irinotecan. Clin Cancer Res. 2012;18:1177–85. doi:10.1158/1078-0432.CCR-11-0564.
85. Diehl F, Li M, Dressman D, He Y, Shen D, Szabo S, Diaz LA, Goodman SN, David KA, Juhl H, Kinzler KW, Vogelstein B. Detection and quantification of mutations in the plasma of patients with colorectal tumors. Proc Natl Acad Sci U S A. 2005;102:16368–73. doi:10.1073/pnas.0507904102.
86. Taly V, Pekin D, Benhaim L, Kotsopoulos SK, Le Corre D, Li X, Atochin I, Link DR, Griffiths AD, Pallier K, Blons H, Bouché O, Landi B, Hutchison JB, Laurent-Puig P. Multiplex picodroplet digital PCR to detect KRAS mutations in circulating DNA from the plasma of colorectal cancer patients. Clin Chem. 2013;59:1722–31. doi:10.1373/clinchem.2013.206359.
87. Diehl F, Schmidt K, Choti MA, Romans K, Goodman S, Li M, Thornton K, Agrawal N, Sokoll L, Szabo SA, Kinzler KW, Vogelstein B, Diaz LA. Circulating mutant DNA to assess tumor dynamics. Nat Med. 2008;14:985–90. doi:10.1038/nm.1789.
88. Olsson E, Winter C, George A, Chen Y, Howlin J, Tang M-HE, Dahlgren M, Schulz R, Grabau D, van Westen D, Fernö M, Ingvar C, Rose C, Bendahl P-O, Rydén L, Borg Å, Gruvberger-Saal SK, Jernström H, Saal LH. Serial monitoring of circulating tumor DNA in patients with primary breast cancer for detection of occult metastatic disease. EMBO Mol Med. 2015;7:1034–47. doi:10.15252/emmm.201404913.
89. Heitzer E, Ulz P, Belic J, Gutschi S, Quehenberger F, Fischereder K, Benezeder T, Auer M, Pischler C, Mannweiler S, Pichler M, Eisner F, Haeusler M, Riethdorf S, Pantel K, Samonigg H, Hoefler G, Augustin H, Geigl JB, Speicher MR. Tumor-associated copy number changes in the circulation of patients with prostate cancer identified through whole-genome sequencing. Genome Med. 2013c;5:30. doi:10.1186/gm434.
90. Newman AM, Bratman SV, To, J, Wynne JF, Eclov NCW, Modlin LA, Liu CL, Neal JW, Wakelee HA, Merritt RE, Shrager JB, Loo BW, Alizadeh AA, Diehn M. An ultrasensitive method for quantitating circulating tumor DNA with broad patient coverage. Nat Med. 2014;20:548–54. doi:10.1038/nm.3519.
91. Forshew T, Murtaza M, Parkinson C, Gale D, Tsui DWY, Kaper F, Dawson S-J, Piskorz AM, Jimenez-Linan M, Bentley D, Hadfield J, May AP, Caldas C, Brenton JD, Rosenfeld N. Noninvasive identification and monitoring of cancer mutations by targeted deep sequencing of plasma DNA. Sci Transl Med. 2012;4:136ra68. doi:10.1126/scitranslmed.3003726.
92. Chan KCA, Jiang P, Zheng YWL, Liao GJW, Sun H, Wong J, Siu SSN, Chan WC, Chan SL, Chan ATC, Lai PBS, Chiu RWK, Lo YMD. Cancer genome scanning in plasma: detection of tumor-associated copy number aberrations, single-nucleotide variants, and tumoral heterogeneity by massively parallel sequencing. Clin Chem. 2013b;59:211–24. doi:10.1373/clinchem.2012.196014.
93. Chan KCA, Jiang P, Chan CWM, Sun K, Wong J, Hui EP, Chan SL, Chan WC, Hui DSC, Ng SSM, Chan HLY, Wong CSC, Ma BBY, Chan ATC, Lai PBS, Sun H, Chiu RWK, Lo YMD. Noninvasive detection of cancer-associated genome-wide hypomethylation and copy

number aberrations by plasma DNA bisulfite sequencing. Proc Natl Acad Sci U S A. 2013a;110:18761–8. doi:10.1073/pnas.1313995110.
94. Heitzer E, Auer M, Ulz P, Geigl JB, Speicher MR. Circulating tumor cells and DNA as liquid biopsies. Genome Med. 2013b;5:73. doi:10.1186/gm477.
95. Heidary M, Auer M, Ulz P, Heitzer E, Petru E, Gasch C, Riethdorf S, Mauermann O, Lafer I, Pristauz G, Lax S, Pantel K, Geigl JB, Speicher MR. The dynamic range of circulating tumor DNA in metastatic breast cancer. Breast Cancer Res. 2014;16:421. doi:10.1186/s13058-014-0421-y.
96. Belic J, Koch M, Ulz P, Auer M, Gerhalter T, Mohan S, Fischereder K, Petru E, Bauernhofer T, Geigl JB, Speicher MR, Heitzer E. Rapid identification of plasma DNA samples with increased ctDNA levels by a modified FAST-SeqS approach. Clin Chem. 2015;61:838–49. doi:10.1373/clinchem.2014.234286.
97. Frattini M, Gallino G, Signoroni S, Balestra D, Lusa L, Battaglia L, Sozzi G, Bertario L, Leo E, Pilotti S, Pierotti MA. Quantitative and qualitative characterization of plasma DNA identifies primary and recurrent colorectal cancer. Cancer Lett. 2008;263:170–81. doi:10.1016/j.canlet.2008.03.021.
98. Kim K, Shin DG, Park MK, Baik SH, Kim TH, Kim S, Lee S. Circulating cell-free DNA as a promising biomarker in patients with gastric cancer: diagnostic validity and significant reduction of cfDNA after surgical resection. Ann Surg Treat Res. 2014;86:136–42. doi:10.4174/astr.2014.86.3.136.
99. Neumann L, Freund KG, Sørenson KH. Long-term results of Charnley total hip replacement. Review of 92 patients at 15 to 20 years. J Bone Joint Surg Br. 1994;76:245–51.
100. Sozzi G, Conte D, Mariani L, Lo Vullo S, Roz L, Lombardo C, Pierotti MA, Tavecchio L. Analysis of circulating tumor DNA in plasma at diagnosis and during follow-up of lung cancer patients. Cancer Res. 2001;61:4675–8.
101. Dawson S-J, Tsui DWY, Murtaza M, Biggs H, Rueda OM, Chin S-F, Dunning MJ, Gale D, Forshew T, Mahler-Araujo B, Rajan S, Humphray S, Becq J, Halsall D, Wallis M, Bentley D, Caldas C, Rosenfeld N. Analysis of circulating tumor DNA to monitor metastatic breast cancer. N Engl J Med. 2013;368:1199–209. doi:10.1056/NEJMoa1213261.
102. Sorensen BS, Wu L, Wei W, Tsai J, Weber B, Nexo E, Meldgaard P. Monitoring of epidermal growth factor receptor tyrosine kinase inhibitor-sensitizing and resistance mutations in the plasma DNA of patients with advanced non-small cell lung cancer during treatment with erlotinib: monitoring EGFR mutations in plasma DNA. Cancer. 2014;120:3896–901. doi:10.1002/cncr.28964.
103. Jiang T, Ren S, Zhou C. Role of circulating-tumor DNA analysis in non-small cell lung cancer. Lung Cancer. 2015;90:128–34. doi:10.1016/j.lungcan.2015.09.013.
104. Oxnard GR, Paweletz CP, Kuang Y, Mach SL, O'Connell A, Messineo MM, Luke JJ, Butaney M, Kirschmeier P, Jackman DM, Jänne PA. Noninvasive detection of response and resistance in EGFR-mutant lung cancer using quantitative next-generation genotyping of cell-free plasma DNA. Clin Cancer Res. 2014;20:1698–705. doi:10.1158/1078-0432.CCR-13-2482.
105. Qiu M, Wang J, Xu Y, Ding X, Li M, Jiang F, Xu L, Yin R. Circulating tumor DNA is effective for the detection of EGFR mutation in non-small cell lung cancer: a meta-analysis. Cancer Epidemiol Biomarkers Prev. 2015;24:206–12. doi:10.1158/1055-9965.EPI-14-0895.
106. Shaib W, Mahajan R, El-Rayes B. Markers of resistance to anti-EGFR therapy in colorectal cancer. J Gastrointest Oncol. 2013;4:308–18. doi:10.3978/j.issn.2078-6891.2013.029.
107. Diaz LA, Williams RT, Wu J, Kinde I, Hecht JR, Berlin J, Allen B, Bozic I, Reiter JG, Nowak MA, Kinzler KW, Oliner KS, Vogelstein B. The molecular evolution of acquired resistance to targeted EGFR blockade in colorectal cancers. Nature. 2012;486:537–40. doi:10.1038/nature11219.
108. Azad AA, Volik SV, Wyatt AW, Haegert A, Le Bihan S, Bell RH, Anderson SA, McConeghy B, Shukin R, Bazov J, Youngren J, Paris P, Thomas G, Small EJ, Wang Y, Gleave ME, Collins CC, Chi KN. Androgen receptor gene aberrations in circulating cell-free DNA: biomarkers of therapeutic resistance in castration-resistant prostate cancer. Clin Cancer Res. 2015;21:2315–24. doi:10.1158/1078-0432.CCR-14-2666.

109. Mohan S, Heitzer E, Ulz P, Lafer I, Lax S, Auer M, Pichler M, Gerger A, Eisner F, Hoefler G, Bauernhofer T, Geigl JB, Speicher MR. Changes in colorectal carcinoma genomes under anti-EGFR therapy identified by whole-genome plasma DNA sequencing. PLoS Genet. 2014;10:e1004271. doi:10.1371/journal.pgen.1004271.
110. Heitzer E, Ulz P, Geigl JB, Speicher MR. Non-invasive detection of genome-wide somatic copy number alterations by liquid biopsies. Mol Oncol. 2016;10:494–502. doi:10.1016/j.molonc.2015.12.004.
111. Tie J, Kinde I, Wang Y, Wong HL, Roebert J, Christie M, Tacey M, Wong R, Singh M, Karapetis CS, Desai J, Tran B, Strausberg RL, Diaz LA, Papadopoulos N, Kinzler KW, Vogelstein B, Gibbs P. Circulating tumor DNA as an early marker of therapeutic response in patients with metastatic colorectal cancer. Ann Oncol. 2015;26:1715–22. doi:10.1093/annonc/mdv177.
112. Kinde I, Wu J, Papadopoulos N, Kinzler KW, Vogelstein B. Detection and quantification of rare mutations with massively parallel sequencing. Proc Natl Acad Sci U S A. 2011;108:9530–5. doi:10.1073/pnas.1105422108.
113. Murtaza M, Dawson S-J, Tsui DWY, Gale D, Forshew T, Piskorz AM, Parkinson C, Chin S-F, Kingsbury Z, Wong ASC, Marass F, Humphray S, Hadfield J, Bentley D, Chin TM, Brenton JD, Caldas C, Rosenfeld N. Non-invasive analysis of acquired resistance to cancer therapy by sequencing of plasma DNA. Nature. 2013;497:108–12. doi:10.1038/nature12065.
114. Arroyo JD, Chevillet JR, Kroh EM, Ruf IK, Pritchard CC, Gibson DF, Mitchell PS, Bennett CF, Pogosova-Agadjanyan EL, Stirewalt DL, Tait JF, Tewari M. Argonaute2 complexes carry a population of circulating microRNAs independent of vesicles in human plasma. Proc Natl Acad Sci. 2011;108:5003–8. doi:10.1073/pnas.1019055108.
115. Zhang J, Li S, Li L, Li M, Guo C, Yao J, Mi S. Exosome and exosomal microRNA: trafficking, sorting, and function. Genomics Proteomics Bioinformatics. 2015;13:17–24. doi:10.1016/j.gpb.2015.02.001.
116. Tsui NBY, Ng EKO, Lo YMD. Stability of endogenous and added RNA in blood specimens, serum, and plasma. Clin Chem. 2002;48:1647–53.
117. Kodahl AR, Lyng MB, Binder H, Cold S, Gravgaard K, Knoop AS, Ditzel HJ. Novel circulating microRNA signature as a potential non-invasive multi-marker test in ER-positive early-stage breast cancer: a case control study. Mol Oncol. 2014;8:874–83. doi:10.1016/j.molonc.2014.03.002.
118. Ng EKO, Tsui NBY, Lam NYL, Chiu RWK, Yu SCH, Wong SCC, Lo ESF, Rainer TH, Johnson PJ, Lo YMD. Presence of filterable and nonfilterable mRNA in the plasma of cancer patients and healthy individuals. Clin Chem. 2002;48:1212–7.
119. Al-Nedawi K, Meehan B, Micallef J, Lhotak V, May L, Guha A, Rak J. Intercellular transfer of the oncogenic receptor EGFRvIII by microvesicles derived from tumour cells. Nat Cell Biol. 2008;10:619–24. doi:10.1038/ncb1725.
120. Iero M, Valenti R, Huber V, Filipazzi P, Parmiani G, Fais S, Rivoltini L. Tumour-released exosomes and their implications in cancer immunity. Cell Death Differ. 2008;15:80–8. doi:10.1038/sj.cdd.4402237.
121. Kosaka N, Iguchi H, Ochiya T. Circulating microRNA in body fluid: a new potential biomarker for cancer diagnosis and prognosis. Cancer Sci. 2010;101:2087–92. doi:10.1111/j.1349-7006.2010.01650.x.
122. Skog J, Würdinger T, van Rijn S, Meijer DH, Gainche L, Sena-Esteves M, Curry WT, Carter BS, Krichevsky AM, Breakefield XO. Glioblastoma microvesicles transport RNA and proteins that promote tumour growth and provide diagnostic biomarkers. Nat Cell Biol. 2008;10:1470–6. doi:10.1038/ncb1800.
123. Taylor DD, Gercel-Taylor C. MicroRNA signatures of tumor-derived exosomes as diagnostic biomarkers of ovarian cancer. Gynecol Oncol. 2008;110:13–21. doi:10.1016/j.ygyno.2008.04.033.
124. Webber J, Steadman R, Mason MD, Tabi Z, Clayton A. Cancer exosomes trigger fibroblast to myofibroblast differentiation. Cancer Res. 2010;70:9621–30. doi:10.1158/0008-5472.CAN-10-1722.

125. Thakur BK, Zhang H, Becker A, Matei I, Huang Y, Costa-Silva B, Zheng Y, Hoshino A, Brazier H, Xiang J, Williams C, Rodriguez-Barrueco R, Silva JM, Zhang W, Hearn S, Elemento O, Paknejad N, Manova-Todorova K, Welte K, Bromberg J, Peinado H, Lyden D. Double-stranded DNA in exosomes: a novel biomarker in cancer detection. Cell Res. 2014;24:766–9. doi:10.1038/cr.2014.44.
126. Vickers KC, Palmisano BT, Shoucri BM, Shamburek RD, Remaley AT. MicroRNAs are transported in plasma and delivered to recipient cells by high-density lipoproteins. Nat Cell Biol. 2011;13:423–33. doi:10.1038/ncb2210.
127. Mitchell PS, Parkin RK, Kroh EM, Fritz BR, Wyman SK, Pogosova-Agadjanyan EL, Peterson A, Noteboom J, O'Briant KC, Allen A, Lin DW, Urban N, Drescher CW, Knudsen BS, Stirewalt DL, Gentleman R, Vessella RL, Nelson PS, Martin DB, Tewari M. Circulating microRNAs as stable blood-based markers for cancer detection. Proc Natl Acad Sci. 2008;105:10513–8. doi:10.1073/pnas.0804549105.
128. Moldovan L, Batte K, Wang Y, Wisler J, Piper M. Analyzing the circulating microRNAs in exosomes/extracellular vesicles from serum or plasma by qRT-PCR. In: Kosaka N, editor. Circulating microRNAs, Springer protocols. New York: Humana Press; 2013. p. 129–45.
129. Wang K, Yuan Y, Cho J-H, McClarty S, Baxter D, Galas DJ, Ahuja SK. Comparing the MicroRNA spectrum between serum and plasma. PLoS One. 2012;7:e41561. doi:10.1371/journal.pone.0041561.
130. Sourvinou IS, Markou A, Lianidou ES. Quantification of circulating miRNAs in plasma. J Mol Diagn. 2013;15:827–34. doi:10.1016/j.jmoldx.2013.07.005.
131. Grasedieck S, Schöler N, Bommer M, Niess JH, Tumani H, Rouhi A, Bloehdorn J, Liebisch P, Mertens D, Döhner H, Buske C, Langer C, Kuchenbauer F. Impact of serum storage conditions on microRNA stability. Leukemia. 2012;26:2414–6. doi:10.1038/leu.2012.106.
132. Mraz M, Malinova K, Mayer J, Pospisilova S. MicroRNA isolation and stability in stored RNA samples. Biochem Biophys Res Commun. 2009;390:1–4. doi:10.1016/j.bbrc.2009.09.061.
133. Bravo V, Rosero S, Ricordi C, Pastori RL. Instability of miRNA and cDNAs derivatives in RNA preparations. Biochem Biophys Res Commun. 2007;353:1052–5. doi:10.1016/j.bbrc.2006.12.135.
134. Moldovan L, Batte KE, Trgovcich J, Wisler J, Marsh CB, Piper M. Methodological challenges in utilizing miRNAs as circulating biomarkers. J Cell Mol Med. 2014;18:371–90. doi:10.1111/jcmm.12236.
135. Ma W, Wang M, Wang Z-Q, Sun L, Graber D, Matthews J, Champlin R, Yi Q, Orlowski RZ, Kwak LW, Weber DM, Thomas SK, Shah J, Kornblau S, Davis RE. Effect of long-term storage in TRIzol on microarray-based gene expression profiling. Cancer Epidemiol Biomarkers Prev. 2010;19:2445–52. doi:10.1158/1055-9965.EPI-10-0565.
136. Kim Y-K, Yeo J, Kim B, Ha M, Kim VN. Short structured RNAs with low GC content are selectively lost during extraction from a small number of cells. Mol Cell. 2012;46:893–5. doi:10.1016/j.molcel.2012.05.036.
137. Pritchard CC, Cheng HH, Tewari M. MicroRNA profiling: approaches and considerations. Nat Rev Genet. 2012;13:358–69. doi:10.1038/nrg3198.
138. Hydbring P, Badalian-Very G. Clinical applications of microRNAs. F1000Research. 2013. doi:10.12688/f1000research.2–136.v3.
139. Griffiths-Jones S, Saini HK, van Dongen S, Enright AJ. miRBase: tools for microRNA genomics. Nucleic Acids Res. 2008;36:D154–8. doi:10.1093/nar/gkm952.
140. Lawrie CH, Gal S, Dunlop HM, Pushkaran B, Liggins AP, Pulford K, Banham AH, Pezzella F, Boultwood J, Wainscoat JS, Hatton CSR, Harris AL. Detection of elevated levels of tumour-associated microRNAs in serum of patients with diffuse large B-cell lymphoma. Br J Haematol. 2008;141:672–5. doi:10.1111/j.1365-2141.2008.07077.x.
141. Volinia S, Calin GA, Liu C-G, Ambs S, Cimmino A, Petrocca F, Visone R, Iorio M, Roldo C, Ferracin M, Prueitt RL, Yanaihara N, Lanza G, Scarpa A, Vecchione A, Negrini M, Harris CC, Croce CM. A microRNA expression signature of human solid tumors defines cancer gene targets. Proc Natl Acad Sci U S A. 2006;103:2257–61. doi:10.1073/pnas.0510565103.

142. Boeri M, Verri C, Conte D, Roz L, Modena P, Facchinetti F, Calabro E, Croce CM, Pastorino U, Sozzi G. MicroRNA signatures in tissues and plasma predict development and prognosis of computed tomography detected lung cancer. Proc Natl Acad Sci. 2011;108:3713–8. doi:10.1073/pnas.1100048108.
143. Keller A, Leidinger P, Bauer A, ElSharawy A, Haas J, Backes C, Wendschlag A, Giese N, Tjaden C, Ott K, Werner J, Hackert T, Ruprecht K, Huwer H, Huebers J, Jacobs G, Rosenstiel P, Dommisch H, Schaefer A, Müller-Quernheim J, Wullich B, Keck B, Graf N, Reichrath J, Vogel B, Nebel A, Jager SU, Staehler P, Amarantos I, Boisguerin V, Staehler C, Beier M, Scheffler M, Büchler MW, Wischhusen J, Haeusler SFM, Dietl J, Hofmann S, Lenhof H-P, Schreiber S, Katus HA, Rottbauer W, Meder B, Hoheisel JD, Franke A, Meese E. Toward the blood-borne miRNome of human diseases. Nat Methods. 2011;8:841–3. doi:10.1038/nmeth.1682.
144. Tiberio P, Callari M, Angeloni V, Daidone MG, Appierto V, Tiberio P, Callari M, Angeloni V, Daidone MG, Appierto V. Challenges in using circulating miRNAs as cancer biomarkers. Biomed Res Int. 2015;2015:e731479. doi:10.1155/2015/731479.

Computational Pathology 13

Peter J. Schüffler, Qing Zhong, Peter J. Wild, and Thomas J. Fuchs

Contents

Abstract

Computational pathology offers a comprehensive framework for advanced study design in a wide range of research questions, as well as for standardized pipeline development for fast and reproducible computer-assisted routine diagnostics. This new field emerges at the border of pathology and computer science and

P.J. Schüffler • T.J. Fuchs (✉)
Memorial Sloan Kettering Cancer Center, 417 E 68th St, New York, NY 10065, USA
e-mail: schueffp@mskcc.org; fuchst@mskcc.org

Q. Zhong • P.J. Wild
Department of Pathology and Molecular Pathology, University Hospital Zurich, Schmelzbergstr. 12, Zurich 8091, Switzerland
e-mail: qing.zhong@usz.ch; peter.wild@usz.ch

J. Haybaeck (ed.), *Mechanisms of Molecular Carcinogenesis – Volume 2*,
DOI 10.1007/978-3-319-53661-3_13

shows high potential to revolutionize established workflows in research and clinic, since not only computational models get faster and more efficient than before but also since an incredible amount of training data is being generated in modern hospitals which is mandatory for the training of informed and validated models.

We review the field of computational pathology and illustrate on two research examples how it will contribute to an accurate, objective, and reproducible study design comprising informed data acquisition, advanced pattern recognition, and transparent model validation.

13.1 Introduction

In the intersection of pathology and computer science, statistics and computational biology emerged as an exciting new interdisciplinary field of natural sciences, which aims to fuse techniques from both disciplines for advanced medical research (see Fig. 13.1). This chapter outlines this field of *computational pathology* with its promises and limits and illustrates two typical applications of computational pathology in nowadays cancer research.

Early work in *computational pathology* started in 2008 by Fuchs et al. [1], who further defined the term in 2011 as a holistic strategy for probabilistic pathology research. As sketched in Fig. 13.2, it comprises strategic planning for *data acquisition* and labeling (top row), computer vision methods and supervised and unsupervised machine learning approaches for *image analysis* and outcome prediction (middle row), as well as model validation and *survival analysis* for a given target outcome on large patient cohorts (bottom row) [2]. In this sense, computational pathology

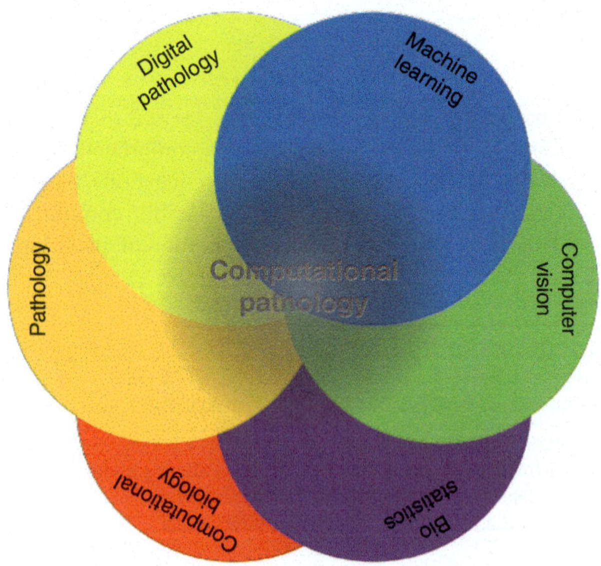

Fig. 13.1 Computational pathology finds itself at the overlap of pathology, digital pathology, computer science with mainly machine learning and computer vision, statistics, and computational biology

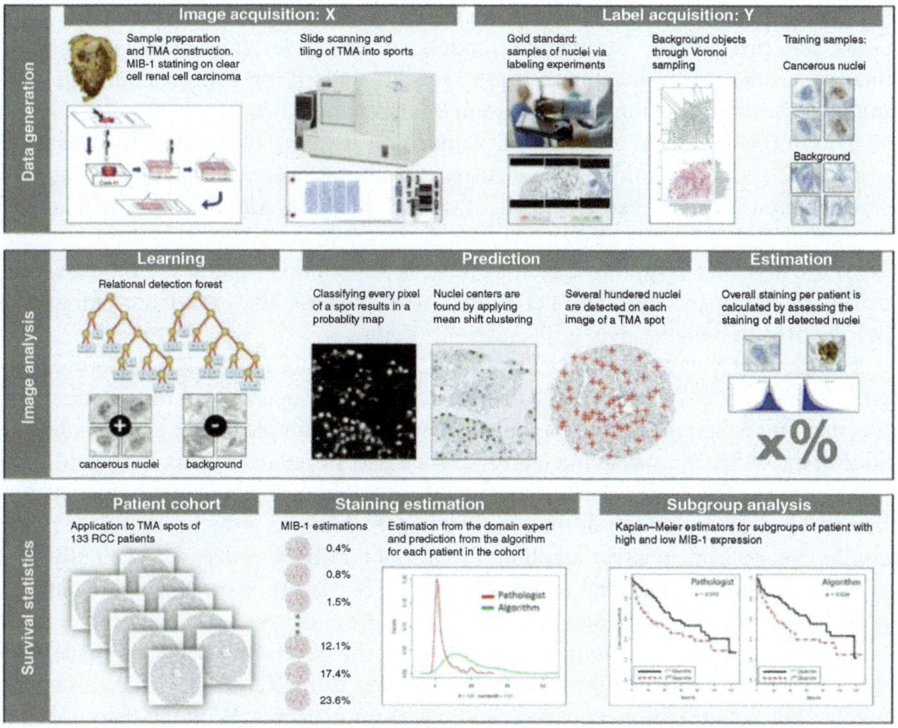

Fig. 13.2 Definition of computational pathology: "Computational pathology investigates a complete probabilistic treatment of scientific and clinical workflows in general pathology, i.e. it combines experimental design (*top*), statistical pattern recognition (*middle*) and survival analysis (*bottom*) within a unified framework to answer scientific and clinical questions in pathology" [2]

provides a rich collection of tools and techniques for a comprehensive study design for sophisticated research questions in pathology. Other definitions came up in the following years, and all share the idea of computational pathology to be a holistic approach from data generation and processing, over automated learning and modeling of data from multiple sources for the best possible medical decision [3, 4].

A big breakthrough for computational pathology was the facilitated digitizing of pathology tissue slides by high-resolution slide scanners that serve simultaneously as digital microscopes, scanners, and data servers. Whereas classical pathology was (and still is) completely analogue with by-eye examinations with the microscope and with handwritten medical reports, the digitizing of these scans and reports allowed for the first time for access of advanced computer vision and machine learning algorithms to those data [2, 5]. The research field of *digital pathology* investigates new methods for quality control and standardization of digital pathology slides [6, 7], as well as new methodologies for image acquisition and registration [8, 9].

Machine learning enables us to learn distinct patterns in large collections of data, which might be nonobvious or nonintuitive by the human eye. Particularly in cancer research, where large data sets can get very complex, machine learning helps to structure and model these data to make informed, objective, and reproducible decisions and predictions. Further, as models of the underlying biology might be unknown, data-driven machine learning reveals underlying patterns, thus supporting hypothesis generation and validation on real patient data.

The improvement of a manifold of machine learning techniques in the recent years (e.g., reducing computational costs, enhancing computational power), as well as the increase of scanned tissue slices over the last decade, led to an increase of computational pathology-driven research design in many institutes.

The University Hospital Zurich (USZ), as an example, increased the research-driven scanning job numbers per month over the last 10 years from below 500 before 2011 to recently over 1500 scans or 50/day.

Further, the Memorial Sloan Kettering Cancer Center (MSKCC) performs digital scanning not only for research projects but also for operational routine diagnostics, leading to much higher scanning volume. For example, every second-opinion slide at the MSKCC is systematically scanned and stored over years for recall. Also other tissue slides are scanned. Therefore, the number of digital slides at this institution reaches 17000/month at the moment and is going to grow up to 40000 in the coming years. This massive amount of data can hardly be processed manually, and at the same time, it embodies a rich source of necessary (potential) training data for generic and widely validated algorithms.

A major goal in computational pathology is the development of automated and standardized methods for the holistic assessment of digital tissue images, such as whole-slide images, biopsy images, and tissue microarrays. With the help of state-of-the-art algorithms from computer vision and machine learning, the analysis of these highly complex image data is aimed to be more objective and reproducible while at the same time even faster and cheaper compared to manual examination by trained human specialists.

For pathologists, on the other hand, computational pathology will be a tool for computer-assisted and computer-informed decision support [10]. The increasing complexity of medical data urges computer-assisted technologies for fast and accurate assessment of individual cases. With the tools of computational pathology helping in routine diagnostics, pathologists will have more time to concentrate on their underlying research questions and on the interesting borderline cases.

Computational pathology has high impact in current medical research and manifold applications, two of which we outline in the next sections: an automated staining estimation pipeline for renal cell carcinoma and computational ISH assessment for tumor heterogeneity detection.

13.2 Staining Estimation Pipeline for Clear Cell Renal Cell Carcinoma

Clear cell renal cell carcinoma (ccRCC) is a special form of cancer of the kidney where the cancer cells have a mostly clear cytoplasm. The proliferation rate of those cancer cells is an important prognostic indicator for survival of the patient [11]. To

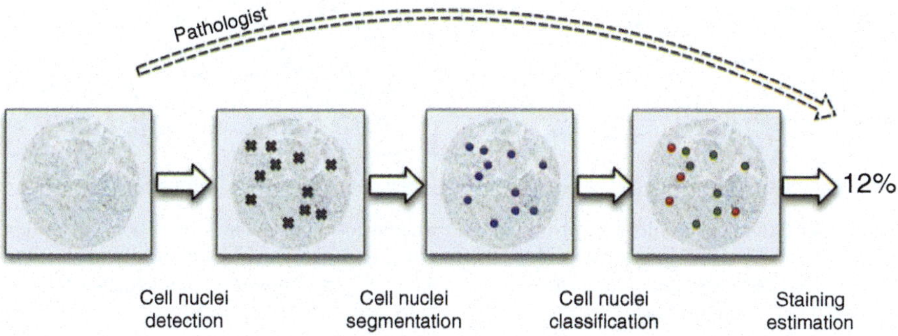

Fig. 13.3 Automated staining estimation pipeline for ccRCC. Starting from a TMA sample, a pathologist would estimate the relative amount of MIB-1-positive (proliferating) cancer cells (*dashed arrow*). The corresponding computational approach partitions this workflow into the four subtasks: (1) nucleus detection (Sect. 13.2.1), (2) nucleus segmentation (Sect. 13.2.2), (3) classification into cancer/noncancer (Sect. 13.2.3), and (4) staining estimation (Sect. 13.2.4) [13]

support this hypothesis, a tissue microarray (TMA) of 133 ccRCC samples has been stained with the marker MIB-1 binding at the proliferation-associated nuclear protein Ki-67. This immunostaining reveals proliferating cells with MIB-1-positive nuclei. A hematoxylin counterstaining shows negative nuclei of the cells in the surrounding tissue. Together with the clinical records of survival, the hypothesis can be tested.

After the staining, the *staining estimation* is the process to determine the relative amount of stained cancer cells in a given specimen. An important aspect is that the estimate is only relevant on the subset of cancer cells and not on other cells (normal, connective tissue, and others) which might be present in the same specimen. The differentiation of cancer cells from others cannot be done by staining and is usually visually done by experienced pathologists counting or estimating the relative amount of stained cancer cells. It has been shown that such estimates can be very variable among and within experts [2].

To better standardize this task, a computational pipeline for the staining estimation has been proposed by Fuchs et al. [1] and Schüffler et al. [13, 14] as outlined in Fig. 13.3. This pipeline partitions the staining estimation into the four steps: (1) nucleus detection (Sect. 13.2.1), (2) nucleus segmentation (Sect. 13.2.2), (3) classification into cancer/noncancer (Sect. 13.2.3), and (4) staining estimation (Sect. 13.2.4).

13.2.1 Nucleus Detection

Cell nuclei in immunohistochemically stained tissue images are bluish or dark brownish objects (depending on their stain) in an environment of bluish cytoplasm, cell membranes, and other cellular and subcellular structures. The vast amount of different nuclear shapes and sizes, the over- and under-staining of parts of the image,

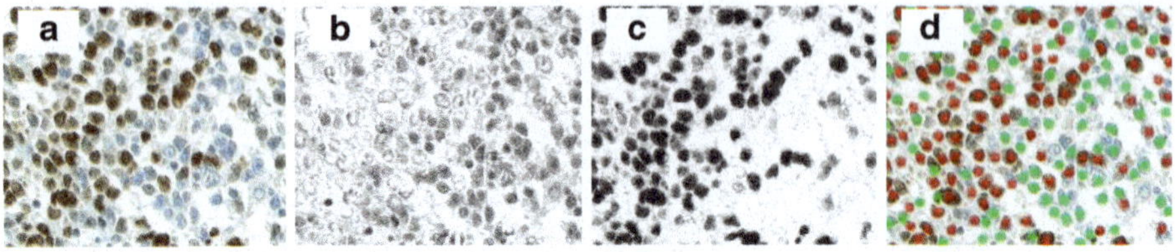

Fig. 13.4 Example of color deconvolution and watershed-based nucleus detection. (**a**) Original MIB-1 staining ccRCC image patch. (**b**) Hematoxylin channel after color deconvolution. (**c**) DAB channel after color deconvolution. (**d**) Detected nuclei on both channels (*green*, hematoxylin channel; *red*, DAB channel) [14]

and the variability of image and tissue quality among samples make automated object detection problematic. Further, the underlying tissue slice is three-dimensional, and the cell nuclei can be out of focus or cut through their body, which complicates automated nucleus detection further.

An unsupervised method for cell nucleus detection is color deconvolution followed by watershed segmentation (see Fig. 13.4) [14]. The color deconvolution algorithm [15] separates the bluish-stained image channel from the brown nucleus channel (Fig. 13.4b, c), at the same time reducing differently stained background structures. The resulting channel images can be interpreted as probability maps for cell nuclei. After smoothing the individual channel images, a watershed-based maximum finder will find the centroids of the nuclei (Fig. 13.4d) [14].

Recent supervised approaches incorporated random forests and specialized forms of them, such as, e.g., randomized tree ensembles [16], to generate advanced probability maps for nucleus detection. The idea is to classify every pixel in a window-shifting approach into foreground (cell nucleus) and background (other structures). Challenging is then the development of highly predictive image characteristics, such as intensity-based or shape-based features for accurate prediction.

To overcome the design of handcraft features for pixel-wise classification, newer approaches for cell detection use deep learning and convolutional neural networks (CNN) [17]. Here, high-order features are learned by a network of connected *neurons* which are hierarchically stacked in successive layers. Although CNN showed promising results a bunch of image-based classification tasks, their drawback still is the relatively large amount of required training samples and high computation time during training due to the immense number of fitted parameters.

13.2.2 Nucleus Segmentation

For subsequent nucleus classification, a proper nucleus segmentation is important to separate the nucleus from its surrounding tissue as well as to collect nucleus shape-based features. Shape features might play an important role as RCC cells tend to have irregular shapes, whereas healthy cells usually are roundish and smooth.

Several ways for nucleus segmentation exist, such as pixel-wise clustering or pixel-wise classification. We outline here an energy minimization method with

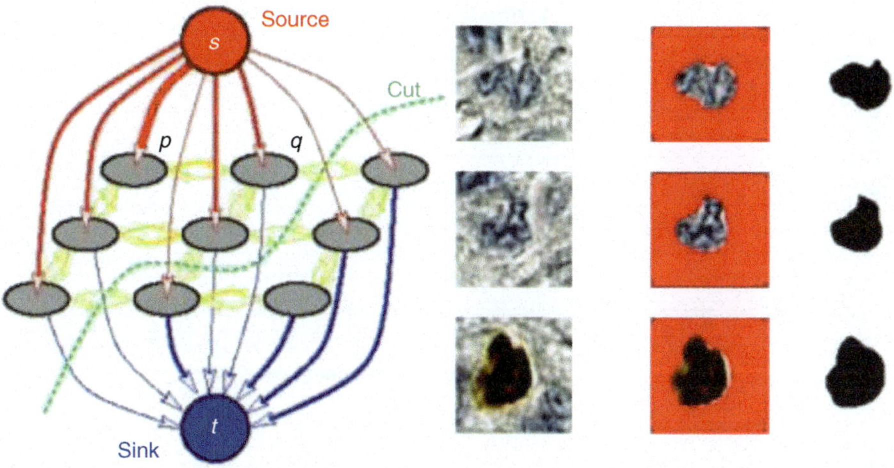

Fig. 13.5 Graph cuts for nucleus segmentation. *Left*: each image pixel is connected to a source node and a sink node. The weights are given by the intensity. A minimum cut partitions the graph into foreground and background. *Right*: three examples of segmented nucleus patches [19]

graph cuts [19]. For this, the image pixels of a smoothed gray-scaled nucleus patch are interpreted as a rectangular graph with every pixel being connected to its four neighbors. The weights of the connections are given by the intensity difference. Also, every pixel is connected to a foreground node, called *source*, and a background node, called *sink*. The weights of the pixels to these two nodes are given by the pixel intensity (see Fig. 13.5) and roundish shape priors [13].

The graph is then partitioned by a minimum cut separating source from sink, such that every pixel is either assigned to the foreground node or to the background node. Figure 13.5 shows three examples for nucleus segmentation with graph cuts.

13.2.3 Nucleus Classification

For nucleus classification, intensity features and shape features were extracted and concatenated. Intensity features comprise foreground intensity, background intensity, and pyramid histogram of oriented gradients [20]. Shape-based features utilize the segmentation line of nucleus calculating the Freeman chain code and the 1D-signature [21]. The classification is then performed with, e.g., support vector machines [22, 23] or random forests [24]. Importantly, we compare the classification system with the interobserver error, thus interpreting the algorithm as additional "expert" which should be compared with other, real experts.

13.2.3.1 Active Learning

Active learning is a scenario where the classifier can select training samples of higher value for the classification performance compared to, e.g., redundant training samples [25], and has been used in medical image processing for classification and

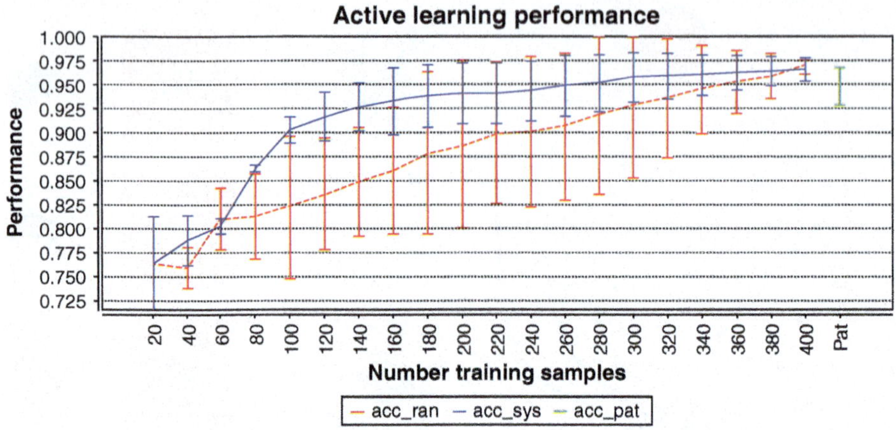

Fig. 13.6 Active learning for nucleus classification (*blue*) reduces the number of samples needed for high accuracy compared to random sampling (*red*). Test performed on three TMA images with 400 nuclei. The best accuracy (0.97) is comparable to the inter-pathologist error of two experts (0.95, *green*) [24]

segmentation [26, 27]. An active learning classifier can score the classification confidence for an unknown sample. If the confidence is too low, the sample needs to be labeled by an expert. For nucleus classification, where hundreds or thousands of nuclei would have to be labeled by pathologists for supervised learning, active learning is a great tool to reduce the number of expensive expert annotations.

To illustrate that, it has been shown on the example of three TMA images that in a simple active learning approach where training samples were successively added selected by (low) class probability, a user would have to label less than 45% of the nuclei than he or she would have to label when added in a randomized manner, in order to achieve the best classification result (Fig. 13.6) [14]. In total, the total classification accuracy (0.97 ± 0.15) is within the range of that of two pathologists (0.95 ± 0.02) who labeled all cells of the same images independently.

13.2.4 Staining Estimation

After all nuclei are classified into cancerous and benign, the stained cancerous nuclei have to be counted. To separate stained nuclei from unstained ones, a simple red/blue relation can be used: whenever a nucleus patch contains more red than blue, it is considered stained. Of course, color models can also be learned on the data [1]. Note that if color deconvolution has been used for nucleus detection, the staining information of a nucleus arises also from the channel on which the nucleus has been found.

13.2.4.1 Survival Analysis

Survival analysis on a larger patient cohort is an important validation of the computational staining estimation pipeline. Since MIB-1 staining is correlated with survival in RCC [11], we expect the algorithm to separate RCC patients according to their prognosis by the predicted MIB-1 staining estimation. The pipeline was trained on

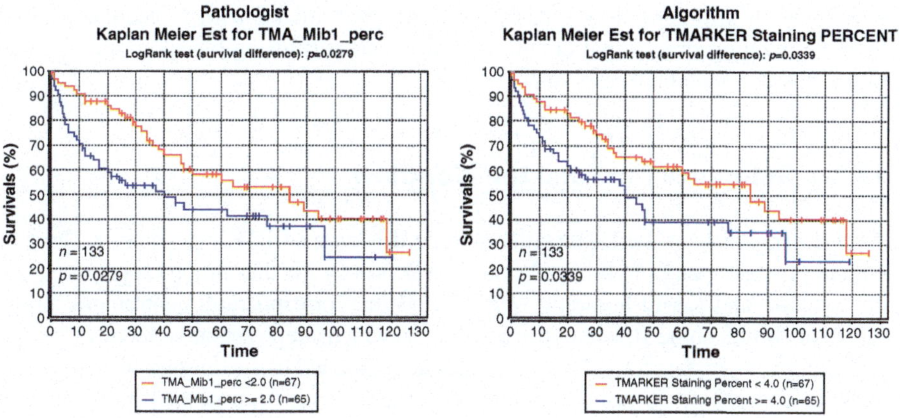

Fig. 13.7 Kaplan–Meier estimates for the cumulative survival rates after months after diagnosis. *Left*: the pathologist's staining estimation separates 133 ccRCC patients into a high MIB-1 (*blue*) and a low MIB-1 (*red*) group, with significantly different prognosis ($p = 0.03$). *Right*: the same task performed by the staining estimation pipeline, trained on eight fully labeled TMA images ($p = 0.03$) [28]

eight ccRCC TMA images whose nuclei were fully annotated by two pathologists. The algorithm was then tested on TMA images of a cohort of 133 ccRCC patients. Figure 13.7 shows with two Kaplan–Meier estimates that the algorithm was equally well able to identify the low and high MIB-1 expression groups as a trained human expert was. Patients with a high MIB-1 expression rate show a significantly lower survival time, where the difference in survival is significant ($p = 0.03$) [28].

13.3 ISHProfiler

Recent large-scale genome analyses and molecular profiling of human tissue samples have uncovered extensive genetic alterations and tumor heterogeneity in most tumor entities, demonstrating its significant role in cancer treatment and personalized medicine [18, 29–31]. Yet, such studies often excluded corresponding histomorphology and failed to predict mutations at single-cell level. Fluorescence in situ hybridization is a well-established method that measures genetic variations with detailed tissue morphology. However, the labor-intensive counting of signals under a fluorescence microscope is prone to interobserver variability [2, 32], hampers a streamlined assessment of the genetic status, and impedes a systematic quantification of tumor heterogeneity. These obstacles have prevented tumor heterogeneity to be properly evaluated in research and clinical practice, leading to a poor understanding of cancer evolution. The integrative method ISHProfiler alleviates these limitations by combining an image-based computational workflow with a dual-color chromogenic and silver in situ hybridization (DISH) assay that accurately detects copy number variation (CNV) at single-cell resolution, expressively visualizes multilevel heterogeneity, and objectively quantifies heterogeneous genetic alterations of various genes in diverse human tumors [33]. The ISHProfiler is objective and generic and can be potentially extended for profiling heterogeneous allelic gains and losses of any gene in any tissue specimen hybridized with molecular probes, thereby enabling precise patient stratification and permitting broad applications in tissue-based biomedical research.

13.3.1 Reference Data and the Workflow

As reference data for the ISHProfiler and the detection of PTEN deletion, a TMA of human prostate cancer hybridized with DISH probes for the tumor suppressor gene phosphatase and tensin homolog (*PTEN*) and the corresponding centromeric probe (CEP) of chromosome 10 was constructed, from which a representative set of 71 benign and malignant prostate formalin-fixed, paraffin-embedded tissue samples were selected. This set consists of 38 primary acinar adenocarcinomas from radical prostatectomy (RPE) specimens, 10 castration resistant prostate cancers (CRPCs), 6 PC lymph node metastases, 1 distant metastasis, and 16 benign prostatic hyperplasias (BPHs).

ISHProfiler comprises three major algorithmic steps: First, each tissue core was digitized and preprocessed (white balancing, deconvolution, and contrast modification) using the scanner's default auto-correction settings. Images were then resized by bicubic interpolation to 4096 × 4096 pixels for efficient tiling ($4096 = 2^{12}$) and served as input data. Second, DISH signals were detected by the circular Hough transform [12]. Third, a SVM model [22, 23] was trained and fivefold cross validated on the basis of an independent training set of *PTEN* DISH signals with expert annotations. The final model was used to classify the signals into five classes: *PTEN*, CEP10, mixed class *PTEN* + CEP10, background noise, and cell stains. *PTEN* deletion was defined if the division of all *PTEN* by all CEP10 signals in a single tissue core is less than or equal to 60% (Fig. 13.8).

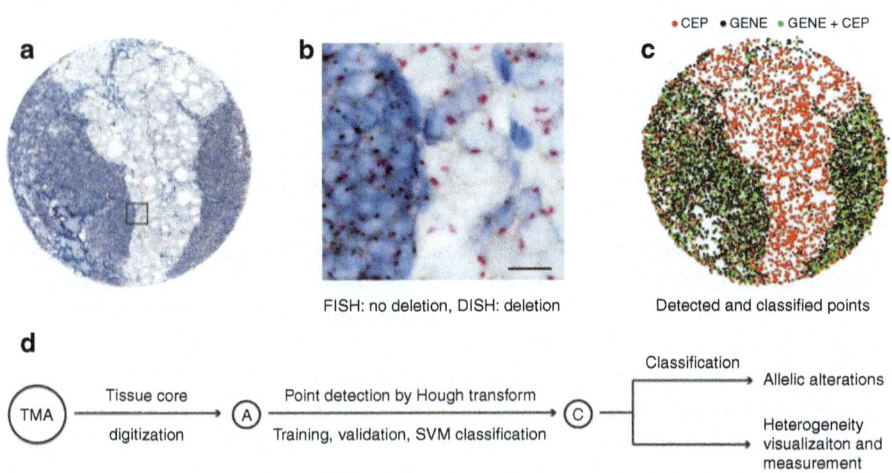

Fig. 13.8 (**a**) Example of a PC lymph node metastasis showing cellular heterogeneity for *PTEN* status. (**b**) Zoomed image showing PC with *PTEN* deletion (*right side*) and lymph node structures without *PTEN* deletion (*left side*). Scale bar, 10 μm. (**c**) Detected and classified *PTEN* gene and CEP points are displayed as a signal color map. (**d**) Computational workflow ISHProfiler. *Circled letters* correspond to the respective results shown in (**a**, **c**) [18]

13.3.2 Visualization and Quantification of Multilevel Heterogeneity

Since *PTEN* deletion has been shown to be heterogeneous [34], dichotomization of *PTEN* status into deletion and non-deletion using a single-valued threshold is arbitrary, thus cannot reflect tumor heterogeneity. To demonstrate a straightforward visual categorization of heterogeneous *PTEN* status, we applied ISHProfiler to reference data of 71 tissue cores and generated respective signal color maps, in which *PTEN* and CEP10 were illustrated as colored squares (Fig. 13.8). By grouping representative signal color maps in terms of a classification tree (Fig. 13.9), two major

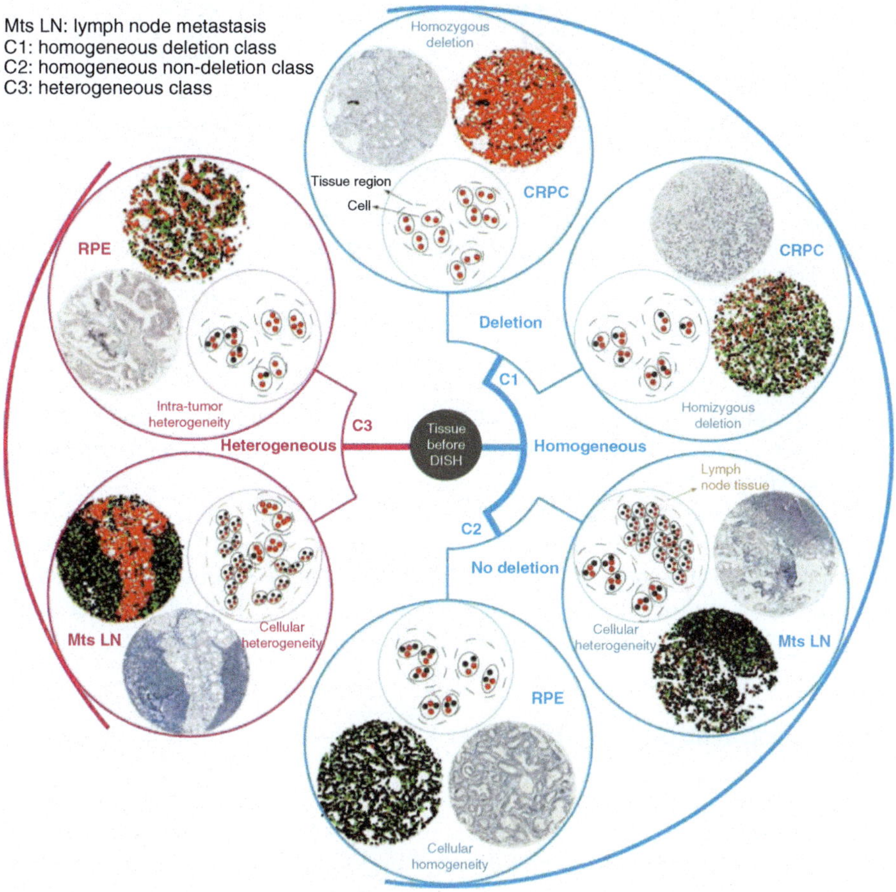

Fig. 13.9 Visualization of multilevel heterogeneity by ISHProfiler. Heterogeneity of *PTEN* CNV of different tissue cores can be visually classified into two major groups (homogeneous and heterogeneous classes), three subclasses (C1 and C2, homogeneous deletion and non-deletion, and C3, heterogeneous class), or six subgroups (homozygous deletion, hemizygous deletion, cellular homogeneity, cellular heterogeneity with either homogeneous or heterogeneous genetic status, and intra-tumor heterogeneity). In each *circle*, an original tissue core, a signal color map, and a sketch are illustrated [18]

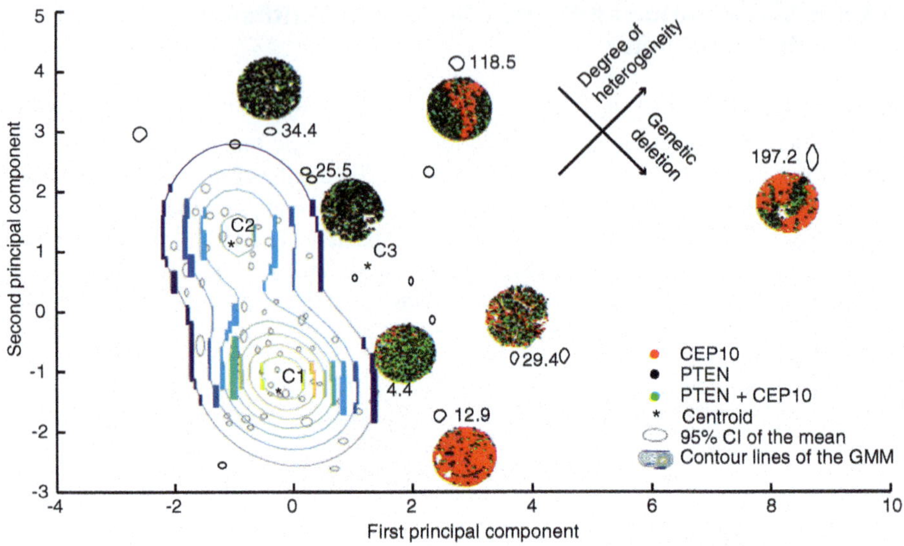

Fig. 13.10 Each tissue core is depicted as an *ellipse* in the two-dimensional PCA subspace ($n = 71$), superimposed with selected signal color maps shown in Fig. 13.9, where the axes of the ellipses indicate the 95% confidence interval (CI) of an experiment with a total of $n = 100$ repetitions by varying the number of random points from 201 to 300. The number is the Mahalanobis distance of each point to the respective centroids that are illustrated as *stars* [18]

classes of homogeneous and heterogeneous events, three subclasses, and six prototypes can be defined, revealing multilevel heterogeneity (cellular, inter-, and intra-tumor heterogeneity).

By incorporating the randomized local ratio and the randomized local density based on the detected *PTEN* and CEP10 signals, followed by feature extraction and principal component analysis (PCA), Gaussian mixture modeling computationally uncovered the multilevel heterogeneity among and within individual tissue cores (Fig. 13.10), matching the proposed visual categorization (Fig. 13.9).

13.3.3 Applications of ISHProfiler

ISHProfiler, integrated into the open source software TMARKER [14], uses supervised machine learning and statistical methods to generate computational models of CNV based on the classification of detected molecular signals, without relying on computationally intensive algorithms for single-cell recognition [35, 36]. It achieves classification accuracy similar to that of manual assessment, while the evaluation time is tremendously reduced, outperforming manual assessment by at least four orders of magnitude. Thus, ISHProfiler can also analyze large whole-tissue slides with unprecedented throughput. For investigation of the complex intra-tumor CNV landscape of the *PTEN* deletion on a *PTEN* DISH whole slide (with 108,000 × 138,000

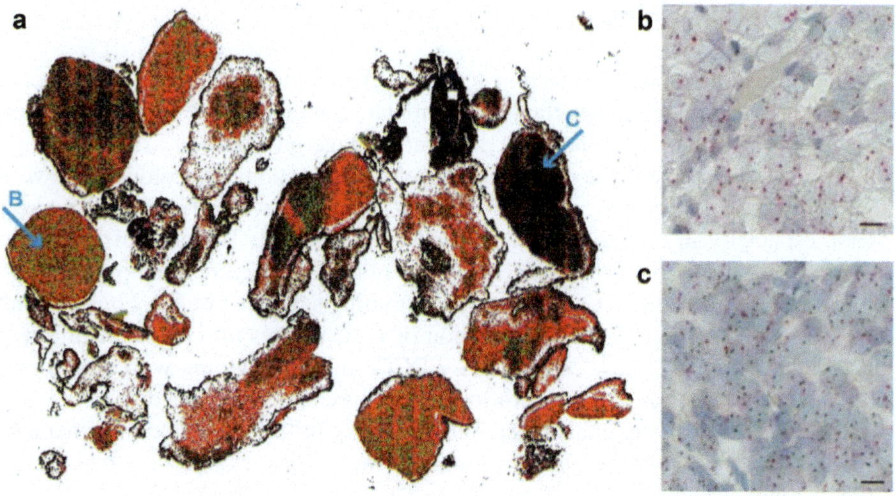

Fig. 13.11 (**a**) A signal color map of detected *PTEN* and CEP10 signals generated for a whole-slide image of CRPC. (**b**, **c**) Zoomed *PTEN* DISH images of areas marked with the *arrows* in (**a**). Scale bar, 10 μm [18]

pixels) of a transurethrally resected CRPC, ISHProfiler classified more than one million molecular signals and generated a signal color map (Fig. 13.11) that considerably agreed with a serial section that was immunohistochemically stained with anti-PTEN antibody.

ISHProfiler can be applied to additional cancer entities and genetic loci. For measuring amplification of 19q12, including *CCNE1* and *URI* in ovarian and endometrial cancers, ISHProfiler obtained computational results that closely matched manual assessment [37]. Moreover, ISHProfiler workflow successfully detected a heterogeneous *HER2* gene amplification in a well-differentiated gastric adenocarcinoma. The amplified regions agreed with the overexpression of HER2 detected by immunohistochemical staining.

13.4 Future of Computational Pathology

Computational pathology has gained more and more impact in research and in daily clinic especially in the recent years. This is mainly due to two reasons: First, computational methods and algorithms are getting more efficient and faster than before. Second, hospitals are collecting huge amounts of image data in their archives which are easier scanned and digitized than before. As a result, highly specialized classifier systems have for the first time access to hundreds of thousands potential training images, which is mandatory for higher robustness at better accuracy. The much broader validated algorithms are coming closer to clinical application than, for example, highly specified prototypes from smaller research studies.

13.4.1 Deep Learning

Computational methods and machine learning are gaining more and more impact in pathology research. Although classical computational image pipelines have already shown considerable performance in automatic image processing [2, 8, 13, 33, 35, 36, 38], a further boost to the new research field computational pathology came with the revival of deep learning and convolutional neural networks [17] in the present years, especially in biological and medical tasks [39–42]. This sort of classifiers seems to have superior performance in image-related tasks compared to other classification systems at the cost of many more training samples needed. For example, it has been shown that tissue classification of RCC images into the three different types clear cell RCC, clear cell papillary RCC, and oncocytoma with neural networks outperforms classical handcrafted feature-based methods in a direct comparison [43]. Modern tissue analysis studies regularly utilize deep learning methods [44–48], and these methods keep promise to outperform existing state-of-the-art classification systems.

13.4.2 Gathering of Training Data

Computational pathology is generally limited by the expensive acquisition on training labels. In contrast to classical imaging tasks, such as object detection or face recognition, where a vast amount of training labels can, e.g., be collected by nonexpert users on social networks, computational pathology relies on accurate labels by trained expert in highly specific medical tasks. Also, it is often not clear which image features exactly distinguish one medical class from another [49]. Therefore, there is an urgent need of new simplified methods for gathering expert-labeled data for digitized pathology slides. Research is ongoing on how to track pathologists in their clinical work flows when analyzing tissue slides with device tracking systems [50, 51] or without any human interaction [52].

13.4.3 Computational Pathology and Big Data

As pathology exclusively enables the research of localized spacial disease patterns and morphology in tissue specimen, it is a key component in medical cancer research and adds to other methods, such as genetics or microbiology [53]. Together with clinical data and health records, all these fields contribute to a holistic view and understanding of the cancer biology but also accumulate to highly complex data relationships, manifold diversity, and huge data amounts (so-called *Big Data*) [54]. The integration of these data from multiple sources is necessary for informed diagnoses by pathologists and decisions of medical doctors [55]. Computational methods are therefore needed to identify and quantify the necessary bits of information needed for high efficient personalized precision medicine.

Acknowledgment We thank Norbert Wey and Monika Bieri from University Hospital Zurich for their manifold support and their careful evaluation of the scanning statistics at the hospital. This chapter was partly funded through the NIH/NCI Cancer Center Support Grant P30 CA008748.

References

1. Fuchs TJ, Wild PJ, Moch H, Buhmann JM. Computational pathology analysis of tissue microarrays predicts survival of renal clear cell carcinoma patients. In: Proceedings of the international conference on Medical Image Computing and Computer-Assisted Intervention MICCAI, Lecture Notes in Computer Science, vol. 5242. Berlin: Springer; 2008. p. 1–8.
2. Fuchs TJ, Buhmann JM. Computational pathology: challenges and promises for tissue analysis. Comput Med Imaging Graph. 2011;35(78):515–30.
3. Louis DN, Feldman M, Carter AB, Dighe AS, Pfeifer JD, Bry L, Almeida JS, Saltz J, Braun J, Tomaszewski JE, Gilbertson JR, Sinard JH, Gerber GK, Galli SJ, Golden JA, Becich MJ. Computational pathology: a path ahead. Arch Pathol Lab Med. 2016;140(1):41–50.
4. Louis DN, Gerber GK, Baron JM, Bry L, Dighe AS, Getz G, Higgins JM, Kuo FC, Lane WJ, Michaelson JS, Le LP, Mermel CH, Gilbertson JR, Golden JA. Computational pathology: an emerging definition. Arch Pathol Lab Med. 2014;138(9):1133–8.
5. Gurcan M, Boucheron L, Can A, Madabhushi A, Rajpoot N, Yener B. Histopathological image analysis: a review. IEEE Rev Biomed Eng. 2009;2:147–71.
6. Bautista PA, Hashimoto N, Yagi Y. Color standardization in whole slide imaging using a color calibration slide. J Pathol Inform. 2014;5(1):4.
7. Bautista PA, Yagi Y. Staining correction in digital pathology by utilizing a dye amount table. J Digit Imaging. 2015;28(3):283–94.
8. Giesen C, Wang HA, Schapiro D, Zivanovic N, Jacobs A, Hattendorf B, Schüffler PJ, Grolimund D, Buhmann JM, Brandt S, Varga Z, Wild PJ, Gunther D, Bodenmiller B. Highly multiplexed imaging of tumor tissues with subcellular resolution by mass cytometry. Nat Methods. 2014;11:417–22.
9. Hashimoto N, Bautista PA, Haneishi H, Snuderl M, Yagi Y. Development of a 2d image reconstruction and viewing system for histological images from multiple tissue blocks: towards high-resolution whole-organ 3d histological images. Pathobiology. 2016;83(2–3):127–39.
10. Balis U, Hipp J, Flotte T, Monaco J, Cheng J, Madabhushi A, Yagi Y, Rodriguez-Canales J, Emmert-Buck M, Dugan M, Hewitt S, Toner M, Tompkins R, Lucas D, Gilbertson J. Computer aided diagnostic tools aim to empower rather than replace pathologists: lessons learned from computational chess. J Pathol Inform. 2011;2(1):25.
11. Kramer BA, Gao X, Davis M, Hall M, Holzbeierlein J, Tawfik O. Prognostic significance of ploidy, MIB-1 proliferation marker, and p53 in renal cell carcinoma. J Am Coll Surg. 2005;201(4):565–70.
12. Duda RO, Hart PE. Use of the Hough transformation to detect lines and curves in pictures. Commun ACM. 1972;15(1):11–5.
13. Schüffler PJ, Fuchs TJ, Ong CS, Roth V, Buhmann JM. Computational TMA analysis and cell nucleus classification of renal cell carcinoma. In: Proceedings of the 32nd DAGM conference on Pattern Recognition. Berlin: Springer; 2010. p. 202–11.
14. Schüffler PJ, Fuchs TJ, Ong CS, Wild P, Buhmann JM. TMARKER: a free software toolkit for histopathological cell counting and staining estimation. J Pathol Inform. 2013;4(2):2.
15. Ruifrok AC, Johnston DA. Quantification of histochemical staining by color deconvolution. Anal Quant Cytol Histol. 2001;23:291–9.
16. Fuchs TJ, Haybaeck J, Wild PJ, Heikenwalder M, Moch H, Aguzzi A, Buhmann JM. Randomized tree ensembles for object detection in computational pathology. In: Proceedings of the 5th international symposium on Advances in Visual Computing: Part I, ISVC '09. Berlin, Heidelberg, Las Vegas, Nevada: Springer; 2009. p. 367–78.
17. LeCun Y, Bengio Y, Hinton G. Deep learning. Nature. 2015;521(7553):436–44.

18. Gerlinger M, Rowan AJ, Horswell S, Larkin J, Endesfelder D, Gronroos E, Martinez P, Matthews N, Stewart A, Tarpey P, Varela I, Phillimore B, Begum S, McDonald NQ, Butler A, Jones D, Raine K, Latimer C, Santos CR, Nohadani M, Eklund AC, Spencer-Dene B, Clark G, Pickering L, Stamp G, Gore M, Szallasi Z, Downward J, Futreal PA, Swanton C. Intratumor heterogeneity and branched evolution revealed by multiregion sequencing. N Engl J Med. 2012;366(10):883–92.
19. Boykov Y, Veksler O, Zabih R. Fast approximate energy minimization via graph cuts. IEEE Trans Pattern Anal Mach Intell. 2001;23(11):1222–39.
20. Bosch A, Zisserman A, Munoz X. Representing shape with a spatial pyramid kernel. In: Proceedings of the 6th ACM international conference on Image and Video Retrieval, CIVR '07. New York: ACM; 2007. p. 401–8.
21. Gonzalez RC, Woods RE, Eddins SL. Digital image processing using MATLAB. 2nd ed. Knoxville: Gatesmark Pub, S.I.; 2009.
22. Cortes C, Vapnik V. Support-vector networks. Mach Learn. 1995;20(3):273–97.
23. Friedman J, Hastie T, Tibshirani R. The elements of statistical learning, Springer Series in Statistics, vol. 1. New York: Springer; 2001.
24. Breiman L. Random forests. Mach Learn. 2001;45(1):5–32.
25. Cohn DA, Ghahramani Z, Jordan MI. Active learning with statistical models. J Artif Intell Res. 1996;4(1):129–45.
26. Mahapatra D, Schüffler PJ, Tielbeek JAW, Makanyanga JC, Stoker J, Taylor SA, Vos FM, Buhmann JM. Active learning based segmentation of Crohn's disease using principles of visual saliency. In: 2014 IEEE 11th international symposium on Biomedical Imaging (ISBI); 2014. p. 226–9.
27. Mahapatra D, Schüffler PJ, Tielbeek JAW, Vos FM, Buhmann JM. Semi-supervised and active learning for automatic segmentation of Crohn's disease. In: Mori K, Sakuma I, Sato Y, Barillot C, Navab N, editors. Medical image computing and computer-assisted intervention, Lecture Notes in Computer Science, vol. 8150. Berlin: Springer; 2013. p. 214–21.
28. Schüffler PJ. Machine learning approaches for structure analysis in medical image data; 2014.
29. Marusyk A, Almendro V, Polyak K. Intra-tumour heterogeneity: a looking glass for cancer? Nat Rev Cancer. 2012;12:323–34.
30. Swanton C. Intratumor heterogeneity: evolution through space and time. Cancer Res. 2012;72(19):4875–82.
31. The Cancer Genome Atlas Research Network, et al. The Cancer Genome Atlas Pan-Cancer analysis project. Nat Genet. 2013;45:1113–20.
32. Zhong Q, Busetto AG, Fededa JP, Buhmann JM, Gerlich DW. Unsupervised modeling of cell morphology dynamics for time-lapse microscopy. Nat Methods. 2013;9:711–3.
33. Zhong Q, Rüschoff JH, Guo T, Gabrani M, Schüffler PJ, Rechsteiner M, Liu Y, Fuchs TJ, Rupp NJ, Fankhauser C, Buhmann JM, Perner S, Poyet C, Blattner M, Soldini D, Moch H, Rubin MA, Noske A, Rüschoff J, Haffner MC, Jochum W, Wild PJ. Image-based computational quantification and visualization of genetic alterations and tumour heterogeneity. Sci Rep. 2016;6:24146.
34. Krohn A, Freudenthaler F, Harasimowicz S, Kluth M, Fuchs S, Burkhardt L, Stahl P, C Tsourlakis M, Bauer M, Tennstedt P, Graefen M, Steurer S, Sirma H, Sauter G, Schlomm T, Simon R, Minner S. Heterogeneity and chronology of pten deletion and erg fusion in prostate cancer. Mod Pathol. 2014;27:1612–20.
35. Beck AH, Sangoi AR, Leung S, Marinelli RJ, Nielsen TO, Vijver MJvd, West RB, Rijn Mvd, Koller D. Systematic analysis of breast cancer morphology uncovers stromal features associated with survival. Sci Transl Med. 2011;3(108):108ra113.
36. Yuan Y, Failmezger H, Rueda OM, Ali HR, Grf S, Chin SF, Schwarz RF, Curtis C, Dunning MJ, Bardwell H, Johnson N, Doyle S, Turashvili G, Provenzano E, Aparicio S, Caldas C, Markowetz F. Quantitative image analysis of cellular heterogeneity in breast tumors complements genomic profiling. Sci Transl Med. 2012;4(157):157ra143.
37. Noske A, Henricksen LA, LaFleur B, Zimmermann AK, Tubbs A, Singh S, Storz M, Fink D, Moch H. Characterization of the 19q12 amplification including /CCNE1/ and /URI/ in different epithelial ovarian cancer subtypes. Exp Mol Pathol. 2015;98(1):47–54.

38. Zerhouni E, Prisacari B, Zhong Q, Wild P, Gabrani M. Deciphering protein signatures using color, morphological, and topological analysis of immunohistochemically stained human tissues. In Proc. SPIE 9791, Medical Imaging 2016: Digital Pathology, 97910T (March 23, 2016); 2016.
39. Greenspan H, van Ginneken B, Summers RM. Guest editorial deep learning in medical imaging: overview and future promise of an exciting new technique. IEEE Trans Med Imaging. 2016;35(5):1153–9.
40. Rampasek L, Goldenberg A. TensorFlow: biology's gateway to deep learning? Cell Syst. 2016;2(1):12–4.
41. Rueckert D, Glocker B, Kainz B. Learning clinically useful information from images: past, present and future. Med Image Anal. 2016;33:13–8.
42. Zerhouni E, Prisacari B, Zhong Q, Wild P, Gabrani M. A computational framework for disease grading using protein signatures. In: 2016 IEEE 13th international symposium on Biomedical Imaging (ISBI); 2016. p. 1401–4.
43. Schüffler PJ, Sarungbam J, Muhammad H, Reznik E, Tickoo SK, Fuchs TJ. Mitochondria-based renal cell carcinoma subtyping: learning from deep vs. flat feature representations. J Mach Learn Res Proc. 2016.
44. Bauer S, Carion N, Schüffler P, Fuchs T, Wild P, Buhmann JM. Multi-organ cancer classification and survival analysis; 2016. arXiv:1606.00897 [cs, q-bio, stat].
45. Litjens G, Snchez CI, Timofeeva N, Hermsen M, Nagtegaal I, Kovacs I, Hulsbergen van de Kaa C, Bult P, van Ginneken B, van der Laak J. Deep learning as a tool for increased accuracy and efficiency of histopathological diagnosis. Sci Rep. 2016;6:26286.
46. Sirinukunwattana K, Raza SEA, Tsang YW, Snead DRJ, Cree IA, Rajpoot NM. Locality sensitive deep learning for detection and classification of nuclei in routine colon cancer histology images. IEEE Trans Med Imaging. 2016;35(5):1196–206.
47. Xu J, Xiang L, Liu Q, Gilmore H, Wu J, Tang J, Madabhushi A. Stacked sparse autoencoder (SSAE) for nuclei detection on breast cancer histopathology images. IEEE Trans Med Imaging. 2016;35(1):119–30.
48. Zhou Y, Chang H, Barner K, Spellman P, Parvin B. Classification of histology sections via multispectral convolutional sparse coding. Conf Comput Vis Pattern Recognit Workshops. 2014;2014:3081–8.
49. Levenson RM, Krupinski EA, Navarro VM, Wasserman EA. Pigeons (*Columba livia*) as trainable observers of pathology and radiology breast cancer images. PLoS One. 2015;10(11):e0141357.
50. Braxton M, Bruny T, Elmore J, Gagnon S, Raghunath V, Reisch L, Shapiro L, Weaver D, Allison K. Mouse cursor movement and eye tracking data as an indicator of pathologists attention when viewing digital whole slide images. J Pathol Inform. 2012;3(1):43.
51. Mercan E, Aksoy S, Shapiro LG, Weaver DL, Bruny TT, Elmore JG. Localization of diagnostically relevant regions of interest in whole slide images: a comparative study. J Digit Imaging. 2016;29(4):496–506.
52. Schaumberg AJ, Sirintrapun SJ, Al-Ahmadie HA, Schüffler PJ, Fuchs TJ. Deep-scope: nonintrusive whole slide saliency annotation and prediction from pathologists at the microscope. In: 13th international conference on Computational Intelligence Methods for Bioinformatics and Biostatistics. Stirling, UK; 2016.
53. Nawaz S, Yuan Y. Computational pathology: exploring the spatial dimension of tumor ecology. Cancer Lett. 2016;380(1):296–303.
54. Zhong Q, Barnert R, Ratsch G, Fuchs TJ, Wild PJ. Big Data in der Medizin. Leading Opinions: Hämatologie & Onkologie; 2016. p. 102–5.
55. Yagi Y, Riedlinger G, Xu X, Nakamura A, Levy B, Iafrate AJ, Mino-Kenudson M, Klepeis VE. Development of a database system and image viewer to assist in the correlation of histopathologic features and digital image analysis with clinical and molecular genetic information. Pathol Int. 2016;66(2):63–74.

In Vitro Three-Dimensional Cell Cultures as Tool for Precision Medicine

14

Alessandra Silvestri, Dirk Schumacher, Maxine Silvestrov, Reinhold Schäfer, Christoph Reinhard, Jens Hoffmann, Karsten Boehnke, and Christian R.A. Regenbrecht

Contents

A. Silvestri, PhD • M. Silvestrov • C.R.A. Regenbrecht, PhD (✉)
CPO-Cellular Phenomics and Oncology Berlin-Buch GmbH,
Robert-Rössle-Str. 10, Berlin 13125, Germany
e-mail: christian.regenbrecht@cellphenomics.com

D. Schumacher • R. Schäfer, PhD
Institute of Pathology and Comprehensive Cancer Center, Charité-Universitätsmedizin Berlin,
Charitéplatz 1, Berlin 10117, Germany

German Cancer Consortium (DKTK), German Cancer Research Center (DKFZ),
Im Neuenheimer Feld 280, Heidelberg 69120, Germany

C. Reinhard, PhD
Eli Lilly and Company, Indianapolis 46285, IN, USA

J. Hoffmann, PhD
EPO-Experimental Pharmacology and Oncology Berlin-Buch GmbH,
Robert-Rössle-Str. 10, Berlin 13125, Germany

K. Boehnke, PhD
Quantitative Biology, Eli Lilly and Company,
Avenida de la Industria, 30, Alcobendas 28108, Madrid, Spain

J. Haybaeck (ed.), *Mechanisms of Molecular Carcinogenesis – Volume 2*,
DOI 10.1007/978-3-319-53661-3_14

Abstract

For decades of molecular cancer research, immortal cancer cell lines have served as an easily accessible source for basic cancer biology research and preclinical testing of anticancer drugs. However, numerous studies have suggested that these cell lines poorly recapitulate the diversity, heterogeneity, and resulting drug resistance or relapse in patients. The derivation and (short-term) culture of primary cells from solid tumors have thus gained significant importance in personalized cancer therapy. This chapter focuses on our current understanding and the pros and cons of different preclinical and prospective clinical models of three-dimensional primary tumor cultures. We will discuss cell culture approaches, such as biomimetic scaffolds and growth factor supplemented, chemically defined media for various forms of solid tumors. Complex culture models of primary tumor cells could finally provide a key missing link between compound screening and clinical trials and ultimately will help redefining therapeutic intervention with high translational relevance at the level of the individual patient.

14.1 Introduction

Despite significant advances in our understanding of various aspects of cancer initiation, progression, metastasis, tumor microenvironment, and cancer stem cells, every year more than 1,5 million people are dying of cancer in Europe alone [1]. Therefore a fast and successful translation of research-generated knowledge into the clinic is of utmost importance. Driven by an increased understanding of the underlying biology, evidence-based drug development has led to considerable improvements in cancer management. However, tumor relapse is still frequently observed. In the last decade, many targeted drugs have been successfully developed and are today's clinical standard of care. In a next phase of drug discovery, better predictors for treatment success or failure are needed. Here, in early-phase drug discovery, *in vitro* and *in vivo* models will help finding response-correlated signatures that reliably predict the clinical response of already available and future drugs in more stratified patient cohorts.

14.1.1 The Preclinical Use of Two- and Three-Dimensional Cell Cultures

Experimental models for the study of tumor development and progression, as well as for drug testing, have been pivotal to our understanding of tumor biology, and they represent fundamental tools to improve cancer treatment.

Since 1907, when Harrison and colleagues first developed two-dimensional (2D) culture techniques as a model for the study of mammalian tissues and organs [2], scientists have the possibility of studying cellular physiology in an easy, fast, and relative inexpensive manner. One of the main concerns regarding the use of established cancer cell lines is the accumulation of additional genetic aberrations during *in vitro* passaging, hence limiting their suitability as *in vitro* preclinical models [3, 4] that hardly represent clinical scenarios [5]. Moreover, there is a wide range of variability in patient response to the same drugs used on tumors that carry identical genomic aberrations. Thus, studying drug response in a rather small number of established cell lines may pose an oversimplification and underestimation of the (epi)genetic diversity on the basis of the individual cancer patient [6, 7]. In medical science, models representing physiological and pathological conditions are fundamental tools for drug development. However, the complexity of biological systems requires experimental models to be able to approximate the original structure and functionality of the studied organ, tissue, or cell type. The composition of solid tumors and their (micro)environment is highly heterogeneous, comprising not only tumor cells but also supporting stroma cells, such as fibroblast, endothelial cells and immune cells. These different cell types cooperate in a well-organized manner, and this complexity seems to be a prerequisite for tumor growth and progression [8]. Considering that *in vivo* tissues are integral three-dimensional (3D) structures, the use of 3D *in vitro* techniques to study cellular biology allows for a better representation of the physiological situation. Already in the late 1960s, researchers started using matrix tissue cultures to better recreate the physiological tumor microenvironment *in vitro* [9], and, since then, several research groups have improved this approach. Studies by Inch and Sutherland represent the first application of multicellular 3D spheroid models to study nodular carcinomas [10, 11]. Following these groundbreaking publications, several other groups focused on the development of methods to easily obtain experimental systems for the study of tumors in their natural 3D microenvironment [12–15].

In the last 30 years, an increasing interest of the scientific community in tumor microenvironment has led to substantial improvements in comprehending the function of extracellular matrix proteins and tissues [16–19]. These findings greatly contributed to the advancement and optimization of more complex experimental 3D model systems.

14.1.2 Tumor Heterogeneity

Since the 1970s, several studies have highlighted the presence of distinct subpopulations of cancer cells residing within an individual solid tumor [20–22], characterized by varying tumorigenic potential, the capacity to metastasize, and treatment susceptibility. Analysis of multiple tumors demonstrated both inter- and intra-tumor heterogeneity that may be explained by either of the two predominant theories describing the process of tumorigenesis and progression. The "clonal evolution model" is based on the acquisition of additional mutations within single cells over time, thereby creating more aggressive cellular subclones facilitating tumor progression and metastatic spread [23]. Several recent publications demonstrated the presence of a high degree of genetic heterogeneity inside the same tumor mass [24, 25]. Gerlinger and colleagues [24] analyzed separate tumor regions of primary tumors or the corresponding metastasis from patients with renal carcinoma. By means of exome sequencing, chromosomal aberration analysis, and ploidy profiling, they demonstrated that heterogeneous genetic alterations were present in different regions of a single tumor mass, revealing extensive genetic intra-tumor heterogeneity. Based on these data, they suggested that the presence of cancer cells with divergent molecular characteristics may cause the generation of misleading profiles when only single biopsies are used for analysis. In a similar study, de Bruin and colleagues [25] found further evidence for spatial and temporal (branched) clonal evolution in non-small cell lung cancer (NSCLC), encompassing the de novo acquisition of driver mutations and complex genomic aberrations like translocations or copy number alterations. Besides renal cancer and NSCLC, further evidence of genetic heterogeneity was presented for colorectal cancer (CRC) [26]. The study proposes a so-called Big Bang model of tumor initiation, in which a single clonal expansion occurs early after transition to advanced carcinoma, creating pervasive private (subclone exclusive) alterations that persist within the final neoplasm alongside public (tumor wide) alterations. It also postulates branched evolution, also described in the aforementioned renal carcinoma study, to be a natural result of early clonal expansion. Massive clonal expansion in the course of further progression is described to be rare owing to spatial constraints limiting selective forces that drive progression. Still, additional subclones of potential aggressive nature may evolve, intermix, and persist at low population sizes, providing the "substrate" for a dominant, resistant population following treatment-induced selection. Thus, this model may serve as an interpretation for the heterogeneous nature of (primary) cancers. Finally, the model permits compatibility to the second theory of tumor initiation and progression, the "stem cell hypothesis." This theory argues that tumors harbor a small subpopulation of cancer stem cells considered to drive tumor progression and to be the culprit of relapse, in contrast to the bulk of proliferating tumor cells [27]. In a murine model of intestinal cancer, Barker and colleagues [28] showed that LGR5-positive stem cells of the intestinal crypt could be transformed by solely deleting APC, a tumor suppressor gene of the Wnt pathway. Loss of this gatekeeper led to the formation of macroscopic adenomas. These two theories are not mutually exclusive since hierarchical, stem cell-like, and clonal mechanisms could be present and work synergistically to sustain tumor progression [29].

Besides tumorigenesis and metastatic spread, intra-tumor heterogeneity may also influence therapy resistance, and relapse. Ding and colleagues analyzed somatic mutations in primary tumors and relapse samples from patients with acute myeloid leukemia, demonstrating the association of tumor recurrence with the accrual of additional *de novo* mutations and clonal evolution. Moreover, they showed that the development of new mutations is caused, in part, by chemotherapy, selecting for the most resistant cellular subclone [30].

The high level of heterogeneity observed in tumors has important clinical implications, and it represents a fundamental aspect for the design of drug development, as well as biomarker determination studies. The presence of different cellular subpopulations with divergent molecular characteristics may be one of the causes of the clinical failure of several anticancer therapies. Studies conducted in both *in vitro* and in *in vivo* experiments do not take this complexity into account and thus produce misleading results. Moreover, in the field of precision medicine, the presence of regions in the tumor mass with different characteristics limits the informative value of single needle biopsies or surgical excision, since they cannot accurately represent the complete profile of the entire tumor mass.

The increasing attention to tumor heterogeneity has stimulated researchers to develop new experimental models able to maintain or mimic this complexity. For this reason, in the last decades, the use of 3D multicellular cultures, especially tissue organoids in which the original tissue characteristics, such as morphology, cell polarity, and marker expression, are preserved, has strongly increased. Therefore, these models represent useful tools for drug screening, as well as for predictive biomarkers determination.

These disparities in clinical response and patient-dependent tumor variability are the driving force behind precision medicine [31–33] and provide the impetus to develop methods of generating and culturing primary tumor cells from patients that reflect both clonal subpopulations and retain stemlike characteristics, thus enabling a more effective bench to bedside translation.

In the era of personalized therapy, researchers need a repertoire of patient-derived primary tumor cells that can generate high-fidelity data for translating *in vitro* findings to *in vivo* models and ultimately to clinical settings. This will provide a more refined database compared to a tissue bank.

14.2 Preclinical Models for the Study of Tumor Development and Progression: Comparison of Three-Dimensional Cell Cultures to Xenografts and Two-Dimensional Cell Lines

14.2.1 *In Vivo* Preclinical Models

Using mouse models to study human tumor growth dates back to the late 1960s [34]. With the development of immune-deficient mouse strains like nude or severe combined immunodeficiency (SCID) [35] in the mid-1980s, rodent models became common use in basic research and preclinical studies. These models have been successfully used to study drug tolerability and sensitivity of pharmaceuticals with different modes of action and, however, are not yet the bona fide model system representing human cancers. Currently, efforts are being made to improve predictivity of xenograft models by using orthotopic implantation, which is technically demanding but resembles a closer link to the environment of the donor tissue. In addition, cocultures with human immune and stromal cells are under evaluation. To improve predictivity of preclinical data for clinical trials in recent years, large panels of patient-derived xenografts (PDX) were developed and investigated [36, 37]. Because PDX involve transplanting cancer patient tissue directly into immune-compromised mice, genetic information and immunohistological markers are correlative to the patient and can be applied to evaluate novel anticancer drugs [38] and personalized cancer therapies in preclinical phase II studies' design. The advantages of PDX models can be summarized as follows: (1) preserve and stabilize genetic, histological, and phenotypic features of the tumor, (2) maintain stromal and stem cell components of the tumor, (3) facilitate biomarker assessment, and (4) can be used to predict the response to anticancer drugs. Working with PDX model needs approval of an institutional review board for utilization of patient-derived tumor tissue. Unfortunately, this model is technically challenging, expensive, and time-consuming. The freshly excised primary human tumors need to be delivered from the operating room to the laboratory within 1 day, while, simultaneously, a sample of the primary human tumors should be examined by a pathologist. Therefore, it is necessary to have good infrastructure in place to ensure smooth transition of the samples from the surgeon, via the pathologist to the researcher. Only then, the original primary human tumors can be compared with tumor tissues of the passaged tumor graft. Notwithstanding these efforts, the take rate of PDX between 20 and 70% [39–41] is sometimes a limiting factor. Additionally, each passage of tumor tissue has to be compared with the original tissue with regard to its key immunohistochemical features. From the second passage, nude mice can be utilized, and pieces of tumor tissues can be frozen in liquid nitrogen for later use. Once these hurdles are taken, PDX models are available for preclinical validation of anticancer drug sensitivity and prediction of patient prognosis. PDX are certainly an extremely promising model for personalized cancer therapy. Accordingly, several global research centers are establishing resource libraries of PDX models. Development and standardization of animal models can increase the predictability of the anticancer drug response and be utilized as a tool for preclinical assessment of anticancer drugs.

14.2.2 *In Vitro* 2D Models

Since 1951, when for the first time tumor cells were isolated and propagated as a stable cell line (HeLa) [42], numerous tumor cell lines have been established and are commercially available as models for different cancer types characterized by distinctive genetic backgrounds. Another important advantage of stable cell lines is the fact that they can be easily manipulated to study specific cellular functions involved in tissue homeostasis and malignant transformation. The long-standing experience with such cells caused the fact that, today, early-phase drug discovery relies heavily on screening based on 2D *in vitro* models, as they are relatively easy to perform in an automated fashion, and only confident results delivered by these models warrant approval of animal use in later stages. Unfortunately, since in conventional 2D cultures cells are forced to grow in monolayers, the 3D architecture of the tumor tissue is completely lost, and, therefore, the complex cross talk between tumor and microenvironment cannot be maintained (Fig. 14.1a). Moreover,

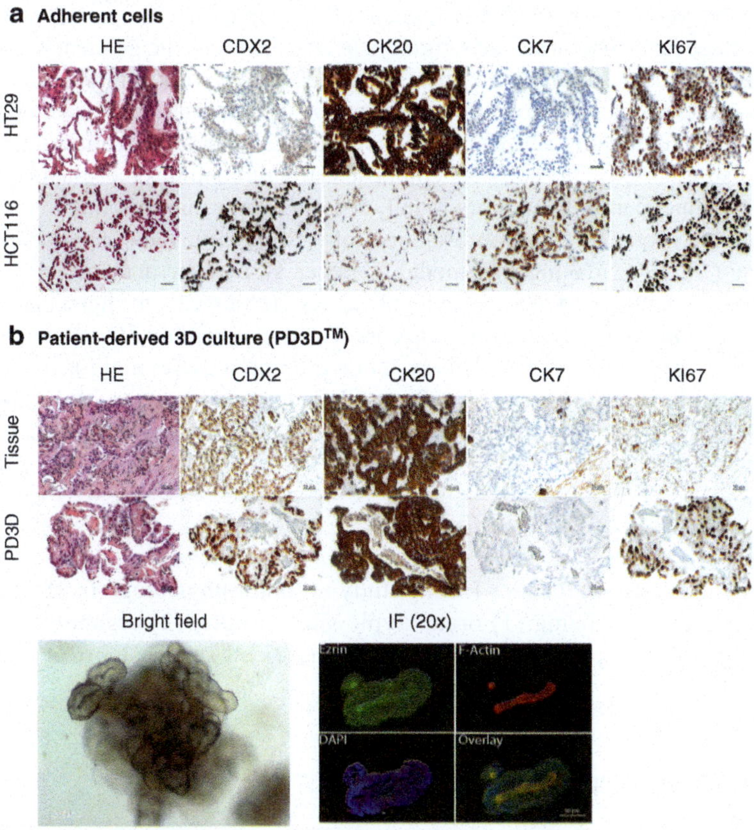

Fig. 14.1 Analysis of morphology and marker expression in colorectal cancer cell lines, tumor tissue, and patient-derived 3D (PD3D) cell culture. (**a**) Analysis of morphology and marker expression in HT29 and HCT116 colorectal cancer cell lines by eosin staining (HE) and immunohistochemistry, respectively. (**b**) Comparison of morphology and marker expression in tumor tissues and in the corresponding PD3D cell culture directly isolated from fresh tumor material

monolayer cell cultures are often characterized by a different morphology resulting from alterations of the normal cellular processes compared to the tissues of origin.

14.2.3 *In Vitro* 3D Models

Limitations of 2D cell cultures are partially overcome by 3D cell cultures that allow for a more accurate mimicry of the native cancer tissue since it is possible to preserve cellular morphology and heterogeneity characterizing physiological tissues. This is a fundamental requirement because both morphology and cell–environment cross talk strongly influence gene expression and therefore cell behavior [43–45]. Moreover, the use of multicellular tumor models could mimic tissue architecture to investigate not only tumor cell activity but also the influence of tumor microenvironment on malignant transformation.

Another very important feature of 3D cultures is the presence of gradients of soluble factors. Due to the 3D geometry, nutrients, oxygen, growth factors, as well as CO_2 and wastes are exchanged between cells and microenvironment in a way that is very close to the one observed in tumors *in vivo*. The specific geometry of 3D cell cultures induces the formation of concentric cell layers with different phenotypes. For instance, in spheroid models bigger than 500 μm, an external layer of proliferating cells, a middle quiescent viable cell zone, and an internal necrotic core can easily be distinguished. This complex composed structure mimics the situation *in vivo* where the inner part of the tumors usually receives fewer nutrients due to the devascularization. Moreover, this conformation reflects the diffusion of chemicals inside the tumor *in vivo*, making this model a proper system for drug efficacy studies [46–48]. Recently, Rodenhizer and colleagues developed an engineered model where scaffold–tumor composite strips are rolled to assemble 3D structures mimicking *in vivo* tumor geometry. This system represents a useful tool for studying the influence of hypoxic gradients on cell growth, metabolic pathways, and response to therapy [49].

14.3 Models of 3D Cell Cultures

Currently, several approaches for the study of tumor physiology in 3D are used, the most important being (1) organotypic slice culture, (2) organotypic cocultures, (3) cellular spheroids generated by single cell suspensions, or (4) tissue organoids.

14.3.1 Organotypic Slice Culture

The easiest way to maintain tissue architecture *ex vivo* is to cultivate the whole organ or fragments/slices of it in the so-called organotypic slice culture (Fig. 14.2a). This system allows studying the normal/altered physiology of the organ maintaining the original tissue organization. Several studies have been done on various

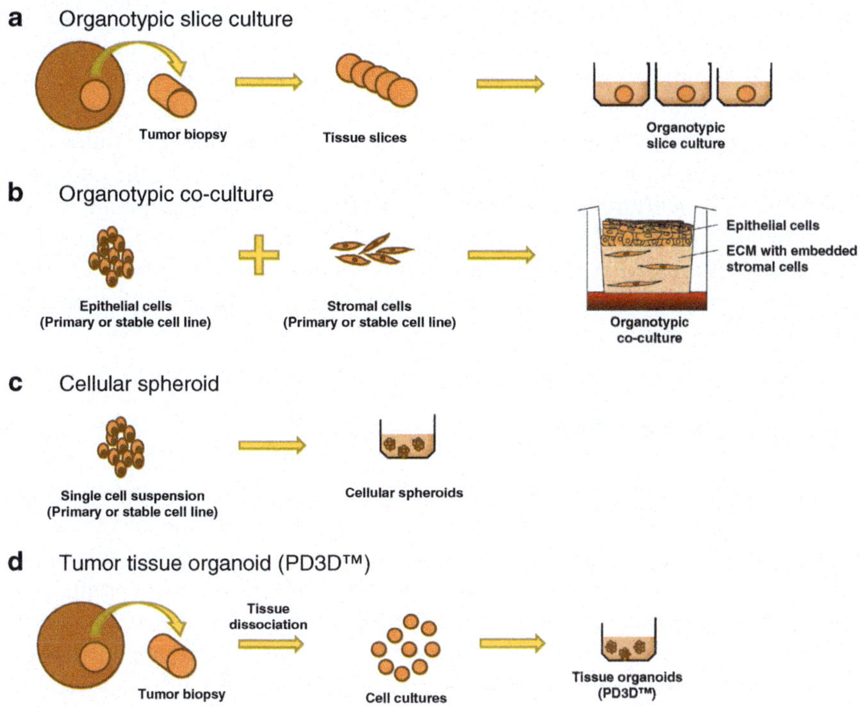

Fig. 14.2 Models of 3D cell cultures. (**a**) Organotypic slice culture: the whole organ or fragments of it are cultivated *ex vivo*. (**b**) Organotypic coculture: epithelial cells are cultured on an extracellular matrix with embedded stromal cells. (**c**) Cellular spheroids: 3D structures obtained by aggregation of single cells from primary or stable cell lines. (**d**) Tumor tissue organoids: multicellular structures directly isolated from fresh tissue without prior cell enrichment

tumor types, such as breast [50, 51], ovarian [52], and prostate cancer [53]. In these studies, drug uptake, proliferation, and cell death induction, as well as protein or gene expression alteration, have been considered. Unfortunately, tissue slices can be maintained in culture only for short periods of time without observing an alteration in cell viability and tissue architecture and can therefore be used only for short-term experiments. Moreover, several factors during the collection of the tissue, such as intraoperative handling and pathology processing, can strongly influence tissue slice characteristics, requiring the need for a high number of experimental replicates to obtain unbiased results. The short-term culturing and the inability of the cultures to expand restrict the use of this model to low-throughput studies in which the number of drugs and analytical endpoints to be analyzed is limited.

14.3.2 Organotypic Cocultures

Organotypic coculture models or raft cultures have been established for decades. The air–liquid interface allows for stratified tissue organization of epithelial cells growing on extracellular matrix gels with embedded stromal cells (Fig. 14.2b).

These complex cocultures permit the reconstruction of an *in vivo*-like microenvironment and provided important insights into the regulation of epidermal differentiation and epithelial–mesenchymal interactions [54–58]. To further optimize the support of long-term epithelial growth and differentiation, the composition of the stromal equivalents has been modified by using various extracellular matrix approaches mimicking the microenvironment [59–63]. However, the complexity of organotypic cocultures precluded their validation and establishment as robust drug-screening models. In particular, the application as 384-well-based assays and the ease of molecular readouts and rapid imaging analysis remain enormous challenges for both organotypic slice cultures and organotypic coculture models.

14.3.3 Cellular Spheroids

Although organotypic slice cultures and organotypic coculture models are very complex and innovative tools, the most frequently used 3D models are cellular spheroids (Fig. 14.2c). These structures are obtained by aggregation of single cell suspension of stable cell lines or isolated primary cells. Cellular spheroids can be composed of a single cell type (homotypic) or different cell types (heterotypic), such as epithelial cells, endothelial cells, and/or fibroblasts plated together. The latter approach is particularly interesting when the aim of the analysis is to study the interaction between tumor cells and microenvironment and its influence on tumor progression. Several studies have demonstrated that heterotypic spheroids are useful tools for studying tumor–stroma interactions when epithelial cells and fibroblasts are plated together [64, 65]. Fong and colleagues recently developed a 3D hydrogel system where prostate cancer cells and osteoblastic cells were grown together to recapitulate the tumor cell–microenvironment interaction within the bone metastatic microenvironment [66].

Moreover, the introduction of macrophages or other immune cells into the 3D culture enables the study of the effects of the immune system on tumor progression [67]. In addition, tumor cell invasion and metastatic potential can be thoroughly investigated by coculturing epithelial cells with fibroblasts and endothelial cells [68].

Not only stable cell lines can be used to generate spheroids but also isolated stem cells and iPS form 3D structures in specific culture conditions. Ricci-Vitiani and colleagues demonstrated that $CD133^{+}$ cells isolated from colon cancer are able to form tumor spheres *in vitro* and that these spheroids retain the ability to engraft and produce tumors with the same characteristics of the original mass [69]. Weiswald et al. demonstrated that it is possible to obtain *ex vivo* colospheres from cancer cells but not from normal colonic mucosa and that sphere-forming capacity was associated with tumor aggressiveness [70]. Moreover, they showed that colospheres matched the gene expression profile of the parental tissue [71].

14.3.4 Tumor Tissue Organoids (Patient-Derived 3D Cell Cultures, PD3D™)

Tumor tissue organoids (Fig. 14.2d) are novel multicellular systems freshly isolated from primary organs by mechanical and enzymatic dissociation. In this system, 3D organoids are composed of cell clusters that develop and maintain the complex composition of the tissue of origin. These structures are able to self-organize and to reproduce the exact original tissue architecture and marker expression (Fig. 14.1b). Since this model represents a useful tool to study tumor development and progression, as well as to test drug efficacy, several groups focus on the development of protocols for growing and expanding tumor organoids from primary tissues for different cancer types [72–74]. Moreover, it has been shown that $Lgr5^+$ cells isolated from different tissues, such as the stomach [75], liver [76], and intestine [77], can be cultured *in vitro* and are able to form long-term organoids that reproduce the structure of the tissue of origin.

Recent works of Clevers [78] that originate from the early matrix-associated models established in the 1960s [9] show that organoids represent a promising model for patient stratification and oncogenic therapeutic development since they are easy to establish and have an almost unlimited proliferative potential. Clevers and his group demonstrated that Lgr5 stem cells from multiple organs are able to form epithelial organoids that retain tissue identity [79]. Moreover, they showed that this approach can be applied not only to primary tumors but also to secondary metastatic masses [80], making this model a suitable tool for personalized anticancer treatment. Lately it has been reported that further improved patient-derived 3D cell culture models are of high value for drug screening in colon cancer patients [81].

In this book chapter, we will focus on tumor organoids to better understand their advantages and their potential application in the field of precision medicine.

14.4 Methods for 3D Spheroid Culture Development and Growth

14.4.1 Spontaneous Cell Aggregation

14.4.1.1 Nonadhesive Surfaces: Liquid Overlay Techniques

Different techniques for spheroid formation have been developed. Tissue organoids and single cell suspension spheroids can be mainly developed using similar experimental techniques. The most frequently used approaches in spheroid culturing are those based on spontaneous cell aggregation with liquid overlay techniques. Several tumor cell types are able to adhere spontaneously to each other or to other cell types forming cell aggregates like tissue spheroids. In order for this to be possible, aggregation forces between cells have to be stronger than the adhesive forces on the surface of the culture plate (Fig. 14.3Aa). Several surface modifications are possible to avoid cell adhesion and to allow spontaneous spheroid formation. Plate surface coating with substances that prevent cell adhesion, such as agar or agarose,

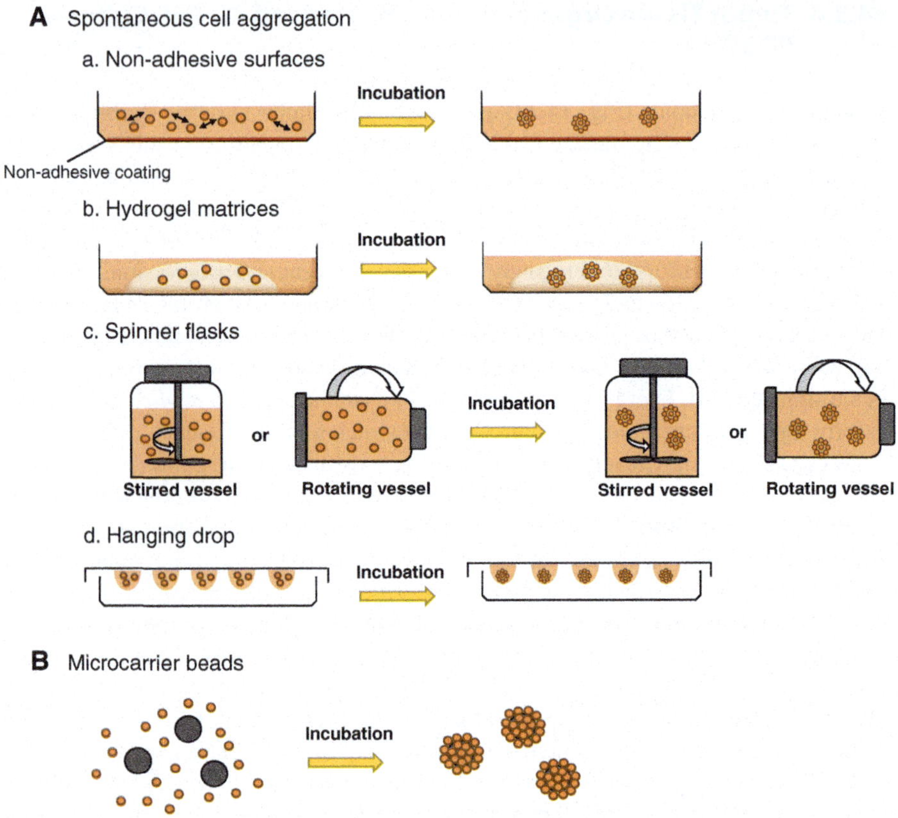

Fig. 14.3 Methods for 3D spheroid development and growth. (**A**) Spontaneous cell aggregation. Several cell types can spontaneously aggregate forming spheroids when adhesive forces between cells are stronger than adhesive forces to the surfaces of the culture plate. (**a**) Nonadhesive surfaces: when plate surfaces are modified to reduce cell adhesion, cells aggregate, forming 3D structures. (**b**) Hydrogel matrices: natural or synthetic matrices can be used to better mimic the extracellular microenvironment. Spheroids can form by aggregation of single cells or by monoclonal cell growth. (**c**) Spinner flasks: for high-scale spheroid production, stirred or rotating vessels are used. The constant movement of the vessel prevents cell adhesion to the surface of the plate allowing 3D spheroid formation. (**d**) Hanging drop: cells are seeded in small drops of medium where, due to gravity forces, cells aggregate at the tip of the drop. (**B**) Microcarrier beads: natural or synthetic solid materials can be used to facilitate spheroid formation. Cells adhere to the bead surface and proliferate forming spheroidal structures

polyHEMA, proteoglycans, or positively charged polystyrene, has been one of the first approaches in the development of cellular spheroids [12, 82–85]. More recently, the development of cell culture products with nonadhesive surfaces has allowed researchers to obtain cellular spheroids more easily and with a higher level of reproducibility. Methods for developing multicellular spheroids based on nonadhesive surfaces are inexpensive, easy to produce, and suitable for several cell types. Unfortunately, this approach leads to the formation of spheroids with variable size and shape, and it does not allow an upscaled production.

14.4.1.2 Cell Embedding in Hydrogel Matrices

To better mimic the extracellular microenvironment, cells can also be embedded in natural or synthetic hydrogels, such as collagen or laminin matrices or other more complex matrix gels, such as Matrigel™ [86] or Puramatrix™ [87]. These are useful materials for 3D spheroid culturing since they not only prevent cell adhesion to plate surface, but they also allow recreating the physiological situation that is present *in vivo* (Fig. 14.3Ab). Based on the process that has to be studied, researchers can choose between matrices with different characteristics: Collagen I gels model the ECM of connective tissue, basement membrane extracts, such as Matrigel, recapitulate the tumor epithelial microenvironment, while fibronectin-rich matrices mimic the microenvironment during wound healing. Moreover, the possibility to easily control the density and the flexibility of these materials makes them a very useful tool to model *in vivo* tissues.

14.4.1.3 Spinner Flask, Gyratory Rotator, and Microgravity System

For high-scale spheroid formation, several systems based on the application of external forces have been developed. Spinner flasks are bioreactors in which the continuous mixing of the medium prevents cell adhesion to the flask surface, allowing spheroid formation [88] (Fig. 14.3Ac). A similar system is represented by the gyratory rotators in which cells are maintained in suspension thanks to the constant rotatory movement of the flask. These methods allow for a good control of spheroid size by adjusting cell density, rotatory velocity, and incubation time. One of the limitations of these systems is the presence of high shear forces generated by the movement that can affect cell physiology. To overcome this problem, a microgravity system has been developed in which cells are maintained in a rotatory, weightless microenvironment. This method, developed by the National Aeronautic and Space Agency (NASA), allows for spontaneous formation of 3D spheroids with homogeneous size and shape [46]. The major advantages of the rotary cell culture system (RCCS) lie in the possibility of obtaining spheroids with larger size, characterized by a higher level of differentiation. Moreover, the RCCS system allows the coculturing of different cell types with the production of heterogeneous spheroids that very well mimic the epithelial architecture of the tissue *in vivo*. One of the limitations of this system is the need for specialized, expensive equipment that makes the use of this technology limited to dedicated laboratories.

Since rotatory systems allow massive production of spheroids, this is the method of choice to generate high amounts of homogeneous cellular spheroids for further applications. Indeed, when the goal is to study single spheroid formation and behavior, the static systems based on liquid overlay techniques are preferred since they permit an accurate analysis of cellular characteristics.

14.4.1.4 Hanging Drop Method

With the aim to develop an easier methodology for obtaining 3D cell cultures, Kelm and colleagues applied the hanging drop method used for culturing stem cell embryoid bodies to spheroid formation [89]. In this method, the cell suspension is deposited as drops of typically 20 μL onto the underside of a plate lid. Subsequently, the lid is carefully inverted, with the drops being kept in place because of surface tension. Caused by gravity, the cells inside the drop descend to the tip of the drop aggregating as spheroids of well-defined and homogeneous size (Fig. 14.3Ad). This ability of obtaining spheroids with constant dimension within a series of experiments is important to obtain reliable results. Reproducibility of the hanging drop system is also promoted by its technical conditions that are completely free of contact with any artificial support of surface or matrices.

Although this method is characterized by several advantages, it is not applicable to high-throughput drug screenings since the massive production of spheroids is difficult and the maximum volume per drop is only 50 μL, including the drug test medium. Furthermore, the system is difficult to handle, since quality of the spheroids strongly depends on technical skills, in particular the capability of maintaining structured spheroids during all the passages that involve medium removal and change. In summary, the hanging drop method is inexpensive using standard plates but difficult to handle and reproduce. In contrast, the use of commercially available specialized plates makes the system easy to perform but leads to an increase in costs [90].

14.4.2 Microcarrier Beads and Scaffold-Based Culture with Natural or Synthetic Solid Materials

To facilitate spheroid formation for cells that do not spontaneously aggregate, microcarrier beads are used since they allow attachment and aggregation of cells, thus facilitating cell–cell interaction and the consequent formation of spheroidal structures (Fig. 14.3B). Several bead types are available, differing in size, composition, and surface coating. Due to the wide surface area, these carriers allow obtaining high cell densities. Once cells adhere to the surface of the carriers, they proliferate creating mini-spheroids that, in turn, can aggregate one to each other, forming spheroids with bigger dimensions. Another advantage of this method is represented by the possibility of growing on the same bead different cell types, thus obtaining heterogeneous spheroids in which intercellular cross talk is mimicked [91]. This makes them a useful tool for studying the interactions between tumor cells and microenvironment. Unfortunately, due to the large surface of the beads, the spheroids produced by this method are mainly composed of the microcarrier support, making them poorly comparable to the physiological situation.

Another system used to facilitate spheroid formation in cells that do not spontaneously aggregate is the use of natural or synthetic prefabricated solid scaffolds. Cells are seeded on these porous supports, and they migrate along the surface and aggregate creating 3D structures. Growing spheroids increase in dimension and fill the interstices within the scaffold, creating well-organized 3D spheroids [92]. Several different materials can be used, such as collagen, chitosan, or synthetic polymers like D,D,–L,L polylactic acid. These materials very well support 3D long-term cell growth and differentiation, allowing cells to more closely mimic *in vivo* cell morphogenesis and differentiation. Moreover, the availability of different scaffolds with various density, porosity, and flexibility provides the possibility of easily finding the best option for each cell line and experimental need. Due to the nature of the scaffold material the results may vary depending on the polymer's composition.

14.4.3 Pluripotent Stem Cells for Modeling Human Cancer

Since 2008, when Yoshiki Sasai discovered that stem cells can be guided to form three-dimensional structures of neural cells [93], scientists are aiming at rebuilding organ parts or even complete organs from pluripotent stem cells. Until today, different protocols have been established to derive micro-stomachs, brains, breasts, kidneys, and many others from human or induced pluripotent stem cells [94].

More recently, this feature of pluripotent stem cells was applied to model gastric cancer in a three-dimensional fashion. McCracken and colleagues induced human pluripotent stem cells to form gastric tissue, growing as 3D structures in Matrigel [95]. These organoids were next infected with *Helicobacter pylori*, a bacterium linked to chronic gastritis and an increased risk of developing adenocarcinoma of the stomach [96]. After infection, significant epithelial responses were observed, including robust phosphorylation of c-Met and increased cell proliferation.

Huang et al. published a similar but more specific approach to mirror ductal pancreatic cancer in a three-dimensional fashion [97]. They subsequently induced human embryonic stem cells to differentiate into defined endoderm, polarized organoids growing on Matrigel, and finally to pancreatic progenitors. In order to guide the cells to the specific fate, different growth factor and nutrients critical for pancreas development were subsequently added to the culture medium. After that, differentiated organoids were transduced to model pancreatic cancer to express mutant variants of the KRAS or TP53 gene, the most frequently observed mutations in pancreatic adenocarcinomas. Injected into mice, transduced organoids showed abnormal ductal architecture and nuclear morphology consistent with neoplastic transformation, showing the feasibility of the protocol to model human cancer.

Although this kind of protocol is rather difficult to establish, stem cells, once established, represent an unlimited source of cancer organoids bearing known mutant variations of specific onco- or tumor suppressor genes. This offers a state-of-the-art methodology to study the impact of specific mutations and their combination to drug response and tumor behavior.

14.4.4 Selection of the Most Appropriate Model

Since several 3D models with different characteristics are available, it is fundamental to choose the most appropriate approach based on the experimental aim of the study. Every 3D culture method has its own characteristics and therefore can meet different experimental needs (Table 14.1).

Table 14.1 Summary of advantages and disadvantages of 3D culturing methods/models based on the final experimental application

Application	Method/model	Advantages	Disadvantages
Massive spheroid production	Spinner flasks/ gyratory rotators	Simple to perform	Need of specialized equipment
		Spheroids with controllable and reproducible size	High shear forces
		Long-term coculture	
	NASA microgravity system	Simple to perform	Need of specialized equipment
		Spheroids with controllable and reproducible size	Expensive
		Long-term coculture possible	
		Better differentiation	
Morphological studies	Hanging drop method	Inexpensive	Labor intensive
		Easy to perform	
		Spheroids with controllable and reproducible size	
	Liquid overlay technique	Suitable for many cell line	Limited amount of spheroids
		Easy to perform	Variability in spheroid size
		Suitable for coculture experiments	
Tumor–microenvironment cross talk studies	Heterotypic coculture spheroids	Increased level of complexity that better mimics *in vivo* tumors	Difficult to determine the influence of the single components
		Influence of stroma, angiogenesis, immune system	Labor intensive
			Experimental variability
Cell migration and invasion analysis	Hydrogel matrices	Good 3D extracellular support	Labor intensive
		Flexible, possibility of modifying matrix stiffness based on experimental needs	Experimental variability
			Expensive

(continued)

Table 14.1 (continued)

Application	Method/model	Advantages	Disadvantages
Low-throughput drug screening	Tissue slices	Direct analysis of drug efficacy on *ex vivo* tumor tissue	Limited amount of starting material
		Influence of tumor microenvironment	Low reproducibility
			Only few analytical techniques can be applied
	Tissue organoids	Closely recapitulate *in vivo* tumors	Absence of tumor microenvironment
		Cell expansion allows increasing the number of drugs and drug concentrations that can be tested	
		Maintenance of cell heterogeneity	
		Based on the 3D culture method, several analytical assays can be applied	
High-throughput drug screening (HTS)	Cellular spheroids	Highly pure cell population	Loss of tumor heterogeneity
		Stem cell sensitivity analysis	
		Easy to handle	
	Tissue organoids	Closely recapitulate *in vivo* tumors	Not all the 3D techniques are applicable to HTS
		Cell expansion allows increasing the number of drugs and drug concentrations that can be tested	Absence of tumor microenvironment
		Maintenance of cell heterogeneity	
		Based on the 3D culture method, several analytical assays can be applied	

(continued)

Table 14.1 (continued)

Application	Method/model	Advantages	Disadvantages
3D culture of cells with low aggregation capacities	Microcarrier beads/solid scaffolds	Good 3D extracellular support	Variability between scaffolds
		Availability of materials with different chemistry, porosity, and density	Biomechanical influence on tumor cell behavior
		More physiological cell morphology and differentiation than 2D systems	
Precision medicine	Tissue slices	Direct analysis of *ex vivo* tumor tissue	Limited amount of material for the analysis
		Reduced time needed for testing (no need of 3D culture formation and expansion)	Low reproducibility
	Tissue organoids	Closely recapitulate *in vivo* tumors	Limited amount of starting material
		Cell expansion allows increasing the number of drugs and drug concentrations that can be tested	Absence of tumor microenvironment
		Maintenance of cell heterogeneity	
		Based on the 3D culture method, several analytical assays can be applied	

First, it is important to consider whether the goal is the massive production of cellular spheroids for further analyses or if the main focus is to study cellular morphology and behavior. In the first case, it is possible to use methods, such as spinner flasks and gyratory rotators, that allow for the production of great amounts of spheroids with controllable and reproducible size but that do not permit any kind of direct analysis. Indeed, when the goal is to proceed with morphological studies or molecular analysis, low-scale production methods have to be preferred since they permit the "on-site," direct study of the spheroid characteristics of interest.

One of the most important aspects to take into account is the level of *in vivo* complexity that has to be maintained. For example, if the main focus is on the role of tumor microenvironment and how the cross talk between tumor and stroma cells influences tumor progression, it is useful to recapitulate this heterogeneity by using 3D spheroids composed of different cell types, such as tumor epithelial cells, fibroblast, and endothelial cells.

On the other hand, if a high-throughput approach is required, for example, in drug-screening experiments, it is very important to use a method that is easy to manipulate and to analyze, therefore being applicable to 384-well plate format and characterized by a high level of reproducibility. When the readout of the analysis is based on optical imaging, polymers and scaffolds need to be used carefully because layer depth and transparency need to be taken into account. Moreover, when the goal is to work with human tissue to specifically analyze the patient's tumor profile in order to determine the best therapeutic approach, methods that allow working with limited amounts of starting material have to be chosen.

14.4.5 Innovative *In Vitro* Approaches for Precision Medicine: Organ-on-Chip

Over the last few years, efforts have been made to develop new, more complex *in vitro* systems that efficiently mimic human organs. One of the more innovative models is the "organ-on-chip," bioengineered devices that accurately simulate structure and function of human organs, developed to study physiological and pathological conditions. These models are composed of several interconnected compartments that allow for cell to cell crosstalk in a dynamic microenvironment where signaling molecules and metabolites can be exchanged through a perfusion system.

Several organ-on-chip models have been developed for liver [98–100], gut [101, 102], lung [103, 104], and heart [105, 106]. These models are very promising tools not only for the study of tissue development and pathological progression but also for testing drug efficacy and toxicity. Perfusion systems can in fact be used to deliver therapeutic agents simulating blood supply and replicating drug delivery *in vivo*. The development of more precise organ-on-chip models will improve the clinical translation of drug efficacy testing and reduce the use of animals in preclinical analysis.

Organ-on-chip can be developed using cells grown in 2D or 3D structures. Moreover, it is possible to introduce different cell types, for example, tumor epithelial cells and endothelial cells, which together with a perfusion system allow for exchange of signaling molecules and metabolites. Due to their dynamic characteristics and the possibility of building different organs on the same chip [107], these models allow for the study of both drug efficacy and mechanisms of tumor cell dissemination through the bloodstream and colonization of secondary tissues.

Another aspect that makes organ-on-chip very attractive for antitumor research is the possibility of introducing immune cells in the system developing the so-called immunocompetent cancer chips. These models have been originally established to study inflammatory diseases in several organs, such as gut [101, 102], lung [108, 109], and skin [110]. Considering the increasing interest in immune system response, modulation for cancer therapy and the limitations correlated with the use of animal models in this field, immunocompetent microfluidic *in vitro* systems represent an attractive solution for drug development and efficacy testing.

Finally, the promising application of induced human pluripotent stem cells (iPS) in the development of differentiated tissues gives rise to the possibility of building organ-on-chip using PS cells directly isolated from cancer patients. The improvement of these multi-organ human-on-chip models, matching the patient genetic profile, will therefore strongly improve precision medicine.

14.5 Patient Tumor-Derived Organoids as Tool for Drug Screening, Precision Medicine, and Biomarker Discovery

14.5.1 Drug Screening

The determination of drug efficacy, as well as the discovery of new drug targets, strongly requires experimental models that recapitulate the physiological condition as much as possible. It is generally accepted that *in vivo* animal models represent the best approach to studying drug sensitivity and toxicity since they allow for the analysis of treatment effect on an entire organism. However, as highlighted above, the use of animal models is often a limiting condition due to ethical and economic aspects, as well as to the high level of variability that makes data more difficult to validate. Moreover, due to a different physiology and metabolism, drug sensitivity testing in animal models is rarely predictive of real clinical efficacy. In addition, both the high number of new drug candidates that need to be tested and the increasing use of combination therapies that require the testing of simultaneous treatment with different drugs have increased the need for high-throughput screening methods. For all these reasons, in the last decades, the use of animal models for drug testing has decreased with a constant increment in the development and use of *in vitro*/*ex vivo* approaches. In this context, the advances in the field of 3D cultures have allowed researchers and pharmaceutical companies to test new compounds in a system that is closer to the physiological situation compared to standard 2D models, increasing the probability of success in translating the results into animal and human trials.

Several works have been published in which protocols for high-throughput drug screening with 3D cultures have been developed [111–113]. Friedrich et al. described an easy-handling protocol for spheroid establishment by liquid overlay technique on agar-coated plates and their treatment [114]. This protocol requires 96 h for spheroid formation and 72 h incubation with the drugs. Following the treatment, several analytical endpoints have been considered: spheroid integrity was analyzed by phase contrast imaging, cell integrity and viability were calculated by an acid phosphatase assay that was established by the same group [115] and that does not require spheroid dissociation, and spheroid growth delay was determined by analyzing spheroid volume kinetics. They also provided a list of human carcinoma cell lines partly selected by the NCI-DTP 60-cell line panel that can be used for drug screening producing treatable spheroids with a high level of reproducibility.

To improve screening capacity, Lovitt et al. developed a 3D cancer model in a miniaturized 384-well plate format growing cells onto a Matrigel layer [116]. They tested proliferation and cell viability in different cell lines and following diverse treatments, and they demonstrated the high reproducibility of the assay. In particular, they analyzed spheroid size and cellular metabolic activity using resazurin. For both analyses, different technologies have been used to read assay output demonstrating the versatility of the system and its applicability to different laboratory setups. Boehnke et al. describe that high-throughput drug screening of 3D tumor cultures in 384-well format is feasible and represents a robust tool for future drug discovery [117].

Several new approaches have been proposed for spheroid analysis after drug treatment. Klein et al. described the use of optical coherence tomography for visualizing, monitoring, and quantifying growth and treatment response dynamics [118]. With this technology, it is possible to monitor—constantly and without any external perturbation of the system—spheroid growth after treatment over the course of hours or days.

Most of the current 3D models used for drug screening are homogeneous systems composed of one single cell type that is usually cancer epithelial cells. Considering the fundamental role of the microenvironment in tumor development and progression, the last few years have seen a development of several 3D heterocellular systems in which fibroblast, endothelial, and/or immune cells were introduced. These models are more difficult to handle and to analyze due to the presence of components with different characteristics and behavior, but they are able to better mimic the *in vivo* situation, representing a step forward in the development of 3D *in vitro* systems that can replace, at least in part, the use of animal models and that can be applied to treatment selection. Kenny and colleagues have recently published a paper in which they showed that heterogeneous 3D culture of tumor microenvironment can be adapted for quantitative high-throughput screening [119]. In particular, they developed a heterotypic coculture assay containing human fibroblasts, mesothelial cells, and extracellular matrix to reproduce the superficial tissue layers encountered by ovarian cancer cells during metastasis. Applying this system to 384- and 1536-well format screenings, they identified compounds able to inhibit the early steps of ovarian cancer metastasis. The efficacy of these compounds was confirmed in *in vivo* assays with two different cell lines, demonstrating the predictive value of this system for *in vivo* efficacy determination.

In summary, the application of 3D culture systems as disease-specific human drug-screening models has enormous potential to connect compound screening and clinical trials. Importantly, a successful establishment of novel and complex technologies in the drug discovery pipeline requires accurate assay validation and implementation possibilities for automatic robotic platforms into the workflow. In addition, novel experimental setups and readouts should allow for the evaluation of compound combination assays. The goals of multidrug-screening platforms are to achieve enhanced potency by exploiting greater-than-additivity therapeutic effects, to lower the doses of drugs with non-overlapping toxicity, and to delay drug resistance. The evaluation of drug combinations requires the selection of the proper assay format and, more importantly, the assessment of different algorithms to obtain precise results of drug–drug interaction studies [120–122]. Moreover, the potential value of pharmacodynamically synergistic drug combinations involves the investigation of the modes of action from the perspectives of coordinated molecular interactions and network regulation [123].

14.5.2 Precision Medicine

In the last decades, the presence of patient-specific differences in drug response including efficacy and side effects has prompted many researchers to focus on the determination of the key molecules that play a role in this phenomenon, leading to the foundation of "pharmacogenomics." Moreover, with advances in the development and use of so-called molecular therapeutics, drugs that act by targeting specific molecules known to be determinants of cancer growth and spreading, the study of the differences between individuals at the molecular level has become one of the main interests in the cancer field. The National Academy of Science has defined "precision medicine" as "the use of genomic, epigenomic, exposure and other data to define individual patterns of disease, potentially leading to better individual treatment" [124]. The possibility of stratifying patients for the most appropriate treatment could lead to an important reduction of side effects, as well as to an improvement of patients' quality of life.

Due to the great number of available anticancer drugs and the possibility of drug combination(s), there is a strong need for new, easy-to-handle, high-throughput assays for the evaluation of treatment efficacy in individual tumors. 3D patient-derived organotypic cultures represent a useful tool for this application since they are close enough to the physiological situation to efficiently mimic the tumor behavior *in vivo*, and, compared to animal models, they allow for screening of a greater number of single as well as of combination drugs (Fig. 14.4). On a 384-well plate, it is possible to test three pairs of compounds in a full 6×6 matrix per patient within the same time frame as single drugs. An analogous xenograft experiment would require an unfeasible number of animals. Also, 3D *in vitro* assays allow obtaining drug response information in a few weeks compared to the months required for *in vivo* experiments (Fig. 14.5).

Several recent articles demonstrated the feasibility of using patient-derived tumor tissues for the determination of the most appropriate treatment. Vaira and colleagues isolated *ex vivo* organotypic slice cultures of human tumor specimens, and they showed how epithelial–stromal relationship, tumor tissue morphology, proliferation, and viability were preserved after culturing. Moreover, at the molecular level, pathway activity was maintained. In particular, they focus on the activation of the PI3K/Akt pathway, and they demonstrated that treating tumor slices with PI3K inhibitor LY294002 induced a partial decrease in tissue proliferation and increase in apoptosis with a general decrement in cell viability over time with increasing concentration of the drug. Even if limited to one treatment, this analysis showed the utility of this approach in predicting patient tumor sensitivity to drugs [125].

Other groups applied this approach to study tumor response to chemotherapeutic drugs [126] and irradiation treatment [127] in head and neck squamous carcinomas

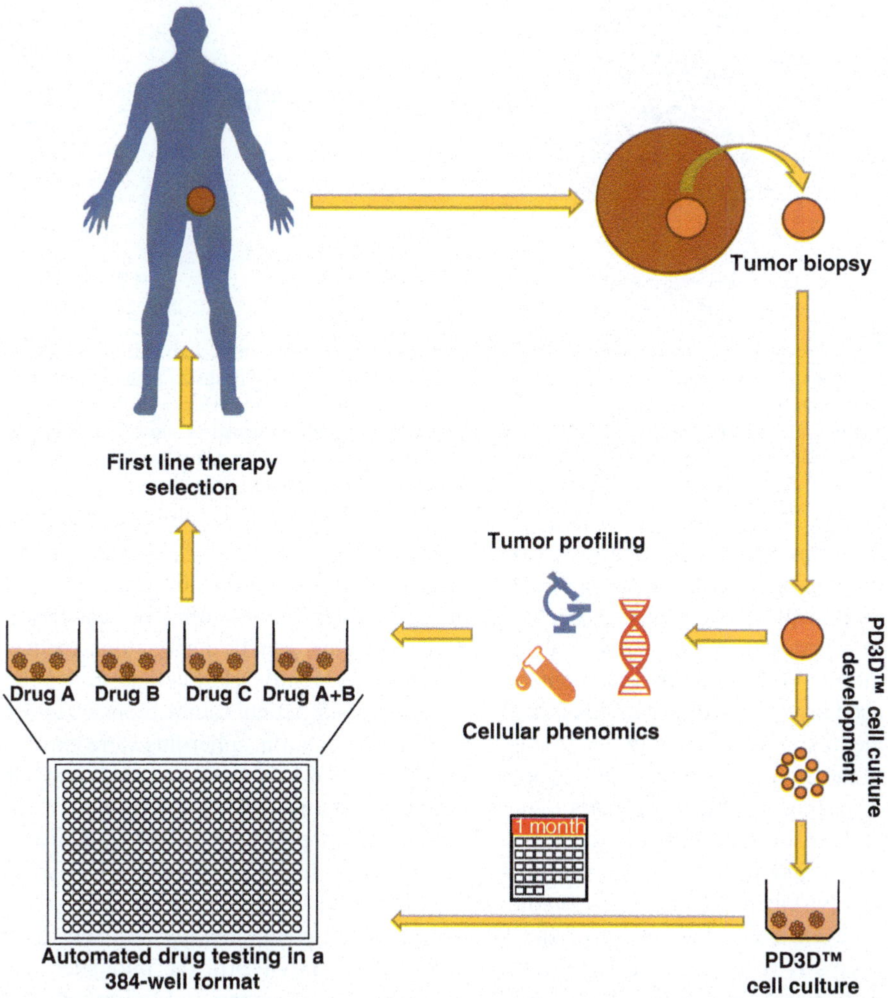

Fig. 14.4 Schematic representation of PD3D cell cultures as future tool for treatment selection. PD3D cell cultures are grown directly from fresh patient material and expanded. Based on tumor profiling, a panel of available drugs is selected and tested on PD3D cultures as single treatment or in combination. Efficacy data obtained *in vitro* on 3D spheroids could in the future be translated into clinic to improve therapy selection

and glioblastoma multiforme, respectively. Patient-derived tissue slices before and after treatment were histologically analyzed for the expression of markers for proliferation (Ki67), apoptosis/cell death (cleaved-caspase 3/propidium iodide), and DNA double-strand breaks (γH2AX).

Not only tissue slices but also 3D tumor spheroids represent feasible tools for personalized therapy [72, 74]. Gao et al. developed a protocol for isolating 3D spheroids from prostate cancer specimens, and they showed that these organoids

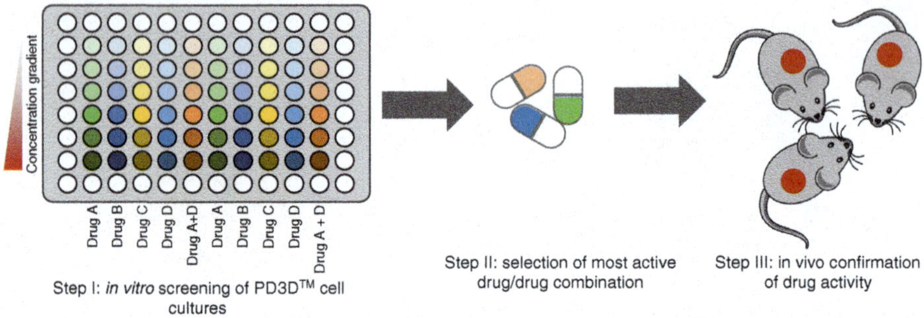

Fig. 14.5 *In vitro* prescreening of candidate drugs for selecting the most active drugs/drug combinations. *In vitro* screening of PD3D cell cultures allows for testing drugs as single agents or in combination in a system that better mimic physiological conditions. The efficacy of the most active compounds selected by the *in vitro* prescreening can then be confirmed *in vivo*. This approach allows for testing of a greater number of drugs/drugs combinations, for reducing the time needed to obtain drug efficacy data, and for strongly decreasing the number of animals used

recapitulate the molecular diversity of prostate cancer subtypes [128]. The possibility of generating a collection of patient-derived tumor spheroids that maintain inter-individual heterogeneity makes this model a useful tool for selecting the best therapy in a patient-specific manner, as well as for studying genetic lesions that mediate tumor progression and resistance to treatment. Recently, Halfter and colleagues used breast cancer organoids directly generated from patient tumor biopsies to predict response to neoadjuvant therapy [129]. After isolation, spheroids were exposed to the equivalent therapeutic agents, and cell survival was measured. Survival data were then correlated to pathological complete response (pCR) determined at surgery. Even if the authors used a limited number of samples with the subsequent need for an independent validation of the results, they showed that this model was a highly sensitive and specific predictor of pCR in the cohort of patients analyzed.

Even if currently none of the available *in vitro* chemotherapy and sensitivity assays is recommended for the use in clinical settings [130], the published data demonstrate that 3D *in vitro* approaches represent a promising tool to improve personalized therapy and to help researchers in understanding the mechanisms of tumor resistance.

14.5.2.1 Treatment Tailoring After Recurrence

Currently, the main interest of pharmaceutical companies is the discovery of tumor targets for the development of new cytotoxic drugs. Since tumor recurrence is often caused by drug resistance, it is fundamental not only to expand the panel of available anticancer drugs but also to better understand the mechanisms that allow cells to bypass drug activity, with the aim to develop new methods for circumventing tumor resistance. An interesting characteristic of cellular spheroids is that cancer cells grown in 3D show a similar resistance behavior observed in patient tumors. This is due to the fact that cells grown as spheroids produce more extracellular matrix, and the crosstalk between tumor cells and microenvironment makes cells less sensitive to anticancer treatment [131, 132]. Based on this feature, 3D cancer spheroids represent a very useful tool for the study of drug resistance. Moreover, the possibility of culturing 3D organotypic spheroids directly from the patient tumor allows the use of this system for direct treatment selection after recurrence. It is in fact possible to grow 3D organoids from the relapsed tumor and test the post-recurrence cells with a complete drug panel, allowing the selection of a second-line therapy (Fig. 14.6a). The possibility of expanding the organotypic cultures makes them a useful tool also for determining tumor resistance *in vitro*, therefore potentially predicting the relapse in the patient. After first-line treatment selection, 3D organoids can be treated with the same regimen of the patient recreating the selective pressure to which tumor cells are subjected during therapy. If cells develop resistance, it is possible to retest the cells with a second panel of drugs, thus allowing the selection of second-line therapy. This approach would strongly improve the management of tumor patients since it would be possible to earlier predict resistance to treatment and to select a new therapy (Fig. 14.6b).

Moreover, the analysis of cellular spheroids from tumors before and after recurrence can help to understand the mechanisms of resistance giving important information that can be used for the improvement of the current treatment regimens, for the development of new anticancer drugs, and also for the identification of predictive biomarkers.

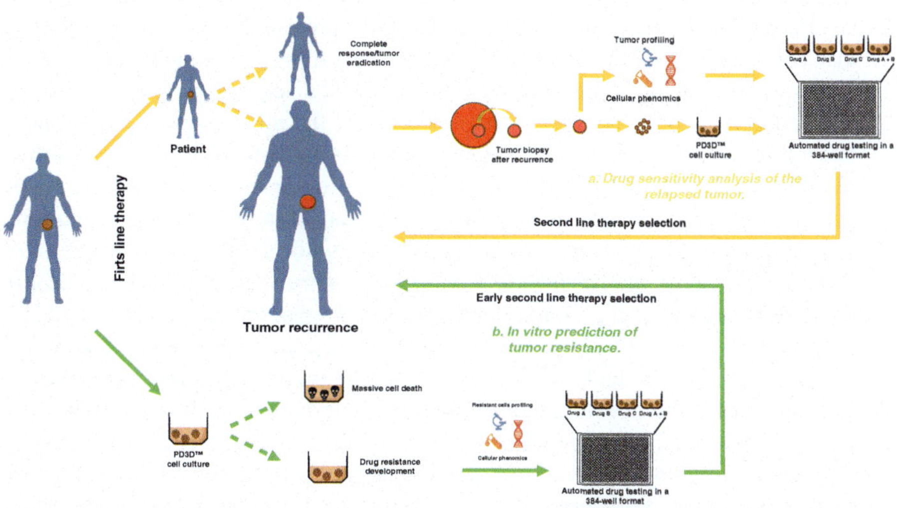

Fig. 14.6 Schematic representation of PD3D cell cultures as tool for treatment selection after recurrence and for prediction of tumor resistance. (**a**) After recurrence, PD3D cells can be grown from the relapsed tumor and tested with a second panel of drugs to determine second-line therapy. (**b**) During first line therapy, PD3D cells grown from the original tumor can be treated *in vitro* with the same regiment of the patient. If cells develop resistance before clinical relapse is detectable, a second panel of drugs can be screened to early select a new therapy

14.5.3 Biomarker Discovery

3D cultures derived from patients' tumor specimens can be applied to both drug screening and treatment selection to improve anticancer therapy efficacy. Moreover, since this model maintains the molecular characteristics of the tumor of origin, it represents a potential tool for biomarker discovery. Currently, new prognostic and predictive biomarkers are determined by direct analysis of tumor specimens. Limitations to this approach are the accessibility of primary material, as well as the variability in handling this material, which could cause alterations in cellular phenomics, thus possibly strongly influencing the results of the analysis. 3D *in vitro* cell cultures allow bypassing these limitations. By growing cells *in vitro*, it is in fact possible to expand tumor cells, thus increasing the amount of tissue available for the analysis, and, simultaneously, to maintain intact the intracellular activity, thus avoiding alterations at the molecular profile level. The generation of large collections of patient-derived 3D cultures may therefore represent a potential biobank for population studies in which these *in vitro* models are used as representative of the original tumor for biomarker determination. Moreover, the possibility of storing 3D cultures for long periods of time allows to create a collection of tumor specimens with long-term follow-up that can be used for retrospective studies.

Conclusions

In the last decades, the commonly used "trial-and-error" approach for treatment determination has partly been replaced by more precise, marker-guided methods that, considering the individual molecular profile, allow decreasing side effects and improving treatment efficacy. To be useful in clinical settings, prognostic and predictive assays need to mimic as close as possible the physiological conditions to allow the direct translation of results to patient management. In this perspective, 3D cultures directly generated from fresh patient tumor tissue have a great potential for improving cancer patients outcome. As shown in this chapter, 3D cultures can already be used to study the mechanisms of tumor development, progression, and resistance to therapy but also as a direct tool for drug screening. After appropriate clinical studies, in the future, it may be possible that PD3D cultures could also have an impact on clinical decision-making in cancer medicine and treatment.

Together with the lack of efficacy, one of the main causes of drug failure in clinical trials is the development of previously undefined toxic side effects. For this reason, there is an urgent need for new models translating the *in vitro* toxicity after acute and chronic exposure to chemicals into clinical trials. 3D cultures represent a promising approach in this field. Some recent publications have in fact demonstrated the feasibility of using 3D models for the determination of both nephro- and hepatic-toxicity [133–136]. Together with supporting the determination of drug toxicity, this approach could also help to highlight and explain the interindividual differences in drug transformation and activation.

Due to their versatility, 3D models represent an efficient tool for translational research, and the constant improvement in 3D culture technologies increases the potential of these *in vitro* methodologies in supporting clinical decisions.

References

1. Ferlay J, Steliarova-Foucher E, Lortet-Tieulent J, Rosso S, Coebergh JWW, Comber H, et al. Cancer incidence and mortality patterns in Europe: estimates for 40 countries in 2012. Eur J Cancer. 2013;49(6):1374–403.
2. Harrison R. Observations on the living developing nerve fiber. Anat Rec. 1907;1:116–28.
3. Gillet JP, Calcagno AM, Varma S, Marino M, Green LJ, Vora MI, et al. Redefining the relevance of established cancer cell lines to the study of mechanisms of clinical anti-cancer drug resistance. Proc Natl Acad Sci U S A. 2011;108(46):18708–13.
4. Gazdar AF, Gao B, Minna JD. Lung cancer cell lines: useless artifacts or invaluable tools for medical science? Lung Cancer. 2010;68(3):309–18.
5. Kirk R. Genetics: personalized medicine and tumour heterogeneity. Nat Rev Clin Oncol. 2012;9(5):250.
6. Lima SC, Hernandez-Vargas H, Herceg Z. Epigenetic signatures in cancer: implications for the control of cancer in the clinic. Curr Opin Mol Ther. 2010;12(3):316–24.
7. Toyota M, Issa JP. Epigenetic changes in solid and hematopoietic tumors. Semin Oncol. 2005;32(5):521–30.
8. Calon A, Lonardo E, Berenguer-llergo A, Espinet E, Hernando-momblona X, Iglesias M, et al. Stromal gene expression defines poor-prognosis subtypes in colorectal cancer. Nat Genet. 2015;47(February):320–9.
9. Kalus M, Delmonte L, Ghidoni JJ, Liebelt RA. Transplantation of mouse mammary carcinoma through matrix tissue cultures. Tex Rep Biol Med. 1968;26(4):517–24.
10. Inch WR, McCredie JA, Sutherland RM. Growth of nodular carcinomas in rodents compared with multi-cell spheroids in tissue culture. Growth. 1970;34(3):271–82.
11. Sutherland RM, McCredie JA, Inch WR. Growth of multicell spheroids in tissue culture as a model of nodular carcinomas. J Natl Cancer Inst. 1971;46(1):113–20.
12. Yuhas JM, Li AP, Martinez AO, Ladman AJ. A simplified method for production and growth of multicellular tumor spheroids. Cancer Res. 1977;37(10):3639–43.
13. Yuhas JM, Tarleton AE, Harman JG. In vitro analysis of the response of multicellular tumor spheroids exposed to chemotherapeutic agents in vitro or in vivo. Cancer Res. 1978;38(11 Pt 1):3595–8.
14. Yang J, Richards J, Bowman P, Guzman R, Enami J, McCormick K, et al. Sustained growth and three-dimensional organization of primary mammary tumor epithelial cells embedded in collagen gels. Proc Natl Acad Sci U S A. 1979;76(7):3401–5.
15. Mauchamp J, Margotat A, Chambard M, Charrier B, Remy L, Michel-Bechet M. Polarity of three-dimensional structures derived from isolated hog thyroid cells in primary culture. Cell Tissue Res. 1979;204(3):417–30.
16. Bhowmick NA, Neilson EG, Moses HL. Stromal fibroblasts in cancer initiation and progression. Nature. 2004;432(7015):332–7.
17. Sternlicht MD, Lochter A, Sympson CJ, Huey B, Rougier JP, Gray JW, et al. The stromal proteinase MMP3/stromelysin-1 promotes mammary carcinogenesis. Cell. 1999;98(2):137–46.
18. Paszek MJ, Zahir N, Johnson KR, Lakins JN, Rozenberg GI, Gefen A, et al. Tensional homeostasis and the malignant phenotype. Cancer Cell. 2005;8(3):241–54.
19. Wiseman BS, Sternlicht MD, Lund LR, Alexander CM, Mott J, Bissell MJ, et al. Site-specific inductive and inhibitory activities of MMP-2 and MMP-3 orchestrate mammary gland branching morphogenesis. J Cell Biol. 2003;162(6):1123–33.

20. Heppner GH, Miller BE. Tumor heterogeneity: biological implications and therapeutic consequences. Cancer Metastasis Rev. 1983;2(1):5–23.
21. Fidler IJ, Kripke ML. Metastasis results from preexisting variant cells within a malignant tumor. Science. 1977;197(4306):893–5.
22. Fidler IJ. Tumor heterogeneity and the biology of cancer invasion and metastasis. Cancer Res. 1978;38(9):2651–60.
23. Nowell PC. The clonal evolution of tumor cell populations. Science. 1976;194(4260):23–8.
24. Gerlinger M, Rowan AJ, Horswell S, Larkin J, Endesfelder D, Gronroos E, et al. Intratumor heterogeneity and branched evolution revealed by multiregion sequencing. N Engl J Med. 2012;366(10):883–92.
25. de Bruin EC, McGranahan N, Mitter R, Salm M, Wedge DC, Yates L, et al. Spatial and temporal diversity in genomic instability processes defines lung cancer evolution. Science. 2014;346(6206):251–6.
26. Sottoriva A, Kang H, Ma Z, Graham TA, Salomon MP, Zhao J, et al. A Big Bang model of human colorectal tumor growth. Nat Genet. 2015;47(3):209–16.
27. Reya T, Morrison SJ, Clarke MF, Weissman IL. Stem cells, cancer, and cancer stem cells. Nature. 2001;414(6859):105–11.
28. Barker N, Ridgway RA, van Es JH, van de Wetering M, Begthel H, van den Born M, et al. Crypt stem cells as the cells-of-origin of intestinal cancer. Nature. 2009;457(7229):608–11.
29. Shackleton M, Quintana E, Fearon ER, Morrison SJ. Heterogeneity in cancer: cancer stem cells versus clonal evolution. Cell. 2009;138(5):822–9.
30. Ding L, Ley TJ, Larson DE, Miller CA, Koboldt DC, Welch JS, et al. Clonal evolution in relapsed acute myeloid leukaemia revealed by whole-genome sequencing. Nature. 2012;481(7382):506–10.
31. Schilsky RL. Personalized medicine in oncology: the future is now. Nat Rev Drug Discov. 2010;9(5):363–6.
32. Mitsiades CS, Davies FE, Laubach JP, Joshua D, San Miguel J, Anderson KC, et al. Future directions of next-generation novel therapies, combination approaches, and the development of personalized medicine in myeloma. J Clin Oncol. 2011;29(14):1916–23.
33. Trusheim MR, Burgess B, Hu SX, Long T, Averbuch SD, Flynn AA, et al. Quantifying factors for the success of stratified medicine. Nat Rev Drug Discov. 2011;10(11):817–33.
34. Rygaard J, Povsen CO. Heterotransplantation of a human malignant tumour to "nude" mice. 1969. APMIS. 2007;115:604–6.
35. Bosma GC, Custer RP, Bosma MJ. A severe combined immunodeficiency mutation in the mouse. Nature. 1983;301:527–30.
36. Cook N, Jodrell DI, Tuveson DA. Predictive in vivo animal models and translation to clinical trials. Drug Discov Today. 2012;17(5–6):253–60.
37. Tentler JJ, Tan AC, Weekes CD, Jimeno A, Leong S, Pitts TM, et al. Patient-derived tumour xenografts as models for oncology drug development. Nat Rev Clin Oncol. 2012;9(6):338–50.
38. Moro M, Bertolini G, Tortoreto M, Pastorino U, Sozzi G, Roz L. Patient-derived xenografts of non small cell lung cancer: resurgence of an old model for investigation of modern concepts of tailored therapy and cancer stem cells. J Biomed Biotechnol. 2012;2012:568567.
39. Fichtner I, Rolff J, Soong R, Hoffmann J, Hammer S, Sommer A, et al. Establishment of patient-derived non-small cell lung cancer xenografts as models for the identification of predictive biomarkers. Clin Cancer Res. 2008;14(20):6456–68.
40. Zhuo Y, Wu Y, Guo A, Chen S, Su J. Establishment of patient-derived lung cancer xenograft models and study for its biological characteristics. Zhongguo Fei Ai Za Zhi. 2010;13(6):568–74.
41. Jin K, He K, Han N, Li G, Wang H, Xu Z, et al. Establishment of a PDTT xenograft model of gastric carcinoma and its application in personalized therapeutic regimen selection. Hepatogastroenterology. 2011;58(110–111):1814–22.
42. Scherer WF, Syverton JT, Gey GO. Studies on the propagation in vitro of poliomyelitis viruses. IV. Viral multiplication in a stable strain of human malignant epithelial cells (strain HeLa) derived from an epidermoid carcinoma of the cervix. J Exp Med. 1953;97(5):695–710.

43. Kenny PA, Lee GY, Myers CA, Neve RM, Semeiks JR, Spellman PT, et al. The morphologies of breast cancer cell lines in three-dimensional assays correlate with their profiles of gene expression. Mol Oncol. 2007;1(1):84–96.
44. Silberstein GB. Tumour-stromal interactions. Role of the stroma in mammary development. Breast Cancer Res. 2001;3(4):218–23.
45. Schmeichel KL, Bissell MJ. Modeling tissue-specific signaling and organ function in three dimensions. J Cell Sci. 2003;116(Pt 12):2377–88.
46. Curcio E, Salerno S, Barbieri G, De Bartolo L, Drioli E, Bader A. Mass transfer and metabolic reactions in hepatocyte spheroids cultured in rotating wall gas-permeable membrane system. Biomaterials. 2007;28(36):5487–97.
47. Mueller-Klieser W. Method for the determination of oxygen consumption rates and diffusion coefficients in multicellular spheroids. Biophys J. 1984;46(3):343–8.
48. Alvarez-Perez J, Ballesteros P, Cerdan S. Microscopic images of intraspheroidal pH by 1H magnetic resonance chemical shift imaging of pH sensitive indicators. MAGMA. 2005;18(6):293–301.
49. Rodenhizer D, Gaude E, Cojocari D, Mahadevan R, Frezza C, Wouters BG, et al. A three-dimensional engineered tumour for spatial snapshot analysis of cell metabolism and phenotype in hypoxic gradients. Nat Mater. 2016;15:227–34.
50. van der Kuip H, Murdter TE, Sonnenberg M, McClellan M, Gutzeit S, Gerteis A, et al. Short term culture of breast cancer tissues to study the activity of the anticancer drug taxol in an intact tumor environment. BMC Cancer. 2006;6:86.
51. Milani C, Welsh J, Katayama ML, Lyra EC, Maciel MS, Brentani MM, et al. Human breast tumor slices: a model for identification of vitamin D regulated genes in the tumor microenvironment. J Steroid Biochem Mol Biol. 2010;121(1–2):151–5.
52. Estes JM, Oliver PG, Straughn Jr JM, Zhou T, Wang W, Grizzle WE, et al. Efficacy of anti-death receptor 5 (DR5) antibody (TRA-8) against primary human ovarian carcinoma using a novel ex vivo tissue slice model. Gynecol Oncol. 2007;105(2):291–8.
53. Kiviharju-af Hallstrom TM, Jaamaa S, Monkkonen M, Peltonen K, Andersson LC, Medema RH, et al. Human prostate epithelium lacks Wee1A-mediated DNA damage-induced checkpoint enforcement. Proc Natl Acad Sci U S A. 2007;104(17):7211–6.
54. Mackenzie IC, Fusenig NE. Regeneration of organized epithelial structure. J Invest Dermatol. 1983;81:189s–94s.
55. Stark HJ, Baur M, Breitkreutz D, Mirancea N, Fusenig NE. Organotypic keratinocyte cocultures in defined medium with regular epidermal morphogenesis and differentiation. J Invest Dermatol. 1999;112:681–91.
56. Szabowski A, Maas-Szabowski N, Andrecht S, Kolbus A, Schorpp-Kistner M, Fusenig NE, et al. c-Jun and JunB antagonistically control cytokine-regulated mesenchymal-epidermal interaction in skin. Cell. 2000;103(5):745–55.
57. Barton CE, Johnson KN, Mays DM, Boehnke K, Shyr Y, Boukamp P, et al. Novel p63 target genes involved in paracrine signaling and keratinocyte differentiation. Cell Death Dis. 2010;1(9):e74.
58. Commandeur S, van Drongelen V, de Gruijl FR, El Ghalbzouri A. Epidermal growth factor receptor activation and inhibition in 3D in vitro models of normal skin and human cutaneous squamous cell carcinoma. Cancer Sci. 2012;103(12):2120–6.
59. El Ghalbzouri A, Commandeur S, Rietveld MH, Mulder AA, Willemze R. Replacement of animal-derived collagen matrix by human fibroblast-derived dermal matrix for human skin equivalent products. Biomaterials. 2009;30(1):71–8.
60. Boehnke K, Mirancea N, Pavesio A, Fusenig NE, Boukamp P, Stark HJ. Effects of fibroblasts and microenvironment on epidermal regeneration and tissue function in long-term skin equivalents. Eur J Cell Biol. 2007;86(11–12):731–46.
61. Muffler S, Stark H-J, Amoros M, Falkowska-Hansen B, Boehnke K, Bühring H-J, et al. A stable niche supports long-term maintenance of human epidermal stem cells in organotypic cultures. Stem Cells. 2008;26(10):2506–15.
62. Shahabeddin L, Berthod F, Damour O, Collombel C. Characterization of skin reconstructed on a chitosan-cross-linked collagen-glycosaminoglycan matrix. Skin Pharmacol. 1990;3(2):107–14.

63. Laplante AF, Germain L, Auger FA, Moulin V. Mechanisms of wound reepithelialization: hints from a tissue-engineered reconstructed skin to long-standing questions. FASEB J. 2001;15(13):2377–89.
64. Ronnov-Jessen L, Petersen OW, Koteliansky VE, Bissell MJ. The origin of the myofibroblasts in breast cancer. Recapitulation of tumor environment in culture unravels diversity and implicates converted fibroblasts and recruited smooth muscle cells. J Clin Invest. 1995;95(2):859–73.
65. Barcellos-Hoff MH, Ravani SA. Irradiated mammary gland stroma promotes the expression of tumorigenic potential by unirradiated epithelial cells. Cancer Res. 2000;60(5):1254–60.
66. Fong ELS, Wan X, Yang J, Morgado M, Mikos AG, Harrington DA, et al. A 3D in vitro model of patient-derived prostate cancer xenograft for controlled interrogation of in vivo tumor-stromal interactions. Biomaterials. 2016;77:164–72.
67. Sacks PG, Taylor DL, Racz T, Vasey T, Oke V, Schantz SP. A multicellular tumor spheroid model of cellular immunity against head and neck cancer. Cancer Immunol Immunother. 1990;32(3):195–200.
68. Correa de Sampaio P, Auslaender D, Krubasik D, Failla AV, Skepper JN, Murphy G, et al. A heterogeneous in vitro three dimensional model of tumour-stroma interactions regulating sprouting angiogenesis. PLoS One. 2012;7(2):e30753.
69. Ricci-Vitiani L, Lombardi DG, Pilozzi E, Biffoni M, Todaro M, Peschle C, et al. Identification and expansion of human colon-cancer-initiating cells. Nature. 2007;445(7123):111–5.
70. Weiswald LB, Richon S, Validire P, Briffod M, Lai-Kuen R, Cordelieres FP, et al. Newly characterised ex vivo colospheres as a three-dimensional colon cancer cell model of tumour aggressiveness. Br J Cancer. 2009;101(3):473–82.
71. Weiswald LB, Richon S, Massonnet G, Guinebretiere JM, Vacher S, Laurendeau I, et al. A short-term colorectal cancer sphere culture as a relevant tool for human cancer biology investigation. Br J Cancer. 2013;108(8):1720–31.
72. Kondo J, Endo H, Okuyama H, Ishikawa O, Iishi H, Tsujii M, et al. Retaining cell-cell contact enables preparation and culture of spheroids composed of pure primary cancer cells from colorectal cancer. Proc Natl Acad Sci U S A. 2011;108(15):6235–40.
73. Endo H, Okami J, Okuyama H, Kumagai T, Uchida J, Kondo J, et al. Spheroid culture of primary lung cancer cells with neuregulin 1/HER3 pathway activation. J Thorac Oncol. 2013;8(2):131–9.
74. Ashley N, Jones M, Ouaret D, Wilding J, Bodmer WF. Rapidly derived colorectal cancer cultures recapitulate parental cancer characteristics and enable personalized therapeutic assays. J Pathol. 2014;234(1):34–45.
75. Barker N, Huch M, Kujala P, van de Wetering M, Snippert HJ, van Es JH, et al. Lgr5(+ve) stem cells drive self-renewal in the stomach and build long-lived gastric units in vitro. Cell Stem Cell. 2010;6(1):25–36.
76. Huch M, Dorrell C, Boj SF, van Es JH, Li VS, van de Wetering M, et al. In vitro expansion of single Lgr5+ liver stem cells induced by Wnt-driven regeneration. Nature. 2013;494(7436):247–50.
77. Sato T, Vries RG, Snippert HJ, van de Wetering M, Barker N, Stange DE, et al. Single Lgr5 stem cells build crypt-villus structures in vitro without a mesenchymal niche. Nature. 2009;459(7244):262–5.
78. Sachs N, Clevers H. Organoid cultures for the analysis of cancer phenotypes. Curr Opin Genet Dev. 2014;24:68–73.
79. van de Wetering M, Francies HE, Francis JM, Bounova G, Iorio F, Pronk A, et al. Prospective derivation of a living organoid biobank of colorectal cancer patients. Cell. 2015;161(4):933–45.
80. Weeber F, van de Wetering M, Hoogstraat M, Dijkstra KK, Krijgsman O, Kuilman T, et al. Preserved genetic diversity in organoids cultured from biopsies of human colorectal cancer metastases. Proc Natl Acad Sci. 2015;112(43):201516689.
81. Schuette et al. Molecular dissection of colorectal cancer in pre-clinical models identifies biomarkers predicting sensitivity to EGFR inhibitors. Nature Communications. 8:14262. doi: 10.1038/ncomms14262, www.nature.com/naturecommunications, http://dx.doi.org/10.1038/ncomms14262

82. Koide N, Sakaguchi K, Koide Y, Asano K, Kawaguchi M, Matsushima H, et al. Formation of multicellular spheroids composed of adult rat hepatocytes in dishes with positively charged surfaces and under other nonadherent environments. Exp Cell Res. 1990;186(2):227–35.
83. Bae SI, Kang GH, Kim YI, Lee BL, Kleinman HK, Kim WH. Development of intracytoplasmic lumens in a colon cancer cell line cultured on a non-adhesive surface. Cancer Biochem Biophys. 1999;17(1–2):35–47.
84. Emfietzoglou D, Kostarelos K, Papakostas A, Yang WH, Ballangrud A, Song H, et al. Liposome-mediated radiotherapeutics within avascular tumor spheroids: comparative dosimetry study for various radionuclides, liposome systems, and a targeting antibody. J Nucl Med. 2005;46(1):89–97.
85. Lin RZ, Chang HY. Recent advances in three-dimensional multicellular spheroid culture for biomedical research. Biotechnol J. 2008;3(9–10):1172–84.
86. Hughes CS, Postovit LM, Lajoie GA. Matrigel: a complex protein mixture required for optimal growth of cell culture. Proteomics. 2010;10(9):1886–90.
87. Abu-Yousif AO, Rizvi I, Evans CL, Celli JP, Hasan T. PuraMatrix encapsulation of cancer cells. J Vis Exp. 2009;(34). pii:1692, doi:10.3791/1692.
88. Sutherland RM, Inch WR, McCredie JA, Kruuv J. A multi-component radiation survival curve using an in vitro tumour model. Int J Radiat Biol Relat Stud Phys Chem Med. 1970;18(5):491–5.
89. Kelm JM, Timmins NE, Brown CJ, Fussenegger M, Nielsen LK. Method for generation of homogeneous multicellular tumor spheroids applicable to a wide variety of cell types. Biotechnol Bioeng. 2003;83(2):173–80.
90. Breslin S, O'Driscoll L. Three-dimensional cell culture: the missing link in drug discovery. Drug Discov Today. 2013;18(5–6):240–9.
91. Johns RA, Tichotsky A, Muro M, Spaeth JP, Le Cras TD, Rengasamy A. Halothane and isoflurane inhibit endothelium-derived relaxing factor-dependent cyclic guanosine monophosphate accumulation in endothelial cell-vascular smooth muscle co-cultures independent of an effect on guanylyl cyclase activation. Anesthesiology. 1995;83(4):823–34.
92. Bell E. Strategy for the selection of scaffolds for tissue engineering. Tissue Eng. 1995;1(2):163–79.
93. Eiraku M, Watanabe K, Matsuo-Takasaki M, Kawada M, Yonemura S, Matsumura M, et al. Self-organized formation of polarized cortical tissues from ESCs and its active manipulation by extrinsic signals. Cell Stem Cell. 2008;3(5):519–32.
94. Willyard C. The boom in mini stomachs, brains, breasts, kidneys and more. Nature. 2015;523(7562):520–2.
95. Blaser MJ, Perez-Perez GI, Kleanthous H, Cover TL, Peek RM, Chyou PH, et al. Infection with *Helicobacter pylori* strains possessing cagA is associated with an increased risk of developing adenocarcinoma of the stomach. Cancer Res. 1995;55(10):2111–5.
96. McCracken KW, Cata EM, Crawford CM, Sinagoga KL, Schumacher M, Rockich BE, et al. Modelling human development and disease in pluripotent stem-cell-derived gastric organoids. Nature. 2014;516(7531):400–4.
97. Huang L, Holtzinger A, Jagan I, BeGora M, Lohse I, Ngai N, et al. Ductal pancreatic cancer modeling and drug screening using human pluripotent stem cell- and patient-derived tumor organoids. Nat Med. 2015;21(11):1364–71.
98. Sivaraman A, Leach JK, Townsend S, Iida T, Hogan BJ, Stolz DB, et al. A microscale in vitro physiological model of the liver: predictive screens for drug metabolism and enzyme induction. Curr Drug Metab. 2005;6(6):569–91.
99. Lee PJ, Hung PJ, Lee LP. An artificial liver sinusoid with a microfluidic endothelial-like barrier for primary hepatocyte culture. Biotechnol Bioeng. 2007;97(5):1340–6.
100. Wagner I, Materne E-M, Brincker S, Süssbier U, Frädrich C, Busek M, et al. A dynamic multi-organ-chip for long-term cultivation and substance testing proven by 3D human liver and skin tissue co-culture. Lab Chip. 2013;13(18):3538–47.

101. Kim HJ, Huh D, Hamilton G, Ingber DE. Human gut-on-a-chip inhabited by microbial flora that experiences intestinal peristalsis-like motions and flow. Lab Chip. 2012;12(12):2165.
102. Kim HJ, Ingber DE. Gut-on-a-Chip microenvironment induces human intestinal cells to undergo villus differentiation. Integr Biol (Camb). 2013;5(9):1130–40.
103. Zhang L, Wang J, Zhao L, Meng Q, Wang Q. Analysis of chemoresistance in lung cancer with a simple microfluidic device. Electrophoresis. 2010;31(22):3763–70.
104. Tavana H, Zamankhan P, Christensen PJ, Grotberg JB, Takayama S. Epithelium damage and protection during reopening of occluded airways in a physiologic microfluidic pulmonary airway model. Biomed Microdevices. 2011;13(4):731–42.
105. Li XJ, Li PCH. Contraction study of a single cardiac muscle cell in a microfluidic chip. Methods Mol Biol. 2006;321(1):199–225.
106. Grosberg A, Alford PW, McCain ML, Parker KK. Ensembles of engineered cardiac tissues for physiological and pharmacological study: Heart on a chip. Lab Chip. 2011;11(24):4165.
107. Maschmeyer I, Lorenz AK, Schimek K, Hasenberg T, Ramme AP, Hübner J, et al. A four-organ-chip for interconnected long-term co-culture of human intestine, liver, skin and kidney equivalents. Lab Chip. 2015;15(12):2688–99.
108. Huh D, Matthews BD, Mammoto A, Montoya-Zavala M, Hsin HY, Ingber DE. Reconstituting organ-level lung functions on a chip. Science. 2010;328(5986):1662–8.
109. Huh D, Leslie DC, Matthews BD, Fraser JP, Jurek S, Hamilton GA, et al. A human disease model of drug toxicity-induced pulmonary edema in a lung-on-a-chip microdevice. Sci Transl Med. 2012;4(159):159ra147.
110. Ataç B, Wagner I, Horland R, Lauster R, Marx U, Tonevitsky AG, et al. Skin and hair on-a-chip: in vitro skin models versus ex vivo tissue maintenance with dynamic perfusion. Lab Chip. 2013;13(18):3555–61.
111. Tung YC, Hsiao AY, Allen SG, Torisawa YS, Ho M, Takayama S. High-throughput 3D spheroid culture and drug testing using a 384 hanging drop array. Analyst. 2011;136(3):473–8.
112. Di Z, Klop MJ, Rogkoti VM, Le Devedec SE, van de Water B, Verbeek FJ, et al. Ultra high content image analysis and phenotype profiling of 3D cultured micro-tissues. PLoS One. 2014;9(10):e109688.
113. Cavnar SP, Salomonsson E, Luker KE, Luker GD, Takayama S. Transfer, imaging, and analysis plate for facile handling of 384 hanging drop 3D tissue spheroids. J Lab Autom. 2014;19(2):208–14.
114. Friedrich J, Seidel C, Ebner R, Kunz-Schughart LA. Spheroid-based drug screen: considerations and practical approach. Nat Protoc. 2009;4(3):309–24.
115. Friedrich J, Eder W, Castaneda J, Doss M, Huber E, Ebner R, et al. A reliable tool to determine cell viability in complex 3-d culture: the acid phosphatase assay. J Biomol Screen. 2007;12(7):925–37.
116. Lovitt CJ, Shelper TB, Avery VM. Miniaturized three-dimensional cancer model for drug evaluation. Assay Drug Dev Technol. 2013;11(7):435–48.
117. Boehnke K, Iversen PW, Schumacher D, Lallena MJ, Haro R, Amat J, Haybaeck J, Liebs S, Lange M, Schäfer R, Regenbrecht CRA, Reinhard C, Velasco JA. Assay Establishment and Validation of a High-Throughput Screening Platform for Three-Dimensional Patient-Derived Colon Cancer Organoid Cultures. J Biomol Screen. 2016;21(9):931–41. doi:10.1177/1087057116650965, jbx.sagepub.com, http://jbx.sagepub.com/content/early/2016/05/26/1087057116650965.full.
118. Klein OJ, Jung YK, Evans CL. Longitudinal, quantitative monitoring of therapeutic response in 3D in vitro tumor models with OCT for high-content therapeutic screening. Methods. 2014;66(2):299–311.
119. Kenny HA, Lal-Nag M, White EA, Shen M, Chiang CY, Mitra AK, et al. Quantitative high throughput screening using a primary human three-dimensional organotypic culture predicts in vivo efficacy. Nat Commun. 2015;6:6220.
120. Fitzgerald JB, Schoeberl B, Nielsen UB, Sorger PK. Systems biology and combination therapy in the quest for clinical efficacy. Nat Chem Biol. 2006;2(9):458–66.

121. Zhao L, Wientjes MG, Au JLS. Evaluation of combination chemotherapy: integration of nonlinear regression, curve shift, isobologram, and combination index analyses. Clin Cancer Res. 2004;10(23):7994–8004.
122. Lehár J, Krueger AS, Avery W, Heilbut AM, Johansen LM, Price ER, et al. Synergistic drug combinations tend to improve therapeutically relevant selectivity. Nat Biotechnol. 2009;27(7):659–66.
123. Jia J, Zhu F, Ma X, Cao Z, Li Y, Chen YZ. Mechanisms of drug combinations: interaction and network perspectives. Nat Rev Drug Discov. 2009;8(2):111–28.
124. National Research Council (US) Committee on A Framework for Developing a New Taxonomy of Disease. Toward precision medicine: building a knowledge network for biomedical research and a new taxonomy of disease. Washington, DC: National Academies Press; 2011.
125. Vaira V, Fedele G, Pyne S, Fasoli E, Zadra G, Bailey D, et al. Preclinical model of organotypic culture for pharmacodynamic profiling of human tumors. Proc Natl Acad Sci U S A. 2010;107(18):8352–6.
126. Gerlach MM, Merz F, Wichmann G, Kubick C, Wittekind C, Lordick F, et al. Slice cultures from head and neck squamous cell carcinoma: a novel test system for drug susceptibility and mechanisms of resistance. Br J Cancer. 2014;110(2):479–88.
127. Merz F, Gaunitz F, Dehghani F, Renner C, Meixensberger J, Gutenberg A, et al. Organotypic slice cultures of human glioblastoma reveal different susceptibilities to treatments. Neuro Oncol. 2013;15(6):670–81.
128. Gao D, Vela I, Sboner A, Iaquinta PJ, Karthaus WR, Gopalan A, et al. Organoid cultures derived from patients with advanced prostate cancer. Cell. 2014;159(1):176–87.
129. Halfter K, Ditsch N, Kolberg H-C, Fischer H, Hauzenberger T, von Koch FE, et al. Prospective cohort study using the breast cancer spheroid model as a predictor for response to neoadjuvant therapy—the SpheroNEO study. BMC Cancer. 2015;15(1):519.
130. Burstein HJ, Mangu PB, Somerfield MR, Schrag D, Samson D, Holt L, et al. American Society of Clinical Oncology clinical practice guideline update on the use of chemotherapy sensitivity and resistance assays. J Clin Oncol. 2011;29(24):3328–30.
131. Longati P, Jia X, Eimer J, Wagman A, Witt MR, Rehnmark S, et al. 3D pancreatic carcinoma spheroids induce a matrix-rich, chemoresistant phenotype offering a better model for drug testing. BMC Cancer. 2013;13:95.
132. Hazlehurst LA, Landowski TH, Dalton WS. Role of the tumor microenvironment in mediating de novo resistance to drugs and physiological mediators of cell death. Oncogene. 2003;22(47):7396–402.
133. Astashkina AI, Mann BK, Prestwich GD, Grainger DW. A 3-D organoid kidney culture model engineered for high-throughput nephrotoxicity assays. Biomaterials. 2012;33(18):4700–11.
134. DesRochers TM, Suter L, Roth A, Kaplan DL. Bioengineered 3D human kidney tissue, a platform for the determination of nephrotoxicity. PLoS One. 2013;8(3):e59219.
135. Hrach J, Mueller SO, Hewitt P. Development of an in vitro liver toxicity prediction model based on longer term primary rat hepatocyte culture. Toxicol Lett. 2011;206(2):189–96.
136. Fey SJ, Wrzesinski K. Determination of drug toxicity using 3D spheroids constructed from an immortal human hepatocyte cell line. Toxicol Sci. 2012;127(2):403–11.

Stratifying Cancer Therapies by Molecular Interactions and Imaging

15

Myria Galazi, Gregory Weitsman, James Monypenny, Oana Coban, Hanna Milewicz, Valenti Gomez, Francesca Ciccarelli, and Tony Ng

Contents

M. Galazi • G. Weitsman • J. Monypenny • O. Coban • H. Milewicz • F. Ciccarelli
Division of Cancer Studies, Richard Dimbleby Department of Cancer Research, New Hunt's House, King's College London, London SE1 1UL, UK

V. Gomez
UCL Cancer Institute, University College London, Paul O'Gorman Building, London, WC1E 6DD, UK

T. Ng (✉)
Division of Cancer Studies, Richard Dimbleby Department of Cancer Research, New Hunt's House, King's College London, London SE1 1UL, UK

UCL Cancer Institute, University College London, Paul O'Gorman Building, London, WC1E 6DD, UK

Breast Cancer Now Research Unit, Department of Research Oncology, Guy's Hospital, King's College London School of Medicine, London SE1 9RT, UK
e-mail: tony.ng@kcl.ac.uk

J. Haybaeck (ed.), *Mechanisms of Molecular Carcinogenesis – Volume 2*,
DOI 10.1007/978-3-319-53661-3_15

Abstract

Accumulated knowledge generated by years of fundamental research and more recently the implementation of high-throughput sequencing analysis and genomic technologies have led to the identification of novel molecular events that are critical oncogenic drivers amenable to targeted therapy. As a result, in the past decade, we have observed the introduction of molecularly targeted therapies for the treatment of cancers within clinical trials that have then subsequently gained approval for use in routine clinical practice. Some of these agents have demonstrated dramatic efficacies, not previously observed in the treatment of metastatic cancers, such as malignant melanoma, non-small cell lung cancer (NSCLC), colorectal cancer and breast cancer.

15.1 Introduction

Accumulated knowledge generated by years of fundamental research and more recently the implementation of high-throughput sequencing analysis and genomic technologies have led to the identification of novel molecular events that are critical oncogenic drivers amenable to targeted therapy. As a result, in the past decade, we have observed the introduction of molecularly targeted therapies for the treatment of cancers within clinical trials that have then subsequently gained approval for use in routine clinical practice. Some of these agents have demonstrated dramatic efficacies, not previously observed in the treatment of metastatic cancers, such as malignant melanoma, non-small cell lung cancer (NSCLC), colorectal cancer and breast cancer.

The timeline of translating these molecular agents from the laboratory bench into clinical practice is accelerating, increasing the need for diagnostics/biomarkers to prescreen and stratify patients that are likely to respond and to detect early signs of drug response or acquired resistance. In addition, despite the profound initial responses seen with the introduction of these therapies, acquired resistance by cancer cells is increasingly observed. This necessitates a greater understanding of the cellular mechanisms underlying the plasticity and adaptation of cancer cells in response to targeted therapies.

15.2 Matching Molecule-Targeted Agents to Mutations

The complexity of neoplastic disease depends on a variety of biological processes. Sustained mitogenic signalling is arguably the most fundamental trait of cancer cells. This is conveyed in large by growth factors that bind cell-surface receptors allowing activation of their tyrosine kinase domains and activation of branched intracellular signalling pathways that regulate cell cycle and growth. In addition, activating somatic mutations within tumours lead to constitutive signalling of important cellular pathways, such as the mitogen-activated protein (MAP) kinase and the phosphoinositide 3-kinase (PI3K) pathways [1, 2]. Dependency of a tumour on a specific oncogene renders these malignancies sensitive to inhibitors. Finally, the disruption of negative feedback loops within these pathways enhances proliferative signalling further—for example, the loss of function mutations in PTEN that amplify PI3K signalling [3].

15.2.1 The Ras-Raf-MEK-ERK Signalling Pathway

The Ras-Raf-MEK-ERK signalling pathway is instrumental in linking growth factor-dependent receptor activation to a variety of fundamental cellular processes, including proliferation, cell cycle arrest, apoptosis and migration. In response to extracellular receptor activation and subsequent recruitment and activation of the monomeric GTPase Ras, a protein kinase signalling cascade is triggered that leads to the phosphorylation of ERK (extracellular signal-regulated kinases). Phosphorylated ERKs translocate to the nucleus where they themselves phosphorylate transcription factors, thereby regulating their activity [4].

BRAF is a gene that encodes the B-Raf protein, a key component of the Ras-Raf-MEK-ERK pathway. It is a proto-oncogene that is frequently mutated in melanoma (50–60%) and at a lower frequency in other human cancers (10% of colorectal and 6% of lung cancers). This occurs mainly by somatic missense point mutations, all of which lie within the kinase domain.

Since the initial identification of BRAF mutations in cancer, about 300 distinct missense mutations have been identified in tumour samples and cancer cell lines [5]. Most occur in the activation loop (A-loop) near Val 600, or in the Gly-Ser-Gly-Ser-Phe-Gly phosphate-binding loop (P-loop) [6]. These mutations lead to A-loop or P-loop phosphorylation, or disrupt the A-loop–P-loop interaction that stabilises the inactive conformation of B-Raf, promoting the kinase-active form of the protein. The most common *BRAF* mutation leads to the substitution of Glu for Val at amino acid 600 (V600E) and accounts for 80% of the mutations seen in this gene. This substitution mimics phosphorylation of the activation loop, thereby inducing constitutive B-Raf kinase activity [7].

In clinical trials of *BRAF* V600E melanoma patients, B-Raf inhibitors (vemurafenib and dabrafenib) have induced high rates of response and longer survival [8, 9].

Despite this, resistance to B-Raf inhibitors has been documented both in the primary (intrinsic) and secondary (acquired) setting. A significant proportion (16%) of patients with *BRAF*-mutated malignant melanoma have shown primary resistance to dabrafenib [10], and the durability and survival benefit of the patients who have shown initial response to either of these agents seems to be limited to about 1 year [10–12].

In contrast, in colorectal cancers with the same *BRAF* V600E mutation, Raf inhibitor monotherapy has proven disappointingly ineffective, with response rates of only approximately 5% [13]. This disparity in B-Raf inhibition between different tumour types poses a great challenge to our current understanding of inhibitor insensitivity.

Primary resistance to B-Raf inhibition has been shown to occur through activation of signalling through wild-type RAF by relieving feedback mechanisms [14]. Moreover, the cancer type and the signalling pathways that are primarily involved in cancer cell proliferation can interact and predict efficacy to B-Raf inhibition. For example, the decreased efficacy of vemurafenib in BRAF-mutated metastatic colorectal cancer (mCRC) is mainly attributed to a feedback increase in epidermal growth factor receptor (EGFR) signalling when B-Raf is inhibited [15].

Hypotheses for mechanisms of acquired resistance to B-Raf inhibition generally include secondary mutations in *BRAF* V600E, MAP kinase pathway reactivation and activation of alternative survival pathways. Reactivation of the MAP kinase signalling pathway occurs through upregulation of KRAS, NRAS, MEK1 and MEK2 [16–18]. For example, sequencing of *BRAF* V600E-positive melanoma cell lines and patient tumours with acquired secondary resistance to vemurafenib demonstrated NRAS upregulation and subsequent MAP kinase pathway reactivation that was also sensitive to MEK inhibition [17]. Additionally, the identification of a *BRAF* V660E splice variant, p61BRAF(V600E) [19], which demonstrates enhanced homodimerisation in vemurafenib-resistant melanoma cell lines, promoted an increase in ERK signalling. This splice variant was also found in the tumours of patients with acquired resistance to vemurafenib, supporting the hypothesis that failure of B-Raf inhibitors occurs due to alternative, Ras-independent mechanisms of Raf activation. The p61BRAF(V600E) is Ras independent as it lacks exons 4–8, a region that encompasses the RAS-binding domain [19].

Furthermore, whole exome sequencing (WES) in colorectal cancer from paired pretreatment and post-progression biopsies, from patients with *BRAF*-mutant tumours and initial clinical response or prolonged stable disease, demonstrated MAP kinase pathway alterations unique to the resistant tumour, including *KRAS* amplification and *MEK1* mutation [20]. Indeed, the continued MAP kinase pathway dependence of *BRAF*-mutant colorectal cancer cells with these molecular alterations that promotes acquired resistance to B-Raf inhibition is supported by the fact that they remain sensitive to ERK inhibitors [20]. Amplified or mutant *KRAS* produces a constitutively GTP-bound and active form of the protein that promotes stimulus-independent and persistent activation of downstream effectors, leading to the proliferation and survival of cancer cells predominantly through the Ras-Raf-MEK-ERK pathway [21].

Acquired resistance to B-Raf inhibition has also been observed to occur via a MAP kinase pathway-independent mechanism in approximately 30% of melanoma patients [22]. Activation of receptor tyrosine kinases (RTKs) provides one such mechanism of BRAF inhibitor resistance. Increased expression of either platelet-derived growth factor receptor-β (PDGFRβ) or insulin-like growth factor 1 receptor (IGF1R) was identified in cultured cells and in specimens from patients with vemurafenib-resistant melanomas. PDGFRβ or IGF1R signalling allows for the activation of MAPK pathway-independent pro-survival signalling pathways, such as the PI3K–AKT axis, which render cells resistant to the effects of B-Raf inhibition [17, 23]. EGF signalling also confers resistance to B-Raf inhibition and induces melanoma invasion through Src-dependent pathways. Inhibition of the EGF receptor and Src resensitises treatment-resistant BRAF-mutant melanoma cells to vemurafenib and blocks their invasiveness [24]. In addition, activation of HER3 signalling has been identified as an adaptive mechanism of resistance in a subset of patients with melanoma. It is thought that B-Raf kinase inhibition promotes the upregulation of the transcription factor forkhead box protein D3 (FOXD3), which, in turn, directs increased expression of HER3 and allows for enhanced HER2–HER3 signalling [25].

Identifying and understanding these resistance mechanisms in B-Raf inhibitor-resistant cancers will be instrumental for the development of additional rational pathway-targeted therapeutic combinations to achieve longer-lasting responses and better patient survival. In addition, a deeper understanding into the Ras-Raf-MEK-ERK pathway and the identification of new biomarkers to predict response will be essential for future drug development in RAF-driven cancers.

15.2.2 The Epidermal Growth Factor Receptor (EGFR/ErbB) Family

EGFR (also known as ErbB or HER1) is the archetypal member of the epidermal growth factor family of receptor tyrosine kinases (RTKs). Upon ligand binding with peptide growth factors of the EGF family, the receptor undergoes dimerisation that stimulates its intrinsic intracellular protein tyrosine kinase activity. This leads to further activation of downstream pathways, such as Ras-Raf-MEK-ERK and PI3K-AKT-mTOR, which drive DNA synthesis and cell proliferation [26].

EGFR and EGF-like peptides are often overexpressed in many human carcinomas, such as lung, colorectal, head and neck cancers and glioblastomas. Amplification of the EGFR gene and mutations of the EGFR catalytic tyrosine kinase domain have been demonstrated to occur in carcinoma patients. The most common EGFR-activating mutations, exon 19 deletion and exon 21 point mutation (L858R), predict sensitivity to EGFR tyrosine kinase inhibitors [27, 28]. The distribution of these mutations around the 'catalytic kinase domain' is distinct in NSCLC in contrast to EGFR mutations in glioblastomas located in the extracellular portion of the protein [29]. Additional mutations in the EGFR gene have been identified but these are rarer, and their impact on predicting sensitivity to anti-EGFR therapy or clinical prognosis still needs to be elucidated [30].

The complex interactions of EGFR with the other epidermal growth factor family receptors and ligands, especially its ability to form, and heterodimers allow for sustained activation of downstream signalling pathways despite inhibition, leading to cancer cell proliferation posing further therapeutic challenges [31].

Erlotinib and gefitinib reversibly inhibit the EGFR tyrosine kinase domain by competitively binding with ATP. Afatinib is an irreversible inhibitor of EGFR and

HER2 receptor tyrosine kinases. Cetuximab, a chimeric mouse-human IgG1 antibody, and panitumumab, a fully humanised IgG2 antibody, are monoclonal antibodies that block ligand binding to the extracellular domain of EGFR [32].

Gefitinib, erlotinib and afatinib have been used in the subset of patients with NSCLC (10–35%) who have activating EGFR gene mutations. Several randomised phase III trials in patients with advanced NSCLC and activating EGFR mutation have shown the efficacy of these agents in terms of response rate (RR) and progression-free survival (PFS) when compared with standard platinum-based chemotherapy [33–40]. In addition, cetuximab is effective in a subset of KRAS wild-type metastatic colorectal cancers. Response, however, is transient, and secondary resistance occurs, limiting the clinical benefit of these anti-EGFR agents.

The resistance mechanisms to EGFR small molecule inhibitors or antibodies are due to novel secondary EGFR mutations and activation of bypass signalling pathways (such as MET amplification, KRAS activation) [41].

The EGFR T790M mutation in exon 20 of the EGFR gene is found in approximately 60% of EGFR TKI-treated patients with NSCLC as a form of acquired resistance. It leads to increase in kinase activity for ATP, therefore reducing antagonism by the ATP-competitive tyrosine kinase inhibitors [42]. EGFR inhibitors that target the T790M mutation have been identified. These novel third-generation EGFR inhibitors, such as rociletinib (CO-1686) and AZD9291, have shown clinical responses in approximately 60% of EGFR-mutant patients with T790M-mediated secondary resistance [43, 44]. Similarly, the EGFR extracellular domain S492R mutation reduces cetuximab binding in colorectal cancer [45].

Amplification of the MET gene is observed in up to a fifth of patients with EGFR TKI-resistant NSCLC. In an EGFR-mutant gefitinib-resistant lung cancer cell line, activated MET led to phosphorylation of the p85 subunit of PI3K allowing interaction with phosphorylated HER3 and downstream AKT signalling through this MET/HER3/PI3K axis. Therefore, the MET-mediated increase in PI3K signalling rescues the EGFR inhibitor-dependent inactivation of PI3K. In the same study, dual inhibition of EGFR and MET signalling in xenograft models showed tumour regression [46].

In both NSCLC and colorectal cancer, KRAS and EGFR mutations generally occur in a mutually exclusive pattern, and the presence of one predicts non-response to therapy directed against the other [47]. The KRAS gene encodes for the GTP-binding protein Ras. Once mutated, Ras becomes constitutively active. Deep sequencing and BEAMing (beads, emulsions, amplification and magnetic, a technique to detect with a 5% mutated/wild-type allele sensitivity) in metastatic colorectal cell lines and in patient biopsies ($n = 8$) pre- and post-cetuximab treatment (who had initially responded and then developed resistance) demonstrated KRAS mutations. This emergence of cetuximab-resistant populations derives either from selection of pre-existing KRAS clones or of 'de novo' acquisition of a KRAS mutation under the pressure of cetuximab treatment. Co-administration of cetuximab and selective inhibitors of MEK kinase in these resistant cell clones showed that they become sensitive to treatment [48].

Although the spectrum of KRAS mutations affects the same three-dimensional area of the folded protein, molecular studies suggest that different amino acid substitutions may define the biological properties of the KRAS protein in terms of binding affinity for downstream effectors. A recent study indicated that NSCLC cell lines with G12C and G12V amino acid substitutions appear to have lower levels of phosphorylated AKT to those of cell lines carrying other KRAS mutations [49]. Similarly, levels of phosphorylated MEK did differ between KRAS G12C or G12V and other KRAS mutations or wild-type KRAS, which suggests that tumours harbouring G12C or G12V variants may have greater dependency upon mitogen-activated protein/extracellular signal-regulated kinase (MAP/ERK) signalling for proliferation. Although experimental, these results seem to suggest that MEK inhibitors may be more active against tumours with KRAS G12C or G12V variants, a hypothesis that will be prospectively evaluated in the ongoing SELECT-1 study (https://clinicaltrials.gov/ct2/show/NCT01750281).

15.2.3 PI3K/AKT Pathway

Phosphoinositide 3-kinases (PI3Ks) belong to a family of lipid kinases that phosphorylate 3-hydroxyl group of phosphoinositides. Importantly, phosphorylation of phosphatidylinositol-4,5-biphosphate by PI3K results in the production of phosphatidylinositol-3,4,5-triphosphate (PIP_3), which is critical in recruiting AKT for activation of growth, proliferation and survival. PTEN (phosphatase and tensin homolog) is a tumour suppressor that negatively regulates PIP_3 [50].

Activating mutations in PIK3CA, the gene encoding the p110a catalytic subunit of PI3K, is frequently mutated in cancer. PI3KCA amplification, PTEN loss, AKT mutations and receptor tyrosine kinase amplification promote tumorigenesis by upregulating the PI3K/AKT signalling axis [51].

The complexity of the PI3K pathway, however, leads to negative feedback release as a result of targeted inhibition, leading to the activation of compensatory signalling pathways. These comprise the FOXO-dependent feedback reactivation of receptor tyrosine kinases (such as HER2, HER3, IGF1R and insulin receptor) and downstream kinases, including ERK, MYC amplification and NOTCH or Wnt-b-catenin pathway activation [52–54].

mTOR inhibitors were the first to enter the clinic (rapamycin) and have been effective in some malignancies, for example, renal cell carcinoma, when administered as single agents [55]. Dual PI3K-mTOR and AKT inhibitors have also been tested within clinical trials in various tumour types and have produced variable results. A comprehensive list of these is sited in the supplementary table of the review article by Fruman et al. with the relevant citations [56]. Because of the complexity of the PI3K pathway described above, inhibitors of the PI3K/AKT/mTOR pathway are used in combination with other targeted treatments as well, for example, with trastuzumab or lapatinib [57].

Furthermore, in the case of hormone-sensitive malignancies, such as prostate and breast cancer, PI3K pathway alterations are triggered as resistance mechanisms to hormone therapies. In PTEN-deficient prostate cancer, PI3K and androgen receptor (AR) regulate one another by a reciprocal feedback mechanism. AR inhibition activates AKT signalling, and PI3K inhibition results in feedback signalling to HER2–HER3 and AR reactivation [58, 59]. The effect of combined targeted inhibition of the PI3K/AKT/mTOR pathway and androgen deprivation therapy is being investigated within clinical trials. Finally, the positive results of the combination of the aromatase inhibitor exemestane and everolimus in oestrogen receptor-positive breast cancer have led to the design of many other clinical trials of PI3K/AKT/mTOR pathway inhibitors with antioestrogen therapies [60].

15.2.4 Other Agents

Less frequent genetic alterations include the ALK (anaplastic lymphoma kinase) fusion gene rearrangement. This is present in a 4–5% of NSCLCs and its presence is mutually exclusive for EGFR or KRAS mutations. Crizotinib is an ALK inhibitor that demonstrated significantly improved RR and PFS in ALK-positive NSCLC and is now approved for use in these patients [61, 62].

Agents targeting the Ras-Raf-MEK-ERK pathway such as the MEK inhibitor selumetinib are being evaluated as therapeutics in KRAS-mutant NSCLC within clinical trials (https://clinicaltrials.gov/ct2/show/NCT01750281).

Finally, the MET receptor tyrosine kinase is overexpressed in 25–75% of NSCLC and has been associated with poorer outcomes. Tivantinib, a novel MET tyrosine kinase inhibitor, and onartuzumab, a monoclonal antibody against the MET proto-oncogene product, have been evaluated within clinical trials with contradictory clinical outcomes that might though reflect the fact that patients were recruited without any stratification based to MET expression [63–65].

15.2.5 Redesigning Cancer Clinical Trials

The identification of pathways involved in carcinogenesis, metastasis and drug resistance as well as the increasing availability of technologies allowing molecular profiling of tumours has stimulated the development of programmes towards personalised cancer care through the design of novel clinical trials in which patients with targetable genomic or molecular aberrations are allocated to the relevant agent.

The Phase I Trial by The University of Texas MD Anderson Cancer Center assigned patients ($N = 1144$) with various types of advanced or metastatic cancers refractory to standard therapy and with a molecular aberration for which a matched targeted agent was available (including PIKCA, mTOR, BRAF, MEK, multikinases, KIT, EGFR and RET). From the patients screened, the 175 with molecular aberrations and matched therapy had an overall response rate of 27% (complete response, CR 2%; partial response, PR 25%) vs. 5% in the 116 patients treated with non-matched therapy and time to treatment failure (TTF) of 5.2 months in the matched targeted therapy group vs. 2.2 in the non-matched group ($p < 0.0001$) [66].

More clinical trials are being designed and started recruitment. FOCUS4 is a molecularly stratified, multi-arm, multistage (MAMS) design and multisite randomised trial for patients with metastatic colorectal cancer (mCRC). It aims to recruit up to 643 patients who will initially receive standard treatment. Biomarker testing will be performed on original tumour specimens, and molecular stratification will be performed whilst patients are still responding or have stable disease to standard therapy at 16 weeks. The agents available within the FOCUS4 trial include PIK3CA, AKT, MEK, HER1, HER2 and HER3 inhibitors, and they will be used to assess primary outcomes of progression-free survival, and secondary outcomes are safety, toxicity, tumour response and quality of life.

In early 2015, the National Cancer Institute (NCI) will be launching the NCI-MATCH trial planning to enrol at least 1000 patients, from the initial 3000 screened, for a targeted combination therapy independent of tumour histology. Similarly, the NCI-MPACT trial is randomly assigning patients with a known genetic mutation to pathway-specific targeted therapy vs. treatment not known to be pathway specific.

Despite these efforts to match targeted treatments to the correct patients most likely to respond and derive benefit, we are still being faced with challenges:

- How do we know that the mutations identified are the drivers of cancer cell proliferation?
- How do we treat patients with ≥1 targetable molecular aberration?
- What do we do when rare and non-targetable mutations are identified?
- Do we still treat patient with progressive disease and non-targetable mutations based on the knowledge of responders from the same histologic group?

Finally, histopathological information remains a clinically significant variable despite the growing interest in moving beyond this with the development and use of targeted therapies. The above and many more other trials have demonstrated how mutations segregate along histological lines, e.g. EGFR mutations and NSCLC. In addition, the data with targeted agents so far indicates that the disease-specific presence of a targetable mutation also determines whether the target represents a clinically relevant driver mutation (e.g. sensitivity of V600E BRAF-mutant melanoma to BRAF inhibition, compared to colorectal cancers with the same mutation).

15.3 Target Concentration Alone Not Predictive of Clinical Response to Targeted Therapies

Agents targeting the ErbB/HER receptors have been used for various cancers. The initial strategy was to use agents with high specificity towards one target in combination with standard chemotherapy. Indeed, the use of cetuximab for lung cancer in combination with cisplatin and vinorelbine showed statistically significant survival benefit in patients with high expression levels of EGFR [67]. Data from the S0342 study suggested that such effect could be attributed to high copy numbers of EGFR [68]. However, in the BMS099 trial, addition of cetuximab to taxane/carboplatin did not provide any additional benefit [69] despite high expression levels of EGFR by immunohistochemistry (IHC) and fluorescence in situ hybridisation (FISH) analyses.

The efficacy of cetuximab in colorectal cancer patients has been assessed in the subgroup of patients with wild-type kRas since multiple studies have shown that activating kRas mutations (which propagate the signals downstream of EGFR) render tumours insensitive to anti-EGFR therapy [70–74]. In early studies, cetuximab showed increase in overall survival [75, 76], associated with high EGFR copy number [77] or high EGFR mRNA [78] in patients with wild-type kRas protein. However, in recent phase III trials, this outcome was not achieved (COIN trial) [79], and in the New EPOC trial [78–80], the use of cetuximab was disadvantageous. A confounding factor could be the type of chemotherapy regimen in that a more favourable clinical outcome was obtained with EGFR treatment when this was combined with irinotecan-containing regimen, e.g. the phase 3 CRYSTAL trial [81, 82] cf. oxaliplatin-based chemotherapy in COIN and new EPOC. Similarly, in head and neck tumours, cetuximab addition to platinum chemotherapy led to statistically significant increase in overall survival [83], which was not associated with high expression levels of EGFR [84].

The use of anti-EGFR inhibitors also showed variable results. Erlotinib alone [85] or addition of erlotinib [86] or gefitinib [87] to chemotherapy did not improve outcomes in patients with wild-type EGFR. However, treatment with erlotinib alone [85] or addition of gefitinib to cisplatin/docetaxel [36] or to carboplatin/paclitaxel [88] significantly increased progression-free survival (PFS) in patients harbouring EGFR activation mutation. This supports current selection criteria for use of these inhibitors only in patients with tumours sensitive to the EGFR signalling due to activating mutations.

This pattern of contrasting responses to anti-EGFR therapy and the diverse predictive power of EGFR expression levels for treatment stratification could be due to multiple reasons. A great variation in molecular diagnosis of EGFR levels utilising FISH assay was found among five international centres from Belgium, Italy, Switzerland and the USA, the within-subject coefficient of variation reaching ~57% [89]. These discrepancies are mainly attributed to technical issues of the quality of the slide, the equipment used and the personnel difference in interpretation, rather than tumour heterogeneity. Additionally, other genes similarly to kRas acting downstream of EGFR can affect the response to these agents. Gene mutations, such as BRaf and PTEN [90], presence of ligands EGF and TGF-α in serum [91] and epiregulin and amphiregulin in tumour [72] or high levels of pERK1/2 and pAkt [87] can influence response rates to anti-EGFR agents.

Furthermore, acquired resistance due to the changes in trafficking and degradation of the receptor, or strong activation of other members of the family (HER2, HER3), or even other receptor tyrosine kinase (RTK) like cMet can have a significant role [92]. Prolonged inhibition of EGFR leads to increased EGFR expression and its association with HER2, HER3 and cMet RTKs, leading to transactivation of HER2 and HER3 which could be blocked by pertuzumab. Upregulation of HER3 and its dimerisation with EGFR was also shown in triple-negative breast cancer patients following cetuximab and panitumumab treatment [93].

HER2 overexpression is known to have strong prognostic [94] and predictive value [93, 95] in breast cancer as well as in advanced gastric cancer [96]. The predictive value of HER2 overexpression became questionable in breast cancer patients after results of the National Surgical Adjuvant Breast and Bowel Project (NSABP) trial when effectiveness of trastuzumab was shown in patients with low expression levels of HER2 [97]. This trial also showed discrepancies with standardised methods for HER2 amplification detection [98] (patients with low HER2 expression levels were initially scored as HER2 positive and were treated with trastuzumab) [99, 100].

15.4 Tumour Evolution and Protein Network Rewiring: Outside the Mutation Box

The cancer cell is an integrated system of processes that form a tightly interconnected network rather than an ensemble of independent genes and proteins. As a result, the effect of genetic aberrations should be interpreted taking into account their impact on this system's equilibrium [101]. The emerging heterogeneity of cancer genomic landscape has been used to question the usefulness of large-scale screenings, the main concern being that the discovery of rare mutations adds very little to the overall knowledge of cancer genetics. Approaches that focus on the identification of global features, more than to the study of single genes, show instead that a comprehensive catalogue of cancer genetic determinants is instrumental to trace recurrent patterns in their system-level and evolutionary properties [101].

Studying cancer genes thereby requires more systematic approaches to help understand better the genetic determinants of cancer. A multidimensional analysis that combines sequence similarity, functional annotations, protein–protein interactions and molecular pathways in examining the genes mutated in breast and colorectal cancers has helped to clarify the preferred evolutionary 'nodes' (i.e. where tumour cells are likely to mutate within the cancer genome) by overlaying the mutation landscapes on top of the protein interaction network [102]. For instance, in breast cancer, over half of the mutated proteins were predicted to be involved in a large interaction cluster that is centred upon TP53, BRCA1, PIK3R1 and NF-κB. Systems biology approaches examine the effects of mutations on the complex cellular network and protein–protein interactions, and such analyses can reveal where perturbations are likely to have a maximum impact [103]. Pathway crosstalk/compensatory activities [104, 105] constitute a major mechanism of resistance to targeted therapy. An example of this is the MAPK and PI3K signalling pathway cascades that are both activated by a diverse family of growth factors and other stimuli. Both these pathways are mutated in cancer and have important roles in cell proliferation, differentiation, metabolism and survival. Aberrant signalling and intensive crosstalk between these pathways are critical in growth and survival. In general, PI3K positively regulates MAPK cascade, facilitating maximal ERK responses and physiological stimuli, whereas activated ERK negatively controls the PI3K/AKT pathway [106, 107]. Experimental data and computational simulations show that this dynamic crosstalk, however, is context dependent, influenced dramatically by the concentrations of growth factors and the levels of receptors and scaffolding proteins such as GAB and IRS [107, 108]. This crosstalk leads to activation of compensatory signalling, allowing cancer cells to evade apoptosis if only a single pathway is targeted therapeutically.

Within these cross-compensatory pathways, the formation of proteins into stable complexes plays a fundamental role in creating feedback regulation (e.g. TSC1/TSC2 and TORC1 and 2). The clinical utility of our knowledge of protein complexes and protein–protein interactions has been hampered by the lack of technologies that can probe such interactions in clinical samples.

15.5 Currently Available Techniques to Quantify Single Protein Pair Interactions in Patient Samples

Tumour biopsy tissue is important to provide us additional information about the molecular characteristics of an individual tumour beyond the histological description. IHC and FISH are the most commonly used techniques to detect expression levels of single proteins of interest in formalin-fixed, paraffin-embedded (FFPE) tissue. Rewiring of ErbB receptor complexes has been documented in models of resistance and recently in clinical studies (e.g. through EGFR-HER dimer formation in breast cancer patients undergoing neoadjuvant treatment with cetuximab/panitumumab) [93]. Expression level alone is not sufficient to monitor this process, whereby ErbB inhibition leads to protein network rewiring within the cancer cells, thereby giving rise to potential resistance mechanisms in response to targeted cancer therapies. However, probing protein complexes in FFPE tissues is a challenge.

Collaborative enzyme enhanced reactive immunoassay utilises capture antibody to detect a protein or its modification (i.e. phosphorylation) with antibodies conjugated to glucose oxidase (GO) and horseradish peroxidase (HRP). The extremely high sensitivity of the assay can be used to detect proteins in single tumour cells and has the potential for use to detect protein–protein interactions [109]. Another method utilises two antibodies specific to the same protein or two different proteins [110] conjugated to reported tag and biotin, respectively. Both antibodies are used simultaneously on FFPE tissue after antigen retrieval. Photosensitiser methylene blue conjugated to streptavidin binds to biotin to give rise to singlet oxygen under illumination with 670 nm light, which, in turn, will cleave thioether bond between first antibody and the reporter tag. The latter is collected in the buffer and analysed using capillary electrophoresis. This method allows detection of total or modified protein and protein complexes with resolution distance 30–100 nm [110]. The HER2 or HER2 homodimer levels detected with VeraTag showed 98% concordance with IHC and FISH in 237 breast cancer samples [111]. Furthermore, the levels of HER1–HER2 complex strongly correlated with expression levels of HER1 or HER2 [112]. Unfortunately, these methods do not allow for cellular resolution and provide an average value for the whole sample.

Proximity ligation assay (PLA) was developed by Soderberg et al. [113] for detection of two proteins in close proximity. Two antibodies are directed against different proteins modified by covalent attachment of the 5″ end of two different oligonucleotides with complementary sequence for linear linker. When the antibodies are applied on the sample, the linear linker will bind two oligonucleotides only when they are in close proximity and create circular DNA which serves as a template for rolling circle amplification (RCA). Addition of fluorescent probes complementary to the sequence of DNA after RCA reaction allows visualisation of the proteins. The calculated distance between proteins could be below 30 nm [113]. This method was modified for commercialisation and use in general laboratories by addition of secondary antibodies linked with oligonucleotides. The Duolink kit (Sigma) only requires two antibodies of interest (mouse and rabbit host). The drawback of the commercial kit over the original method [113] is that the addition of secondary antibodies increases distance between proteins beyond 30 nm. The use of this technique in cancer tissue [114, 115] found strong correlation between complex level and concentration of the interacting proteins. The method was further developed to detect multiple complexes in the same sample utilising different probes simultaneously [116]. The utility of multiplex imaging of proteins will be discussed in the next section.

Förster resonance energy transfer-fluorescence lifetime imaging microscopy (FRET-FLIM) [117–119] is the accepted gold standard technique for measuring protein proximity, typically within the <10 nm range. By measuring the donor lifetime in the presence and in the absence of acceptor, one can accurately determine the distance between the donor- and acceptor-labelled proteins. Unlike intensity-based FRET methods which cannot distinguish between an increase in FRET efficiency (i.e. tighter coupling) and an increase in FRET population (concentration of FRET species), FRET-FLIM can resolve this by using multicomponent analysis. For cell-based research, proteins of interest with attached fluorescent proteins (i.e. GFP-mRFP1, donor–acceptor) can be visualised to demonstrate protein–protein interaction in time and following stimulation.

To adopt this method for imaging endogenous proteins in cells or tissue samples (Fig. 15.1), antibodies are directly labelled with fluorophores, one acting as donor (shorter emission wavelength, i.e. Alexa546) and one as acceptor (longer emission wavelength, i.e. Cy5). As an example, this technique has been applied to quantify the homo- and heterodimeric interactions between ErbB family members stained with fluorescently labelled antibodies against specific targets that are endogenously expressed [93, 120–122]. The challenge in the tissue application of this technique is the interfering endogenous and preparation-induced fluorescence in stromal and epithelial components of the tissue [123, 124]. An improved FLIM histology technique was developed [125] specifically to circumvent the problem

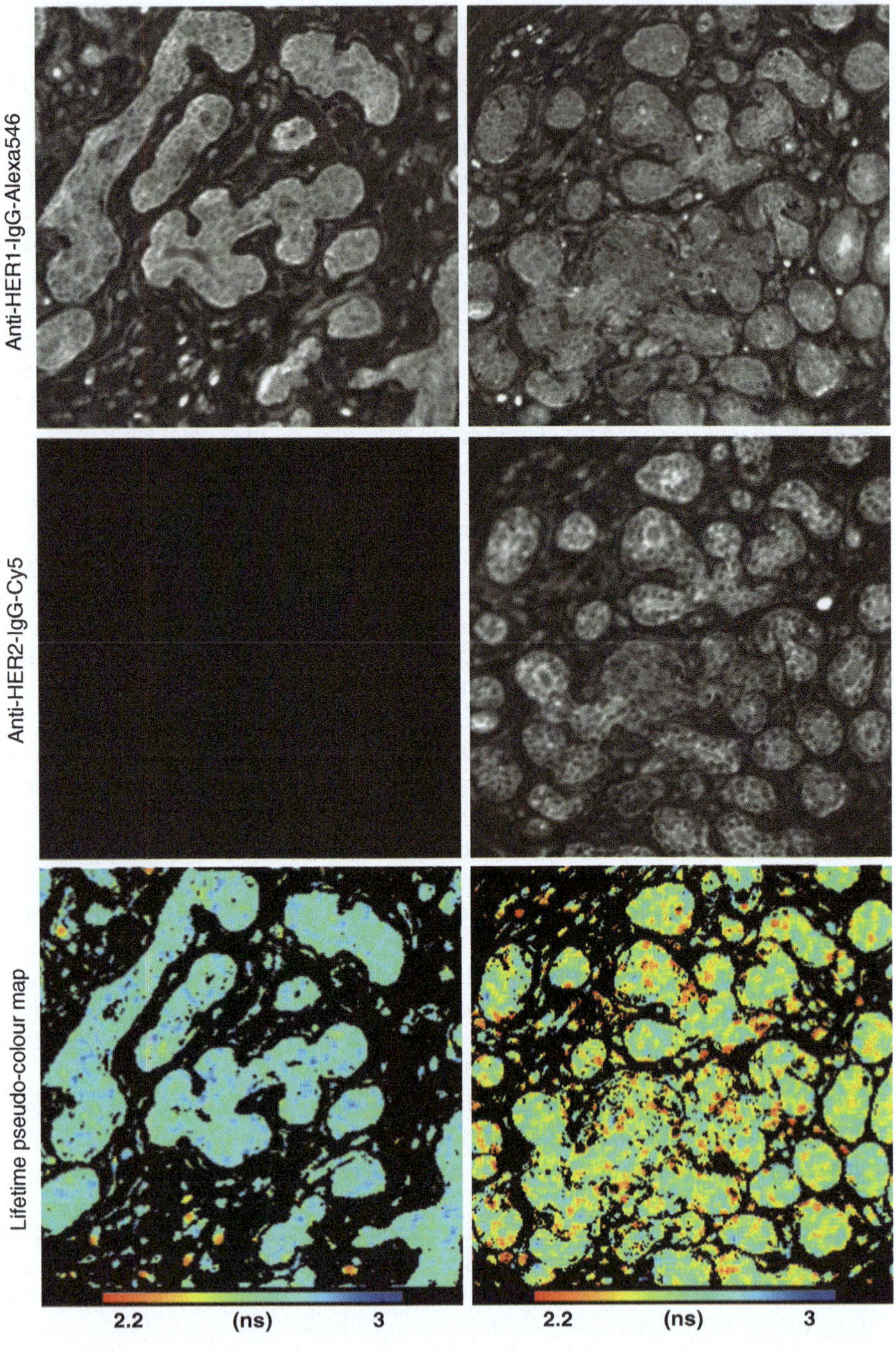

Fig. 15.1 Breast cancer sample stained with anti-HER1-IgG-Alexa546 and anti-HER2-Cy5 antibodies. The shift from *blue pallet* to *yellow/red* indicates decrease in lifetime of Alexa546, which is consistent with FRET between two fluorophores and indicates interaction between HER1 and HER2 proteins

of contaminating autofluorescence in FFPE tissues. With this technique we were able to quantify with high accuracy endogenous protein–protein interactions in archived pathological tissues [126]. Also, the EGFR-HER3 heterodimer assay was successfully applied for validation of mechanism for resistance development in patients with triple-negative breast cancer [93] treated with anti-EGFR antibodies. We found that anti-EGFR antibodies increase dimer formation and signalling through HER3 receptor, which is consistent with resistance to cetuximab observed in cell lines [92].

15.6 Highly Multiplexed Imaging of Tumour Tissues

Individual measurement of protein levels can no longer predict treatment outcome [94]. We now understand that multiple proteins may play a role in disease progression [72, 87, 90, 91] and treatment adaptation [92, 93]. Hence there is a need to develop new imaging techniques to capture disease heterogeneity. Using standard IHC utilising monochromatic dye, the expression of multiple proteins in patient samples is challenging. This can be overcome by sequential staining of the same tissue sample after removal of previous staining [127]. Up to six proteins can be detected this way, but it is limited by tissue degradation and reduced image quality. Using the same principle, changing the detection mode to fluorescence and developing special chemistry to inactivate fluorescent dye up to 60 proteins were imaged in whole tissue samples [128]. More than 700 patient samples analysed with this technique allowed for the mapping of cellular mTORC1 and MAPK signal transduction patterns in tissues with subcellular resolution. Furthermore, it allowed for combination of protein images with DNA FISH analysis in single tissue samples. The principle in both studies is similar to the MELC technology [129], the sequential imaging of the same field of view of tissue stained with different primary antibodies and corresponding fluorescently labelled secondary antibody combined with photobleaching cycles in between staining. The number of proteins imaged can improve dramatically from 49 [129] to 2100 [130] and allow for generation of combinatorial molecular phenotype maps for further analysis of protein networks.

An alternative approach utilising mass spectrometry [131] for detection of bound antibody to the target is a rasterising oxygen duoplasmatron ion beam liberating ions from lanthanide-conjugated antibodies. Up to ten different proteins were imaged simultaneously in human FFPE breast cancer samples. Another study showed even higher multiplexing capability when tissue stained with 32 rare earth metal-conjugated antibodies [132] were hit with high-resolution UV laser resulting in ablation of metals from antibodies spot by spot and line by line. By mass cytometry, proteins with cellular resolution of 1 μm were imaged. Both studies generated multicolour composite images, and quantitative capacity in a dynamic range greatly improved over quantitative immunofluorescence. There is a great potential to develop both methods for practical applications to understand protein networking in human samples.

15.7 Single-Molecule Imaging to Quantify ErbB Kinetics

The advent of next-generation sequencing (NGS) [133] has revealed genetic mutations that are widely heterogeneous among tumour subclones, even within the same patient tumour [134, 135], and the development of targeted therapies is expanding. Molecular imaging can provide detailed information on the temporal and spatial distribution of specific aberrations within a tumour, but identifying reliable markers for predicting response to therapies is an ongoing effort. As previously discussed, EGFR overexpression has been linked to the progression of many human cancers [129, 130, 136, 137], but it does not reliably correlate with response to treatment [138]. Even though the molecular concentration of EGFR has no predictive power, EGFR still has the potential of a biomarker candidate, and alternative read-outs have been explored. Ligand-induced homo- and heterodimer formation of ErbB receptors results in different biological outcomes irrespective of recruitment and activation of similar effector proteins. The use of quantitative, single-molecule-based imaging techniques to probe the effect of drug treatment on EGFR will complement the NGS approach, by providing a functional read-out that can potentially detect genetic variants (of different allelic fractions) which break through as the tumour subclones evolve.

Recently, we have shown the effect of EGFR inhibitors, gefitinib and lapatinib, on the EGFR homodimerisation kinetics in a basal-like breast cancer cell line, HCC1954 [121]. Gefitinib confers EGFR stability and allows EGFR to form a higher fraction of homodimers. This ability of gefitinib to modulate EGFR homodimerisation is likely to be important to cellular signalling by enhancing cell proliferation, contrary to the intended role of TKIs. Sensitivity to gefitinib was restored when using a double-site dimerisation-deficient EGFR mutant (the aforementioned EGFR I706Q, V948R mutation which has recently been shown by us to perturb the ability of EGFR to dimerise through either the C lobe (activator) or N lobe (receiver)) [122]. This suggests that the modest drug-induced increase in stability of the EGFR homodimer may have a significant biological impact on the tumour cell's proliferation potential.

15.8 Exosomes for Monitoring Proteomic and RNAome Changes in Response to Therapies

Currently available diagnostic and prognostic tools for monitoring molecular changes in cancer consist of invasive methods requiring tumour tissue. Of particular interest in this context are exosomes that are secreted by a variety of cell types and tissues in the body, including cancer cells. In the last decade, exosomes gained enormous attention in the field of cancer due to their potential to reflect tumorigenesis [139] and the proteome and RNAome of their cellular origin. In addition, exosome analysis is ideally suited for monitoring the evolving tumour longitudinally, in terms of its whole transcriptome, miRNome and proteome profiles [140] and can be used to monitor tumour progression in response to cancer therapy [141]. The research of exosomes is still in its early days from the clinical perspective, but the importance of exosome for monitoring the response to anticancer treatments is increasing.

Exosomes are vesicles with a size of 30–120 nm that are shed extracellularly. They can be released from various cells, such as epithelial, immune and inflammatory cells, as well as body fluids, including saliva, breast milk, urine and blood endothelial. Their cargo contains proteins, mRNA and miRNA, DNA and lipids that can be exchanged between donor and recipient cells. The number of identified miRNA, proteins and lipids in exosomes is growing and comprehensively comprised in the ExoCarta database [142, 143]. The assembly of their cargo and their selective molecular sorting is still poorly understood. Exosomes are characterised by canonical proteins, such as CD63, TSG101 and CD9, but also cell-specific content reflecting their origin.

Exosomes are released by cells under both physiological and pathological conditions, including cancers. Therefore, the biogenesis of exosomes may represent a response to different biological processes such as ligand stimulation [144] or cell stress [145], but also it might represent a particular pathological phenotype. Particularly in cancer, the biogenesis of exosomes might reflect changes in the microenvironment, hypoxia or malignant transformation. The unique ability of exosomes to selectively pack biological information into their cargo and to transfer molecules horizontally allows for a crosstalk between cells distant in the body. In this context, the exosomal cargo contains also antigens, diverse receptors, oncogenes and other proteins that have been shown to contribute to cancer invasion, cancer progression, angiogenesis, immune suppression and metastasis. The transfer of exosomes by recipient cells may affect the biological function of target cells in many different manners. The uptake of exosomes by target cells may be mediated through receptor ligand interaction, phagocytosis or fusion with the cell membrane. This results in the delivery of the exosomal cargo, including transcription factors, mRNA, non-coding RNA and miRNA as well as oncogenes, into recipient cells that may activate downstream pro-survival signalling processes [146, 147].

15.8.1 Multifaceted Function of Exosomes in Cancer

Cancer-derived exosomes may facilitate the oncogenic transfer proteins or miRNA between cancer cells. In EGFR-associated carcinogenesis, it has been demonstrated in glioma cells that the oncogenic form of EGFR (EGFRvIII) carrying an activating mutation may stimulate the formation of microvesicles and thus the increase in exosome production [148]. Furthermore, these exosome cargos carrying the EGFRvIII mutation can be delivered to cells lacking the oncogenic receptor and inducing oncogenic activity by activating AKT/MAPK signalling pathways. In another study, it has been shown that exosomes extracted from mutant KRAS in colon cancer cells confer a higher invasive potential on target cells than from exosomes derived from KRAS wild-type cells [144]. Taking this work further, the authors showed by proteomic analysis that mutant KRAS affects the composition of these exosomes and carry many tumour-promoting proteins, including KRAS and EGFR that can also be detected in transformed recipient cells [149].

The potential of exosomes to initiate angiogenesis has been shown by shed microvesicles from diverse human cancer cells harbouring oncogenic activated EGFR. These cells were taken up by endothelial cells and lead to autocrine expression of VEGF [150]. The question of how tumour cells may enhance angiogenesis under hypoxic conditions leads to the findings in another study that this is mediated through exosomes by activating the development of endothelial cells [151]. Other cell players involved in angiogenesis and activated by exosomes are myofibroblasts [152, 153]. Recent findings in a prostate cancer model suggest that prostate cancer-derived exosomes expressing TGFβ1 can trigger the differentiation of fibroblast to myofibroblasts which are pro-angiogenic [153, 154].

Metastasis is the predominant cause of mortality in cancer patients, and many efforts are made to understand the mechanisms leading to the spread of tumours to distant organs in the body. In this context, cancer-derived exosomes have been shown to contribute to the formation of the pre-metastatic niche that is associated with cancer progression and correlates with the development of aggressive cancers. It has been shown that exosomes originating from human and mice melanoma cells have the ability to modulate bone marrow progenitor cells and may give rise to a pro-metastatic behaviour [155]. Another finding shows that HIF and RAB22A-expressing breast cancer cells stimulate the production of exosomes under hypoxic conditions, and these exosomes lead to spontaneous metastasis to the lungs [156]. These findings suggest that oncogenic transfer of molecules by exosomes enhances carcinogenesis and has pro-metastatic features; however, it is not clear whether these changes remain permanent.

Both de novo and acquired resistance to anticancer therapies remains a major challenge in cancer treatment [157]. Various studies have demonstrated the impact of exosomes on chemotherapies and influence on the development of resistance to cancer treatment. For instance, in HER2-overexpressing breast cancer cells, the resistance to trastuzumab was linked to HER2-overexpressing exosomes [158]. In another study, the development of cisplatin resistance has been proved to be mediated by exosomes released from ovarian cancer cells [159].

In addition, exosomes have the potential to modulate the immune system by stimulating an immunosuppressive tumour environment [147, 160, 161]. For example, tumour-derived microvesicles from cancer patients' sera are capable of inducing immune suppression by promoting T regulatory cell expansion and apoptosis of antitumour $CD8^+$ effector T cells [162]. The analysis of microvesicles from cancer patients' sera might therefore serve as surrogate to monitor for changes in the immunological milieu in cancers. Also, tumour-derived exosomes can induce myeloid-derived suppressor cells (MDSC) that have an immune suppressive effect on cytotoxic $CD8^+$ T cells by modulating the differentiation pathway of myeloid cells to MDSC pathway, therefore promoting a pro-tumorigenic phenotype [163]. Taken together, exosomes have the potential to manipulate a number of biological processes in cancers leading to cell proliferation [164], angiogenesis, promote matrix remodelling, induce metastasis and contribute to drug resistance.

15.8.2 Proteome and RNAome Changes in Cancer-Derived Exosomes

A number of studies observed a modulated exosome composition of proteins, mRNA and miRNA in cancer cell-derived exosomes compared to healthy cells. Not surprisingly, many research efforts are made into analysing the proteome and RNAome of cancer exosomes on a large scale to provide insights into exosomes. RNAome and proteome analyses of exosomes can provide molecular signatures in cancers and identify novel biomarkers in circulating blood of cancer patients. The ExoCarta database comprises all entries of proteins and miRNAs identified in exosomes. For instance, EGFR has been identified in exosomes in several proteomic experiments in colon, bladder and pancreatic cancer studies [165–168]. Other proteomic studies of cancer-derived exosomes demonstrate the presence of proteins promoting angiogenesis and cell and cell motility/metastasis and the related process of pre-metastatic niche priming, including Met, S100A8 and S100A9 [169–171]. In addition to protein analysis, lipidomic profiling has also been performed on exosomes derived from prostate cancer cell lines with distinct AR phenotypes [172].

In the context of HER2 overexpression in solid tumours, quantitative proteomic analysis has been carried out on exosomes from wild-type- or HER2-overexpressing mammary epithelial cells. An altered extracellular vesicular proteomic content was observed in response to HER2 overexpression pointing towards a potential interplay between HER receptor signalling and cargo protein selection or packaging into exosomes [173].

MicroRNA (miRNA) is a major component of exosomes. The biogenesis of miRNA is tightly controlled and its expression deregulated in cancers [174]. There is evidence that exosomes transfer mRNA and miRNA to recipient cells and mediate intercellular communication [175, 176]. Skog et al. provided the first major evidence for the transport of miRNA by exosomes. The messengers encoding for EGFRvIII protein and miR-21 were delivered by glioblastoma-derived exosomes into normal recipient cells and were then translated into functional signals, stimulating proliferation of cancer cells [164]. In a different study, Zhou et al. have shown that cancer-secreted exosomes carrying miR-105 destroy vascular endothelial barriers to promote metastasis [177].

Different findings demonstrate that miRNAs from exosomes also have the potential to modulate the immune regulation. MicroRNA derived from exosomes of lung cancer cells could bind to Toll-like receptors in macrophages, leading to TLR-mediated NFkB activation and release of pro-metastatic inflammatory cytokines [178]. Furthermore, it has been suggested that miRNA-containing macrovesicles secreted by macrophages regulate the invasiveness of breast cancer cells through exosomes [179]. In this regard, the identification of novel miRNA, isolated from circulating exosomes, is of considerable interest to be used as novel biomarkers for anticancer therapies.

15.8.3 Clinical Relevance of Exosomes for Monitoring Anticancer Therapy

Exosomes released into blood, urine and body fluids offer a great opportunity to access biological information about the cancer without performing invasive procedures. In addition, exosome analysis from patients' blood offers almost a continuous and longitudinal access to information on the status of the tumour, in order to track its time-dependent and treatment-responsive evolution and heterogeneity [180].

Sufficient evidence exists that biological information retrieved from cancer cells is reflected in cancer-derived exosomes. Many known cellular proteins that are currently used as biomarkers for cancer diagnosis have been identified in exosomes [181]. EGFR, HER2, CEA and many others can be detected in exosomes or microvesicles [182, 183].

Many researchers have observed that oncogenic pathways stimulate the production of exosomes, and indeed, a number of cancer cells produce elevated levels of exosomes. The concentration of exosomes in the blood of cancer patients is often increased compared to exosomes from the blood of healthy donors. Quantification of exosomes in the plasma from melanoma patients using ELISA test showed a significantly increased number of exosomes in melanoma patients as compared to healthy donors [184]. Similarly, a significant higher mean concentration of exosome and exosomal miRNA was detected in plasma of lung cancer patients compared to that of control individuals [185].

In the context of changes in the exosome concentration, several studies have shown that oncogenes, tumour suppressor genes or miRNA may influence the exosome production [141, 186]. Apart from blood, other exosomal fluids might be of interest as an alternative diagnostic biomarker, such as urine exosomes in prostate cancer [187].

One major advantage of exosomes for using exosomes as biomarkers is their stability. They are known to be stable at 4 °C in the short-term and at −80 °C for long-term storage. At the current state, one of the challenges for using exosomes to monitor cancer progression and responses to cancer treatment is the potential need to isolate cancer cell-derived exosomes from a mixed population of circulating exosomes from different cells in the blood [147]. Commercial companies are making huge efforts to develop novel assays to improve the current diagnostic tools available, of which several are under clinical validation.

15.9 Super-resolution Imaging to Quantify ErbB Oligomer Heterogeneity

Resolving the spatial distribution of EGFR on the plasma membrane is very important as several studies have suggested the existence of oligomers larger than homo- and heterodimers between different ErbB species. Several studies reported on the existence of higher-order ErbB oligomers and their functional relevance to signalling [188–190]. Understanding the heterogeneous/variable downstream consequences (in terms of signalling output) of these ErbB receptor associations has been aided by systems modelling [191]. In cancers, the influence of the stoichiometry and nanoscale proximity of ErbB receptors within these oligomers is still not well understood but may be concerned with and serve as a surrogate biomarker for the signal amplification in the ErbB network and the determination of cell fates.

Fluorescence microscopy is a sensitive and powerful tool extensively used to visualise complex dynamic structures and functions occurring within different types of cells and tissues. The main disadvantage is poor, diffraction-limited resolution which is typically an order of magnitude larger than distances between interacting proteins. To gain a better understanding of the biological mechanisms, a plethora of super-resolution fluorescence techniques were developed, which achieve resolution below the diffraction limit [192, 193]. Nanometre resolution imaging opened the possibility to discover new surrogate/correlative markers of cellular function, by providing precise stoichiometry and nanoscale proximity parameters within, e.g. the aforementioned ErbB oligomeric protein network. Super-resolution imaging techniques have been applied to visualise protein trafficking and the protein architecture in various cell types. The spatial organisation of a protein inside a cell is often directly linked to its function. Dual-colour imaging permits nanoscale co-localisation and the direct probing of protein–protein interactions [117]. Quantitative single-molecule localisation microscopy allows for counting the number of molecules of a protein species in a single cell [194], studying the heterogeneity in protein spatial organisation [195–198] and co-localisation analysis [199, 200]. The unique advantages of high molecular specificity coupled with nanometre resolutions make super-resolution imaging a superior tool to access detailed spatial information available otherwise only by electron microscopy (EM). Recent efforts lead to great progress in quantitative single-molecule localisation microscopy and highlighted its potential to address biological questions in the sheer complexity of the cellular environment [201].

Super-resolution fluorescence microscopy offers the advantage of simpler optical set-ups and multicolour imaging which can be performed in live cells if desirable. It can achieve spatial resolution of 10–20 nm, depending on the photostability, brightness and switching properties of the fluorophores. QD blinking can be successfully used as an alternative to photoswitching of organic fluorophores or photoactivatable proteins [202]. High-resolution imaging methods are important to investigate the significance of receptor nanoscale organisation in regulating its function. Detailed spatial information on the EGFR distribution is critical to improve our understanding of the mechanisms controlling protein interactions and EGFR-driven cell signalling and cancer progression. Quantitative data on the size, composition and subcellular localisation of receptor clusters is not accessible with diffraction-limited optical microscopy. It requires high-resolution imaging techniques, such as near-field scanning optical microscopy (NSOM) [203], atomic force microscopy, transmission EM [190, 204] or the more recently developed single-molecule localisation-based super-resolution methods.

15.10 Monitoring the Occurrence of EMT by Sampling Circulating Tumour Cells

Over the last 20 years, evidence has accumulated to support the role of epithelial to mesenchymal transition (EMT) as a means through which solid tissue epithelial cancers invade and metastasise [205]. Remarkable phenotype plasticity of epithelial cells underlies morphogenesis, epithelial repair and tumour invasiveness. During progressive dedifferentiation in epithelial cancer, the conversion from multicellular growth and invasion towards mesenchymal single-cell migration is termed the epithelial–mesenchymal transition and has been implicated in the process whereby carcinoma cells disseminate from the local environment and metastasise to a secondary site [206, 207].

EMT occurs through events that alter the integrity of cell–cell junctions and loss of apico-basal polarity, eventually resulting in the loss of contact between neighbouring cells and the ability to facilitate movement. These changes in cell morphology and function are accompanied by changes in epithelial cancer cell protein expression profiles. The loss of E-cadherin is a characteristic feature of EMT. This occurs due to reduced expression, loss of function cadherin mutations and catenin-signalling pathways and protease function deregulation leading to cadherin degradation. De novo expression of vimentin is also observed [208].

The process of malignant EMT is under the strict control of growth factors (EGF, HGF, IGF and TNFα) and downstream transcription factors (Snail1 and Snail2) [206]. The transforming growth factor (TGF-β) is the most extensively studied inducer of EMT. Members of the TGF-β family of growth factors can initiate and maintain EMT in a variety of biological systems and pathophysiological context by activating major signalling pathways and transcriptional regulators integrated in extensive signalling networks [209].

EMT is therefore considered a significant step in the invasive cascade of tumours, and some reports suggest that it is manifested as and can therefore be monitored by quantifying mesenchymal circulating tumour cells (CTCs). It marks the transition from a collective to a single-cell migration mechanism and represents phenotypic and functional plasticity of tumour cells. Once the tumour has reached the dedifferentiated stage of single-cell dissemination, metastatic spread is increased, and the process is associated with poor prognosis [210].

15.10.1 Collective Migration in Cancer

Epithelial cancer cells disseminate from the primary tumour either as individual cells or as cell clusters or strands. Collective cell movement is this important mechanism of cancer cell local invasion and distant spread characterised in epithelial cancers of high and intermediate differentiation grade (e.g. lobular breast cancer, epithelial prostate cancer and NSCLC) and melanoma [211–213]. In highly differentiated epithelial cancer, tubular and glandular structures may be retained in the invading zone, whereas with the loss of apical–basal polarity and dedifferentiation, cell strands can extend within the tissue.

Protruding sheets and strands of cells that maintain contact with the primary site can generate local invasion, such as in invasive epithelial cancers, whereas detached cell clusters or 'nests' can extend along interstitial tissue gaps as well as along perineural structures [212]. These cell clusters allow metastasis also through lymphatics and have been isolated from peripheral blood. Collective dissemination of cancer cells is thought to be more effective in increasing the probability of tumour cell survival during metastasis [214].

The integrin family of transmembrane receptors is involved in the regulation of cell cluster migration. The importance of integrins of the β1 family in providing attachment and dynamic force generation was shown for neoplastic cancer cell clusters. Actin is also required along cell–cell junctions in sustaining collective integrity [215]. Furthermore, migrating cell collectives develop preferential protease expression and proteolytic degradation of extracellular matrix by matrix metalloproteinases (MMPs) upregulated by tumour cells that allow for expansion of cell groups [216]. Motility-inducing chemokines and growth factors induce and maintain migration by signal transduction via the PI3K, RAC and RHO signalling pathways [217].

15.10.2 CTCs in Prognostication, Response Prediction and Treatment Stratification

The presence of viable circulating tumour cells (CTCs) is a prerequisite for establishing distant metastases. CTCs can provide 'liquid, real-time' biopsy to allow exploration of tumorigenesis and metastasis. The detection and enumeration of CTCs has demonstrated clinical utility with respect to prognosis in breast, colorectal and prostate cancers [218–220]. More recent evidence from studies also suggests that CTCs can be used for pharmacodynamic information to guide therapeutic decision making by providing information on drug resistance mechanisms.

Technological advances, in particular cytometric approaches, have led to advances in CTC purification and molecular characterisation. Still, there are numerous challenges and uncertainties with the use of current technologies especially in terms of purifying the CTC detection from a patient blood sample as well as the degree of biological heterogeneity within a CTC population.

Cytometric approaches, mainly the CellSearch system, employ immunostaining to identify CTCs based on the expression of epithelial markers (EpCAM, cytokeratins) and the absence of leucocyte markers (anti-CD45) [221]. More novel methods of CTC detection utilise isolation by size exclusion (isolation by size of epithelial tumour cells; ISET by ScreenCell) that allows enrichment of epithelial cells using filtration that captures tumour cells [222]. The advantage of this technique is that tumour cell loss is minimised as compared to immunolabelling with epithelial cell-specific markers. This is because not all epithelial tumour cells express EpCAM (variable between tumour types and about 70–80%) [223]. In addition, given the knowledge on biological plasticity and EMT, the expression profile of epithelial markers, such as EpCAM and cytokeratins, by CTCs is likely to be modified dynamically during the process of metastasis as this involves phenotypic changes in a subset of cells within the primary tumour, during which epithelial cells become more motile and invasive through epithelial–mesenchymal transition [206]. ISET, therefore, poses as an advantageous method compared to immunostaining.

In using CTCs within studies for cancer prognostication and therapeutic prediction, it is important to take into account the process of incomplete EMT as suggested in the literature [224]. Collective migration of tumour cells in the circulation during cancer progression, as tumour microemboli (CTM), suggests that these CTCs may manifest a broad and heterogeneous spectrum of phenotypic changes, hence limiting the applicability of immunostaining techniques as CTC detection approaches [206, 213, 225][206, 213, 225]. In addition EMT, as a favourable process of cancer metastasis, only takes place in a small percentage of tumour cells. In a pilot study in lung cancer patients using ISET, inter- and intra-patient heterogeneity was seen by the variable expression of the epithelial markers E-cadherin and cytokeratins, as well as the mesenchymal markers vimentin and N-cadherin in CTCs/CTM. The presence of this heterogeneous tumour cell population in the circulation has not been evaluated for its clinical significance in patient prognostication and stratification [226].

The molecular characteristics of CTCs have also been investigated in trials as predictive biomarkers to guide therapy. HER2 expression between primary tumour and corresponding CTCs in advanced breast cancer demonstrated a difference, suggesting the possibility of clonal selection during treatment and a mechanism of disease progression [227]. In a different study in breast cancer patients, activated EGFR and downstream PI3K/AKT pathway activation was also observed in CTCs [228]. Furthermore, enrichment of CTCs by CellSearch allows CTC analysis by fluorescence in situ hydridisation (FISH) and the CTC-chip technology to detect both the classical EGFR-activating mutation and the acquired T790M resistance mutation. Clearly, the ability to reproduce this within large-scale trials offers the possibility of monitoring changes in epithelial tumour genotypes during the course of treatment for better patient stratification without the need to obtain tumour tissue [229].

With ongoing research in CTC technologies, molecular characterisation of CTCs is possible and possesses unique advantages to study tumour biology and the process of metastasis. From a translational perspective, CTCs can serve as important biomarkers of predictive information for the stratification of patients to therapies, but the CTC methodologies need to have reliable, reproducible and robust assays to make this a clinical reality.

15.11 A Cinderella Story: Signal Transduction-Metabolism Link—Metabolic Changes in Cancer and Their Influence by Protein Interactions Within Signalling Pathways

The ability of cancer cells to undergo major reprogramming of their energy metabolism to fuel continuous growth and proliferation as well as dissemination is a new hallmark of carcinogenesis and malignant transformation [230, 231].

15.11.1 Metabolic Requirements of Cancer Cells

Normal cells have as a main source of energy the generation of ATP through the process of oxidative phosphorylation (OXPHOS). Known as the *Warburg effect*, cancer cells produce lactate as a result of the metabolism of glucose through the pathway of glycolysis even in the presence of normal concentrations of oxygen [232, 233].

Despite the fact that the yield of ATP created per glucose consumed is low and regardless of how proliferative cancer cells are, ATP and NADH are not limited in cancer cells [234], since the maintenance of ATP/ADP or NADH/NAD^+ ratios is critical for the cells not to undergo cell cycle arrest, autophagy or apoptosis [235, 236]. The accumulation of glycolysis intermediates allows the cell to divert into anabolic pathways, providing precursors for macromolecular synthesis: acetyl-coA for fatty acids, non-essential amino acids and ribose for nucleotides and NADPH through the pentose phosphate shunt [237].

Due to the lower efficiency of glucose metabolism within tumours, cancer cells have created the need for other energetic sources. Glutamine is a key substrate required for tumour growth as the primary nitrogen donor for the synthesis of nucleotides and non-essential amino acids [238]. Furthermore, by losing the amide and amine groups, glutaminolysis allows the conversion of glutamine to α-ketoglutarate, a tricarboxylic acid that can be derived into the Krebs cycle for obtaining ATP or to acetyl-coA for the lipid synthesis required by the proliferating cancer cell [239].

15.11.2 Molecular Mechanisms Regulating Cancer Cell Metabolism

(a) Growth factors:

PI3K is a major downstream effector of receptor tyrosine kinases (RTKs) and G protein-coupled receptors (GPCR) and transduces signals from various growth factors and cytokines into intracellular messages. In response to extracellular signals, activated PI3K phosphorylates phosphatidylinositol (4,5)-bisphosphate (PIP2) to generate phosphatidylinositol (3,4,5)-triphosphate (PIP3), leading to the recruitment of AKT to the plasma membrane, where it is phosphorylated and activated. The phosphatase PTEN is the most important negative regulator of the PI3K signalling pathway. PTEN dephosphorylates the 3-position phosphate of PIP3 to produce PIP2, thus inhibiting PI3K-dependent AKT activation. Activated AKT can directly activate mTORC1 by phosphorylation and indirectly by phosphorylating TSC2. Therefore, genetic alterations in PI3K/AKT/mTOR and PTEN lead to increased and sustained activation of this signalling pathway and are common features in many cancers [2, 240].

(b) Oxygen availability—HIF1:

Hypoxia occurs frequently in human cancers, inducing a broad metabolic reprogramming coordinated at the transcriptional level by HIF-1, which functions as a master regulator to balance oxygen supply and demand. HIF-1 reprograms glucose metabolism through transactivation of PDK1 (TCA cycle inhibitor) and activation of glycolytic components, such as glucose receptors (GLUT) and glycolytic enzymes (HK, PKM2 and LDHA) [241].

mTOR activity stimulates translation, stabilisation and transactivation of HIF1A, thus collaborating to the metabolic switch in response to hypoxia [242]. Sustained activation of mTOR and loss of PTEN increase dramatically the levels and activity of HIF-1, which, in turn, correlates with cancer progression and increased risk of mortality in many human cancers [243].

(c) Amino acid availability:
The proto-oncogene c-myc enables cancer cells to maximise glutamine uptake from the extracellular space through upregulation of the glutamine importer ASCT2 and conversion into glutamic acid via overexpression of glutaminase (GLS). Once glutamine enters the cell, it can be metabolised through glutaminolysis to provide NADPH or exported. The fraction of glutamine that is shuttled out in exchange for essential amino acids (i.e. leucine) directly activates mTOR for the initiation of protein translation and cell growth [244]. C-myc is one of the major transcription factors driving the key mechanisms in tumorigenesis (cell cycle, apoptosis, differentiation, migration, etc.), being mutated or amplified in almost all types of human cancers, a phenomenon that also leads to the increase in mTOR signalling and constitutive changes in cell metabolism [245].

(d) Energy:
AMPK is activated in response to low cellular energy (high AMP/ATP ratio). Once activated it downregulates protein synthesis and stimulates ATP generation via fatty acid oxidation (Inoki et al., 2003). In addition, AKT activates mTORC1 by regulation of cellular energy maintaining a high ATP level that causes a decrease in the AMP/ATP ratio and inhibits AMPK pathway [246]. Loss of AMPK activity has been observed in several tumour types and can cooperate with oncogenic drivers to reprogram tumour cell metabolism and enhance cell growth and proliferation [247].

15.11.3 Imaging Cancer Cell Metabolism

Due to their altered metabolism, malignant tumours can be detected with high sensitivity and specificity by imaging their increased metabolic rates for glucose, amino acids or lipids. Metabolic imaging provides essential data regarding the functional state of the tumour (or even within it) that cannot be approached with conventional genetic or biochemical techniques, such as immunohistochemistry. These approaches are of importance in the stratification of cancer patients and prediction of treatment response:

(a) Glucose metabolism imaging:
FDG is a modified glucose molecule with the radionuclide fluorine-18 (^{18}F) in place of the hydroxyl group on the 2 carbon. The uptake of ^{18}FDG is identical to the glucose, but this analogue cannot be metabolised due to the lack of the hydroxyl

group and therefore accumulates in the cell. The method of imaging with FDG is positron emission through the decay of ^{18}F. ^{18}FDG-PET has been used in numerous PET diagnostic imaging procedures, helping to define tumour volumes and localisation to complement computed tomography (CT) or magnetic resonance imaging (MRI). Nevertheless, PET images lack the resolution necessary to identify distributions of FDG within a tumour other than large areas of increased or decreased uptake. Hence, novel fluorescent alternatives have been developed in the recent years to improve the resolution of the glucose analogues [248].

Nuclear magnetic resonance spectroscopy (NMR spectroscopy, MRS) exploits the magnetic properties of certain atomic nuclei. MRS provides information about the structure, dynamics and concentration of molecules. ^{13}C nuclear magnetic resonance (NMR) spectroscopy allows observation of labelled glucose through glycolysis, TCA or fatty acid synthesis, determining the activity of the different metabolic pathways in the cell. Sensitivity and poor spatial resolution are the main weaknesses of the procedure in vivo. Several approaches have been recently used to solve these limitations: (a) proton (^{1}H) decoupling that combines the high sensitivity of ^{1}H-MRS and uses sophisticated data acquisition schemes that detect only ^{1}H attached to ^{13}C labels within molecules [249] and (b) generation of hyperpolarised ^{13}C agents by DNP (dynamic nuclear polarization) that increase the signal by more than 10,000-fold [250]. By monitoring the fluxes of pyruvate through the cell in preclinical models of prostate cancer, it has been shown the potential of this technique on understanding prostate cancer and the diseases of other organs [251].

Furthermore, glucoCEST is a system that uses natural, non-radioactive glucose at physiologically reasonable quantities. It is visualised by MRI through a mechanism known as chemical exchange saturation transfer (CEST) [252, 253]. In this context, intracellular lactate and lactate excretion measured by ^{1}H-MRS can help to reflect the metabolic status of a specific tissue [254, 255].

(b) Amino acid metabolism imaging:
Altered nutrient uptake can also be assessed with radiolabelled and cold approaches. Increased glutamine metabolism shown by many tumours was the basis for the design of several glutamine analogues as PET radiotracers: ^{18}F-, ^{11}C- and ^{13}N-glutamine or glutamate compounds have been recently developed. Magnetic resonance spectroscopy can also be used to evaluate glutamine and its metabolites within tumours (e.g. conversion of ^{13}C-glutamine to ^{13}C-glutamate) [256].

(c) Imaging of hypoxia and REDOX status within tumours:
The approaches previously described are based on the different uptake of metabolites between the cancer and the normal tissue, but the differential processes occurring within the tumour environment also can be assessed by measuring the presence of hypoxic regions and the balance between anabolic and catabolic reactions based on quantification of nicotinamide adenine dinucleotides (NAD and NADP). Tumour hypoxia is a common feature in most of the cancers, which correlates with disease progression, malignancy and poor

prognosis; hence the visualisation and quantification of the hypoxic regions within tumours have diagnostic importance. Quantification of pO_2 has been studied with various techniques: oxygen electrodes, biomarkers (e.g. pimonidazole) and blood oxygen level-dependent (BOLD) or tissue oxygen level-dependent (TOLD) MRI. Further developments in imaging techniques that utilise radiotracers combined with PET/SPECT (Fleming et al., 2015) and fluorescent probes based on Förster resonance energy transfer (FRET) are very powerful prognostic tools [257].

Conclusion

In summary, we have provided a precise synopsis of the pathophysiological processes of cancer growth, metastasis and metabolism through to the various methodological approaches to assess the molecular network organisation of tumour cells. The ultimate translational goal is to monitor the temporal evolution of these genomic events and molecular mechanisms within the tumours of patients through imaging and circulating biomarkers. This is in line with our recently outlined C2c (turning cancer to chronic disease) strategy of integrating non-invasive imaging with circulating biomarkers to track tumour evolution [140, 258]. The future aim is to instigate or change treatment once pathway-specific changes in the tumour's genome and/or proteome are detected, preferably early on during the course of targeted treatment. The introduction of such technologies in routine clinical practice will promote the provision of personalised treatment in cancer and undoubtedly improve clinical outcomes.

References

1. Davies MA, Samuels Y. Analysis of the genome to personalize therapy for melanoma. Oncogene. 2010;29:5545–55.
2. Yuan TL, Cantley LC. PI3K pathway alterations in cancer: variations on a theme. Oncogene. 2008;27:5497–510.
3. Jiang BH, Liu LZ. PI3K/PTEN signaling in angiogenesis and tumorigenesis. Adv Cancer Res. 2009;102:19–65.
4. Chen RH, Sarnecki C, Blenis J. Nuclear localization and regulation of erk- and rsk-encoded protein kinases. Mol Cell Biol. 1992;12:915–27.
5. Forbes SA, Bindal N, Bamford S, Cole C, Kok CY, Beare D, Jia M, Shepherd R, Leung K, Menzies A, Teague JW, Campbell PJ, Stratton MR, Futreal PA. COSMIC: mining complete cancer genomes in the Catalogue of Somatic Mutations in Cancer. Nucleic Acids Res. 2011;39:D945–50.
6. Garnett MJ, Marais R. Guilty as charged: B-RAF is a human oncogene. Cancer Cell. 2004;6:313–9.
7. Davies H, Bignell GR, Cox C, Stephens P, Edkins S, Clegg S, Teague J, Woffendin H, Garnett MJ, Bottomley W, Davis N, Dicks E, Ewing R, Floyd Y, Gray K, Hall S, Hawes R, Hughes J, Kosmidou V, Menzies A, Mould C, Parker A, Stevens C, Watt S, Hooper S, Wilson R, Jayatilake H, Gusterson BA, Cooper C, Shipley J, Hargrave D, Pritchard-Jones K, Maitland N, Chenevix-Trench G, Riggins GJ, Bigner DD, Palmieri G, Cossu A, Flanagan A, Nicholson A, Ho JW, Leung SY, Yuen ST, Weber BL, Seigler HF, Darrow TL, Paterson H, Marais R, Marshall CJ, Wooster R, Stratton MR, Futreal PA. Mutations of the BRAF gene in human cancer. Nature. 2002;417:949–54.

8. Hauschild A, Grob JJ, Demidov LV, Jouary T, Gutzmer R, Millward M, Rutkowski P, Blank CU, Miller Jr WH, Kaempgen E, Martin-Algarra S, Karaszewska B, Mauch C, Chiarion-Sileni V, Martin AM, Swann S, Haney P, Mirakhur B, Guckert ME, Goodman V, Chapman PB. Dabrafenib in BRAF-mutated metastatic melanoma: a multicentre, open-label, phase 3 randomised controlled trial. Lancet. 2012;380:358–65.
9. Long GV, Menzies AM, Nagrial AM, Haydu LE, Hamilton AL, Mann GJ, Hughes TM, Thompson JF, Scolyer RA, Kefford RF. Prognostic and clinicopathologic associations of oncogenic BRAF in metastatic melanoma. J Clin Oncol. 2011;29:1239–46.
10. Ascierto PA, Minor D, Ribas A, Lebbe C, O'hagan A, Arya N, Guckert M, Schadendorf D, Kefford RF, Grob JJ, Hamid O, Amaravadi R, Simeone E, Wilhelm T, Kim KB, Long GV, Martin AM, Mazumdar J, Goodman VL, Trefzer U. Phase II trial (BREAK-2) of the BRAF inhibitor dabrafenib (GSK2118436) in patients with metastatic melanoma. J Clin Oncol. 2013;31:3205–11.
11. Chapman PB, Hauschild A, Robert C, Haanen JB, Ascierto P, Larkin J, Dummer R, Garbe C, Testori A, Maio M, Hogg D, Lorigan P, Lebbe C, Jouary T, Schadendorf D, Ribas A, O'day SJ, Sosman JA, Kirkwood JM, Eggermont AM, Dreno B, Nolop K, Li J, Nelson B, Hou J, Lee RJ, Flaherty KT, Mcarthur GA. Improved survival with vemurafenib in melanoma with BRAF V600E mutation. N Engl J Med. 2011;364:2507–16.
12. Sosman JA, Kim KB, Schuchter L, Gonzalez R, Pavlick AC, Weber JS, Mcarthur GA, Hutson TE, Moschos SJ, Flaherty KT, Hersey P, Kefford R, Lawrence D, Puzanov I, Lewis KD, Amaravadi RK, Chmielowski B, Lawrence HJ, Shyr Y, Ye F, Li J, Nolop KB, Lee RJ, Joe AK, Ribas A. Survival in BRAF V600-mutant advanced melanoma treated with vemurafenib. N Engl J Med. 2012;366:707–14.
13. Kopetz S, Desai J, Chan E, Hecht JR, O'dwyer PJ, Lee RJ, Nolop KB, Saltz L. PLX4032 in metastatic colorectal cancer patients with mutant BRAF tumors. J Clin Oncol. 2010;28
14. Lito P, Pratilas CA, Joseph EW, Tadi M, Halilovic E, Zubrowski M, Huang A, Wong WL, Callahan MK, Merghoub T, Wolchok JD, de Stanchina E, Chandarlapaty S, Poulikakos PI, Fagin JA, Rosen N. Relief of profound feedback inhibition of mitogenic signaling by RAF inhibitors attenuates their activity in BRAFV600E melanomas. Cancer Cell. 2012; 22:668–82.
15. Corcoran RB, Ebi H, Turke AB, Coffee EM, Nishino M, Cogdill AP, Brown RD, Della Pelle P, Dias-Santagata D, Hung KE, Flaherty KT, Piris A, Wargo JA, Settleman J, Mino-Kenudson M, Engelman JA. EGFR-mediated re-activation of MAPK signaling contributes to insensitivity of BRAF mutant colorectal cancers to RAF inhibition with vemurafenib. Cancer Discov. 2012;2:227–35.
16. Corcoran RB, Dias-Santagata D, Bergethon K, Iafrate AJ, Settleman J, Engelman JA. BRAF gene amplification can promote acquired resistance to MEK inhibitors in cancer cells harboring the BRAF V600E mutation. Sci Signal. 2010;3:ra84.
17. Nazarian R, Shi H, Wang Q, Kong X, Koya RC, Lee H, Chen Z, Lee MK, Attar N, Sazegar H, Chodon T, Nelson SF, Mcarthur G, Sosman JA, Ribas A, Lo RS. Melanomas acquire resistance to B-RAF(V600E) inhibition by RTK or N-RAS upregulation. Nature. 2010;468:973–7.
18. Van Allen EM, Wagle N, Sucker A, Treacy DJ, Johannessen CM, Goetz EM, Place CS, Taylor-Weiner A, Whittaker S, Kryukov GV, Hodis E, Rosenberg M, Mckenna A, Cibulskis K, Farlow D, Zimmer L, Hillen U, Gutzmer R, Goldinger SM, Ugurel S, Gogas HJ, Egberts F, Berking C, Trefzer U, Loquai C, Weide B, Hassel JC, Gabriel SB, Carter SL, Getz G, Garraway LA, Schadendorf D. The genetic landscape of clinical resistance to RAF inhibition in metastatic melanoma. Cancer Discov. 2014;4:94–109.
19. Poulikakos PI, Persaud Y, Janakiraman M, Kong X, Ng C, Moriceau G, Shi H, Atefi M, Titz B, Gabay MT, Salton M, Dahlman KB, Tadi M, Wargo JA, Flaherty KT, Kelley MC, Misteli T, Chapman PB, Sosman JA, Graeber TG, Ribas A, Lo RS, Rosen N, Solit DB. RAF inhibitor resistance is mediated by dimerization of aberrantly spliced BRAF(V600E). Nature. 2011;480:387–90.

20. Ahronian LG, Sennott EM, Van Allen EM, Wagle N, Kwak EL, Faris JE, Godfrey JT, Nishimura K, Lynch KD, Mermel CH, Lockerman EL, Kalsy A, Gurski Jr JM, Bahl S, Anderka K, Green LM, Lennon NJ, Huynh TG, Mino-Kenudson M, Getz G, Dias-Santagata D, Iafrate AJ, Engelman JA, Garraway LA, Corcoran RB. Clinical acquired resistance to RAF inhibitor combinations in BRAF-mutant colorectal cancer through MAPK pathway alterations. Cancer Discov. 2015;5:358–67.
21. Roberts PJ, Stinchcombe TE. KRAS mutation: should we test for it, and does it matter? J Clin Oncol. 2013;31:1112–21.
22. Shi H, Hugo W, Kong X, Hong A, Koya RC, Moriceau G, Chodon T, Guo R, Johnson DB, Dahlman KB, Kelley MC, Kefford RF, Chmielowski B, Glaspy JA, Sosman JA, van Baren N, Long GV, Ribas A, Lo RS. Acquired resistance and clonal evolution in melanoma during BRAF inhibitor therapy. Cancer Discov. 2014;4:80–93.
23. Villanueva J, Vultur A, Lee JT, Somasundaram R, Fukunaga-Kalabis M, Cipolla AK, Wubbenhorst B, Xu X, Gimotty PA, Kee D, Santiago-Walker AE, Letrero R, D'andrea K, Pushparajan A, Hayden JE, Brown KD, Laquerre S, Mcarthur GA, Sosman JA, Nathanson KL, Herlyn M. Acquired resistance to BRAF inhibitors mediated by a RAF kinase switch in melanoma can be overcome by cotargeting MEK and IGF-1R/PI3K. Cancer Cell. 2010;18:683–95.
24. Girotti MR, Pedersen M, Sanchez-Laorden B, Viros A, Turajlic S, Niculescu-Duvaz D, Zambon A, Sinclair J, Hayes A, Gore M, Lorigan P, Springer C, Larkin J, Jorgensen C, Marais R. Inhibiting EGF receptor or SRC family kinase signaling overcomes BRAF inhibitor resistance in melanoma. Cancer Discov. 2013;3:158–67.
25. Abel EV, Basile KJ, Kugel 3rd CH, Witkiewicz AK, Le K, Amaravadi RK, Karakousis GC, Xu X, Xu W, Schuchter LM, Lee JB, Ertel A, Fortina P, Aplin AE. Melanoma adapts to RAF/MEK inhibitors through FOXD3-mediated upregulation of ERBB3. J Clin Invest. 2013;123:2155–68.
26. Ciardiello F, Tortora G. EGFR antagonists in cancer treatment. N Engl J Med. 2008;358:1160–74.
27. Bell DW, Gore I, Okimoto RA, Godin-Heymann N, Sordella R, Mulloy R, Sharma SV, Brannigan BW, Mohapatra G, Settleman J, Haber DA. Inherited susceptibility to lung cancer may be associated with the T790M drug resistance mutation in EGFR. Nat Genet. 2005;37:1315–6.
28. Oxnard GR, Miller VA, Robson ME, Azzoli CG, Pao W, Ladanyi M, Arcila ME. Screening for germline EGFR T790M mutations through lung cancer genotyping. J Thorac Oncol. 2012;7:1049–52.
29. Lynch TJ, Bell DW, Sordella R, Gurubhagavatula S, Okimoto RA, Brannigan BW, Harris PL, Haserlat SM, Supko JG, Haluska FG, Louis DN, Christiani DC, Settleman J, Haber DA. Activating mutations in the epidermal growth factor receptor underlying responsiveness of non-small-cell lung cancer to gefitinib. N Engl J Med. 2004;350:2129–39.
30. Eberhard DA, Johnson BE, Amler LC, Goddard AD, Heldens SL, Herbst RS, Ince WL, Janne PA, Januario T, Johnson DH, Klein P, Miller VA, Ostland MA, Ramies DA, Sebisanovic D, Stinson JA, Zhang YR, Seshagiri S, Hillan KJ. Mutations in the epidermal growth factor receptor and in KRAS are predictive and prognostic indicators in patients with non-small-cell lung cancer treated with chemotherapy alone and in combination with erlotinib. J Clin Oncol. 2005;23:5900–9.
31. Yarden Y. The EGFR family and its ligands in human cancer. signalling mechanisms and therapeutic opportunities. Eur J Cancer. 2001;37(Suppl 4):S3–8.
32. Chong CR, Janne PA. The quest to overcome resistance to EGFR-targeted therapies in cancer. Nat Med. 2013;19:1389–400.
33. Han J-Y, Park K, Kim S-W, Lee DH, Kim HY, Kim HT, Ahn MJ, Yun T, Ahn JS, Suh C, Lee J-S, Yoon SJ, Han JH, Lee JW, Jo SJ, Lee JS. First-SIGNAL: first-line single-agent Iressa versus gemcitabine and cisplatin trial in never-smokers with adenocarcinoma of the lung. J Clin Oncol. 2012;30:1122–8.
34. Inoue A, Kobayashi K, Maemondo M, Sugawara S, Oizumi S, Isobe H, Gemma A, Harada M, Yoshizawa H, Kinoshita I, Fujita Y, Okinaga S, Hirano H, Yoshimori K, Harada T, Saijo Y,

Hagiwara K, Morita S, Nukiwa T, North-East Japan Study Group. Updated overall survival results from a randomized phase III trial comparing gefitinib with carboplatin–paclitaxel for chemo-naïve non-small cell lung cancer with sensitive EGFR gene mutations (NEJ002). Ann Oncol. 2013;24:54–9.
35. Maemondo M, Inoue A, Kobayashi K, Sugawara S, Oizumi S, Isobe H, Gemma A, Harada M, Yoshizawa H, Kinoshita I, Fujita Y, Okinaga S, Hirano H, Yoshimori K, Harada T, Ogura T, Ando M, Miyazawa H, Tanaka T, Saijo Y, Hagiwara K, Morita S, Nukiwa T. Gefitinib or chemotherapy for non-small-cell lung cancer with mutated EGFR. N Engl J Med. 2010;362:2380–8.
36. Mitsudomi T, Morita S, Yatabe Y, Negoro S, Okamoto I, Tsurutani J, Seto T, Satouchi M, Tada H, Hirashima T, Asami K, Katakami N, Takada M, Yoshioka H, Shibata K, Kudoh S, Shimizu E, Saito H, Toyooka S, Nakagawa K, Fukuoka M, West Japan Oncology, G. Gefitinib versus cisplatin plus docetaxel in patients with non-small-cell lung cancer harbouring mutations of the epidermal growth factor receptor (WJTOG3405): an open label, randomised phase 3 trial. Lancet Oncol. 2010;11:121–8.
37. Mok TS, Wu Y-L, Thongprasert S, Yang C-H, Chu D-T, Saijo N, Sunpaweravong P, Han B, Margono B, Ichinose Y, Nishiwaki Y, Ohe Y, Yang J-J, Chewaskulyong B, Jiang H, Duffield EL, Watkins CL, Armour AA, Fukuoka M. Gefitinib or carboplatin-paclitaxel in pulmonary adenocarcinoma. N Engl J Med. 2009;361:947–57.
38. Sequist LV, Yang JC-H, Yamamoto N, O'byrne K, Hirsh V, Mok T, Geater SL, Orlov S, Tsai C-M, Boyer M, Su W-C, Bennouna J, Kato T, Gorbunova V, Lee KH, Shah R, Massey D, Zazulina V, Shahidi M, Schuler M. Phase III study of afatinib or cisplatin plus Pemetrexed in patients with metastatic lung adenocarcinoma with EGFR mutations. J Clin Oncol. 2013;31:3327–34.
39. Wu Y-L, Zhou C, Hu C-P, Feng J, Lu S, Huang Y, Li W, Hou M, Shi JH, Lee KY, Xu C-R, Massey D, Kim M, Shi Y, Geater SL. Afatinib versus cisplatin plus gemcitabine for first-line treatment of Asian patients with advanced non-small-cell lung cancer harbouring EGFR mutations (LUX-Lung 6): an open-label, randomised phase 3 trial. Lancet Oncol. 2014;15:213–22.
40. Zhou C, Wu Y-L, Chen G, Feng J, Liu X-Q, Wang C, Zhang S, Wang J, Zhou S, Ren S, Lu S, Zhang L, Hu C, Hu C, Luo Y, Chen L, Ye M, Huang J, Zhi X, Zhang Y, Xiu Q, Ma J, Zhang L, You C. Erlotinib versus chemotherapy as first-line treatment for patients with advanced EGFR mutation-positive non-small-cell lung cancer (Optimal, CTONG-0802): a multicentre, open-label, randomised, phase 3 study. Lancet Oncol. 2011;12:735–42.
41. Wheeler DL, Dunn EF, Harari PM. Understanding resistance to EGFR inhibitors-impact on future treatment strategies. Nat Rev Clin Oncol. 2010;7:493–507.
42. Yu HA, Arcila ME, Rekhtman N, Sima CS, Zakowski MF, Pao W, Kris MG, Miller VA, Ladanyi M, Riely GJ. Analysis of tumor specimens at the time of acquired resistance to EGFR-TKI therapy in 155 patients with EGFR-mutant lung cancers. Clin Cancer Res. 2013;19:2240–7.
43. Janne PA, Yang JC, Kim DW, Planchard D, Ohe Y, Ramalingam SS, Ahn MJ, Kim SW, Su WC, Horn L, Haggstrom D, Felip E, Kim JH, Frewer P, Cantarini M, Brown KH, Dickinson PA, Ghiorghiu S, Ranson M. AZD9291 in EGFR inhibitor-resistant non-small-cell lung cancer. N Engl J Med. 2015;372:1689–99.
44. Sequist LV, Soria JC, Goldman JW, Wakelee HA, Gadgeel SM, Varga A, Papadimitrakopoulou V, Solomon BJ, Oxnard GR, Dziadziuszko R, Aisner DL, Doebele RC, Galasso C, Garon EB, Heist RS, Logan J, Neal JW, Mendenhall MA, Nichols S, Piotrowska Z, Wozniak AJ, Raponi M, Karlovich CA, Jaw-Tsai S, Isaacson J, Despain D, Matheny SL, Rolfe L, Allen AR, Camidge DR. Rociletinib in EGFR-mutated non-small-cell lung cancer. N Engl J Med. 2015;372:1700–9.
45. Montagut C, Dalmases A, Bellosillo B, Crespo M, Pairet S, Iglesias M, Salido M, Gallen M, Marsters S, Tsai SP, Minoche A, Seshagiri S, Serrano S, Himmelbauer H, Bellmunt J, Rovira A, Settleman J, Bosch F, Albanell J. Identification of a mutation in the extracellular domain of the Epidermal Growth Factor Receptor conferring cetuximab resistance in colorectal cancer. Nat Med. 2012;18:221–3.

46. Engelman JA, Zejnullahu K, Mitsudomi T, Song Y, Hyland C, Park JO, Lindeman N, Gale CM, Zhao X, Christensen J, Kosaka T, Holmes AJ, Rogers AM, Cappuzzo F, Mok T, Lee C, Johnson BE, Cantley LC, Janne PA. MET amplification leads to gefitinib resistance in lung cancer by activating ERBB3 signaling. Science. 2007;316:1039–43.
47. Takeda M, Okamoto I, Fujita Y, Arao T, Ito H, Fukuoka M, Nishio K, Nakagawa K. De novo resistance to epidermal growth factor receptor-tyrosine kinase inhibitors in EGFR mutation-positive patients with non-small cell lung cancer. J Thorac Oncol. 2010;5:399–400.
48. Misale S, Yaeger R, Hobor S, Scala E, Janakiraman M, Liska D, Valtorta E, Schiavo R, Buscarino M, Siravegna G, Bencardino K, Cercek A, Chen CT, Veronese S, Zanon C, Sartore-Bianchi A, Gambacorta M, Gallicchio M, Vakiani E, Boscaro V, Medico E, Weiser M, Siena S, Di Nicolantonio F, Solit D, Bardelli A. Emergence of KRAS mutations and acquired resistance to anti-EGFR therapy in colorectal cancer. Nature. 2012;486:532–6.
49. Ihle NT, Byers LA, Kim ES, Saintigny P, Lee JJ, Blumenschein GR, Tsao A, Liu S, Larsen JE, Wang J, Diao L, Coombes KR, Chen L, Zhang S, Abdelmelek MF, Tang X, Papadimitrakopoulou V, Minna JD, Lippman SM, Hong WK, Herbst RS, Wistuba II, Heymach JV, Powis G. Effect of KRAS oncogene substitutions on protein behavior: implications for signaling and clinical outcome. J Natl Cancer Inst. 2012;104:228–39.
50. Cantley LC. The phosphoinositide 3-kinase pathway. Science. 2002;296:1655–7.
51. Samuels Y, Wang Z, Bardelli A, Silliman N, Ptak J, Szabo S, Yan H, Gazdar A, Powell SM, Riggins GJ, Willson JK, Markowitz S, Kinzler KW, Vogelstein B, Velculescu VE. High frequency of mutations of the PIK3CA gene in human cancers. Science. 2004;304:554.
52. Chakrabarty A, Sanchez V, Kuba MG, Rinehart C, Arteaga CL. Feedback upregulation of HER3 (ErbB3) expression and activity attenuates antitumor effect of PI3K inhibitors. Proc Natl Acad Sci U S A. 2012;109:2718–23.
53. Lee T, Yao G, Nevins J, You L. Sensing and integration of Erk and PI3K signals by Myc. PLoS Comput Biol. 2008;4:e1000013.
54. Wang X, Yue P, Chan CB, Ye K, Ueda T, Watanabe-Fukunaga R, Fukunaga R, Fu H, Khuri FR, Sun SY. Inhibition of mammalian target of rapamycin induces phosphatidylinositol 3-kinase-dependent and Mnk-mediated eukaryotic translation initiation factor 4E phosphorylation. Mol Cell Biol. 2007;27:7405–13.
55. Faivre S, Kroemer G, Raymond E. Current development of mTOR inhibitors as anticancer agents. Nat Rev Drug Discov. 2006;5:671–88.
56. Fruman DA, Rommel C. PI3K and cancer: lessons, challenges and opportunities. Nat Rev Drug Discov. 2014;13:140–56.
57. Rodon J, Dienstmann R, Serra V, Tabernero J. Development of PI3K inhibitors: lessons learned from early clinical trials. Nat Rev Clin Oncol. 2013;10:143–53.
58. Carver BS, Chapinski C, Wongvipat J, Hieronymus H, Chen Y, Chandarlapaty S, Arora VK, Le C, Koutcher J, Scher H, Scardino PT, Rosen N, Sawyers CL. Reciprocal feedback regulation of PI3K and androgen receptor signaling in PTEN-deficient prostate cancer. Cancer Cell. 2011;19:575–86.
59. Mellinghoff IK, Vivanco I, Kwon A, Tran C, Wongvipat J, Sawyers CL. HER2/neu kinase-dependent modulation of androgen receptor function through effects on DNA binding and stability. Cancer Cell. 2004;6:517–27.
60. Mayer IA, Abramson VG, Balko JM, Isakoff SJ, Kuba MG, Sanders M, Forero-Torres A, Yap JT, Van Den Abbeele AD, Li Y, Arteaga CL, Winer EP. SU2C phase Ib study of pan-PI3K inhibitor BKM120 with letrozole in ER+/HER2− metastatic breast cancer (MBC). J Clin Oncol. 2014;32(12);1202–9.
61. Kwak EL, Bang Y-J, Camidge DR, Shaw AT, Solomon B, Maki RG, Ou S-HI, Dezube BJ, Jänne PA, Costa DB, Varella-Garcia M, Kim W-H, Lynch TJ, Fidias P, Stubbs H, Engelman JA, Sequist LV, Tan W, Gandhi L, Mino-Kenudson M, Wei GC, Shreeve SM, Ratain MJ, Settleman J, Christensen JG, Haber DA, Wilner K, Salgia R, Shapiro GI, Clark JW, Iafrate AJ. Anaplastic lymphoma kinase inhibition in non-small-cell lung cancer. N Engl J Med. 2010;363:1693–703.

62. Shaw AT, Kim D-W, Nakagawa K, Seto T, Crinó L, Ahn M-J, De Pas T, Besse B, Solomon BJ, Blackhall F, Wu Y-L, Thomas M, O'byrne KJ, Moro-Sibilot D, Camidge DR, Mok T, Hirsh V, Riely GJ, Iyer S, Tassell V, Polli A, Wilner KD, Jänne PA. Crizotinib versus chemotherapy in advanced ALK-positive lung cancer. N Engl J Med. 2013;368:2385–94.
63. Scagliotti GV, Novello S, Schiller JH, Hirsh V, Sequist LV, Soria JC, von Pawel J, Schwartz B, Von Roemeling R, Sandler AB. Rationale and design of MARQUEE: a phase III, randomized, double-blind study of tivantinib plus erlotinib versus placebo plus erlotinib in previously treated patients with locally advanced or metastatic, nonsquamous, non-small-cell lung cancer. Clin Lung Cancer. 2012;13:391–5.
64. Sequist LV, Von Pawel J, Garmey EG, Akerley WL, Brugger W, Ferrari D, Chen Y, Costa DB, Gerber DE, Orlov S, Ramlau R, Arthur S, Gorbachevsky I, Schwartz B, Schiller JH. Randomized phase II study of erlotinib plus tivantinib versus erlotinib plus placebo in previously treated non-small-cell lung cancer. J Clin Oncol. 2011;29:3307–15.
65. Spigel DR, Ervin TJ, Ramlau RA, Daniel DB, Goldschmidt JH, Blumenschein GR, Krzakowski MJ, Robinet G, Godbert B, Barlesi F, Govindan R, Patel T, Orlov SV, Wertheim MS, Yu W, Zha J, Yauch RL, Patel PH, Phan S-C, Peterson AC. Randomized phase II trial of onartuzumab in combination with erlotinib in patients with advanced non-small-cell lung cancer. J Clin Oncol. 2013;31:4105–14.
66. Tsimberidou AM, Iskander NG, Hong DS, Wheler JJ, Falchook GS, Fu S, Piha-Paul S, Naing A, Janku F, Luthra R, Ye Y, Wen S, Berry D, Kurzrock R. Personalized medicine in a phase I clinical trials program: the MD Anderson Cancer Center initiative. Clin Cancer Res. 2012;18:6373–83.
67. Pirker R, Pereira JR, von Pawel J, Krzakowski M, Ramlau R, Park K, de Marinis F, Eberhardt WE, Paz-Ares L, Storkel S, Schumacher KM, von Heydebreck A, Celik I, O'byrne KJ. EGFR expression as a predictor of survival for first-line chemotherapy plus cetuximab in patients with advanced non-small-cell lung cancer: analysis of data from the phase 3 FLEX study. Lancet Oncol. 2012;13:33–42.
68. Hirsch FR, Herbst RS, Olsen C, Chansky K, Crowley J, Kelly K, Franklin WA, Bunn Jr PA, Varella-Garcia M, Gandara DR. Increased EGFR gene copy number detected by fluorescent in situ hybridization predicts outcome in non-small-cell lung cancer patients treated with cetuximab and chemotherapy. J Clin Oncol. 2008;26:3351–7.
69. Khambata-Ford S, Harbison CT, Hart LL, Awad M, Xu LA, Horak CE, Dakhil S, Hermann RC, Lynch TJ, Weber MR. Analysis of potential predictive markers of cetuximab benefit in BMS099, a phase III study of cetuximab and first-line taxane/carboplatin in advanced non-small-cell lung cancer. J Clin Oncol. 2010;28:918–27.
70. De Roock W, Piessevaux H, De Schutter J, Janssens M, De Hertogh G, Personeni N, Biesmans B, Van Laethem JL, Peeters M, Humblet Y, Van Cutsem E, Tejpar S. KRAS wild-type state predicts survival and is associated to early radiological response in metastatic colorectal cancer treated with cetuximab. Ann Oncol. 2008;19:508–15.
71. Di Fiore F, Blanchard F, Charbonnier F, Le Pessot F, Lamy A, Galais MP, Bastit L, Killian A, Sesboue R, Tuech JJ, Queuniet AM, Paillot B, Sabourin JC, Michot F, Michel P, Frebourg T. Clinical relevance of KRAS mutation detection in metastatic colorectal cancer treated by Cetuximab plus chemotherapy. Br J Cancer. 2007;96:1166–9.
72. Khambata-Ford S, Garrett CR, Meropol NJ, Basik M, Harbison CT, Wu S, Wong TW, Huang X, Takimoto CH, Godwin AK, Tan BR, Krishnamurthi SS, Burris III HA, Poplin EA, Hidalgo M, Baselga J, Clark EA, Mauro DJ. Expression of epiregulin and amphiregulin and K-ras mutation status predict disease control in metastatic colorectal cancer patients treated with cetuximab. J Clin Oncol. 2007;25:3230–7.
73. Lievre A, Bachet JB, Boige V, Cayre A, Le Corre D, Buc E, Ychou M, Bouche O, Landi B, Louvet C, Andre T, Bibeau F, Diebold MD, Rougier P, Ducreux M, Tomasic G, Emile JF, Penault-Llorca F, Laurent-Puig P. KRAS mutations as an independent prognostic factor in patients with advanced colorectal cancer treated with cetuximab. J Clin Oncol. 2008;26:374–9.

74. Lievre A, Bachet JB, Le Corre D, Boige V, Landi B, Emile JF, Cote JF, Tomasic G, Penna C, Ducreux M, Rougier P, Penault-Llorca F, Laurent-Puig P. KRAS mutation status is predictive of response to cetuximab therapy in colorectal cancer. Cancer Res. 2006;66:3992–5.
75. Cunningham D, Humblet Y, Siena S, Khayat D, Bleiberg H, Santoro A, Bets D, Mueser M, Harstrick A, Verslype C, Chau I, Van Cutsem E. Cetuximab monotherapy and cetuximab plus irinotecan in irinotecan-refractory metastatic colorectal cancer. N Engl J Med. 2004;351:337–45.
76. Jonker DJ, O'callaghan CJ, Karapetis CS, Zalcberg JR, Tu D, Au HJ, Berry SR, Krahn M, Price T, Simes RJ, Tebbutt NC, van Hazel G, Wierzbicki R, Langer C, Moore MJ. Cetuximab for the treatment of colorectal cancer. N Engl J Med. 2007;357:2040–8.
77. Cappuzzo F, Finocchiaro G, Rossi E, Janne PA, Carnaghi C, Calandri C, Bencardino K, Ligorio C, Ciardiello F, Pressiani T, Destro A, Roncalli M, Crino L, Franklin WA, Santoro A, Varella-Garcia M. EGFR FISH assay predicts for response to cetuximab in chemotherapy refractory colorectal cancer patients. Ann Oncol. 2008;19:717–23.
78. Grimminger PP, Danenberg P, Dellas K, Arnold D, Rodel C, Machiels JP, Haustermans K, Debucquoy A, Velenik V, Sempoux C, Bracko M, Holscher AH, Semrau R, Yang D, Danenberg K, Lenz HJ, Vallbohmer D. Biomarkers for cetuximab-based neoadjuvant radiochemotherapy in locally advanced rectal cancer. Clin Cancer Res. 2011;17:3469–77.
79. Maughan TS, Adams RA, Smith CG, Meade AM, Seymour MT, Wilson RH, Idziaszczyk S, Harris R, Fisher D, Kenny SL, Kay E, Mitchell JK, Madi A, Jasani B, James MD, Bridgewater J, Kennedy MJ, Claes B, Lambrechts D, Kaplan R, Cheadle JP, Investigators MCT. Addition of cetuximab to oxaliplatin-based first-line combination chemotherapy for treatment of advanced colorectal cancer: results of the randomised phase 3 MRC COIN trial. Lancet. 2011;377:2103–14.
80. Primrose J, Falk S, Finch-Jones M, Valle J, O'reilly D, Siriwardena A, Hornbuckle J, Peterson M, Rees M, Iveson T, Hickish T, Butler R, Stanton L, Dixon E, Little L, Bowers M, Pugh S, Garden OJ, Cunningham D, Maughan T, Bridgewater J. Systemic chemotherapy with or without cetuximab in patients with resectable colorectal liver metastasis: the New EPOC randomised controlled trial. Lancet Oncol. 2014;15:601–11.
81. Van Cutsem E, Kohne CH, Hitre E, Zaluski J, Chang Chien CR, Makhson A, D'haens G, Pinter T, Lim R, Bodoky G, Roh JK, Folprecht G, Ruff P, Stroh C, Tejpar S, Schlichting M, Nippgen J, Rougier P. Cetuximab and chemotherapy as initial treatment for metastatic colorectal cancer. N Engl J Med. 2009;360:1408–17.
82. Van Cutsem E, Kohne CH, Lang I, Folprecht G, Nowacki MP, Cascinu S, Shchepotin I, Maurel J, Cunningham D, Tejpar S, Schlichting M, Zubel A, Celik I, Rougier P, Ciardiello F. Cetuximab plus irinotecan, fluorouracil, and leucovorin as first-line treatment for metastatic colorectal cancer: updated analysis of overall survival according to tumor KRAS and BRAF mutation status. J Clin Oncol. 2011;29:2011–9.
83. Vermorken JB, Mesia R, Rivera F, Remenar E, Kawecki A, Rottey S, Erfan J, Zabolotnyy D, Kienzer HR, Cupissol D, Peyrade F, Benasso M, Vynnychenko I, De Raucourt D, Bokemeyer C, Schueler A, Amellal N, Hitt R. Platinum-based chemotherapy plus cetuximab in head and neck cancer. N Engl J Med. 2008;359:1116–27.
84. Licitra L, Mesia R, Rivera F, Remenar E, Hitt R, Erfan J, Rottey S, Kawecki A, Zabolotnyy D, Benasso M, Storkel S, Senger S, Stroh C, Vermorken JB. Evaluation of EGFR gene copy number as a predictive biomarker for the efficacy of cetuximab in combination with chemotherapy in the first-line treatment of recurrent and/or metastatic squamous cell carcinoma of the head and neck: EXTREME study. Ann Oncol. 2011;22:1078–87.
85. Brugger W, Triller N, Blasinska-Morawiec M, Curescu S, Sakalauskas R, Manikhas GM, Mazieres J, Whittom R, Ward C, Mayne K, Trunzer K, Cappuzzo F. Prospective molecular marker analyses of EGFR and KRAS from a randomized, placebo-controlled study of erlotinib maintenance therapy in advanced non-small-cell lung cancer. J Clin Oncol. 2011;29:4113–20.
86. Garassino MC, Martelli O, Broggini M, Farina G, Veronese S, Rulli E, Bianchi F, Bettini A, Longo F, Moscetti L, Tomirotti M, Marabese M, Ganzinelli M, Lauricella C, Labianca R, Floriani I, Giaccone G, Torri V, Scanni A, Marsoni S, Trialists T. Erlotinib versus docetaxel

as second-line treatment of patients with advanced non-small-cell lung cancer and wild-type EGFR tumours (TAILOR): a randomised controlled trial. Lancet Oncol. 2013;14:981–8.

87. Han SW, Hwang PG, Chung DH, Kim DW, Im SA, Kim YT, Kim TY, Heo DS, Bang YJ, Kim NK. Epidermal growth factor receptor (EGFR) downstream molecules as response predictive markers for gefitinib (Iressa, ZD1839) in chemotherapy-resistant non-small cell lung cancer. Int J Cancer. 2005;113:109–15.
88. Fukuoka M, Wu YL, Thongprasert S, Sunpaweravong P, Leong SS, Sriuranpong V, Chao TY, Nakagawa K, Chu DT, Saijo N, Duffield EL, Rukazenkov Y, Speake G, Jiang H, Armour AA, To KF, Yang JC, Mok TS. Biomarker analyses and final overall survival results from a phase III, randomized, open-label, first-line study of gefitinib versus carboplatin/paclitaxel in clinically selected patients with advanced non-small-cell lung cancer in Asia (IPASS). J Clin Oncol. 2011;29:2866–74.
89. Sartore-Bianchi A, Fieuws S, Veronese S, Moroni M, Personeni N, Frattini M, Torri V, Cappuzzo F, Vander Borght S, Martin V, Skokan M, Santoro A, Gambacorta M, Tejpar S, Varella-Garcia M, Siena S. Standardisation of EGFR FISH in colorectal cancer: results of an international interlaboratory reproducibility ring study. J Clin Pathol. 2012;65:218–23.
90. Laurent-Puig P, Cayre A, Manceau G, Buc E, Bachet JB, Lecomte T, Rougier P, Lievre A, Landi B, Boige V, Ducreux M, Ychou M, Bibeau F, Bouche O, Reid J, Stone S, Penault-Llorca F. Analysis of PTEN, BRAF, and EGFR status in determining benefit from cetuximab therapy in wild-type KRAS metastatic colon cancer. J Clin Oncol. 2009;27:5924–30.
91. Han SW, Oh DY, Im SA, Park SR, Lee KW, Song HS, Lee NS, Lee KH, Choi IS, Lee MH, Kim MA, Kim WH, Bang YJ, Kim TY. Phase II study and biomarker analysis of cetuximab combined with modified FOLFOX6 in advanced gastric cancer. Br J Cancer. 2009;100:298–304.
92. Wheeler DL, Huang S, Kruser TJ, Nechrebecki MM, Armstrong EA, Benavente S, Gondi V, Hsu KT, Harari PM. Mechanisms of acquired resistance to cetuximab: role of HER (ErbB) family members. Oncogene. 2008;27:3944–56.
93. Tao JJ, Castel P, Radosevic-Robin N, Elkabets M, Auricchio N, Aceto N, Weitsman G, Barber P, Vojnovic B, Ellis H, Morse N, Viola-Villegas NT, Bosch A, Juric D, Hazra S, Singh S, Kim P, Bergamaschi A, Maheswaran S, Ng T, Penault-Llorca F, Lewis JS, Carey LA, Perou CM, Baselga J, Scaltriti M. Antagonism of EGFR and HER3 enhances the response to inhibitors of the PI3K-Akt pathway in triple-negative breast cancer. Sci Signal. 2014;7:ra29.
94. Slamon DJ, Clark GM, Wong SG, Levin WJ, Ullrich A, Mcguire WL. Human breast cancer: correlation of relapse and survival with amplification of the HER-2/neu oncogene. Science. 1987;235:177–82.
95. Piccart M, Lohrisch C, Di Leo A, Larsimont D. The predictive value of HER2 in breast cancer. Oncology. 2001;61(Suppl 2):73–82.
96. Shitara K, Yatabe Y, Matsuo K, Sugano M, Kondo C, Takahari D, Ura T, Tajika M, Ito S, Muro K. Prognosis of patients with advanced gastric cancer by HER2 status and trastuzumab treatment. Gastric Cancer. 2013;16:261–7.
97. Pogue-Geile KL, Kim C, Jeong JH, Tanaka N, Bandos H, Gavin PG, Fumagalli D, Goldstein LC, Sneige N, Burandt E, Taniyama Y, Bohn OL, Lee A, Kim SI, Reilly ML, Remillard MY, Blackmon NL, Kim SR, Horne ZD, Rastogi P, Fehrenbacher L, Romond EH, Swain SM, Mamounas EP, Wickerham DL, Geyer Jr CE, Costantino JP, Wolmark N, Paik S. Predicting degree of benefit from adjuvant trastuzumab in NSABP trial B-31. J Natl Cancer Inst. 2013;105:1782–8.
98. Dowsett M, Hanna WM, Kockx M, Penault-Llorca F, Ruschoff J, Gutjahr T, Habben K, Van De Vijver MJ. Standardization of HER2 testing: results of an international proficiency-testing ring study. Mod Pathol. 2007;20:584–91.
99. Paik S, Kim C, Wolmark N. HER2 status and benefit from adjuvant trastuzumab in breast cancer. N Engl J Med. 2008;358:1409–11.
100. Romond EH, Perez EA, Bryant J, Suman VJ, Geyer Jr CE, Davidson NE, Tan-Chiu E, Martino S, Paik S, Kaufman PA, Swain SM, Pisansky TM, Fehrenbacher L, Kutteh LA, Vogel VG, Visscher DW, Yothers G, Jenkins RB, Brown AM, Dakhil SR, Mamounas EP, Lingle

WL, Klein PM, Ingle JN, Wolmark N. Trastuzumab plus adjuvant chemotherapy for operable HER2-positive breast cancer. N Engl J Med. 2005;353:1673–84.
101. Ciccarelli FD. The (r)evolution of cancer genetics. BMC Biol. 2010;8:74.
102. Lin J, Gan CM, Zhang X, Jones S, Sjoblom T, Wood LD, Parsons DW, Papadopoulos N, Kinzler KW, Vogelstein B, Parmigiani G, Velculescu VE. A multidimensional analysis of genes mutated in breast and colorectal cancers. Genome Res. 2007;17:1304–18.
103. Rambaldi D, Giorgi FM, Capuani F, Ciliberto A, Ciccarelli FD. Low duplicability and network fragility of cancer genes. Trends Genet. 2008;24:427–30.
104. Kholodenko BN, Hoek JB, Westerhoff HV, Brown GC. Quantification of information transfer via cellular signal transduction pathways. FEBS Lett. 1997;414:430–4.
105. Nakakuki T, Birtwistle MR, Saeki Y, Yumoto N, Ide K, Nagashima T, Brusch L, Ogunnaike BA, Okada-Hatakeyama M, Kholodenko BN. Ligand-specific c-Fos expression emerges from the spatiotemporal control of ErbB network dynamics. Cell. 2010;141:884–96.
106. Hayashi H, Tsuchiya Y, Nakayama K, Satoh T, Nishida E. Down-regulation of the PI3-kinase/Akt pathway by ERK MAP kinase in growth factor signaling. Genes Cells. 2008; 13:941–7.
107. Kiyatkin A, Aksamitiene E, Markevich NI, Borisov NM, Hoek JB, Kholodenko BN. Scaffolding protein Grb2-associated binder 1 sustains epidermal growth factor-induced mitogenic and survival signaling by multiple positive feedback loops. J Biol Chem. 2006;281:19925–38.
108. Borisov N, Aksamitiene E, Kiyatkin A, Legewie S, Berkhout J, Maiwald T, Kaimachnikov NP, Timmer J, Hoek JB, Kholodenko BN. Systems-level interactions between insulin-EGF networks amplify mitogenic signaling. Mol Syst Biol. 2009;5:256.
109. Kim P, Liu X, Lee T, Liu L, Barham R, Kirkland R, Leesman G, Kuller A, Ybarrondo B, Ng SC, Singh S. Highly sensitive proximity mediated immunoassay reveals HER2 status conversion in the circulating tumor cells of metastatic breast cancer patients. Protein Sci. 2011;9:75.
110. Shi Y, Huang W, Tan Y, Jin X, Dua R, Penuel E, Mukherjee A, Sperinde J, Pannu H, Chenna A, Defazio-Eli L, Pidaparthi S, Badal Y, Wallweber G, Chen L, Williams S, Tahir H, Larson J, Goodman L, Whitcomb J, Petropoulos C, Winslow J. A novel proximity assay for the detection of proteins and protein complexes: quantitation of HER1 and HER2 total protein expression and homodimerization in formalin-fixed, paraffin-embedded cell lines and breast cancer tissue. Diagn Mol Pathol. 2009;18:11–21.
111. Huang W, Reinholz M, Weidler J, Yolanda L, Paquet A, Whitcomb J, Lingle W, Jenkins RB, Chen B, Larson JS, Tan Y, Sherwood T, Bates M, Perez EA. Comparison of central HER2 testing with quantitative total HER2 expression and HER2 homodimer measurements using a novel proximity-based assay. Am J Clin Pathol. 2010b;134:303–11.
112. Defazio-Eli L, Strommen K, Dao-Pick T, Parry G, Goodman L, Winslow J. Quantitative assays for the measurement of HER1-HER2 heterodimerization and phosphorylation in cell lines and breast tumors: applications for diagnostics and targeted drug mechanism of action. Breast Cancer Res. 2011;13:R44.
113. Soderberg O, Gullberg M, Jarvius M, Ridderstrale K, Leuchowius KJ, Jarvius J, Wester K, Hydbring P, Bahram F, Larsson LG, Landegren U. Direct observation of individual endogenous protein complexes in situ by proximity ligation. Nat Methods. 2006;3:995–1000.
114. Mukherjee A, Badal Y, Nguyen XT, Miller J, Chenna A, Tahir H, Newton A, Parry G, Williams S. Profiling the HER3/PI3K pathway in breast tumors using proximity-directed assays identifies correlations between protein complexes and phosphoproteins. PLoS One. 2011;6:e16443.
115. Spears M, Taylor KJ, Munro AF, Cunningham CA, Mallon EA, Twelves CJ, Cameron DA, Thomas J, Bartlett JM. In situ detection of HER2:HER2 and HER2:HER3 protein-protein interactions demonstrates prognostic significance in early breast cancer. Breast Cancer Res Treat. 2012;132:463–70.
116. Leuchowius KJ, Clausson CM, Grannas K, Erbilgin Y, Botling J, Zieba A, Landegren U, Soderberg O. Parallel visualization of multiple protein complexes in individual cells in tumor tissue. Mol Cell Proteomics. 2013;12:1563–71.

117. Fruhwirth GO, Fernandes LP, Weitsman G, Patel G, Kelleher M, Lawler K, Brock A, Poland SP, Matthews DR, Keri G, Barber PR, Vojnovic B, Ameer-Beg SM, Coolen AC, Fraternali F, Ng T. How Forster resonance energy transfer imaging improves the understanding of protein interaction networks in cancer biology. Chemphyschem. 2011;12:442–61.
118. Morris JR, Boutell C, Keppler M, Densham R, Weekes D, Alamshah A, Butler L, Galanty Y, Pangon L, Kiuchi T, Ng T, Solomon E. The SUMO modification pathway is involved in the BRCA1 response to genotoxic stress. Nature. 2009;462:886–90.
119. Ng T, Squire A, Hansra G, Bornancin F, Prevostel C, Hanby A, Harris W, Barnes D, Schmidt S, Mellor H, Bastiaens PIH, Parker PJ. Imaging protein kinase C alpha activation in cells. Science. 1999;283:2085–9.
120. Bublil EM, Pines G, Patel G, Fruhwirth G, Ng T, Yarden Y. Kinase-mediated quasi-dimers of EGFR. FASEB J. 2010;24:4744–55.
121. Coban O, Zanetti-Dominguez LC, Matthews DR, Rolfe DJ, Weitsman G, Barber PR, Barbeau J, Devauges V, Kampmeier F, Winn M, Vojnovic B, Parker PJ, Lidke KA, Lidke DS, Ameer-Beg SM, Martin-Fernandez ML, Ng T. Effect of phosphorylation on EGFR dimer stability probed by single-molecule dynamics and FRET/FLIM. Biophys J. 2015;108:1013–26.
122. Kiuchi T, Ortiz-Zapater E, Monypenny J, Matthews DR, Nguyen LK, Barbeau J, Coban O, Lawler K, Burford B, Rolfe DJ, De Rinaldis E, Dafou D, Simpson MA, Woodman N, Pinder S, Gillett CE, Devauges V, Poland SP, Fruhwirth G, Marra P, Boersma YL, Pluckthun A, Gullick WJ, Yarden Y, Santis G, Winn M, Kholodenko BN, Martin-Fernandez ML, Parker P, Tutt A, Ameer-Beg SM, Ng T. The ErbB4 CYT2 variant protects EGFR from ligand-induced degradation to enhance cancer cell motility. Sci Signal. 2014;7:ra78.
123. Berezin MY, Achilefu S. Fluorescence lifetime measurements and biological imaging. Chem Rev. 2010;110:2641–84.
124. Tadrous PJ, Siegel J, French PM, Shousha S, Lalani El N, Stamp GW. Fluorescence lifetime imaging of unstained tissues: early results in human breast cancer. J Pathol. 2003;199:309–17.
125. Barber PR., Tullis IDC, Rowley MI, Martins CD, Weitsman G, Lawler K, Coffey M, Woodman N, Gillett CE, Ng T, Vojnovic B. The Gray Institute open microscopes applied to radiobiology and protein interaction studies. Proceedings of the SPIE. 2014. In press.
126. Weitsman G, Lawler K, Kelleher MT, Barrett JE, Barber PR, Shamil E, Festy F, Patel G, Fruhwirth GO, Huang L, Tullis ID, Woodman N, Ofo E, Ameer-Beg SM, Irshad S, Condeelis J, Gillett CE, Ellis PA, Vojnovic B, Coolen AC, Ng T. Imaging tumour heterogeneity of the consequences of a PKCalpha-substrate interaction in breast cancer patients. Biochem Soc Trans. 2014;42:1498–505.
127. Glass G, Papin JA, Mandell JW. SIMPLE: a sequential immunoperoxidase labeling and erasing method. J Histochem Cytochem. 2009;57:899–905.
128. Gerdes MJ, Sevinsky CJ, Sood A, Adak S, Bello MO, Bordwell A, Can A, Corwin A, Dinn S, Filkins RJ, Hollman D, Kamath V, Kaanumalle S, Kenny K, Larsen M, Lazare M, Li Q, Lowes C, Mcculloch CC, Mcdonough E, Montalto MC, Pang Z, Rittscher J, Santamaria-Pang A, Sarachan BD, Seel ML, Seppo A, Shaikh K, Sui Y, Zhang J, Ginty F. Highly multiplexed single-cell analysis of formalin-fixed, paraffin-embedded cancer tissue. Proc Natl Acad Sci U S A. 2013;110:11982–7.
129. Schubert W, Bonnekoh B, Pommer AJ, Philipsen L, Bockelmann R, Malykh Y, Gollnick H, Friedenberger M, Bode M, Dress AW. Analyzing proteome topology and function by automated multidimensional fluorescence microscopy. Nat Biotechnol. 2006;24:1270–8.
130. Schubert W, Gieseler A, Krusche A, Hillert R. Toponome mapping in prostate cancer: detection of 2000 cell surface protein clusters in a single tissue section and cell type specific annotation by using a three symbol code. J Proteome Res. 2009;8:2696–707.
131. Angelo M, Bendall SC, Finck R, Hale MB, Hitzman C, Borowsky AD, Levenson RM, Lowe JB, Liu SD, Zhao S, Natkunam Y, Nolan GP. Multiplexed ion beam imaging of human breast tumors. Nat Med. 2014;20:436–42.
132. Giesen C, Wang HA, Schapiro D, Zivanovic N, Jacobs A, Hattendorf B, Schuffler PJ, Grolimund D, Buhmann JM, Brandt S, Varga Z, Wild PJ, Gunther D, Bodenmiller B. Highly

multiplexed imaging of tumor tissues with subcellular resolution by mass cytometry. Nat Methods. 2014;11:417–22.
133. Stead LF, Sutton KM, Taylor GR, Quirke P, Rabbitts P. Accurately identifying low-allelic fraction variants in single samples with next-generation sequencing: applications in tumor subclone resolution. Hum Mutat. 2013;34:1432–8.
134. Burrell RA, Mcgranahan N, Bartek J, Swanton C. The causes and consequences of genetic heterogeneity in cancer evolution. Nature. 2013;501:338–45.
135. Gerlinger M, Rowan AJ, Horswell S, Larkin J, Endesfelder D, Gronroos E, Martinez P, Matthews N, Stewart A, Tarpey P, Varela I, Phillimore B, Begum S, Mcdonald NQ, Butler A, Jones D, Raine K, Latimer C, Santos CR, Nohadani M, Eklund AC, Spencer-Dene B, Clark G, Pickering L, Stamp G, Gore M, Szallasi Z, Downward J, Futreal PA, Swanton C. Intratumor heterogeneity and branched evolution revealed by multiregion sequencing. N Engl J Med. 2012;366:883–92.
136. Ross JS, Fletcher JA. The HER-2/neu oncogene in breast cancer: prognostic factor, predictive factor, and target for therapy. Stem Cells. 1998;16:413–28.
137. Selvaggi G, Novello S, Torri V, Leonardo E, De Giuli P, Borasio P, Mossetti C, Ardissone F, Lausi P, Scagliotti GV. Epidermal growth factor receptor overexpression correlates with a poor prognosis in completely resected non-small-cell lung cancer. Ann Oncol. 2004;15:28–32.
138. Cappuzzo F, Gregorc V, Rossi E, Cancellieri A, Magrini E, Paties CT, Ceresoli G, Lombardo L, Bartolini S, Calandri C, de Rosa M, Villa E, Crino L. Gefitinib in pretreated non-small-cell lung cancer (NSCLC): analysis of efficacy and correlation with HER2 and epidermal growth factor receptor expression in locally advanced or metastatic NSCLC. J Clin Oncol. 2003;21:2658–63.
139. Melo SA, Luecke LB, Kahlert C, Fernandez AF, Gammon ST, Kaye J, Lebleu VS, Mittendorf EA, Weitz J, Rahbari N, Reissfelder C, Pilarsky C, Fraga MF, Piwnica-Worms D, Kalluri R. Glypican-1 identifies cancer exosomes and detects early pancreatic cancer. Nature. 2015;523:177–82.
140. Chowdhury R, Ganeshan B, Irshad S, Lawler K, Eisenblatter M, Milewicz H, Rodriguez-Justo M, Miles K, Ellis P, Groves A, Punwani S, Ng T. The use of molecular imaging combined with genomic techniques to understand the heterogeneity in cancer metastasis. Br J Radiol. 2014;87:20140065.
141. Rak J. Microparticles in cancer. Semin Thromb Hemost. 2010;36:888–906.
142. Mathivanan S, Fahner CJ, Reid GE, Simpson RJ. ExoCarta 2012: database of exosomal proteins, RNA and lipids. Nucleic Acids Res. 2012;40:D1241–4.
143. Simpson RJ, Kalra H, Mathivanan S. ExoCarta as a resource for exosomal research. J Extracell Vesicles. 2012;1
144. Higginbotham JN, Demory Beckler M, Gephart JD, Franklin JL, Bogatcheva G, Kremers GJ, Piston DW, Ayers GD, Mcconnell RE, Tyska MJ, Coffey RJ. Amphiregulin exosomes increase cancer cell invasion. Curr Biol. 2011;21:779–86.
145. Hedlund M, Nagaeva O, Kargl D, Baranov V, Mincheva-Nilsson L. Thermal- and oxidative stress causes enhanced release of NKG2D ligand-bearing immunosuppressive exosomes in leukemia/lymphoma T and B cells. PLoS One. 2011;6:e16899.
146. Kahlert C, Kalluri R. Exosomes in tumor microenvironment influence cancer progression and metastasis. J Mol Med (Berl). 2013;91:431–7.
147. EL Andaloussi S, Mager I, Breakefield XO, Wood MJ. Extracellular vesicles: biology and emerging therapeutic opportunities. Nat Rev Drug Discov. 2013;12:347–57.
148. AL-Nedawi K, Meehan B, Micallef J, Lhotak V, May L, Guha A, Rak J. Intercellular transfer of the oncogenic receptor EGFRvIII by microvesicles derived from tumour cells. Nat Cell Biol. 2008;10:619–24.
149. Demory Beckler M, Higginbotham JN, Franklin JL, Ham AJ, Halvey PJ, Imasuen IE, Whitwell C, Li M, Liebler DC, Coffey RJ. Proteomic analysis of exosomes from mutant KRAS colon cancer cells identifies intercellular transfer of mutant KRAS. Mol Cell Proteomics. 2013;12:343–55.
150. Al-Nedawi K, Meehan B, Kerbel RS, Allison AC, Rak J. Endothelial expression of autocrine VEGF upon the uptake of tumor-derived microvesicles containing oncogenic EGFR. Proc Natl Acad Sci U S A. 2009;106:3794–9.

151. Kucharzewska P, Christianson HC, Welch JE, Svensson KJ, Fredlund E, Ringner M, Morgelin M, Bourseau-Guilmain E, Bengzon J, Belting M. Exosomes reflect the hypoxic status of glioma cells and mediate hypoxia-dependent activation of vascular cells during tumor development. Proc Natl Acad Sci U S A. 2013;110:7312–7.
152. Chowdhury R, Webber JP, Gurney M, Mason MD, Tabi Z, Clayton A. Cancer exosomes trigger mesenchymal stem cell differentiation into pro-angiogenic and pro-invasive myofibroblasts. Oncotarget. 2015;6:715–31.
153. Webber JP, Spary LK, Sanders AJ, Chowdhury R, Jiang WG, Steadman R, Wymant J, Jones AT, Kynaston H, Mason MD, Tabi Z, Clayton A. Differentiation of tumour-promoting stromal myofibroblasts by cancer exosomes. Oncogene. 2015;34:290–302.
154. Webber J, Steadman R, Mason MD, Tabi Z, Clayton A. Cancer exosomes trigger fibroblast to myofibroblast differentiation. Cancer Res. 2010;70:9621–30.
155. Peinado H, Aleckovic M, Lavotshkin S, Matei I, Costa-Silva B, Moreno-Bueno G, Hergueta-Redondo M, Williams C, Garcia-Santos G, Ghajar C, Nitadori-Hoshino A, Hoffman C, Badal K, Garcia BA, Callahan MK, Yuan J, Martins VR, Skog J, Kaplan RN, Brady MS, Wolchok JD, Chapman PB, Kang Y, Bromberg J, Lyden D. Melanoma exosomes educate bone marrow progenitor cells toward a pro-metastatic phenotype through MET. Nat Med. 2012;18:883–91.
156. Wang T, Gilkes DM, Takano N, Xiang L, Luo W, Bishop CJ, Chaturvedi P, Green JJ, Semenza GL. Hypoxia-inducible factors and RAB22A mediate formation of microvesicles that stimulate breast cancer invasion and metastasis. Proc Natl Acad Sci U S A. 2014;111:E3234–42.
157. Meads MB, Gatenby RA, Dalton WS. Environment-mediated drug resistance: a major contributor to minimal residual disease. Nat Rev Cancer. 2009;9:665–74.
158. Ciravolo V, Huber V, Ghedini GC, Venturelli E, Bianchi F, Campiglio M, Morelli D, Villa A, Della Mina P, Menard S, Filipazzi P, Rivoltini L, Tagliabue E, Pupa SM. Potential role of HER2-overexpressing exosomes in countering trastuzumab-based therapy. J Cell Physiol. 2012;227:658–67.
159. Safaei R, Larson BJ, Cheng TC, Gibson MA, Otani S, Naerdemann W, Howell SB. Abnormal lysosomal trafficking and enhanced exosomal export of cisplatin in drug-resistant human ovarian carcinoma cells. Mol Cancer Ther. 2005;4:1595–604.
160. Altevogt P, Bretz NP, Ridinger J, Utikal J, Umansky V. Novel insights into exosome-induced, tumor-associated inflammation and immunomodulation. Semin Cancer Biol. 2014;28:51–7.
161. Robbins PD, Morelli AE. Regulation of immune responses by extracellular vesicles. Nat Rev Immunol. 2014;14:195–208.
162. Wieckowski EU, Visus C, Szajnik M, Szczepanski MJ, Storkus WJ, Whiteside TL. Tumor-derived microvesicles promote regulatory T cell expansion and induce apoptosis in tumor-reactive activated CD8+ T lymphocytes. J Immunol. 2009;183:3720–30.
163. Xiang X, Poliakov A, Liu C, Liu Y, Deng ZB, Wang J, Cheng Z, Shah SV, Wang GJ, Zhang L, Grizzle WE, Mobley J, Zhang HG. Induction of myeloid-derived suppressor cells by tumor exosomes. Int J Cancer. 2009;124:2621–33.
164. Skog J, Wurdinger T, Van Rijn S, Meijer DH, Gainche L, Sena-Esteves M, Curry Jr WT, Carter BS, Krichevsky AM, Breakefield XO. Glioblastoma microvesicles transport RNA and proteins that promote tumour growth and provide diagnostic biomarkers. Nat Cell Biol. 2008;10:1470–6.
165. Adamczyk KA, Klein-Scory S, Tehrani MM, Warnken U, Schmiegel W, Schnolzer M, Schwarte-Waldhoff I. Characterization of soluble and exosomal forms of the EGFR released from pancreatic cancer cells. Life Sci. 2011;89:304–12.
166. Graner MW, Alzate O, Dechkovskaia AM, Keene JD, Sampson JH, Mitchell DA, Bigner DD. Proteomic and immunologic analyses of brain tumor exosomes. FASEB J. 2009;23:1541–57.
167. Mathivanan S, Lim JW, Tauro BJ, Ji H, Moritz RL, Simpson RJ. Proteomics analysis of A33 immunoaffinity-purified exosomes released from the human colon tumor cell line LIM1215 reveals a tissue-specific protein signature. Mol Cell Proteomics. 2010;9:197–208.
168. Welton JL, Khanna S, Giles PJ, Brennan P, Brewis IA, Staffurth J, Mason MD, Clayton A. Proteomics analysis of bladder cancer exosomes. Mol Cell Proteomics. 2010;9:1324–38.

169. Harris DA, Patel SH, Gucek M, Hendrix A, Westbroek W, Taraska JW. Exosomes released from breast cancer carcinomas stimulate cell movement. PLoS One. 2015;10:e0117495.
170. Ji H, Greening DW, Barnes TW, Lim JW, Tauro BJ, Rai A, Xu R, Adda C, Mathivanan S, Zhao W, Xue Y, Xu T, Zhu HJ, Simpson RJ. Proteome profiling of exosomes derived from human primary and metastatic colorectal cancer cells reveal differential expression of key metastatic factors and signal transduction components. Proteomics. 2013;13:1672–86.
171. Park JE, Tan HS, Datta A, Lai RC, Zhang H, Meng W, Lim SK, Sze SK. Hypoxic tumor cell modulates its microenvironment to enhance angiogenic and metastatic potential by secretion of proteins and exosomes. Mol Cell Proteomics. 2010;9:1085–99.
172. Hosseini-Beheshti E, Pham S, Adomat H, Li N, Tomlinson Guns ES. Exosomes as biomarker enriched microvesicles: characterization of exosomal proteins derived from a panel of prostate cell lines with distinct AR phenotypes. Mol Cell Proteomics. 2012;11:863–85.
173. Amorim M, Fernandes G, Oliveira P, Martins-De-Souza D, Dias-Neto E, Nunes D. The overexpression of a single oncogene (ERBB2/HER2) alters the proteomic landscape of extracellular vesicles. Proteomics. 2014;14:1472–9.
174. Calin GA, Croce CM. MicroRNA signatures in human cancers. Nat Rev Cancer. 2006;6:857–66.
175. Montecalvo A, Larregina AT, Shufesky WJ, Stolz DB, Sullivan ML, Karlsson JM, Baty CJ, Gibson GA, Erdos G, Wang Z, Milosevic J, Tkacheva OA, Divito SJ, Jordan R, Lyons-Weiler J, Watkins SC, Morelli AE. Mechanism of transfer of functional microRNAs between mouse dendritic cells via exosomes. Blood. 2012;119:756–66.
176. Valadi H, Ekstrom K, Bossios A, Sjostrand M, Lee JJ, Lotvall JO. Exosome-mediated transfer of mRNAs and microRNAs is a novel mechanism of genetic exchange between cells. Nat Cell Biol. 2007;9:654–9.
177. Zhou W, Fong MY, Min Y, Somlo G, Liu L, Palomares MR, Yu Y, Chow A, O'connor ST, Chin AR, Yen Y, Wang Y, Marcusson EG, Chu P, Wu J, Wu X, Li AX, Li Z, Gao H, Ren X, Boldin MP, Lin PC, Wang SE. Cancer-secreted miR-105 destroys vascular endothelial barriers to promote metastasis. Cancer Cell. 2014;25:501–15.
178. Fabbri M, Paone A, Calore F, Galli R, Gaudio E, Santhanam R, Lovat F, Fadda P, Mao C, Nuovo GJ, Zanesi N, Crawford M, Ozer GH, Wernicke D, Alder H, Caligiuri MA, Nana-Sinkam P, Perrotti D, Croce CM. MicroRNAs bind to Toll-like receptors to induce prometastatic inflammatory response. Proc Natl Acad Sci U S A. 2012;109:E2110–6.
179. Yang M, Chen J, Su F, Yu B, Lin L, Liu Y, Huang JD, Song E. Microvesicles secreted by macrophages shuttle invasion-potentiating microRNAs into breast cancer cells. Mol Cancer. 2011;10:117.
180. Vader P, Breakefield XO, Wood MJ. Extracellular vesicles: emerging targets for cancer therapy. Trends Mol Med. 2014;20:385–93.
181. Inal JM, Kosgodage U, Azam S, Stratton D, Antwi-Baffour S, Lange S. Blood/plasma secretome and microvesicles. Biochim Biophys Acta. 2013;1834:2317–25.
182. Choi DS, Park JO, Jang SC, Yoon YJ, Jung JW, Choi DY, Kim JW, Kang JS, Park J, Hwang D, Lee KH, Park SH, Kim YK, Desiderio DM, Kim KP, Gho YS. Proteomic analysis of microvesicles derived from human colorectal cancer ascites. Proteomics. 2011;11: 2745–51.
183. Toth B, Nieuwland R, Liebhardt S, Ditsch N, Steinig K, Stieber P, Rank A, Gohring P, Thaler CJ, Friese K, Bauerfeind I. Circulating microparticles in breast cancer patients: a comparative analysis with established biomarkers. Anticancer Res. 2008;28:1107–12.
184. Logozzi M, De Milito A, Lugini L, Borghi M, Calabro L, Spada M, Perdicchio M, Marino ML, Federici C, Iessi E, Brambilla D, Venturi G, Lozupone F, Santinami M, Huber V, Maio M, Rivoltini L, Fais S. High levels of exosomes expressing CD63 and caveolin-1 in plasma of melanoma patients. PLoS One. 2009;4:e5219.
185. Rabinowits G, Gercel-Taylor C, Day JM, Taylor DD, Kloecker GH. Exosomal microRNA: a diagnostic marker for lung cancer. Clin Lung Cancer. 2009;10:42–6.
186. Yu X, Harris SL, Levine AJ. The regulation of exosome secretion: a novel function of the p53 protein. Cancer Res. 2006;66:4795–801.

187. Chen M, Wang K, Zhang L, Li C, Yang Y. The discovery of putative urine markers for the specific detection of prostate tumor by integrative mining of public genomic profiles. PLoS One. 2011;6:e28552.
188. Clayton AH, Walker F, Orchard SG, Henderson C, Fuchs D, Rothacker J, Nice EC, Burgess AW. Ligand-induced dimer-tetramer transition during the activation of the cell surface epidermal growth factor receptor-A multidimensional microscopy analysis. J Biol Chem. 2005;280:30392–9.
189. Nagy P, Claus J, Jovin TM, Arndt-Jovin DJ. Distribution of resting and ligand-bound ErbB1 and ErbB2 receptor tyrosine kinases in living cells using number and brightness analysis. Proc Natl Acad Sci U S A. 2010;107:16524–9.
190. Yang S, Raymond-Stintz MA, Ying W, Zhang J, Lidke DS, Steinberg SL, Williams L, Oliver JM, Wilson BS. Mapping ErbB receptors on breast cancer cell membranes during signal transduction. J Cell Sci. 2007;120:2763–73.
191. Birtwistle MR, Hatakeyama M, Yumoto N, Ogunnaike BA, Hoek JB, Kholodenko BN. Ligand-dependent responses of the ErbB signaling network: experimental and modeling analyses. Mol Syst Biol. 2007;3:144.
192. Huang B, Babcock H, Zhuang X. Breaking the diffraction barrier: super-resolution imaging of cells. Cell. 2010a;143:1047–58.
193. Toprak E, Kural C, Selvin PR. Super-accuracy and super-resolution getting around the diffraction limit. Methods Enzymol. 2010;475:1–26.
194. Klein T, Proppert S, Sauer M. Eight years of single-molecule localization microscopy. Histochem Cell Biol. 2014;141:561–75.
195. Greenfield D, Mcevoy AL, Shroff H, Crooks GE, Wingreen NS, Betzig E, Liphardt J. Self-organization of the *Escherichia coli* chemotaxis network imaged with super-resolution light microscopy. PLoS Biol. 2009;7:e1000137.
196. Rossy J, Owen DM, Williamson DJ, Yang Z, Gaus K. Conformational states of the kinase Lck regulate clustering in early T cell signaling. Nat Immunol. 2013;14:82–9.
197. Scarselli M, Annibale P, Radenovic A. Cell type-specific beta2-adrenergic receptor clusters identified using photoactivated localization microscopy are not lipid raft related, but depend on actin cytoskeleton integrity. J Biol Chem. 2012;287:16768–80.
198. Sengupta P, Jovanovic-Talisman T, Skoko D, Renz M, Veatch SL, Lippincott-Schwartz J. Probing protein heterogeneity in the plasma membrane using PALM and pair correlation analysis. Nat Methods. 2011;8:969–75.
199. Lehmann M, Rocha S, Mangeat B, Blanchet F, Uji IH, Hofkens J, Piguet V. Quantitative multicolor super-resolution microscopy reveals tetherin HIV-1 interaction. PLoS Pathog. 2011;7:e1002456.
200. Pertsinidis A, Mukherjee K, Sharma M, Pang ZP, Park SR, Zhang Y, Brunger AT, Sudhof TC, Chu S. Ultrahigh-resolution imaging reveals formation of neuronal SNARE/Munc18 complexes in situ. Proc Natl Acad Sci U S A. 2013;110:E2812–20.
201. Deschout H, Shivanandan A, Annibale P, Scarselli M, Radenovic A. Progress in quantitative single-molecule localization microscopy. Histochem Cell Biol. 2014;142:5–17.
202. Wang Y, Fruhwirth G, Cai E, Ng T, Selvin PR. 3D super-resolution imaging with blinking quantum dots. Nano Lett. 2013;13:5233–41.
203. Abulrob A, Lu Z, Baumann E, Vobornik D, Taylor R, Stanimirovic D, Johnston LJ. Nanoscale imaging of epidermal growth factor receptor clustering: effects of inhibitors. J Biol Chem. 2010;285:3145–56.
204. van Belzen N, Rijken PJ, Hage WJ, de Laat SW, Verkleij AJ, Boonstra J. Direct visualization and quantitative analysis of epidermal growth factor-induced receptor clustering. J Cell Physiol. 1988;134:413–20.
205. Hay ED. An overview of epithelio-mesenchymal transformation. Acta Anat (Basel). 1995;154:8–20.
206. Hugo H, Ackland ML, Blick T, Lawrence MG, Clements JA, Williams ED, Thompson EW. Epithelial—mesenchymal and mesenchymal—epithelial transitions in carcinoma progression. J Cell Physiol. 2007;213:374–83.

207. Thiery JP. Epithelial-mesenchymal transitions in tumour progression. Nat Rev Cancer. 2002;2:442–54.
208. Sternlicht MD, Lochter A, Sympson CJ, Huey B, Rougier JP, Gray JW, Pinkel D, Bissell MJ, Werb Z. The stromal proteinase MMP3/stromelysin-1 promotes mammary carcinogenesis. Cell. 1999;98:137–46.
209. Massague J. How cells read TGF-beta signals. Nat Rev Mol Cell Biol. 2000;1:169–78.
210. Moll R, Mitze M, Frixen UH, Birchmeier W. Differential loss of E-cadherin expression in infiltrating ductal and lobular breast carcinomas. Am J Pathol. 1993;143:1731–42.
211. Bell CD, Waizbard E. Variability of cell size in primary and metastatic human breast carcinoma. Invasion Metastasis. 1986;6:11–20.
212. Day Jr CL, Harrist TJ, Gorstein F, Sober AJ, Lew RA, Friedman RJ, Pasternack BS, Kopf AW, Fitzpatrick TB, Mihm Jr MC. Malignant melanoma. Prognostic significance of "microscopic satellites" in the reticular dermis and subcutaneous fat. Ann Surg. 1981;194:108–12.
213. Friedl P, Hegerfeldt Y, Tusch M. Collective cell migration in morphogenesis and cancer. Int J Dev Biol. 2004;48:441–9.
214. Fidler IJ. The relationship of embolic homogeneity, number, size and viability to the incidence of experimental metastasis. Eur J Cancer. 1973;9:223–7.
215. Hegerfeldt Y, Tusch M, Brocker EB, Friedl P. Collective cell movement in primary melanoma explants: plasticity of cell-cell interaction, beta1-integrin function, and migration strategies. Cancer Res. 2002;62:2125–30.
216. Nabeshima K, Inoue T, Shimao Y, Okada Y, Itoh Y, Seiki M, Koono M. Front-cell-specific expression of membrane-type 1 matrix metalloproteinase and Gelatinase a during cohort migration of Colon carcinoma cells induced by hepatocyte growth factor/scatter factor. Cancer Res. 2000;60:3364–9.
217. Giannelli G, Falk-Marzillier J, Schiraldi O, Stetler-Stevenson WG, Quaranta V. Induction of cell migration by matrix metalloprotease-2 cleavage of laminin-5. Science. 1997;277: 225–8.
218. Cohen SJ, Punt CJ, Iannotti N, Saidman BH, Sabbath KD, Gabrail NY, Picus J, Morse M, Mitchell E, Miller MC, Doyle GV, Tissing H, Terstappen LW, Meropol NJ. Relationship of circulating tumor cells to tumor response, progression-free survival, and overall survival in patients with metastatic colorectal cancer. J Clin Oncol. 2008;26:3213–21.
219. de Bono JS, Scher HI, Montgomery RB, Parker C, Miller MC, Tissing H, Doyle GV, Terstappen LW, Pienta KJ, Raghavan D. Circulating tumor cells predict survival benefit from treatment in metastatic castration-resistant prostate cancer. Clin Cancer Res. 2008;14:6302–9.
220. Hayes DF, Cristofanilli M, Budd GT, Ellis MJ, Stopeck A, Miller MC, Matera J, Allard WJ, Doyle GV, Terstappen LW. Circulating tumor cells at each follow-up time point during therapy of metastatic breast cancer patients predict progression-free and overall survival. Clin Cancer Res. 2006;12:4218–24.
221. Riethdorf S, Fritsche H, Muller V, Rau T, Schindlbeck C, Rack B, Janni W, Coith C, Beck K, Janicke F, Jackson S, Gornet T, Cristofanilli M, Pantel K. Detection of circulating tumor cells in peripheral blood of patients with metastatic breast cancer: a validation study of the CellSearch system. Clin Cancer Res. 2007;13:920–8.
222. Vona G, Sabile A, Louha M, Sitruk V, Romana S, Schutze K, Capron F, Franco D, Pazzagli M, Vekemans M, Lacour B, Brechot C, Paterlini-Brechot P. Isolation by size of epithelial tumor cells : a new method for the immunomorphological and molecular characterization of circulatingtumor cells. Am J Pathol. 2000;156:57–63.
223. Went PT, Lugli A, Meier S, Bundi M, Mirlacher M, Sauter G, Dirnhofer S. Frequent EpCam protein expression in human carcinomas. Hum Pathol. 2004;35:122–8.
224. Christiansen JJ, Rajasekaran AK. Reassessing epithelial to mesenchymal transition as a prerequisite for carcinoma invasion and metastasis. Cancer Res. 2006;66:8319–26.
225. Thiery JP, Sleeman JP. Complex networks orchestrate epithelial-mesenchymal transitions. Nat Rev Mol Cell Biol. 2006;7:131–42.

226. Hou JM, Krebs M, Ward T, Sloane R, Priest L, Hughes A, Clack G, Ranson M, Blackhall F, Dive C. Circulating tumor cells as a window on metastasis biology in lung cancer. Am J Pathol. 2011;178:989–96.
227. Pestrin M, Bessi S, Galardi F, Truglia M, Biggeri A, Biagioni C, Cappadona S, Biganzoli L, Giannini A, Di Leo A. Correlation of HER2 status between primary tumors and corresponding circulating tumor cells in advanced breast cancer patients. Breast Cancer Res Treat. 2009;118:523–30.
228. Kallergi G, Agelaki S, Kalykaki A, Stournaras C, Mavroudis D, Georgoulias V. Phosphorylated EGFR and PI3K/Akt signaling kinases are expressed in circulating tumor cells of breast cancer patients. Breast Cancer Res. 2008;10:R80.
229. Maheswaran S, Sequist LV, Nagrath S, Ulkus L, Brannigan B, Collura CV, Inserra E, Diederichs S, Iafrate AJ, Bell DW, Digumarthy S, Muzikansky A, Irimia D, Settleman J, Tompkins RG, Lynch TJ, Toner M, Haber DA. Detection of mutations in EGFR in circulating lung-cancer cells. N Engl J Med. 2008;359:366–77.
230. Cairns RA, Harris IS, Mak TW. Regulation of cancer cell metabolism. Nat Rev Cancer. 2011;11:85–95.
231. Hanahan D, Weinberg RA. Hallmarks of cancer: the next generation. Cell. 2011;144: 646–74.
232. Warburg O. On respiratory impairment in cancer cells. Science. 1956a;124:269–70.
233. Warburg O. On the origin of cancer cells. Science. 1956b;123:309–14.
234. Christofk HR, Vander Heiden MG, Harris MH, Ramanathan A, Gerszten RE, Wei R, Fleming MD, Schreiber SL, Cantley LC. The M2 splice isoform of pyruvate kinase is important for cancer metabolism and tumour growth. Nature. 2008;452:230–3.
235. Lum JJ, Bauer DE, Kong M, Harris MH, Li C, Lindsten T, Thompson CB. Growth factor regulation of autophagy and cell survival in the absence of apoptosis. Cell. 2005;120:237–48.
236. Shaw RJ, Kosmatka M, Bardeesy N, Hurley RL, Witters LA, Depinho RA, Cantley LC. The tumor suppressor LKB1 kinase directly activates AMP-activated kinase and regulates apoptosis in response to energy stress. Proc Natl Acad Sci U S A. 2004;101:3329–35.
237. Vander Heiden MG, Cantley LC, Thompson CB. Understanding the Warburg effect: the metabolic requirements of cell proliferation. Science. 2009;324:1029–33.
238. Wise DR, Thompson CB. Glutamine addiction: a new therapeutic target in cancer. Trends Biochem Sci. 2010;35:427–33.
239. Hatzivassiliou G, Zhao F, Bauer DE, Andreadis C, Shaw AN, Dhanak D, Hingorani SR, Tuveson DA, Thompson CB. ATP citrate lyase inhibition can suppress tumor cell growth. Cancer Cell. 2005;8:311–21.
240. Mabuchi S, Kuroda H, Takahashi R, Sasano T. The PI3K/AKT/mTOR pathway as a therapeutic target in ovarian cancer. Gynecol Oncol. 2015;137:173–9.
241. Semenza GL. HIF-1 mediates metabolic responses to intratumoral hypoxia and oncogenic mutations. J Clin Invest. 2013;123:3664–71.
242. Hudson CC, Liu M, Chiang GG, Otterness DM, Loomis DC, Kaper F, Giaccia AJ, Abraham RT. Regulation of hypoxia-inducible factor 1alpha expression and function by the mammalian target of rapamycin. Mol Cell Biol. 2002;22:7004–14.
243. Semenza GL. Defining the role of hypoxia-inducible factor 1 in cancer biology and therapeutics. Oncogene. 2010;29:625–34.
244. Nicklin P, Bergman P, Zhang B, Triantafellow E, Wang H, Nyfeler B, Yang H, Hild M, Kung C, Wilson C, Myer VE, Mackeigan JP, Porter JA, Wang YK, Cantley LC, Finan PM, Murphy LO. Bidirectional transport of amino acids regulates mTOR and autophagy. Cell. 2009;136:521–34.
245. Meyer N, Penn LZ. Reflecting on 25 years with MYC. Nat Rev Cancer. 2008;8:976–90.
246. Hahn-Windgassen A, Nogueira V, Chen CC, Skeen JE, Sonenberg N, Hay N. Akt activates the mammalian target of rapamycin by regulating cellular ATP level and AMPK activity. J Biol Chem. 2005;280:32081–9.

247. Faubert B, Vincent EE, Poffenberger MC, Jones RG. The AMP-activated protein kinase (AMPK) and cancer: many faces of a metabolic regulator. Cancer Lett. 2015;356:165–70.
248. Cox BL, Mackie TR, Eliceiri KW. The sweet spot: FDG and other 2-carbon glucose analogs for multi-modal metabolic imaging of tumor metabolism. Am J Nucl Med Mol Imaging. 2015;5:1–13.
249. de Graaf RA, Mason GF, Patel AB, Behar KL, Rothman DL. In vivo 1H-[13C]-NMR spectroscopy of cerebral metabolism. NMR Biomed. 2003;16:339–57.
250. Ardenkjaer-Larsen JH, Fridlund B, Gram A, Hansson G, Hansson L, Lerche MH, Servin R, Thaning M, Golman K. Increase in signal-to-noise ratio of >10,000 times in liquid-state NMR. Proc Natl Acad Sci U S A. 2003;100:10158–63.
251. Kurhanewicz J, Bok R, Nelson SJ, Vigneron DB. Current and potential applications of clinical 13C MR spectroscopy. J Nucl Med. 2008;49:341–4.
252. Walker-Samuel S, Ramasawmy R, Torrealdea F, Rega M, Rajkumar V, Johnson SP, Richardson S, Goncalves M, Parkes HG, Arstad E, Thomas DL, Pedley RB, Lythgoe MF, Golay X. In vivo imaging of glucose uptake and metabolism in tumors. Nat Med. 2013;19:1067–72.
253. Ward KM, Balaban RS. Determination of pH using water protons and chemical exchange dependent saturation transfer (CEST). Magn Reson Med. 2000;44:799–802.
254. Lin G, Andrejeva G, Wong Te Fong AC, Hill DK, Orton MR, Parkes HG, Koh DM, Robinson SP, Leach MO, Eykyn TR, Chung YL. Reduced Warburg effect in cancer cells undergoing autophagy: steady- state 1H-MRS and real-time hyperpolarized 13C-MRS studies. PLoS One. 2014;9:e92645.
255. Melo HJ, Goldman SM, Szejnfeld J, Faria JF, Huayllas MK, Andreoni C, Kater CE. Application of a protocol for magnetic resonance spectroscopy of adrenal glands: an experiment with over 100 cases. Radiol Bras. 2014;47:333–41.
256. Rajagopalan KN, Deberardinis RJ. Role of glutamine in cancer: therapeutic and imaging implications. J Nucl Med. 2011;52:1005–8.
257. Kiyose K, Hanaoka K, Oushiki D, Nakamura T, Kajimura M, Suematsu M, Nishimatsu H, Yamane T, Terai T, Hirata Y, Nagano T. Hypoxia-sensitive fluorescent probes for in vivo real-time fluorescence imaging of acute ischemia. J Am Chem Soc. 2010;132:15846–8.
258. Beck S, Ng T. C2c: turning cancer into chronic disease. Genome Med. 2014;6:38.

Modelling Molecular Mechanisms of Cancer Pathogenesis: Virtual Patients, Real Opportunities

16

Hans Lehrach, Thomas Kessler, Lesley Ogilvie, Moritz Schütte, and Christoph Wierling

Contents

H. Lehrach (✉)
Max Planck Institute for Molecular Genetics (MPIMG), Ihnestrasse 63-73, 14195 Berlin, Germany

Dahlem Centre for Genome Research and Medical Systems Biology (DCGMS), Berlin, Germany

Alacris Theranostics GmbH, Berlin, Germany
e-mail: lehrach@molgen.mpg.de

T. Kessler • C. Wierling
Max Planck Institute for Molecular Genetics (MPIMG), Ihnestrasse 63-73, 14195 Berlin, Germany

Alacris Theranostics GmbH, Berlin, Germany

L. Ogilvie • M. Schütte
Alacris Theranostics GmbH, Berlin, Germany

J. Haybaeck (ed.), *Mechanisms of Molecular Carcinogenesis – Volume 2*,
DOI 10.1007/978-3-319-53661-3_16

Abstract

A combination of decades of cancer research and trillions of dollars has helped to exponentially increase our understanding of the molecular processes involved in cancer pathogenesis [1, 2] and to develop new cancer drugs. However, despite progress in diagnosing and treating cancer, these diseases remain one of the leading causes of morbidity and mortality worldwide, responsible for millions of deaths each year [3]. Moreover, we are still faced, on average, with low drug response rates, very serious treatment side effects and questionable survival benefits. In Europe alone, cancer kills around 4000 people every day [4] and costs billions per year [5]. Worldwide, the number of new cancer cases is increasing every year [4], at least partly due to rapidly ageing populations [6], with the number of new cancer cases projected to reach 25 million per year by 2030, and cure rates for many common forms of cancer stagnating [4]. Due to the high number of nonresponders to existing drugs and spiralling costs of new cancer drugs—costs have almost doubled in the last decade, resulting in a dramatic decrease in the number of new drugs—an individualised approach to the diagnosis and treatment of cancer patients is desperately required.

16.1 Introduction

At the heart of these statistics is complexity. The biology (and medicine) of human beings and their diseases is extremely complex, from the processes occurring in a single cell to the interactions occurring between the ~40 trillion cells in our bodies [7], as well as the multiple external (biological, chemical and physical) influences on the organism. The signalling interactions occurring within and between cells are the key to regulating the processes required for the normal functioning of our bodies. These complex interactions are characterised by non-linearity (e.g. crosstalk and feedback mechanisms) and thus present a real challenge to understanding processes beyond a purely cellular level. In cancer patients, this complexity is magnified at multiple levels: every patient is different (monozygotic twins less so than others). Tumours are formed by random processes from somatic cells with different genomes (and epigenomes) that have never identically occurred before (and will never occur again, not even in the same patient), and even subpopulations of the cells of the same tumour can react very differently to drugs. Each tumour is an organised entity consisting of individual and potentially unique cells formed by random processes which enter into crosstalk with each other, the tumour infrastructure and the micro-environment, across both temporal and spatial scales [8]. The genetic alterations (such as mutations and genomic rearrangements) found in cancer cells dysregulate regular signal transduction, disrupting normal controls of cell replication and apoptosis. These alterations are dynamic, accompanied by multiple genetic and epigenetic changes which allow the tumour to evolve continuously in response to selection pressures, enabling the tumour to, for example, acquire resistance to a specific drug by activating/blocking specific signalling pathways. Even subpopulations of cells from the same tumour can exhibit heterogeneity, responding differently to the same treatment [8–10]. Thus, there is no simple way to predict the outcome of a specific treatment in an individual patient.

Currently, the dominant approach in medicine is statistical: different treatment options are given to large groups of patients, where it is assumed or hoped that the multitude of differences between patients are not relevant to the outcome. Subsequently, the treatments that work best on average are determined ('first-line treatment'), or the patient group is subdivided, usually on the basis of biomarkers ('stratified medicine'), to increase the fraction of patients which respond to a treatment. Stratified medicine (though often called 'personalised medicine') is, however, usually still far removed from a truly personalised treatment for every individual patient. While useful, this approach remains ultimately unsatisfactory in many diseases, such as cancer, where treatment response rates remain limited and are exerting a huge, increasing and unsustainable human and economic burden.

Truly personalised medicine is self-evident in areas of medicine, such as surgery, in which the anatomy determines therapy, based on the detailed characterisation of the individual situation by sophisticated imaging techniques, and the capacity of the human brain to easily interpret spatial data. Such a truly personalised therapy choice has, up to now, not been possible in drug-based therapies, where we have extremely complex situations, no or little data (albeit a situation that is rapidly changing) and no obvious way to draw definitive conclusions from the data.

In essentially all other areas where we face similarly complex situations with costly and/or dangerous consequences, we try to build detailed computer models of the process we want to predict (e.g. the weather) or optimise (e.g. the design and safety testing of cars, planes, but also comparatively trivial systems like washing machines or electric toothbrushes) and make mistakes which, in complex situations, cannot be corrected safely, cheaply and quickly in the computer, ending in potentially disastrous consequences. In addition to the necessary computing power, such simulations typically require two main components, information on the mechanisms underlying the processes in the model and a detailed characterisation of the initial state of the system. In most of the sophisticated cases, pragmatic decisions in modelling are required in order to allow the problems to be computable. Until recently, progress has been made in a number of key areas, which allowed for the design of strategies for diagnosing and treating cancer. Here, we discuss how the intelligent combination of technological developments integrating 'big data' and in silico modelling provides a realistic way forward in the fight against cancer and other diseases.

16.2 Mechanisms, Molecular Bases and Exascales

Over the last decade, progress has been made in three key areas: enhanced understanding of disease and drug mechanisms, technological advances in molecular characterisation of patients and enormous gains in computing power. These key areas are now converging to make the prospect of the routine use of virtual patient models in oncology (and beyond) a reality.

Through intense basic and applied research, we have undoubtedly increased our understanding of the mechanisms of cancer, as well as the drugs that we use to treat patients, for instance, the framework posited by Hanahan and Weinberg [1, 2] in their landmark papers, which define the distinct attributes exhibited by tumours across cancer types that enable them to grow and metastasise, charting a path from healthy cell to its malignant (and metastatic) counterpart. A framework that has revealed that perturbations in a cell's signalling system, due to alterations in key components of the signalling networks, can alter a cell's physiology, leading to excessive cell proliferation. The accumulating knowledge base allows us to move from a purely descriptive stance to a mechanistic paradigm, helping us to understand interactions between the components of biological systems across spatial and temporal scales.

At the same time, we are experiencing major progress in DNA sequencing, functional genomics and other omics technologies, enabling rapid analysis of an individual patient on the molecular level in enormous detail. Such sequencing technologies, originally developed for systematic analyses in high-end research applications, have the potential to become key tools in routine diagnostic applications, by providing the detailed information required for modelling individual patients. The feasibility of such an approach is best illustrated by the tremendous progress in DNA sequencing: a decade ago, the human genome sequence was completed at a cost of $3 billion, by an international project lasting 10 years. Today, sequencing a human genome requires 1 day at a cost that has dropped a million fold. Thus, sequence analysis and patient modelling can now be carried out quickly enough to harmonise with the timeline of clinical treatment decisions (Fig. 16.1). As sequencing costs have plummeted, the large-scale deep molecular characterisation of a range of cancer types has become feasible, triggering an ongoing expansion of the molecular knowledge base on cancer (as well as of healthy and diseased

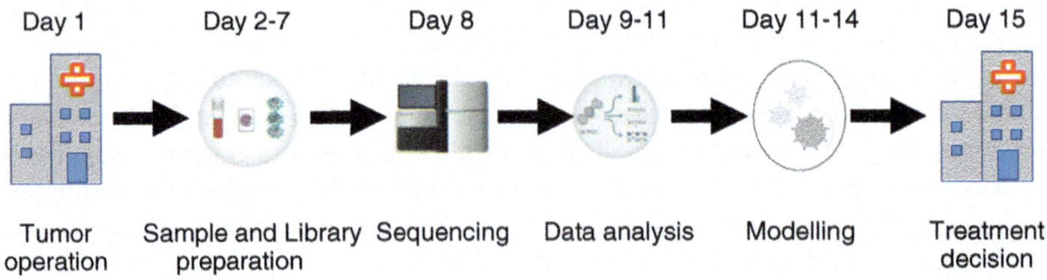

Fig. 16.1 Integrating patient sequencing and modelling into the clinical treatment decision pipeline. Once patient samples are received from the hospital (e.g. tumour and blood), they are processed for sequencing (DNA/RNA extracted, sequencing libraries prepared and sequenced). The complex omics data generated (e.g. genome/exome, transcriptome, possibly proteome) is analysed in detail for tumour-specific alterations, e.g. mutations, gene fusions and copy-number changes. The data is then used to individualise a large-scale mechanistic computational model of cell signalling transduction and associated processes (e.g. ModCell™). The personalised models can then be used to predict how a tumour will respond to a range of drugs, singly or in combination to identify the optimal therapeutic strategy for a specific patient. This information is then given to the oncologist to help them decide on the best treatment strategy for that particular patient

individuals worldwide) through global initiatives, such as the International Cancer Genome Consortium [11], The Cancer Genome Atlas (TCGA; http://cancergenome.nih.gov/), the 1000 Genomes Project (www.1000genomes.org) and the Personal Genomes Project (http://www.personalgenomes.org/). Both the ICGC and TCGA focus on obtaining a comprehensive catalogue of the molecular alterations (e.g. somatic mutations, abnormal expression of genes, epigenetic modifications) in tumour types (and/or subtypes) with significant medical and societal impact to facilitate research into the causes and control of cancer. Currently, the ICGC are cataloguing the genomic, transcriptomic and epigenomic alterations occurring in over 50 tumour types, with a commitment to analyse a further 25,000 tumour genomes. Such 'big data' (sets) help to delineate the molecular basis of cancer, but also further highlight the complexity and heterogeneity associated with cancer of all types [12–14]. The 1000 Genomes Project (www.1000genomes.org) has created a catalogue of the genetic variation (using whole genome and exome sequencing) in over 2504 individuals from a wide range of geographical areas, representing 26 distinct populations. The catalogue is now being maintained and expanded through the International Genome Sample Resource (IGSR), and will cover a wider range of geographical areas and populations. The Personal Genomes Project (www.personalgenomes.org) is set to sequence and analyse the genomes of over 100,000 people across the world, making this data public through open consent.

We are, however, still faced with a number of stumbling blocks when it comes to truly understanding the path from healthy cell to malignant phenotype in terms of how single events combine and how the heterogeneity of tumour cell populations—and their interactions—actually influences the phenotype. The problem is that even if we had this information, we are still far from delineating how these events actually affect the response of a particular patient to a particular drug. In this context, the large-scale and accumulating molecular datasets on tumours are providing molecular portraits which are proving pivotal for understanding tumour behaviour.

The triptych is completed by the ongoing and significant performance gains in computing power (a 1000-fold increase in computing power every decade), with performance hurtling towards exascale systems (machines able to carry out a billion billion calculations per second) in the next 5 years (http://www.top500.org/statistics/perfdevel/). A similar progress is ongoing in improving storage density and network capabilities, enabling collaborative analysis of large datasets, e.g. via cloud computing (see http://www.egi.eu/).

16.3 Computing Complexity

A number of different computational-based approaches are employed in order to analyse these emerging datasets and to widen our understanding of cancer's biological complexity, with statistical and machine learning methodologies, a set of statistical, probabilistic and optimisation techniques, applied most commonly. Statistical approaches have been employed to try and detect molecular signatures or biomarkers of treatment response or likely disease progression and to guide the stratification of patients into molecular subtypes, an important first step in improving outcomes for patients, and to help guide clinical decisions. Such analyses provide information on potential markers of disease and possible drug targets and help to focus on treatments administered. However, although statistical approaches are identifying associations between genetic alterations and disease, they provide no insight into how that alteration is actually driving the effect manifested. Moreover, in the majority of cases, these statistically derived groupings ignore the intrinsic complexity of cancer and the inherent individuality of cancer patients and their disease, a fact unfortunately reflected in the often inadequate response rates to drugs chosen in this way.

As part of a growing momentum in personalised medicine, machine learning approaches are being used to learn and extract information from complex data, such as genomic or transcriptome datasets. The approach can identify hidden patterns in datasets and then 'learn' from these patterns to make predictions on the future state of a system. Such an approach, in the context of systems biology, is suited for exploring the interactions between components of biological systems and could be used to infer functional relationships, for example, machine learning techniques have been shown to improve the understanding of cancer development and progression and facilitate cancer diagnosis and detection and, more recently, prognosis and prediction (see reviews by [15, 16]).

Mechanistic modelling approaches examine the interactions and relationships occurring between the individual components of a complex system, thereby allowing us to understand the system as a whole. The potential of the mechanistic modelling approach in the context of cancer is now emerging more widely, with models simulating disease processes being established. In the first instance, the focus has been on processes that occur at a single biological scale, such as a specific signalling pathway, including the EGFR [17–19], Toll-like receptor [20], Epo [21] and TNFα-mediated NF-κB pathway [22]. However, more global approaches are emerging, which include key signalling and regulatory pathways relevant in cancer, with the capacity to effectively represent key cellular crosstalk mechanisms, the latter being a pivotal feature in predicting patient responses to drugs, as each drug perturbs a wide range of targets and may be involved in a range of biological processes.

Over the last decade, we have focused on the development of an integrated systems biology platform—ModCell™—described in detail elsewhere [23–27], which uses a mechanistic model generated using PyBioS, an integrated, web-based software platform for the design, modelling and simulation of cellular systems [28, 29]. The platform has been used to develop and implement a large mechanistic model of cancer-related signal transduction pathways and cellular processes, as well as functional effects of oncogenes and tumour suppressor genes. The models represent the relevant biological networks of 'normal' human cells, which are then individualised with omics data (e.g. exome/transcriptome/proteome) from a patient sample (e.g. a tumour sample), by mapping of the data, such as relevant germ line or somatic mutations expressed in the mRNA and/or phosphorylation state of critical regulatory proteins, to the functions and parameters linked to the corresponding objects in the model. Drugs or drug combinations are represented by objects, which can, for example, form inactive complexes with the objects representing their target proteins, allowing the systematic prediction of the effect of the drugs on the tumour.

To model the patient, we do, however, have to include additional components influencing the clinical response of the individual patient, for example, the liver, affecting the metabolism and possible activation of the drugs; models representing different, relevant cell types of the body to identify possibly unacceptable side effects; and, ideally, the immune system, able to respond to the tumour and providing targets for a new class of promising therapies, the immunotherapeutics. The virtual tumour patient should therefore consist of all relevant components, represented typically by molecular models interacting through the exchange of signals, e.g. the drug level over time, determined to a large extent by the genome of the patient. The data is then explored using the tools provided by PyBioS [29] and the ModCell™ analysis pipeline, for modelling, simulation, prediction and drug screening of complex molecular systems.

16.4 Critical Factors for Virtual Patient Modelling in Oncology

Although we have the main ingredients to make the prospect of the routine use of virtual patient models in oncology (and beyond) a reality—i.e. enhanced understanding of disease and drug mechanisms, technological advances in molecular characterisation of patients and enormous gains in computing power—through application of the model in preclinical and clinical scenarios, we have identified a range of factors as key to harnessing the real predictive power of such modelling systems.

In this, we are dealing with three basic sources of information:

(a) Data on the individual sample/patient (e.g. tumour exome and transcriptome, patient/germ line exome)
(b) The structure of the model, based on information in pathway databases and publications
(c) Kinetic constants and other quantitative parameters, e.g. the concentration of hard-to-measure components of the model at the beginning of the simulation

16.4.1 Data on the Individual Patient

16.4.1.1 Data Quality and Amount

No matter how advanced a modelling system is, it will depend on the quality and amount of input data. The success of computational modelling relies on (1) high-quality data generation and (2) comprehensive data analysis revealing the essential molecular changes that drive the individual tumour.

One of the key factors affecting data quality and amount is the quality and type of patient sample sequenced. Following surgery, tumour samples are usually fresh frozen or formalin- fixed, paraffin- embedded (FFPE), with fresh frozen material typically giving much better results. While FFPE samples can be used for DNA sequencing, the duration of preservation influences the level of RNA degradation [30].

Tumour purity (i.e. proportion of cancer cells in a sample) will also affect the amount and quality of data generated. Each tumour sample/biopsy will vary in the percentage of cancerous and non-cancerous cells, due to the infiltration of non-cancerous components from the tumour micro-environment, including immune cells, fibroblasts and blood vessels, with the exhibited ratio often varying within and between cell types and tumour origin. Some brain cancer samples, for example, often have a high tumour purity, whereas purity decreases in lung or head and neck cancers [31]. A systematic analysis of the effect of tumour purity on the interpretation of genomic analysis, comprising 9364 tumour samples and 1958 adjacent normal samples across 21 solid tumour types, as expected, showed a confounding effect on quantitative data such as differential gene expression and copy-number analysis [31].

16.4.1.2 Data Type

The tumour-specific model also critically depends on the type of the data available on the tumour and patient. Ideally, a minimum dataset consists of NGS datasets for transcriptome and genome/exome and low coverage genome and transcriptome of the tumour and genome/exome and low coverage genome of the patient. Other less informative datasets (candidate gene sequences, chip-based transcriptome data) can also be used, but with the expected loss of prediction power.

A range of options exist for DNA sequencing. Cancer-specific mutation panels typically target a few to several hundred genes, selected for their known association with cancer or specific cancer types, at very high coverage. These panels are usually designed to detect single-nucleotide variations and small indels (insertions and deletions of DNA) even at very low allele frequency but are limited for the detection of genomic rearrangements.

Whole-exome sequencing (WES) detects mutations within the most well-characterised (~1%) fraction of the genome (exons of most human genes), with approximately 85% of disease-causing mutations being located in coding and functional regions of the genome [32, 33]. Additionally, WES can be used to detect copy-number changes across the genomes. WES therefore reduces the risk of overlooking cancer-relevant genomic alterations in the tumour. The inclusion of data from a control sample (e.g. blood) allows discrimination between somatic (tumour) and germ line (patient genome) genetic alterations and mutations [34].

Whole-genome sequencing (WGS) is the most comprehensive technique enabling detection of genomic rearrangements and interstitial deletions in the range of a few kilobases. The choice of sequencing should, however, also consider the particular cancer type. For mutation-driven cancers, the application of a panel might be sufficient, while for cancers driven by genomic rearrangements, such as mucosal melanoma [35] or leukaemias [36], WGS may be the best choice to capture alterations present.

Transcriptome sequencing (e.g. RNASeq) adds an additional layer to tumour analysis, detecting the expression levels of virtually all genes and enabling the identification of meaningful genetic changes. The following questions arise: Do the genetic changes on the DNA level have an effect on gene function? Are the identified cancer gene variants expressed, e.g. up- or downregulated? Are there any aberrations in the abundance and structure of gene transcripts, i.e. gene products [37]? In addition, RNASeq offers a highly efficient route to identify tumour-specific oncogenic isoforms, e.g. epidermal growth factor receptor variant III (EGFRvIII; [38]), and gene fusions, e.g. TMPRSS2-ERG [39] or BCR-ABL1 [40].

The use of both DNA (exome) and RNA (transcriptome) analyses provides complementary information, both datasets serving as mutual controls, to verify potential 'actionable variants', i.e. changes in a tumour that can be directly linked to treatment options [37, 41, 42]. For instance, a gene amplification should be accompanied by a clear overexpression of the gene, and driver mutations should also be detectable on the RNA level. Such integrated analysis provides a higher level of confidence in the identified alterations, aiding interpretation of novel alterations and translation of results into the right clinical treatment options. Given the significant role that proteins and protein modifications play in regulating the complex biological processes [43–45] being modelled, data on proteins, complexes and modification states can help to enhance the predictive output of models. In particular, proteomics or phosphoproteomics data on a particular tumour sample will provide key information on the results of post-transcriptional regulation in the biological networks [46–51].

Novel sequencing technologies and analysis methodologies, including spatially resolved omics analysis, even at the level of the single cell [47, 48, 52–56], will help to further define the heterogeneity of the tumour landscape and further refine the accuracy of virtual patient modelling applications in oncology.

16.4.2 The Development/Assembly of Computational Models

Systems biology models are considered the key to personalising healthcare [57, 58]. Due to the complexity of changes observed in cancer, mathematical models are essential for the understanding and prediction of the underlying molecular processes [59, 60]. Depending on the goals of the exercise, models can address primarily the large-scale structure and development of tumours on a tissue level or focus on the molecular processes of the tumour cells. Multiscale or hybrid approaches attempt to combine both views [59, 61].

16.4.2.1 Exploiting Existing Data

The development of mechanistic models of disease-related and associated cellular and molecular pathways makes use of existing functional data on the related biological entities, such as genes, proteins or protein complexes, and their interactions. The structure of the model is initially based on information from basic research, available through, for example, pathway databases and the scientific literature. Several manually curated pathway databases exist, including KEGG, Reactome and PID, which cover a substantial amount of searchable cellular signalling pathways, most often depicted as simplified network diagrams [62–64]. A comprehensive list of such databases as well as other publicly available resources useful for the systems biology field is provided by PathGuide (http://www.pathguide.org/; [65]). Helpful tools to aggregate knowledge from these different databases are available from meta-databases such as PathwayCommons [66] and Consensus PathDB [67, 68], which integrate coherent information on functional interactions from diverse data sources.

At least two major benefits are provided by the use of such manually curated databases: (1) the plethora of scientific data that exist for a given pathway in different experimental setups (e.g. different human cell lines, genetic mouse models, xenograft mouse models, etc.) is distilled into a generalised expert view of how the biology of a given pathway 'works' in various biological contexts; (2) updates of the pathway annotations are released in more or less regular intervals that try to incorporate the most recent findings from the literature and thereby improve the pathway annotation. In addition, the information available from these curated pathways is intrinsically prone to be incomplete at any given point in time. One way to try to circumvent such problems is to establish large-scale efforts by the scientific community to recruit expert knowledge on a voluntary basis for the improvement of pathway annotations [69, 70]. Such community efforts can clearly benefit from large-scale projects to extract relevant functional network information from existing datasets [71, 72], and further projects that support a genome-scale integrative interpretation of the current biological knowledge on cellular pathways should be initialised.

16.4.2.2 Reliability and Reproducibility of Existing Data

In recent years, the issue of reproducibility of basic research results has been a common problem, for instance, two thirds of publications on drug targets are not reproducible [73]. This lack of reproducibility does not necessarily result from scientific malpractice but can be based on incompletely understood confounding factors, such as diet formulation in mouse studies or variations in the gut microbiome of the analysed animals [74–76]. Reproducibility of results in biomedical research can also critically depend on the batch of commercially available kits and/or antibodies [77], as well as other factors. Therefore, one important aspect that should be taken into account even during the planning phase of any future large-scale effort to support genome-scale pathway annotations is to, as far as possible, take all measures that maximise the reproducibility of the data. For example, in a simple tandem approach, a rigorous cross-checking of scientific key findings could be performed by an independent lab within the same research consortium, notwithstanding the potentially high costs that may arise for the funding agencies in case of high-throughput techniques, such as NGS or MS. Replicative studies may also be performed by independent labs and posted as such in online repositories (e.g. http://f1000research.com/channels/PRR) or in specialised sections of peer-reviewed journals (http://validation.scienceexchange.com/#/reproducibility-initiative).

A second strategy towards model improvement can, however, be based on comparing an increasing number of predictions with molecular and response data in preclinical and clinical situations. As more and more patients are treated according to predictions of such genome-scale models and the information on treatment success (or failure) is returned, this will increasingly help to eliminate mistakes in the model, complementing the knowledge from basic research by 'reverse engineering' from patient treatment and results from preclinical model systems.

16.4.3 Model Implementation and Optimisation

Mechanistic models of biochemical reaction systems are usually implemented as ordinary differential equation (ODE) systems. These models take into account the reactions occurring, comprising the respective reactants and enzymes, as well as mathematical terms that describe the kinetics of the reactions, based on the reactants' concentrations and reaction-specific kinetic parameters [78]. The initial state of the model is usually determined by the concentrations/amounts of the model species, i.e. reactants, and the kinetic constants or parameters defining the individual characteristics of a reaction kinetic, such as the rate of the reaction or its equilibrium. Thus, the behaviour of the model is defined by its network structure, the stoichiometry of the reactants, the initial state and the kinetic parameters. Moreover, as information on the network structure of cellular reaction systems increases, e.g. generated by new high-throughput methods, such as NGS, quantitative proteomics and metabolomics data, the size and complexity of such models also increases to accommodate this expanding knowledge base.

To ensure that such mechanistic models are predictive, we need a detailed assessment of the most important underlying biological reactions. However, within the large-scale networks generated, much of this information is not easily obtained experimentally, e.g. drug-binding affinities and degradation rates. Moreover, the models become larger, incorporating more signalling pathways and cellular components, and so does the inherent complexity of the model, as well as the associated number of unknown parameters involved in each process. Estimating these parameters accurately is therefore a real challenge to the development of robust and accurate virtual patient models in oncology. This can be addressed by optimisation strategies designed to estimate model parameters or identify regions of the (high-dimensional) parameter space that meet the observations or experimental data or fulfil certain conditions or constraints. A model of cellular signalling pathways which, among other functions, allows us to simulate the effect of a certain mutation or drug treatment on the network, e.g. a BRAF V600E mutation converting BRAF into an oncogene, leads to the activation of the subsequent MAP kinase cascade and downstream transcription factors, such as cMyc, the latter being a driver for cell proliferation [26]. This model can then be used to simulate effects of different drugs, e.g. a drug acting upstream or downstream of the mutation of interest. For example, the in silico proliferation rate, as indicated by the predicted active cMyc concentration, can be correlated with growth rate data from a proliferation assay before and after treatment by different drugs. Simulations of different preclinical models or patients, for which response and/or molecular data are available, define a cost function, i.e. mathematical formula, based on the difference between prediction and observed result. Parameters are then varied systematically to identify parameter vectors giving the best agreement between prediction and observation on a training set and validated on an independent test set. This is, however, a major computational challenge, since at every step and for every result set, different data points on different patients/preclinical models are generated. As models grow in complexity (and parameter number), increasing amounts of data will have to be used to identify regions in the higher and higher dimensional parameter space.

16.5 Future Outlook: From the 'Virtual Patient' to the 'In Silico Self'

Technological progress is ongoing with rapid developments in -omics, and imaging and sensor techniques in a self-learning model will, without doubt, revolutionise medical practice. Incorporation of other factors, such as the role of the liver, allowing individual variations in drug metabolism and activation (due to differences in cytochrome C gene sets), and interactions of tumour cells with the soma and immune systems, into the existing large-scale mechanistic models available, will provide an increased level of predictive accuracy in the context of virtual patient modelling and enhanced predictive capacity. New information types and model components will undoubtedly enhance the accuracy and usefulness of such virtual patient models in oncology. Information types such as that generated by the suite of spatially resolved and single cell analysis approaches will allow tumour heterogeneity to be addressed much more effectively. An important advance for oncology and many other diseases will also be the characterisation and modelling of the deep immune status of a patient. A technology that provides an assessment of an individual's deep immune status, which helps to delineate the potential role of the immune system in a range of disease entities or immune modulation by other factors, such as the microbiome, is urgently needed.

Oncology is, however, only a first step on the way to a truly personalised therapy strategy in many areas of medicine and prevention. A much wider use of computational models for every individual, from pre-birth to old age, will help us to avoid dangers and will contribute to better predicting the response of our bodies to the many challenges we face.

Acknowledgements The authors would like to thank their colleagues at Alacris Theranostics GmbH and the Dahlem Centre for Genome Research and Medical Systems Biology (DCGMS), for fruitful discussions and constructive criticism.

References

1. Hanahan D, Weinberg RA. The hallmarks of cancer. Cell. 2000;100(1):57–70.
2. Hanahan D, Weinberg RA. Hallmarks of cancer: the next generation. Cell. 2011;144(5):646–74.
3. Ferlay J, Soerjomataram I, Dikshit R, Eser S, Mathers C, Rebelo M, Parkin DM, Forman D, Bray F. Cancer incidence and mortality worldwide: sources, methods and major patterns in GLOBOCAN 2012. Int J Cancer. 2015;136(5):E359–86.

4. Stewart BW, Wild CP. World cancer report. IARC Non-serial Publication. WHO Press; 2014. ISBN: 978-92-832-0429-9.
5. Luengo-Fernandez R, Leal J, Gray A, Sullivan R. Economic burden of cancer across the European Union: a population-based cost analysis. Lancet Oncol. 2013;14:1165–74.
6. United Nations, Department of Economic and Social Affairs, Population Division. World population ageing 2013. ST/ESA/SER.A/348. 2013.
7. Bianconi E, Piovesan A, Facchin F, Beraudi A, Casadei R, Frabetti F, Vitale L, Pelleri MC, Tassani S, Piva F, Perez-Amodio S, Strippoli P, Canaider S. An estimation of the number of cells in the human body. Ann Hum Biol. 2013;40(6):463–71. Erratum in: Ann Hum Biol. 2013;40(6):471
8. Meacham CE, Morrison SJ. Tumour heterogeneity and cancer cell plasticity. Nature. 2013;501:328–37.
9. Burrell RA, Swanton C. Tumour heterogeneity and the evolution of polyclonal drug resistance. Mol Oncol. 2014;8:1095–111.
10. Lengauer C, Kinzler KW, Vogelstein B. Genetic instabilities in human cancers. Nature. 1998;396:643–9.
11. International Cancer Genome Consortium, Hudson TJ, et al. International network of cancer genome projects. Nature. 2010;464:993–8.
12. Alexandrov LB, Nik-Zainal S, Wedge DC, Aparicio S, Behjati S, Biankin AV, Bignell GR, Bolli N, Borg A, Borresen-Dale AL, Boyault S, Burkhardt B, Butler AP, Caldas C, Davies HR, Desmedt C, Eils R, Eyfjöro JE, Foekens JA, Graeves M, Hosoda F, Huter B, Ilicic T, Imbeaud S, Imielinks M, Jäger N, Jones DTW, Jones D, Knappskog S, Kool M, Lakhani SR, Lopez-Otin C, Martin S, Munshi NC, Nakamura H, Northcott PA, Pajic M, Papaemmanuil E, Paradiso Angelo, Person JV, Puente XS, Raine K, Ramakrishna Manasa, Richardson AL, Richter J, Rosenstiel P, Schlesner M, Span PN, Teague JW, Totoki Y, Tutt A, Valdes-Mas R, van't Veer L, Vincent-Salomon A, Waddell N, Yates LR, Australian Pancreatic Cancer Genome Initiative, ICGC Breast Cancer Consortium, ICGC MMML-Seq Consortium Zucman-Rossi J, Futreal PA, McDermott U, Lichter P, Meyerson M, Grimmond SM, Siebert R, Campo E, Shibata T, Pflister SM, Campbell PJ, Stratton MR: Signatures of Mutational Processes in Human Cancer. Nature 2013; 500 (7463):415–421.
13. Ciriello G, Gatza ML, Beck AH, Wilkerson MD, Rhie SK, Pastore A, Zhang H, McLellan MM, Yau CY, Kandoth C, Bowlby R, Shen H, Hayat S, Fieldhouse R, Lester SC, Tse GM, Factor RE, Collins LC, Allison KH, Chen YY, Jensen K, Johnson NB, Oesterreich S, Mills GB, Cherniack AD, Robertson G, Benz C, Sander C, Laird PW, Hoadley KA, King TA, TCGA Research Network, Perou CM. Comprehensive molecular portraits of invasive lobular breast cancer. Cell. 2015;63(2):506–19.
14. Jäger N, Schlesner M, Jones DTW, Raffel S, Mallm JP, Junge KM, Weichenhan D, Bauer T, Ishaque N, Kool M, Northcott PA, Korshunov A, Drews RM, Koster J, Versteeg R, Richter J, Hummel M, Mack SC, Taylor MD, Witt H, Swartman B, Schulte-Bockholt D, Sultan M, Yaspo ML, Lehrach H, Hutter B, Brors B, Wolf S, Plass C, Siebert R, Trumpp A, Rippe K, Lehmann I, Lichter P, Pfister SM, Eils R. Hypermutation of the inactive X chromosome is a frequent event in cancer. Cell. 2013;155(3):567–81.
15. Cruz JA, Wishart DS. Applications of machine learning in cancer prediction and prognosis. Cancer Inform. 2007;2:59–77.
16. Kourou K, Exarchos TP, Exarchos KP, Karamouzis MV, Fotiadis DI. Machine learning applications in cancer prognosis and prediction. Comput Struct Biotechnol J. 2015;13:8–17.
17. Oda K, Matsuoka Y, Funahashi A, Kitano H. A comprehensive pathway map of epidermal growth factor receptor signaling. Mol Syst Biol. 2005;1
18. Purvis J, Ilango V, Radhakrishnan R. Role of network branching in eliciting differential short-term signaling responses in the hypersensitive epidermal growth factor receptor mutants implicated in lung cancer. Biotech Prog. 2008;24:540–53.
19. Shih AJ, Purvis J, Radhakrishnan R. Molecular systems biology of ErbB1 signaling: bridging the gap through multiscale modeling and high-performance computing. Mol BioSyst. 2008;4:1151–9.

20. Li F, Thiele I, Jamshidi N, Palsson BO. Identification of potential pathway mediation targets in toll-like receptor signaling. PLoS Comput Biol. 2009;5:e1000292.
21. Bachmann J, Raue A, Schilling M, Becker V, et al. Predictive mathematical models of cancer signalling pathways. J Intern Med. 2012;271:155–65.
22. Cho KH, Shin SY, Lee HW, Wolkenhauer O. Investigations into the analysis and modeling of the TNF alpha-mediated NF-kappa B-signaling pathway. Genome Res. 2003;13(11):2413–22.
23. Henderson D, Ogilvie LA, Hoyle N, Keilholz U, Lange B, Lehrach H, OncoTrack Consortium. Personalized medicine approaches for colon cancer driven by genomics and systems biology: OncoTrack. Biotechnol J. 2014;9(9):1104–14.
24. Ogilvie LA, Wierling C, Kessler T, Lehrach H, Lange BM. Predictive modeling of drug treatment in the area of personalized medicine. Cancer Inform. 2015;14(Suppl. 4):95–103.
25. Röhr C, Kerick M, Fischer A, Kühn A, Kashofer K, Timmermann B, et al. High-throughput miRNA and mRNA sequencing of paired colorectal normal, tumor and metastasis tissues and bioinformatic modeling of miRNA-1 therapeutic applications. PLoS One. 2013;8(7):e67461.
26. Wierling C, Kühn A, Hache H, Daskalaki A, Maschke-Dutz E, Peycheva S, et al. Prediction in the face of uncertainty: a Monte Carlo-based approach for systems biology of cancer treatment. Mutat Res. 2012;746(2):163–70.
27. Wierling C, Kessler T, Ogilvie LA, Lange BM, Yaspo ML, Lehrach H. Network and systems biology: essential steps in virtualising drug discovery and development. Drug Discov Today Technol. 2015;15:33–40.
28. Klipp E, Liebermeister W, Wierling C, Lehrach H, Herwig R. Systems biology: a textbook. Weinheim: Wiley-VCH GmbH & Co. KgaA; 2009.
29. Wierling C, Herwig R, Lehrach H. Resources, standards and tools for systems biology. Brief Funct Genomic Proteomic. 2007;6(3):240–51.
30. Van Allen EM, Wagle N, Stojanov P, Perrin DL, Cibulskis K, Marlow S, Jane-Valbuena J, Friedrich DC, Kryukov G, Carter SL, McKenna A, Sivachenko A, Rosenberg M, Kiezun A, Voet D, Lawrence M, Lichtenstein LT, Gentry JG, Huang FW, Fostel J, Farlow D, Barbie D, Gandhi L, Lander ES, Gray SW, Joffe S, Janne P, Garber J, Macconaill L, Lindeman N, Rollins B, Kantoff P, Fisher SA, Gabriel S, Getz G, Garraway LA. Whole-exome sequencing and clinical interpretation of formalin-fixed, paraffin-embedded tumor samples to guide precision cancer medicine. Nat Med. 2014;20(6):682–8.
31. Aran D, Sirota M, Butte AJ. Systematic pan-cancer analysis of tumour purity. Nat Commun. 2015;6:8971.
32. Botstein D, Risch N. Discovering genotypes underlying human phenotypes: past successes for mendelian disease, future approaches for complex disease. Nat Genet. 2003;33:228–37.
33. Majewski J, Schwartzentruber J, Lalonde E, Montpetit A, Jabado N. What can exome sequencing do for you? J Med Genet. 2011;48:580–9.
34. Jones S, Anagnostou V, Lytle K, Parpart-Li S, Nesselbush M, Riley DR, Shukla M, Chesnick B, Kadan M, Papp E, Galens KG, Murphy D, Zhang T, Kann L, Sausen M, Angiuoli SV, Diaz Jr LA, Velculescu VE. Personalized genomic analyses for cancer mutation discovery and interpretation. Sci Transl Med. 2015;7:283ra53.
35. Furney SJ, Turajlic S, Stamp G, Nohadani M, Carlisle A, Thomas JM, Hayes A, Strauss D, Gore M, van den Oord J, Larkin J, Marais R. Genome sequencing of mucosal melanomas reveals that they are driven by distinct mechanisms from cutaneous melanoma. J Pathol. 2013;230(3):261–9.
36. Roberts KG, Mullighan CG. Genomics in acute lymphoblastic leukaemia: insights and treatment implications. Nat Rev Clin Oncol. 2015;12(6):344–57.
37. Wan MW, Wang J, Gao X, Sklar J. RNA sequencing and its applications in cancer diagnosis and targeted therapy. N A J Med Sci. 2014;7(4):156–62.
38. Gan HK, Cvrljevic AN, Johns TG. The epidermal growth factor receptor variant III (EGFRvIII): where wild things are altered. FEBS J. 2013;280(21):5350–70.
39. Tomlins SA, Rhodes DR, Perner S, Dhanasekaran SM, Mehra R, Sun XW, Varambally S, Cao X, Tchinda J, Kuefer R, Lee C, Montie JE, Shah RB, Pienta KJ, Rubin MA, Chinnaiyan

AM. Recurrent fusion of TMPRSS2 and ETS transcription factor genes in prostate cancer. Science. 2005;310(5748):644–8.
40. Ozsolak F, Milos PM. RNA sequencing: advances, challenges and opportunities. Nat Rev Genet. 2011;12:87–98.
41. Carr TH, McEwen R, Dougherty B, Johnson JH, Dry JR, Lai Z, Ghazoui Z, Laing NM, Hodgson DR, Cruzalegui F, Hollingsworth SJ, Barrett JC. Defining actionable mutations for oncology therapeutic development. Nat Rev Cancer. 2016;16(5):319–29.
42. Wilkerson MD, Cabanski CR, Sun W, Hoadley KA, Walter V, Mose LE, Troester MA, Hammerman PS, Parker JS, Perou CM, Hayes DN. Integrated RNA and DNA sequencing improves mutation detection in low purity tumors. Nucleic Acids Res. 2014;42(13):e107.
43. Bode AM, Dong Z. Post-translational modification of p53 in tumorigenesis. Nat Rev Cancer. 2004;4:793–805.
44. Hitosugi T, Chen J. Post-translational modifications and the Warburg effect. Oncogene. 2014;33(34):4279–85.
45. Markiv A, Rambaruth NDS, Dwek MV. Beyond the genome and proteome: targeting protein modifications in cancer. Curr Opin Pharmacol. 2012;12:408–13.
46. Fredriksson S, Gullberg M, Jarvius J, Olsson C, Pietras K, Gústafsdóttir SM, Ostman A, Landegren U. Protein detection using proximity-dependent DNA ligation assays. Nat Biotechnol. 2002;20:473–7.
47. Geißler D, Stufler S, Löhmannsröben HG, Hildebrandt N. Six-color time-resolved Förster resonance energy transfer for ultrasensitive multiplexed biosensing. J Am Chem Soc. 2013;135:1102–9.
48. Geißler D, Charbonnière LJ, Ziessel RF, Butlin NG, Löhmannsröben HG, Hildebrandt N. Quantum dot biosensors for ultra-sensitive multiplexed diagnostics. Angew Chem Int Ed Engl. 2010;49(8):1396–401.
49. Morgner F, Geißler D, Stufler S, Butlin NG, Löhmannsröben HG, Hildebrandt N. A quantum-dot-based molecular ruler for multiplexed optical analysis. Angew Chem. 2010;49:7570–4.
50. Spurrier B, Ramalingam S, Nishizuka S. Reverse-phase protein lysate microarrays for cell signaling analysis. Nat Protoc. 2008;3:1796–808.
51. Wegner KD, Lindén S, Jin Z, Jennings TL, el Khoulati R, van Bergen en Henegouwen PM, Hildebrandt N. Nanobodies and Nanocrystals: highly sensitive quantum dot-based homogeneous FRET-immunoassay for serum-based EGFR detection. Small. 2014;10(4):734–40.
52. Crosetto N, Bienko M, van Oudenaarden A. Spatially resolved transcriptomics and beyond. Nat Rev Genet. 2015;16:57–66.
53. Mali P, Aach J, Lee J, Levner D, Nip L, Church GM. Barcoding cells using cell-surface programmable DNA-binding domains. Nat Methods. 2013;10(5):403–6.
54. Soderberg O, Gullberg M, Jarvius M, Ridderstråle K, Leuchowius KJ, Jarvius J, Wester K, Hydbring P, Bahram F, Larsson LG, Landegren U. Direct observation of individual endogenous protein complexes in situ by proximity ligation. Nat Methods. 2006;3:995–1000.
55. Weibrecht I, Lundin E, Kiflemariam S, Mignardi M, Grundberg I, Larsson C, Koos B, Nilsson M, Söderberg O. In situ detection of individual mRNA molecules and protein complexes or post-translational modifications using padlock probes combined with the in situ proximity ligation assay. Nat Protoc. 2013;8:355–72.
56. Weibrecht I, Gavrilovic M, Lindbom L, Landegren U, Wählby C, Söderberg O. Visualising individual sequence-specific protein-DNA interactions in situ. New Biotechnol. 2012;29:589–98.
57. Chen R, Snyder M. Systems biology: personalized medicine for the future? Curr Opin Pharmacol. 2012;12(5):623–8.
58. Hood L, Flores M. A personal view on systems medicine and the emergence of proactive P4 medicine: predictive, preventive, personalized and participatory. New Biotechnol. 2012;29(6):613–24.
59. Enderling H, Rejniak KA. Simulating cancer: computational models in oncology. Front Oncol. 2013;3:233.
60. Faratian D, Bown JL, Smith VA, Langdon SP, Harrison DJ. Cancer systems biology. Methods Mol Biol. 2010;662:245–63.

61. Deisboeck TS, Wang Z, Macklin P, Cristini V. Multiscale cancer modeling. Annu Rev Biomed Eng. 2011;13:127–55.
62. Croft D, Mundo AF, Haw R, Milacic M, Weiser J, Wu G, et al. The Reactome pathway knowledgebase. Nucleic Acids Res. 2014;42(Database issue):D472–7.
63. Ogata H, Goto S, Sato K, Fujibuchi W, Bono H, Kanehisa M. KEGG: Kyoto Encyclopedia of Genes and Genomes. Nucleic Acids Res. 1999;27(1):29–34.
64. Schaefer CF, Anthony K, Krupa S, Buchoff J, Day M, Hannay T, Buetow KH. PID: the Pathway Interaction Database. Nucleic Acids Res. 2009;37(Database issue):D674–9.
65. Bader GD, Cary MP, Sander C. Pathguide: a pathway resource list. Nucleic Acids Res. 2006;34(Database issue):D504–6.
66. Cerami EG, Gross BE, Demir E, Rodchenkov I, Babur O, Anwar N, Schultz N, Bader GD, Sander C. Pathway commons, a web resource for biological pathway data. Nucleic Acids Res. 2011;39(Database issue):D685–90.
67. Kamburov A, Wierling C, Lehrach H, Herwig R. ConsensusPathDB—a database for integrating human functional interaction networks. Nucleic Acids Res. 2009;37(Database issue):D623–8.
68. Kamburov A, Pentchev K, Galicka H, Wierling C, Lehrach H, Herwig R. ConsensusPathDB: toward a more complete picture of cell biology. Nucleic Acids Res. 2011;39(Database issue):D712–7.
69. Pico AR, Kelder T, van Iersel MP, Hanspers K, Conklin BR, Evelo C. WikiPathways: pathway editing for the people. PLoS Biol. 2008;6(7):e184.
70. Pratt D, Chen J, Welker D, Rivas R, Pillich R, Rynkov V, Ono K, Miello C, Hicks L, Szalma S, Stojmirovic A, Dobrin R, Braxenthaler M, Kuentzer J, Demchak B, Ideker T. NDEx, the Network Data Exchange. Cell Syst. 2015;1(4):302–5.
71. Hill SM, Heiser LM, Cokelaer T, Unger M, Nesser NK, Carlin DE, Zhang Y, Sokolov A, Paull EO, Wong CK, Graim K, Bivol A, Wang H, Zhu F, Afsari B, Danilova LV, Favorov AV, Lee WS, Taylor D, Hu CW, Long BL, Noren DP, Bisberg AJ, Consortium HPN-DREAM, Mills GB, Gray JW, Kellen M, Norman T, Friend S, Qutub AA, Fertig EJ, Guan Y, Song M, Stuart JM, Spellman PT, Koeppl H, Stolovitzky G, Saez-Rodriguez J, Mukherjee S. Inferring causal molecular networks: empirical assessment through a community-based effort. Nat Methods. 2016;13(4):310–8.
72. Yu MK, Kramer M, Dutkowski J, Srivas R, Licon K, Kreisberg J, Ng CT, Krogan N, Sharan R, Ideker T. Translation of genotype to phenotype by a hierarchy of cell subsystems. Cell Syst. 2016;2(2):77–88.
73. Prinz F, Schlange T, Asadullah K. Believe it or not: how much can we rely on published data on potential drug targets? Nat Rev Drug Discov. 2011;10(9):712.
74. Ericsson AC, Davis JW, Spollen W, Bivens N, Givan S, Hagan CE, McIntosh M, Franklin CL. Effects of vendor and genetic background on the composition of the fecal microbiota of inbred mice. PLoS One. 2015;10(2):e0116704.
75. Jensen MN, Ritskes-Hoitinga M. How isoflavone levels in common rodent diets can interfere with the value of animal models and with experimental results. Lab Anim. 2007;41(1):1–18.
76. Rogers GB, Kozlowska J, Keeble J, Metcalfe K, Fao M, Dowd SE, Mason AJ, McGuckin MA, Bruce KD. Functional divergence in gastrointestinal microbiota in physically-separated genetically identical mice. Sci Rep. 2014;4:5437.
77. Baker M. Reproducibility crisis: blame it on the antibodies. Nature. 2015;521(7552):274–6.
78. Klipp E, Liebermeister L, Wierling C, Kowald A. Systems biology. A Textbook. Weinheim: Wiley-Blackwell; 2016.

The manufacturer's authorised representative in the EU is Springer
Nature Customer Service Centre GmbH, Europaplatz 3, 69115 Heidelberg,
Germany. If you have any concerns regarding our products, please
contact ProductSafety@springernature.com

Printed and bound by CPI Group (UK) Ltd, Croydon, CR0 4YY
01/10/2026
02225869-0005